678·5·033 : 539·4

669-18

R.O. Westcott

T.I.B./Library

B6509

Borrower	Return by
Mr. N. Evans T.I.B.	29-8-80
	29/11/88
T.I.B. Mr N Evans T.I.B.	8.12.88
T.I.B.	
T.I.B.	
T.I.B.	
T.I.B.	

COMPOSITE MATERIALS
AND THEIR USE IN STRUCTURES

MATERIALS SCIENCE SERIES

Advisory Editors

LESLIE HOLLIDAY,

*Visiting Professor, Brunel University,
Uxbridge, Middlesex, Great Britain*

A. KELLY, Sc.D.,

*Deputy Director, National Physical Laboratory,
Teddington, Middlesex, Great Britain*

COMPOSITE MATERIALS AND THEIR USE IN STRUCTURES

JACK R. VINSON

Professor and Chairman of Mechanical and Aerospace Engineering and Professor of Marine Studies, University of Delaware

TSU-WEI CHOU

Associate Professor of Mechanical and Aerospace Engineering and Materials Science, University of Delaware

APPLIED SCIENCE PUBLISHERS LTD

LONDON

APPLIED SCIENCE PUBLISHERS LTD
RIPPLE ROAD, BARKING, ESSEX, ENGLAND

ISBN 0 85334 593 7

WITH 19 TABLES AND 136 ILLUSTRATIONS

© APPLIED SCIENCE PUBLISHERS LTD 1975

Printed in Great Britain by Galliard (Printers) Ltd, Great Yarmouth

This text is dedicated to our wives Trudy and Mei-Sheng for their patience, understanding and encouragement.

PREFACE

Composite materials are not new, since such materials are known to have been used by the ancient Chinese, Israelites and Egyptians, all of whom embedded straw in bricks to improve their structural capabilities. Moreover, reinforced concrete has been a standard materials system for decades. However, with the advent of glass-filament reinforcement in numerous organic matrices the new era of composite materials was born. The most recent advances involve the use of boron and graphite filaments in both organic and metallic matrices. From high-performance composites for use in aerospace vehicle construction to inexpensive composites for low-cost housing the field is rapidly expanding. Man can now 'design' a materials system for the geometry of the structure and the loads applied.

It is most desirable that the present-day engineer understands the materials science aspects and the micromechanics of composite materials. Also, he must realise that new methods of analysis are necessary in order to determine the stresses and deformations in structures involving composite materials, since these structures behave quite differently from those involving isotropic materials.

One thing is generally agreed upon: there is no adequate teaching text for composite materials today. It is hoped by the authors that this book will be sufficient for the purpose. It is emphasised that this book is not an encyclopaedia of all previous research and solutions that are known; it is intended to provide a fundamental understanding of the physical and mathematical aspects of the materials system, and structures comprised of composite materials, at a level that can be understood by students and practising engineers. Although the book deals with fundamentals it is sufficiently comprehensive to ensure that the students are well prepared to pursue advanced literature on the subject. In Chapters 6 and 7, for instance, solutions for only a few illustrative problems are presented, and no attempt has been made to cover even a small fraction of the voluminous literature that has recently developed. However, the methods of analysis discussed in Chapters 6 and 7 all include the features necessary for accurately analysing composite material plates and shells, and thus they provide the background to enable the reader to find solutions to other structural problems. It is our intention that the text should have a proper balance between the materials science and the structural mechanics aspects. Some problems are included, and many more can be easily developed. Directions of future research are also indicated.

This book is intended for mechanical engineers, chemical engineers, civil engineers, aeronautical engineers, ocean engineers, applied mechanicists and materials scientists. It will also be of interest to some physicists, applied mathematicians, chemists and architects.

The authors wish to thank the National Science Foundation (Drs M. P. Gaus and C. J. Astill) and the Air Force Office of Scientific Research (Dr J. Pomerantz and Mr W. J. Walker) and the University of Delaware Research Foundation for their support in developing some of the material included herein. The authors also wish to thank Drs D. O. Harris, R. L. McCullough, R. B. Pipes and C. Calabrese for their extensive reviews and discussions of the manuscript, and our students who have made valuable comments regarding the text, particularly Messrs Y. C. Pan, T. L. Waltz and D. R. Linsenmann.

J. R. VINSON
T. W. CHOU

Newark, Delaware

CONTENTS

ix

Chapter 3
DISPERSION-STRENGTHENED AND DIRECTIONALLY SOLIDIFIED EUTECTIC COMPOSITES

Chapter 4
INTRODUCTION TO PLATE AND SHELL THEORY

Chapter 5
ANISOTROPIC ELASTICITY

Chapter 6
ANALYSIS OF PLATES COMPOSED OF COMPOSITE MATERIALS

Chapter 7
ANISOTROPIC SHELLS

Chapter 8
THE STRENGTH AND FRACTURE OF COMPOSITE MATERIALS

CHAPTER 1

INTRODUCTION

1.1 THE NATURE AND SCOPE OF COMPOSITE MATERIALS

The engineering of modern composite materials has had a significant impact on the technology of design and construction. By combining two or more materials together, we are now able to tailor-make advanced composite materials which are lighter, stiffer and stronger than any other structural materials man has ever used. The history of man-made composite materials can be dated back to ancient Chinese,[1] Egyptians and Israelites.[2] It is interesting to note that they all made bricks by mixing straw with clay. The bamboo used so often in structures by the Chinese is in fact an excellent fibrous composite material. The pattern-welding of sabres developed in ancient China involves the forging together of wrought iron and steel. Laminated composites also were used by the ancient Egyptians. It was recognised that by gluing thin veneers together, the strength of wood was enhanced and the possibility of swelling and shrinkage minimised.

The principle of composite materials can be found in numerous naturally occurring substances. Wood is an organic substance composed primarily of cellulose chains embedded in a lignin matrix at a ratio of about 2 to 1. The bundles of cellulose chains forming walls of the elongated cells are highly crystalline. The cells are held together by the amorphous lignin. The higher the lignin content, the softer and more resilient the combination is. The bond between the fibres and lignin is exceedingly strong, as is evident by the high strength and stiffness of wood.

Bone is another kind of natural composite material.[3] At microscopic level all compact bones are composed of osteones. An osteone is a tubular structure about 1–2 cm in length with an outer diameter of about $250 \, \mu m$ ($\mu m = 1$ micron $= 10^{-4}$ cm). It surrounds a passage of

1

blood vessels which is about 70 μm in diameter. Each osteone is composed of several lamellae of collagen fibres. The collagen fibres in each lamella are parallel and spiral about the axis of the osteone. The direction of the spiral is reversed in each lamella. These strong but soft fibres give the bone its tensile strength. Another important component in compact bones is the inorganic bone mineral which deposits around the periphery of the collagen fibres. These crystalline minerals are bound tightly to the fibres and exhibit good resistance in compressive loadings. The combination of collagen fibres and minerals renders the excellent properties that bones have as load-carrying members in living bodies.

The superior properties of man-made composite materials in structural applications can be best demonstrated by the example of a reinforced concrete beam. Concrete, a relatively inexpensive structural material, is excellent for supporting a compressive load. However, the low resistance of concrete to tension makes it an undesirable material for beam construction. Figure 1.1 indicates that the load-carrying capacity of the concrete beam is restricted by its tensile strength. One way to improve the situation is to strengthen its tensile properties by the use of steel wires. As a result, the majority of tensile stress is now borne by the reinforcing wires, and a heavier load can be applied to the beam without increasing its cross-sectional area. A further improvement to this beam structure can be made by stretching the steel wires when the concrete beam is made. This is the process

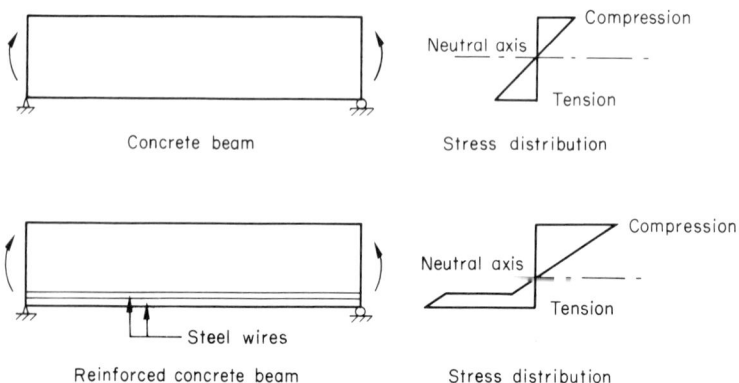

Fig. 1.1.   Load-carrying capacities of non-reinforced and reinforced concrete beams.

used in pre-stressed concrete. By stressing the wire, a compressive force is applied at the lower portion of the beam in Fig. 1.1. Since this portion of the beam will carry tensile stress, the loading capacity of the beam is further improved by pre-stressing the concrete in compression. The combination of steel and concrete has not only made the best use of the strengths of the components but also resulted in properties that cannot be achieved by either component. Furthermore, the desired strength, stiffness and weight of the beam can be tailor-made by adjusting the combination of concrete and steel wires.

In the following sections, the strengthening processes used in the manufacture of alloys are discussed, based upon the understanding of the structure and defects of solids. Finally, the needs for composite materials are dealt with.

1.2 STRUCTURES OF SOLIDS

Solids can be generally categorised, according to their structures, as crystalline and amorphous materials. *Crystals* are characterised by the long-range orders existing in their structures. Whereas in *amorphous solids*, there is a lack of definite structure pattern and the atoms or molecules are arranged in a random manner.[4,5]

Crystals are defined as solids having repetitive three-dimensional arrangements of atoms or ions. Their structures can be most easily classified by using the concept of space lattices. A *space lattice* represents an arrangement of geometrical points in three dimensions, where each point has an identical surrounding. These geometric points are also known as *lattice points.* Since their positions relative to one another are identical throughout the space lattice, it is only necessary to consider a basic structural unit for a particular type of space lattice. The criterion for choosing such a structural unit, a *unit cell*, is that its repetition in three dimensions should create a space lattice. There are 14 space lattices, also called *Bravais lattices*. Their unit cells are shown in Fig. 1.2. The Bravais lattices can be further categorised into seven *crystal systems*. This is done by taking into consideration the relative distances between lattice points and the angles defined by the edges of a unit cell. The geometry of crystal systems is specified in Table 1.1 in the order of increasing degree of symmetry.

The concept of a space lattice, although purely geometric, provides a convenient description of the structure of crystals. A crystal structure

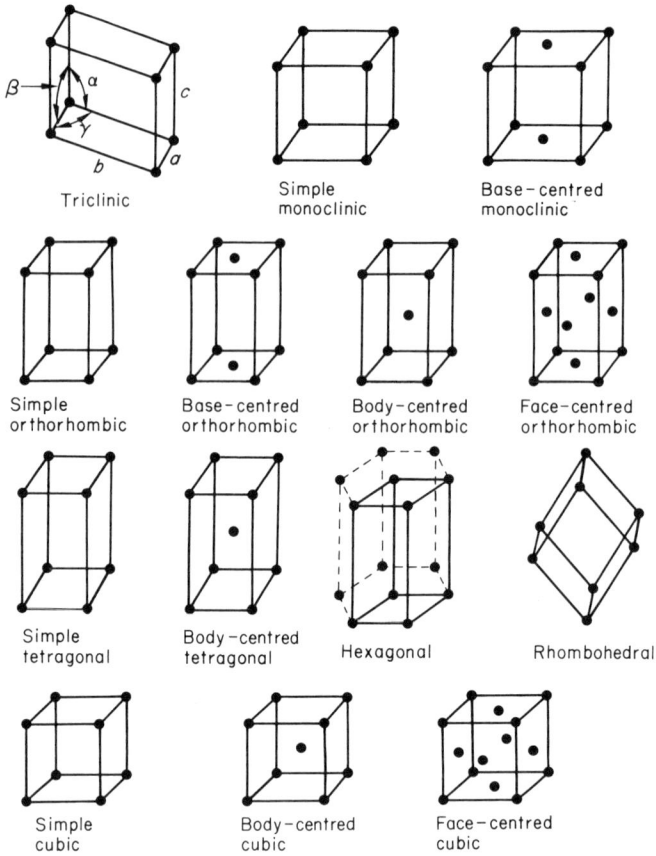

Fig. 1.2. Unit cells of the Bravais lattices. The solid circles indicate the lattice points.

TABLE 1.1

THE CRYSTAL SYSTEMS

Systems	Axes and angles	
Triclinic	$a \neq b \neq c$;	$\alpha \neq \beta \neq \gamma \neq 90°$
Monoclinic	$a \neq b \neq c$;	$\alpha = \beta = 90°, \quad \gamma \neq 90°$
Orthorhombic	$a \neq b \neq c$;	$\alpha = \beta = \gamma = 90°$
Tetragonal	$a = b \neq c$;	$\alpha = \beta = \gamma = 90°$
Hexagonal	$a = b \neq c$;	$\alpha = \beta = 90°, \quad \gamma = 120°$ (or $60°$)
Rhombohedral	$a = b = c$;	$\alpha = \beta = \gamma \neq 90°$
Cubic	$a = b = c$;	$\alpha = \beta = \gamma = 90°$

can be viewed as a space lattice, where the geometrical lattice points are replaced by atoms. In a real crystal, each lattice point can be associated with a single atom, such as in pure elements, or a group of atoms, for example, a molecule. The crystal structures of various materials and their physical properties can be found in References (5) and (6). At room temperature, materials such as aluminium, copper, gold, lead, nickel and silver assume face-centred cubic (FCC) structures; molybdenum, niobium, tantalum, tungsten, and vanadium have body-centred cubic (BCC) structures; beryllium, cadmium, cobalt, magnesium, titanium and zinc are in hexagonal close-packed (HCP) structures. There are also crystals which can exist in different structures at different temperature ranges. A typical crystal of this kind is iron.

The distinction between crystalline and amorphous materials has been made based upon the existence, or the lack of, long-range orders in their structures. A better understanding of the orderliness of structures can be obtained by considering the relative positions of atoms. For example, at temperatures below 910°C and in the range of 1400°C–1534°C, an iron crystal assumes a BCC structure. A unit cell of this structure shows that there are eight atoms surrounding the one at the centre of the cell. They are the first-nearest-neighbours of the central one. By repeating the unit cell in space, it is easily seen that the number of second-nearest-neighbours is six. Since it is not possible to distinguish all the atoms in pure iron, the number of neighbouring atoms at any specified distance from any particular atom remains the same. If iron is heated to the temperature range between 910°C–1400°C, it will assume an FCC structure. Throughout this new structure, the number of nearest-neighbours of different orders for any atom also remains the same. A crystal is thus characterised by its long-range order in structure.

Upon cooling down from the molten state, pure elements can be solidified into crystals. This is due to their relatively simple structures. However, the process of crystallisation can be hindered when the structure of the material is complex and the cooling rate is extremely fast.[7,8] Metallic glasses can be formed by a suitable choice of the component metals in an alloy. Amorphous solids can also be prepared in the form of thin layers by condensation from the vapour state. The method of vapour deposition has been used to produce condensed layers for heavy metal halides such as CuCl, AgCl and AgBr. It is also possible to produce pure non-metallic elements, such as germanium, sulphur, selenium, and silicon, in an amorphous state. For pure metals, it has been found that bismuth and gallium when condensed at helium temperature also exhibit amorphous structures.

However, one of the best-known examples of amorphous solids can be found in oxide glasses. The major component in oxide glasses is quartz (SiO_2). The basic structure of quartz consists of a silicon atom sitting at the centre of a tetrahedron with atomic bonds extending to oxygen atoms at the corners. A two-dimensional view of this structure may look like what is shown in Fig. 1.3(a).[9] The crystal structure of quartz is disturbed when it is made into silica glass (fused quartz). A two-dimensional view of silica glass is shown in Fig. 1.3(b). This network structure is rigid and lacks the regularity of a crystal structure.

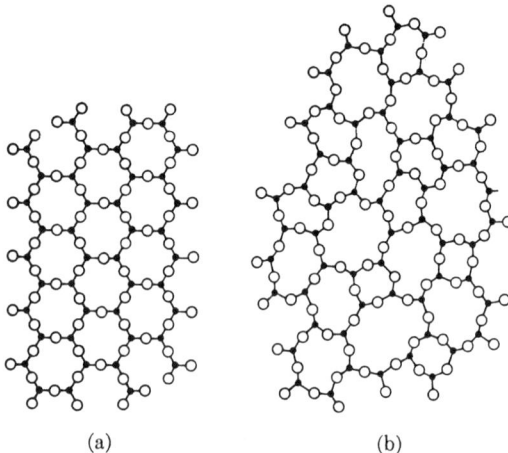

(a)   (b)

Fig. 1.3.   Two-dimensional views of (a) quartz and (b) fused silica. The open and solid circles represent, respectively, the oxygen and silicon atoms. (Repro-duced from Reference 9 by permission of the American Chemical Society.)

The atoms in either crystalline or amorphous structures are bound together by various kinds of inter atomic bonds. They can be generally categorised according to their strengths as primary and secondary bonds. The *metallic, covalent* and *ionic bonds* are all primary bonds. The energy values of primary bonds range from 2 to 8 electron volts ($1 \text{ eV} = 1 \cdot 6 \times 10^{-12} \text{ erg}$). Weaker bonds such as *van der Waals bond*, *hydrogen bond* and *permanent dipole bond* are considered to be secondary bonds, with energy values in the range of $0 \cdot 02$ to $0 \cdot 5 \text{ eV}$.

Metallic bonding involves the sharing of valence electrons by all atoms and is the primary cause of high conductivity of metals. The non-directional nature of this type of bonding provides the deformability of most metals. On the contrary, covalent bonds are highly directional. This type of bond arises from the sharing of valence electrons by adjacent atoms and exists in pure elements as well as molecules. The directionality of covalent bonds in polymers plays a significant role in the configurations of polymer molecules. Ionic bonds, like metallic bonds, are also non-directional. This is due to the fact that a charged ion tends to attract as many ions as possible that have opposite charges. As a result, the packing of ionic crystals is determined by the relative sizes of ions. It is recognised that some pure elements and compounds may have inter-atomic bonds which are characterised as a mixture of the primary bonds.

The various types of atomic bondings and structures in solids can be best illustrated by a discussion of *polymers*. Polymers are characterised by their long molecular chains. These chains consist of one or more atomic groups, the *monomers*, occurring repeatedly. Polymers can be classified according to the mode of polymerisation. For example, *vinyl* polymers are formed when the double bond in a monomer is changed to a single bond, and one single bond on each side is formed with another monomer. In general, the reaction can be represented as

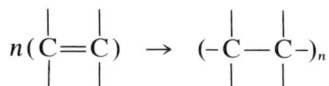

$$n(\overset{|}{\underset{|}{C}}=\overset{|}{\underset{|}{C}}) \quad \rightarrow \quad (-\overset{|}{\underset{|}{C}}-\overset{|}{\underset{|}{C}}-)_n$$

Each dash line in this expression represents a single bond that can be connected to an atom or a group of atoms. The *diene* type polymerisation is characterised by the double bonds in a monomer. The typical reaction then looks like

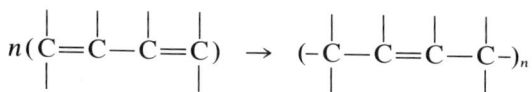

$$n(\overset{|}{\underset{|}{C}}=\overset{|}{C}-\overset{|}{C}=\overset{|}{\underset{|}{C}}) \quad \rightarrow \quad (-\overset{|}{\underset{|}{C}}-\overset{|}{C}=\overset{|}{C}-\overset{|}{\underset{|}{C}}-)_n$$

The mechanisms of polymerisation involve *addition* and *condensation* reactions. Addition polymerisation is achieved by sequential addition of monomers into a chain of many mers. Vinyl type polymers, such as polyethylene $(-H_2C-CH_2-)_n$ and polyvinyl chloride $(-H_2C-CHCl-)_n$ are formed by the addition of ethylene $(H_2C=CH_2)$ and vinyl chloride $(H_2C=CHCl)$, respectively. In addition polymerisation it is necessary to apply heat, pressure, light or catalyst to accelerate the reaction. In order for the addition chain to grow, it is also necessary to stabilise one end of a monomer so that it can attack a neighbouring monomer and continue the polymerisation. The end unit is known as an *initiator*. The peroxides, such as H_2O_2, are commonly used for this purpose. When H_2O_2 dissociates into 2(OH), these atomic groups can combine readily with ions or molecules. The chain reaction in, for instance, polyethylene then looks like

$$HO(C_2H_4\text{---})_n \quad + \quad C_2H_4 \quad \rightarrow \quad HO(C_2H_4\text{---})_{n+1}$$

The time required for polymerisation depends on how readily available the molecules are near the free ends of polymer chains. The movement of polymer molecules is through a diffusion process. It is obvious that when more initiators are used there will be more individual chains with shorter length, and the polymerisation process is faster. Addition polymerisation can also be performed when more than one type of monomer is used. The resulting polymer is called a *copolymer*. In contrast to addition polymerisation, a condensation reaction forms a second molecule as a by-product. Usually the product is water or some simple molecules. Typical examples of condensation polymerisation will be described in Section 2.2.

Polymers with simple structures can be crystallised under suitable conditions. The unit cell in crystallised polyethylene is of the ortho-rhombic type. This is shown in Fig. 1.4. It is also noted that the molecular configuration in polyethylene is not planar. The angles between the carbon bonds are $109 \cdot 5°$. It is possible for the C–C bonds to rotate with respect to each other under thermal agitation. The structure of polyethylene is relatively simple and regular. Hence, polyethylene can be crystallised rather completely. The process of crystallisation becomes difficult if there are bulky side groups attached to the main chain. For example, polyvinyl chloride cannot be as completely crystallised as polyethylene because the chlorine atom is much larger than the hydrogen atom. This process is even more difficult in polystyrene $(-H_2C\text{---}CHC_6H_5-)_n$ where each monomer has a benzene ring attached to it. Another factor which contributes to the difficulty in polymerisation is the rotation of bonds along the molecular chain. It is obvious that if the benzene rings in polystyrene assume random orientations relative to the polymer chain, the chance for crystallisation is slim. On the other hand, if the benzene rings are either

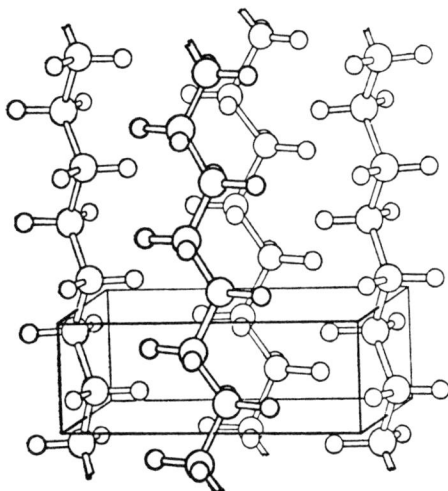

Fig. 1.4. *The crystal structure of polyethylene. A unit cell is shown, whose vertical edge is the fibre repeat-period* (2·53 Å). *The horizontal edges are* 7·40 Å *and* 4·93 Å. *Although four chains are shown it is clear that only two are allocated to one unit cell.*[10,23]

all on one side of a molecular chain or on alternating sides along the chain direction, the possibility of crystallisation is enhanced.

In the vinyl type of polymer, the atomic bonds which bind the atoms together in a molecular chain are of the covalent type. As a result, the strength of polymers is superb along the chain directions. This is the reason why it is desirable to align polymer chains by *crystallisation.* However, the binding forces between polymer chains are due to the van der Waals bonds. The effect of this weak secondary bond can be seen in natural rubber. Natural rubber is a polymer of isoprene

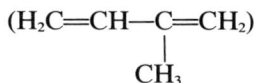

$$(H_2C{=}CH{-}\underset{\underset{\displaystyle CH_3}{|}}{C}{=}CH_2)$$

These two double bonds indicate that polyisoprene is of the diene type. At high temperature the long chains in natural rubber tend to slide over each other and lose their strength. In order to prevent the sliding of chains, a process known as *vulcanisation* has been used very success-fully. In this process sulphur is added to polyisoprene to bind the chains together by forming cross-links:

$$
\begin{array}{cccc}
H & (S)_x & H & H \\
| & | & | & | \\
-C\!-\!\!\!\!\!&\!\!\!\!\!-C\!-\!\!\!\!\!&\!\!\!\!\!-C\!-\!\!\!\!\!&\!\!\!\!\!-C- \\
| & | & | & | \\
H & CH_3 & | & H \\
 & & (S)_x & \\
H & CH_3 & | & H \\
| & | & | & | \\
-C\!-\!\!\!\!\!&\!\!\!\!\!-C\!-\!\!\!\!\!&\!\!\!\!\!-C\!-\!\!\!\!\!&\!\!\!\!\!-C- \\
| & | & | & | \\
H & (S)_x & H & H
\end{array}
$$

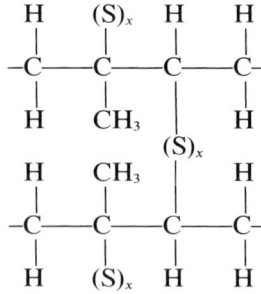

Cross-linking is then a very effective mechanism for the strengthening of polymers. The resulting structure is a rigid three-dimensional network. Another principal strengthening mechanism can be found in ladder-type molecules. The stiffness of this type of polymer is largely due to the links between two molecular chains. The name of ladder

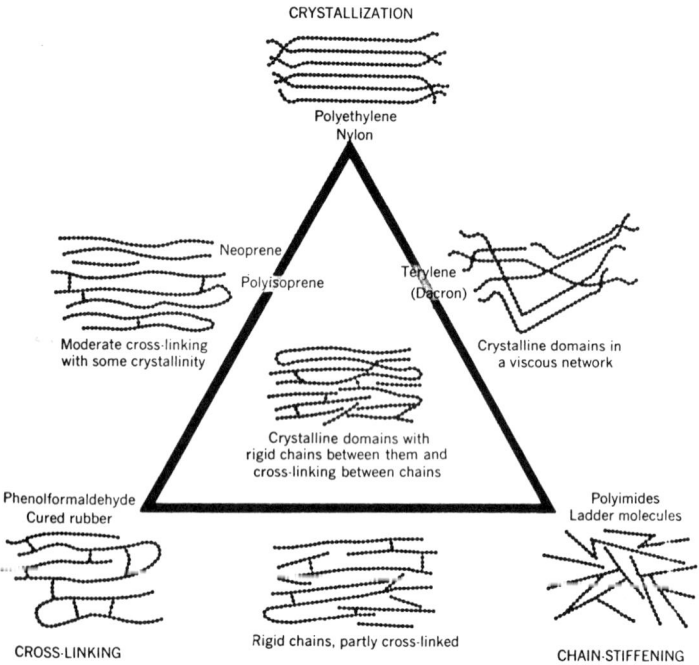

Fig. 1.5. *Schematic illustration of polymer structural configurations due to various strengthening mechanisms. (Reproduced from Reference 11 by permission of Scientific American Inc.)*

molecule is used to simulate the structural configuration. One example of polymer with *chain-stiffening* will be discussed in Section 2.2.

In summary, polymers can be strengthened by crystallisation, cross-linking and chain-stiffening. For certain polymers, the strength is due to a combination of these mechanisms. A schematic illustration of the polymer structure configurations for polymer strengthening is shown in Fig. 1.5.[7,11]

1.3 THE IDEAL COHESIVE STRENGTH OF SOLIDS

The pursuit of greater strength for structural materials has always been a challenge to materials engineers. The maximum attainable strength of a solid affects, among other things, its fracture and deformation behaviours. Hence, it is noteworthy to examine these upper limits of mechanical strength of engineering materials. Theoretical predictions of the ideal cohesive strength of solids have been performed for pure elements and compounds in crystal forms. Some calculations for cohesive strength will be briefly reviewed in this section and then compared with experimental results. The discussion is based upon the work of Macmillan.[12]

Polanyi, and Orowan pioneered the studies of the ideal cohesive tensile strength of solids. Their model considers two parallel atomic planes normal to the loading axis. The equilibrium spacing between the planes is denoted as a. They assumed that in the deformed solid, the uniform tensile stress, σ, and the relative displacement of the planes, u, obey a simple analytical form, such as a sinusoidal function

$$\sigma = \sigma_{\max} \sin \frac{2\pi u}{L} \tag{1.1}$$

L is the wavelength as shown in Fig. 1.6. The ideal cohesive strength is represented by the maximum value of this stress versus the displacement curve. For infinitesimally small strain, the slope of the stress–strain curve should give the correct Young's modulus E, namely

$$\sigma = E\frac{u}{a} \cong \sigma_{\max} \frac{2\pi u}{L} \tag{1.2}$$

and, hence,

$$\sigma_{\max} \cong \frac{E}{2\pi} \frac{L}{a} \tag{1.3}$$

In order to express $\sigma_{\max}$ in terms of physical properties of the solid, it is further assumed that the stored elastic strain energy is converted to

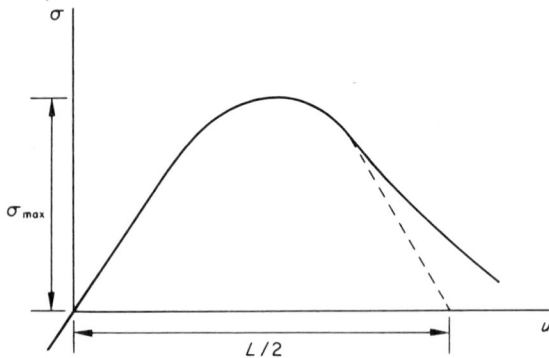

Fig. 1.6.   *Tensile stress: displacement relation in an ideal solid.*

surface energy when the solid is fractured at σ_{max}. Since two new surfaces are created at fracture, the surface energy per unit atomic plane is

$$\gamma = \frac{1}{2} \int_0^{L/2} \sigma_{max} \sin \frac{2\pi u}{L} du$$

$$= \sigma_{max} \frac{L}{2\pi} \qquad (1.4)$$

By combining eqns. (1.3) and (1.4), the ideal cohesive strength is obtained as

$$\sigma_{max} \cong \left(\frac{E\gamma}{a} \right)^{1/2} \qquad (1.5)$$

The ideal cohesive strength in shear was first examined by Frenkel using a method similar to the one just discussed. The repeat distance of atoms in the direction of shear is b. The maximum shear strength is so obtained as to give the correct shear modulus G. It is easily shown (see Problem 1.3) that

$$\tau_{max} = \frac{Gb}{2\pi a} \qquad (1.6)$$

The early work of Polanyi and Frenkel provided convenient evaluation of the ideal cohesive strengths of solids and has been widely quoted. However, their calculations did not take into consideration the crystal structure explicitly. Since then refined calculations have been performed on the basis of inter-atomic forces. Realistic models of inter-atomic forces have been made for van der Waals and ionic solids. Some values of the calculated cohesive strength are shown in Table 1.2. In these models, spherically symmetric, two-body, central

TABLE 1.2

SIMPLE CALCULATIONS OF THE IDEAL COHESIVE STRENGTHS

(Reproduced from Reference 12 by permission of Chapman and Hall.)

Material	Tensile direction	$E \times 10^{-11}$ dynes/cm²	γ ergs/cm²	σ_{max}/E	Shear plane and direction	$G \times 10^{-11}$ dynes/cm²	τ_{max}/G
Ag	$\langle 111 \rangle$	12·1	1130	0·20	$\{111\}\langle 11\bar{2} \rangle$	1·97	0·028–0·039
Al	—	—	—	—	$\{111\}\langle 1\bar{1}0 \rangle$	2·30	0·114
Au	$\langle 111 \rangle$	11·0	1350	0·25	$\{111\}\langle 11\bar{2} \rangle$	1·90	0·028–0·039
C (diamond)	$\langle 111 \rangle$	121·0	5400	0·17	—	—	—
C (graphite)	—	—	—	—	$\{0001\}\langle 1 0\bar{1}0 \rangle$	0·23	0·05
Cu	$\langle 111 \rangle$	19·2	1650	0·20	$\{111\}\langle 11\bar{2} \rangle$	3·08	0·028–0·039
Fe	$\langle 111 \rangle$	26·0	2000	0·18	$\{1\bar{1}0\}\langle 111 \rangle$	6·0	0·11–0·13
Al₂O₃	$\langle 0001 \rangle$	46·0	1000	0·10	$\{0001\}\langle 11\bar{2}0 \rangle$	14·7	0·115
NaCl	$\langle 100 \rangle$	4·4	115	0·10	—	—	—
SiO₂ (glass)	—	7·3	560	0·22	—	—	—

force potentials are usually assumed. It should be noted that these models often ignore the contributions of the kinetic energy of the atoms to the total energy of the system. Furthermore, the assumption is usually made that each atom is at its equilibrium bulk lattice site, and thus surface effects, lattice vacancies and thermal displacements are ignored. Consequently, the theoretical calculations represent a best approximation at $0°K$.

The experimental measurements of very high strengths are usually obtained from three categories of materials. These include: (1) very small whisker single crystals (the preparations and properties of whiskers are discussed in Section 2.3); (2) larger single crystals of materials such as silicon, germanium and alumina; (3) silica glass rods and fibres. Table 1.3[12] is a list of experimental measurements of very great strengths. They represent isolated greatest-strength values recorded in a series of experiments. These strengths are functions of specimen geometry, mode of deformation, temperature and strain rate.

TABLE 1.3

SOME EXPERIMENTAL MEASUREMENTS OF VERY HIGH
STRENGTHS
(Reproduced from Reference 12 by permission of Chapman and Hall.)

Material	$\sigma_{max} \times 10^{-10}$ dynes/cm^2	σ_{max}/E	Remarks
(1) Whiskers			
Ag	1·73	0·040	$\langle 100 \rangle$ tension
Cd	2·80	0·040	$\langle 1\,1\bar{2}0 \rangle$ tension
Cu	2·94	0·015	$\langle 111 \rangle$ tension
Fe	13·10	0·050	$\langle 111 \rangle$ tension
NaCl	1·08	0·025	$\langle 100 \rangle$ tension
(2) Larger single crystals			
Ge	3·82	0·02	Bending
Si	4·14	—	3·8-cm diameter ring pulled in tension
Al$_2$O$_3$	6·85	0·02	Bending
(3) Glass rods and fibres			
SiO$_2$	13·1	0·18	Bending
SiO$_2$	13·8	0·19	Tension

Furthermore, the strength is very sensitive to the surface condition, the presence of dislocations and, hence, is non-reproducible. Since these defects generally occur on a statistical basis, the very great strength of materials is often observed in small specimens.

Knowing the maximum tensile strength, the corresponding shear strength on any crystallographic plane along any arbitrary direction can be evaluated. The theoretical calculations for the ideal shear strength were performed by assuming simple shear conditions. However, experimental high strengths are usually measured in bending or tension. There are both normal and shear stresses acting on a slip system under these conditions. As a result, the shear strength so measured may not reflect the true strength in simple shear.

An examination of Table 1.3 indicates that the experimentally observed high-strength values are consistent with those obtained from theoretical predictions in the order of magnitude. For some materials, the gap between theory and experiment seems to have been closed. However, in others both refined calculations and experimental procedures are needed.

1.4   DEFECTS IN SOLIDS

In the previous sections, the structure and strength of crystals was discussed based upon the assumption that crystals are perfect. This is a highly idealised situation. In fact, nearly all crystals contain some imperfections. It is actually the defect structures that cause the fascinating behaviour and many of the desirable properties of structural materials. It is reasonable to say that materials without imperfections would be too monotonous to study, too brittle to handle, and it would be virtually impossible to manipulate their properties.

Defects in crystals can be classified according to their sizes. The defects having the size of atoms are called point defects. Point defects assume various forms. A *vacancy* is created in a crystal if an atom is missing from its normal atomic site. Vacancies can also exist in groups of two or three. Vacancies are indispensable for the transport of mass in a crystal structure. By exchanging positions with vacancies, atoms can move in a crystal. This is known as a *diffusion* process. Under thermodynamic equilibrium condition, the concentration of vacancy C_v in a crystal varies with temperature in the following manner

$$C_v = C \exp(-E_v/kT) \qquad (1.7)$$

where C is a proportional constant, k is the Boltzmann's constant ($1\cdot38 \times 10^{-16}$ erg/atom°K) and T is the temperature in °K. E_v is the energy needed to form a vacancy and it has values of about 1 eV (for example, $0\cdot9$ eV, $1\cdot0$ eV and $0\cdot75$ eV for copper, gold and aluminium, respectively). Vacancies are usually formed at locations such as free surfaces, grain boundaries and at other types of defects, since the atom that leaves its regular site can be easily accommodated at these locations. It is evident from eqn. (1.7) that the concentration of vacancies increases with temperature. For metals the values of C_v near melting point are between 10^{-3} to 10^{-4}. The concentration of vacancies in aluminium at 330°C, or half of its melting point, is $5\cdot45 \times 10^{-6}$. This is equivalent to $3\cdot3 \times 10^{17}$ vacant sites in a cube of 1 cm³. The high concentration of vacancies at elevated temperature affects the deformation behaviours of metals. The mobility of atoms is greater at elevated temperature due to their having better opportunities of exchanging positions with vacancies. As a result, metals can be deformed at stress levels which are normally insufficient to cause deformation at room temperature. The continuing deformation in materials subjected to constant load or stress is known as *creep*.

Interstitial atoms are another kind of point defect. They are so named because they assume the void spaces in a regular crystal lattice. The types of *interstitials* vary with their origins. If an interstitial atom is of the same kind as the parent crystal, it is called a self interstitial. When a second-phase material is added to a matrix substance, the foreign atoms may occupy either the interstitial positions or the normal sites of the matrix. In the latter case, they are called substitutional impurity atoms. The concentration of interstitial atoms at equilibrium also follows the relation given in eqn. (1.7). The structure of point defects in ionic crystals is more complicated by the fact that the balance of charges needs to be preserved.

The introduction of impurity particles into a matrix material, such as in alloys, generally affects the mechanical properties of the matrix. For instance, a carbon atom dissolved in iron during annealing tends to occupy the void space at the centre of each cubic edge of the FCC unit cell. A calculation of the size of the void indicates that it is smaller than the size of the carbon atom. Naturally, the lattice structure is distorted near the impurity atom. For this particular instance, the lattice distortion is not symmetrical and the stress field induced is very complicated. In order to gain some insight as to the nature of a deformation near a point defect, a simplified model is considered. The void is assumed to be spherically symmetrical with radius r_0. The crystal structure is smeared out and is replaced by a continuous elastic medium. A foreign particle with radius $r_0(1 + \epsilon)$ is now inserted into the hole in the matrix. The constant ϵ measures the misfit between the matrix and the impurity particle. The resulting stress field at a distance r from the point defect is

$$\sigma_{rr} = -\frac{2E\epsilon}{1 + \nu} \left(\frac{r_0}{r}\right)^3$$

$$\sigma_{\theta\theta} = \sigma_{\phi\phi} = \frac{E\epsilon}{1 + \nu} \left(\frac{r_0}{r}\right)^3 \tag{1.8}$$

where E and ν are the Young's modulus and Poisson's ratio respectively of the matrix material, and ϕ and θ are the usual angular co-ordinates used in a spherical system. This stress field is spherically symmetrical and can be one of tension or compression depending upon the relative sizes of the void and the impurity particle. The interaction of impurities with other defects is a major source of strength of alloys.

Defects that possess one long dimension are known as line imperfections, or more commonly as *dislocations*. The role of dislocations in affecting crystal deformation can be understood by examining their behaviour at a microscopic scale. Figure 1.7 shows schematically the cross-sectional views of a crystal deformed under shear load. If the applied stress is within the *elastic limit*, the crystal deforms by changing the inter-atomic bond lengths, and will restore its original shape when the external load is removed (Fig. 1.7(a)). On the other hand, dislocations will propagate in the crystal when the applied stress reaches a threshold value. The passage of dislocations in the crystal will displace the atomic planes relative to each other. The crystal will not return to its original shape upon being relieved of the external

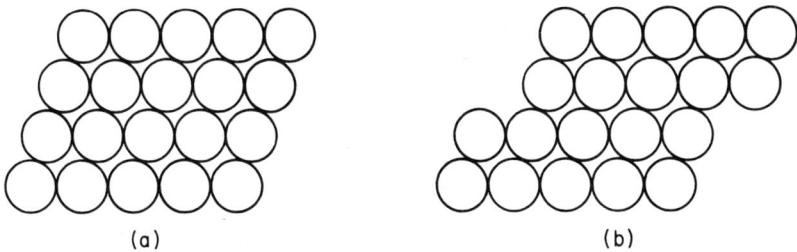

(a)   (b)

Fig. 1.7. (a) *Elastic deformation*; *and* (b) *plastic deformation of crystals.*

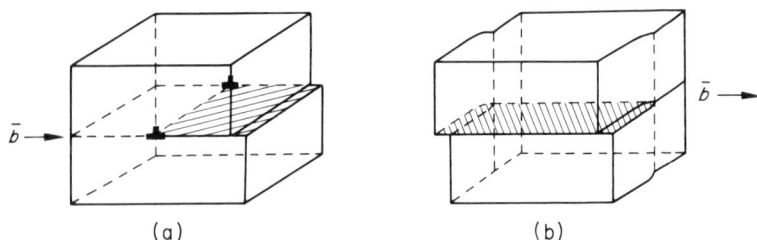

(a)   (b)

Fig. 1.8.   Two kinds of dislocations: (a) edge dislocation; and (b) screw dislocation.

load. Deformation of this kind is termed *plastic deformation*. If a dislocation moves all the way across the width of the crystal, shear steps are created on the free surfaces (Fig. 1.7(b)). It is recognised that dislocations are indispensable for the plastic deformation of crystals.

The nature of line imperfections can be understood by examining the deformation of crystals depicted in Fig. 1.8. Plastic deformations are assumed to have occurred over the shaded planes. These deformations are created by displacing two neighbouring planes relative to each other permanently. The plane across which shear deformation has taken place is called a *slip plane*. Hence, a dislocation can be defined as the boundary which separates the slipped and un-slipped regions of a crystal. The magnitude, together with the direction of shear displacement across the slip plane, is denoted by the *Burgers vector* $\bar{b}$. In Fig. 1.8(a), the Burgers vector is normal to the dislocation which is then called an *edge dislocation*. When a dislocation line lies parallel to the Burgers vector, it is known as a *screw dislocation* (Fig. 1.8(b)). The sense of a Burgers vector is not unique. Depending on the way in which the relative displacement is defined, the directions of the Burgers vectors in Fig. 1.8 may be reversed. However, this will not affect the nature of the dislocations.

When the shear displacement is induced by rigid body rotations of crystal parts above and below a slip plane, the resulting imperfection line is a rotational dislocation, also known as a *disclination*.[13,14,15] Unlike dislocations, disclinations have only been observed in a limited numbers of metals. In view of their rotational nature, disclinations may affect the behaviour of macro-molecular solids where chain rotation is an important deformation configuration.

In order to examine the strengthening mechanisms in composite materials, it is necessary to gain some understanding of the many properties of dislocations. In Section 1.3, where the ideal shear strengths of crystals were evaluated, it was assumed that the crystals contain no imperfections. Shear deformation occurs through the displacement of all the atoms above the slip plane relative to those below by an atomic spacing. This can be done only if all the atomic bonds across the slip plane were broken and subsequently restored, simultaneously. The extremely high stress needed to achieve this kind of deformation actually sets the potential shear strength of a crystal. However, nature has its easier way to deform a crystal. This is achieved by moving dislocations through a crystal. It can be seen from Fig. 1.8 that shear displacements develop within a crystal as dislocations propagate. The atomic bonds are broken and restored in a consecutive manner. The deformation is then non-uniform and highly localised. Obviously, the stress necessary to deform a crystal with dislocations is much less than that for a perfect crystal. Experimental results indicate that aluminium crystals can be deformed at a stress of 3×10^6 dyn/cm^2 which is five orders of magnitude lower than its ideal strength.

The plastic deformations associated with dislocations are evident by the permanent shear displacements left behind along the path of dislocations and at the steps and ledges on free surfaces where dislocations emerge. The internal stress field associated with a dislocation can be found, at least in an approximate manner, by elasticity theory. It is obvious from Fig. 1.8 that a crystal containing dislocations is under severe local stress. For an edge dislocation lying along the z-axis, and having the yz-plane as the extra-half atomic plane, the stress field consists of the following components[16,17]

$$\sigma_x = \frac{-Gb}{2\pi(1-\nu)} \frac{y(3x^2+y^2)}{(x^2+y^2)^2}$$

$$\sigma_y = \frac{Gb}{2\pi(1-\nu)} \frac{y(x^2-y^2)}{(x^2+y^2)^2}$$

$$\sigma_z = \frac{-G\nu by}{\pi(1-\nu)(x^2+y^2)} \qquad (1.9)$$

$$\tau_{xy} = \frac{Gb}{2\pi(1-\nu)} \frac{x(x^2-y^2)}{(x^2+y^2)^2}$$

$$\tau_{xz} = \tau_{yz} = 0$$

where b denotes the magnitude of the dislocation Burgers vector. The origins of these stress components are now briefly examined. The edge dislocation can be formed by joining two crystals along a flat surface. The width of the upper crystal is one atomic spacing larger than the lower one. If these two free surfaces are to be matched exactly, a severe shear stress is developed at this interface, namely, the slip plane of the dislocation. An edge dislocation line can also be pictured as the boundary line where an atomic plane terminates in the crystal. Consequently its elastic field is equivalent to that generated by inserting an extra-half atomic plane into an otherwise perfect crystal. Just as a wedge is pushed through a piece of wood, compressive stresses are induced on both sides of the wedge. Tensile stress exists below the wedge so the wood will be cracked open.

The stress field associated with a screw dislocation is easier to visualise. It is seen from Fig. 1.8(b) that the atomic planes originally perpendicular to the line direction are made continuous when the dislocation is formed. These atomic planes spiral around the dislocation line just as the threads do in a screw. Depending upon the nature of relative shear displacements, and, hence, the sense of spiralling of atomic planes, screw dislocations can be easily classified as left-handed and right-handed. The deformation surrounding a screw dislocation line is cylindrically symmetrical with the dislocation line as the axis of the cylinder. It is consequently desirable to express the non-zero stress component with respect to cylindrical co-ordinate axes as

$$\tau_{z\theta} = \frac{Gb}{2\pi r} \tag{1.10}$$

It can be seen easily from eqns. (1.9) and (1.10) that the intensity of the stress field around a dislocation varies with the distance r from the line in the manner of $1/r$. Hence, at a distance very close to a dislocation line the stresses become unreasonably large due to the elastic singularity. The exact nature of deformation near a dislocation, known as the core region, is not yet clear. Nevertheless, elasticity offers a convenient means for the estimation of dislocation strain energy and interaction between dislocation lines.

The density of dislocations is very high in structural materials. Consider a metallic crystal with a volume of 1 cm^3. In the well annealed condition, the total length of dislocations is in the order of 10^4 cm. This density increases with the amount of plastic deformation. For metals with severe cold-work, the total dislocation length could be as high as 10^{12} cm, or nearly twenty-five times the distance between the earth and the moon. Undoubtedly, all the dislocations are tangled together in random fashion. The densities of dislocations before and after cold-work indicate two interesting facts. First, externally applied stress during cold-working has not sheared dislocations out of the crystal, instead, *dislocation multiplication* has occurred and higher dislocation density has resulted. Secondly, as dislocations multiply the interactions of dislocation stress fields make their movements increasingly difficult (see Problem 1.4). Hence, an increase in applied stress is necessary to bring about further deformation.

The elastic energy associated with a dislocation line can be estimated by considering the total work input in forming such a defect. It can be shown that the energy per unit length of a dislocation is of the order of Gb^2, where G is the shear modulus and b is the magnitude of the Burgers vector. The total energy needed in forming a dislocation is greater in a stronger solid and for a larger shear displacement. By taking, for aluminium, $G = 2 \cdot 8 \times 10^{11}$ dynes/cm^2 and $b = 2 \cdot 9 \times 10^{-8}$ cm, the elastic energy per atomic length of a dislocation is approximately $3 \cdot 95$ eV. This energy is too great to be supplied by thermal activation as in the case of vacancy formation. Hence dislocations can only be created by mechanical force. The energy contribution from the dislocation core is relatively small. Furthermore, it can be shown that the dislocation density for a solid in its equilibrium state should vanish. Structural materials, usually containing a certain amount of dislocations, are apparently not in their equilibrium state from a thermodynamic viewpoint.

Since the total energy of a dislocation line is proportional to its length, a curved dislocation usually tends to shorten its length just as a stretched rubber band does. Consider a straight dislocation segment of length l. The end points of the segment are assumed to have been pinned down on the slip plane. Under applied shear stress the dislocation line bows out and assumes a radius of curvature R. The stress necessary to hold this curvature is inversely proportional to R. This stress reaches a maximum when the dislocation segment is curved into a half-circle (see Problem 1.6)

$$\tau = Gb/l \qquad (1.11)$$

The dislocation segment becomes unstable when it reaches this critical shape and can further bow out without increasing the applied stress. Experimental evidence indicates that under the critical stress given in eqn. (1.11), dislocations can multiply themselves and increase their density.

In the above discussions concerning the dislocation stress field, strain energy, and line tension, the crystal structures of solids were not taken into consideration. Instead, homogeneous isotropic elastic media were assumed for the purpose of easy illustration. However, the smeared-out crystal structure is not a bad assumption in polycrystals and when the degree of crystal anisotropy is relatively small. From the point of view of a continuous medium, the slip plane of a dislocation can be defined as the plane containing both the dislocation line and the Burgers vector. For an edge dislocation, the plane normal to the extra-half atomic layer obviously constitutes the slip plane. As to a screw dislocation, since the line and Burgers vector are collinear there is no unique slip plane. Consequently, a screw dislocation can readily change its direction of slip: a process known as *cross-slip*. However, it is possible for an edge dislocation to move perpendicular to the slip plane in the form of *climb* motion. This can be achieved by diffusing atoms or vacancies to the extra-half atomic plane.

The above definition of dislocation slip plane is originated purely from a consideration of dislocation geometry. However, the slip plane so defined for a dislocation may not be an active slip plane in a real crystal when the crystal structure is taken into consideration. In crystals, the active slip planes for dislocations are the closely packed planes. The reason that shear deformation can take place more easily on these planes is explained by considering, for example, an FCC crystal as shown in Fig. 1.9. Three different atomic planes designated as DBG, ABFE and BEHC are also shown schematically. Plane DBG has the highest density of atoms per unit area and is known as the close-packed-plane in FCC crystals. The densities of atoms in planes ABFE and BEHC are lower than that of DBG and, hence, the void space per unit area is higher. The implication of atomic packing efficiency on shear deformation can be understood by imagining that each plane is made to glide over an identical plane. Apparently, planes that have the highest void space are most difficult to shear as compared to other types of planes with the same structure. A smoother surface of an atomic plane provides less resistance to the shear deformation. Furthermore, it is understood that dislocations also tend to travel along close-packed directions. On the DBG plane, the density of atoms are the highest along the lines DB, BG and GD. Again, it is obvious that shear movements of atoms along close-packed directions will dissipate the lowest energy. Since the inter-atomic distance is less along the close-packed directions than in any other direction, the dislocation Burgers vector and, hence, its strain energy are also minimised.

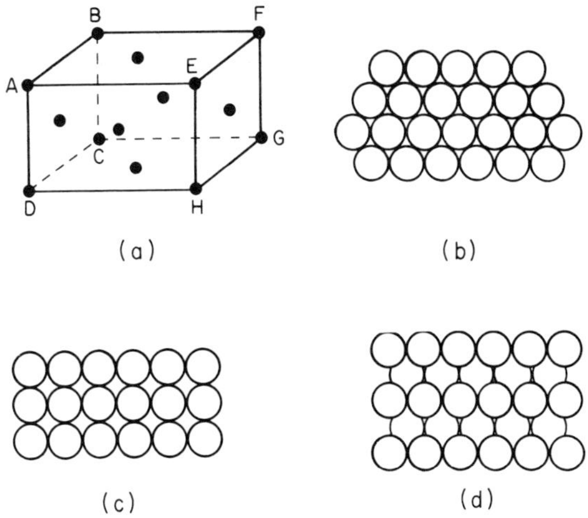

Fig. 1.9.   Atomic planes in an FCC crystal. (a) An FCC crystal; (b) plane DBG;
(c) plane ABFE; (d) plane BEHC.

Crystal imperfections having two long dimensions are called surface defects. They include *free surfaces, grain boundaries, twin boundaries,* and *phase boundaries.* A grain boundary in a polycrystal is an internal boundary which separates regions of crystals which have different orientations. A twin boundary is a low-energy configuration and is formed in such a manner that crystals on both sides of the boundary are mirror images of each other. Phase boundaries separate regions of different structures and compositions and are common in composite materials.

1.5   STRENGTHENING EFFECTS IN METALS

All metallic materials used in structures are polycrystalline materials. Since the grains in a polycrystal have random orientations, polycrystals are usually treated as isotropic, at least on a macroscopic scale. A grain boundary is formed at the location of discontinuity of crystallographic orientations. Naturally, the movements of dislocations are hindered at grain boundaries. Experimental evidence indicates that dislocations generated in the individual grains tend to pile-up against one another in front of these barriers. Plate I[18] demonstrates the *dislocation pile-ups* formed in Cu–4·5% Al. When the distance between dislocations gets very small as in a pile-up, the stress fields of dislocations with the same Burgers vector tend to overlap and to repel

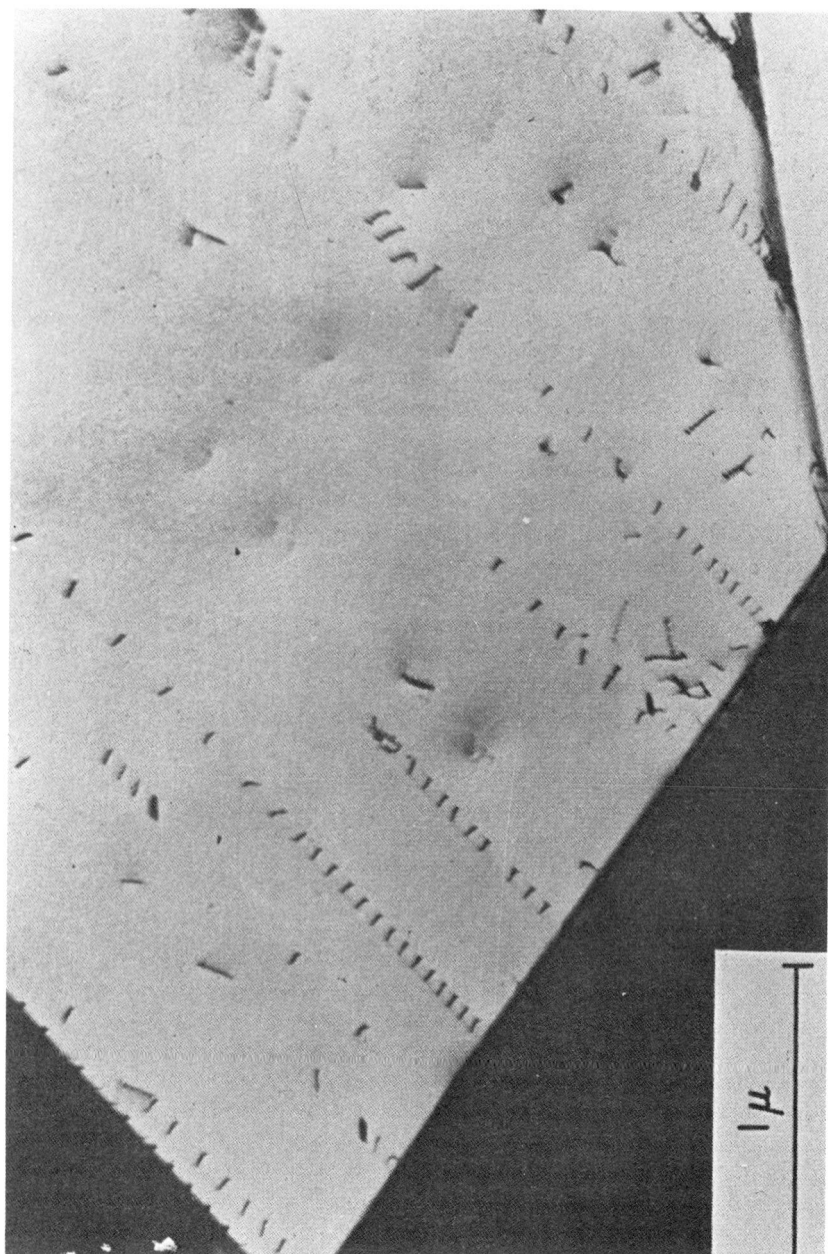

Plate I.   Dislocation pile-ups in a Cu–4·5% Al alloy. (Reproduced from Reference 18 by permission of Interscience Publishers.)

one another (see Problem 1.4). Consequently, external stress is needed to hold the pile-up dislocations together against the barrier. In the meantime, the grain boundary experiences a concentration of stress due to the dislocations. It can be shown that the stress acting on the barrier is proportional to the number of pile-up dislocations and the magnitude of externally applied stress.

The motion of dislocations and the accompanying plastic deformation can be propagated from one grain to another if the local stress concentration is high enough to activate dislocation sources in a neighbouring grain. The strength of crystals can thus be enhanced if the dislocation movement is made more difficult. As far as polycrystalline materials are concerned, a logical way to enhance strength is to implant more barriers, namely, grain boundaries to dislocations. Hall and Petch first realised that the yield strength σ_y of metals increases as the grain diameter d is made smaller. The well-known *Hall–Petch relation* states

$$\sigma_y = \sigma_i + K_y d^{-1/2} \tag{1.12}$$

where σ_i and K_y are constants. In Fe–3% Si, for example, $\sigma_i = 32\cdot4\,\mathrm{kg/mm}^2$ and $K_y = 2\cdot17\,\mathrm{kg/mm}^{3/2}$. The variation of grain size of metals can be controlled by cold work and subsequent heat treatment. The strengthening of polycrystalline materials through the manipulation of grain size is considered to be an effective metallurgical process.

Additional means of strengthening in metals are achieved by adding a second-phase material: a process in metallurgy well-known as *alloying*. The strength of alloys is affected by the manner in which the second-phase solid disperses itself in the matrix. The major hardening phenomena are briefly discussed below.

Solution hardening is effective at low and intermediate temperatures where a second-phase material disperses in the matrix as separate impurity atoms. By taking either substitutional or interstitial positions, the dispersed particles tend to lock dislocations by means of various kinds of interactions. The sources of the interactions may be elastic, chemical, electrical and structural. Among these effects the elastic interaction is considered to be the most effective and is now examined. In the last section, it was noted that impurity atoms occupying interstitial or substitutional positions in alloys induce local stress fields. The magnitude of the internal stress field is proportional to the misfit between a dispersed particle and the site by which it occupies. The distortions surrounding a dispersed particle may be symmetrical, such as substitutional atoms in aluminium and copper; or asymmetrical, such as interstitial carbon atoms in iron and interstitial nitrogen in niobium. The dislocation locking mechanism can be understood from a consideration of the strain energy associated with the alloy. The hydrostatic stress field of the edge dislocation in Fig. 1.8 is compressive above the slip plane and tensile below. Work needs to be done against this stress field when a particle with a volumetric misfit is introduced into a matrix. An oversized particle tends to stay on the tensile side of an edge dislocation so as to relax the hydrostatic dislocation stress.[16] The total elastic energy of the system is thus reduced. The dislocations anchored at the dispersed particles thus experience a retarding force in their movement, and a hardening effect in the alloy is realised. The interaction between point defects and dislocations is certainly more complicated if the nature of crystal structure is taken into consideration.

The dispersed particles in alloys generally have elastic properties different from those of the matrices. This again has the effect of retarding the motion of dislocations. It is recalled that the elastic energy associated with a dislocation is proportional to the shear modulus of the medium. The strain energies of two dislocations situated in homogeneous elastic media with shear moduli G_1 and G_2 are approximately $G_1 b^2$ and $G_2 b^2$, respectively. It is obvious that the dislocations situated in the harder phase have higher strain energies. When a dislocation situated in the medium with modulus G_1 approaches a rigid second phase with modulus $G_2(> G_1)$ as in composite materials, the second phase material at the vicinity of the dislocation is inevitably under the influence of its stress field. As a result, the strain energy associated with the dislocation line is different from that in a homogeneous medium. In this instance, its value lies between the limits of $G_1 b^2$ and $G_2 b^2$. This qualitative argument also explains why dislocations tend to move toward a free surface and away from a rigid second-phase particle. In both cases the total strain energy of the dislocation line is reduced. The interaction of dislocation and solute atoms due to difference in elastic stiffness is believed to be comparable in magnitude to that caused by volume misfit. The extent of solution hardening is certainly affected by the solubility of the hardening constituents.

A supersaturated solid solution occurs when a homogeneous alloy is quenched from elevated temperature. Subsequent annealing of the alloy at intermediate temperature induces precipitation of second-phase particles. The precipitates may have both their composition and structure different from the matrix material. The size and distribution of the particles can be controlled by the temperature and duration of heat treatment. The strengthening in alloys obtained in this manner is called *precipitation hardening*. Precipitated particles, such as in Al–Cu and Al–Ag alloys, provide strong obstacles to the motion of dislocations for reasons discussed above. Furthermore, interfaces between the precipitates and matrix materials may have variable degrees of coherency. Local stresses may be induced at a coherent interface. The dislocation movements are retarded at coherent interfaces as well as at incoherent interfaces where there are discontinuities in atomic planes.

It is now obvious that the second-phase materials in the form of either individual or cluster of atoms in alloys all act as barriers to the motion of dislocations. The stress necessary to cause large-scale plastic deformation by bulging dislocations through the barriers is inversely proportional to the distance between the dispersed particles as given in eqn. (1.11). Hence, the distribution of the alloy constituents has a profound effect on its strength. For instance, in alloys hardened by precipitates, it is desirable to reduce the inter-precipitate distance. However, for a constant volume fraction of the alloying material, this may lead to a reversion to solution hardening. Furthermore, prolonged heat treatment may cause the precipitates to cluster together. The increases in precipitate size and their separation distances weaken the hardening effect.

Metallic structural materials are also strengthened by plastic deformation, namely, an increment in plastic strain is accompanied by a corresponding increment in applied stress. This behaviour, known as *strain hardening* or work hardening, is evident by the positive slope of a stress–strain curve in the plastic range. Strain hardening is generally attributed to the ever-increasing density of dislocations generated during plastic deformations. The rate of work-hardening is directly affected by the dispersion of second-phase particles and the history of plastic deformation of the material. Experimental evidence indicates that the strain-hardening rate is much enhanced if the dispersed particles do not deform plastically with the matrix. The introduction of high dislocation density through plastic deformation also tends to increase the rate of strain-hardening. The process known as *ausforming* is based upon this idea and has been employed in producing high-strength steel. One of the high-temperature forms of steel is called austenite or γ-iron. When austenite is quenched rapidly, it transforms into a hard and brittle phase known as martensite. In ausforming, the quenching is interrupted at a temperature just above the martensitic transformation temperature. The steel is then plastically deformed during this interruption before it is cooled down to room temperature. The martensitic structure formed at low temperature is thus coupled with high dislocation density as a result of plastic deformation. Alloys work-hardened by severe plastic deformation show a distinct dislocation structure called *cell structure*. The wall of a cell is composed of tangled dislocations, while the centre of a cell is almost free of dislocations. Plate II[19] indicates the cell structure to be found in silver. In alloys containing hard particles it can be seen that the particles are predominantly in the cell walls and the size of the cell structure is related to the spacing of the particles. It has been suggested that the hardening of some FCC and BCC metals is due to the bulging of dislocations through openings in the cell walls. Hence, the stress necessary to move dislocations in this stage of the deformation is inversely proportional to the mean dislocation spacing in the cell wall.

1.6   WHY COMPOSITE MATERIALS?

It has always been the hope of metallurgists to be able to produce structural materials possessing both great strength and extreme ductility. Great strength offers high load-carrying capacity. Good ductility prevents sudden and catastrophic failures. However, good strength and ductility are generally incompatible. A glass fibre can have a tensile strength of 500 000 psi ($3 \cdot 45 \times 10^9$ N/m^2) which is much higher than that of a copper wire 60 000 psi ($4 \cdot 14 \times 10^8$ N/m^2). However, a copper wire can withstand severe plastic deformations before being broken, while glass fibres are notorious for their poor qualities in this respect.[20]

As a compromise in the pursuit of strength and ductility, metals have been used as a major form of structural material. With the rapid advancement of technology, the demand for stronger, lighter and tougher structural materials is ever growing. It is, therefore, essential to examine the existing metallic materials, not only for their capacity at the present time, but also their potential for future improvement.

In Fig. 1.10[21] the variation of tensile yield strength with temperature is plotted for the strongest alloys. To realise their potential, the ratios of measured strengths and theoretical strengths are plotted with respect to the equivalent temperature, i.e. temperature/melting point. At low equivalent temperatures, titanium and iron in BCC structure, having strengths of nearly 30 per cent of their theoretical strengths, are

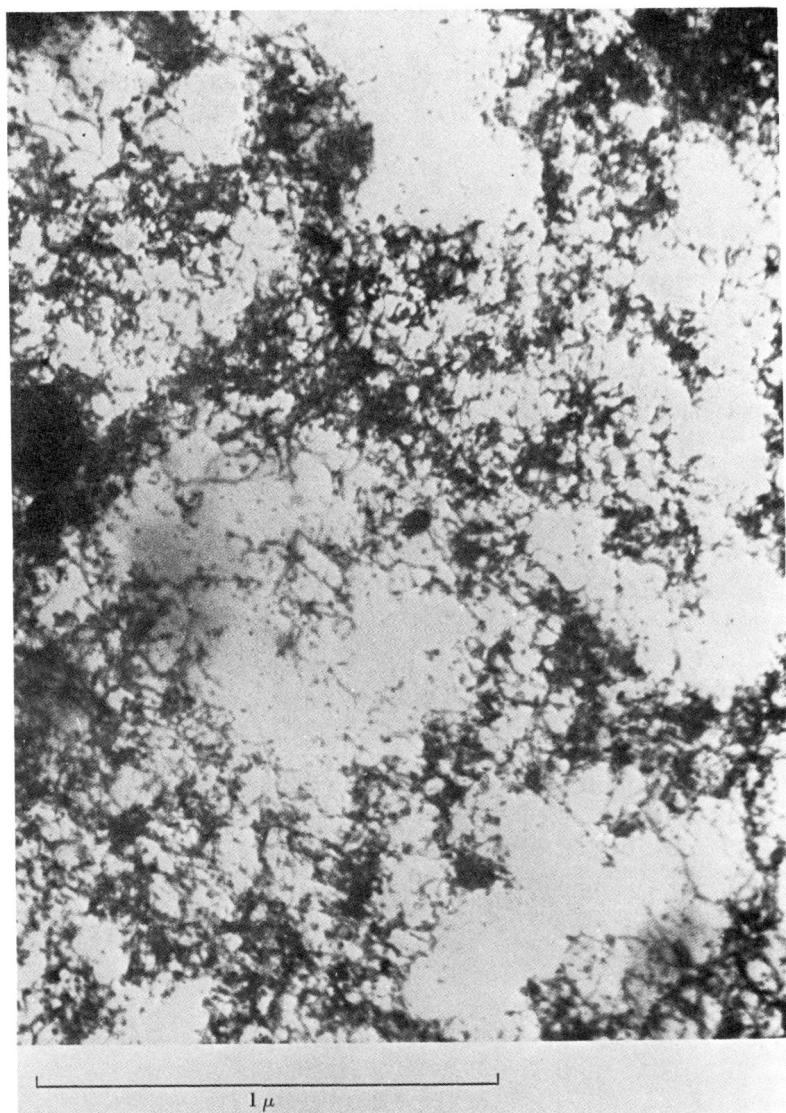

1 μ

Plate II. *Cell structure in silver strained 25 per cent in tension. (Reproduced from Reference 19 by permission of Taylor and Francis Ltd.)*

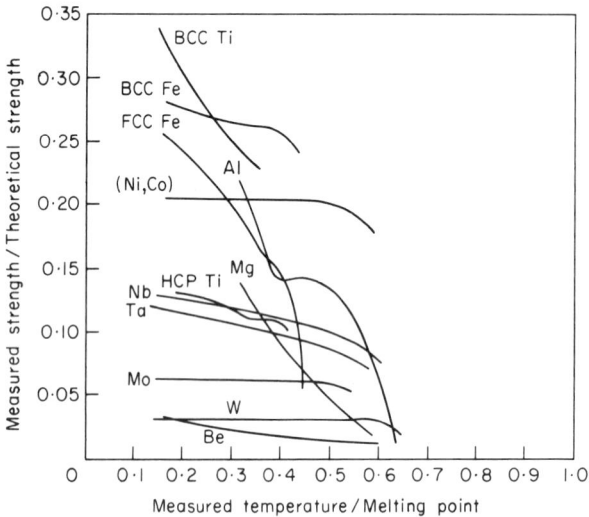

Fig. 1.10. Maximum measured tensile yield strength divided by the theoretical maximum shown as a function of the equivalent temperature. (Reproduced from Reference 21 by permission of John Wiley and Sons.)

far superior to other alloys. Their great strengths are generally attributed to the strong and stable precipitates formed as a consequence of thermomechanical treatments. It is also noted that aluminium and (nickel, cobalt) super alloys have performed reasonably well on an equivalent temperature scale. However, the comparison is disappointing for refractory alloys of tungsten, molybdenum, tantalum, niobium, and beryllium. Refractory metals having high melting temperatures in the range of 2400°C–3400°C, show relatively poor performance in regard to their theoretical strength.[21]

To answer the question of the potential strength of metallic alloys, it is necessary to recall the origin of their strengths. It is generally accepted that the strength of alloys comes mainly from the interaction of dislocations among themselves and with alloying elements. The cell structures formed in cold-worked alloys contain impurity particles predominantly in the cell walls. The size of the cell structure is affected by the spacing of the particles and the extent of cold work. The limiting cell size is about 0·5 μm. Within a cell, dislocations are pinned down at a distance of about one-tenth of the cell diameter. According to eqn. (1.11), the theoretical strength of steel and aluminium may be attained if dislocation barriers are spaced at 15 to 25 atomic distances apart. However, it is really not desirable to have alloying particles actually dispersed in such a fine manner, since dislocations are almost impossible to move at this spacing of barriers and the alloys will have negligible ductility. Ausformed steels have tensile yield strengths close to $3.45 \times 10^9 \, \text{N/m}^2$ (one half million psi). This gives an l/b value of about 50 and a desirable combination of strength and toughness. It is also obvious from the above consideration that ordinary commercial steels with low yield strengths certainly have room for future improvement.[21]

The hard particles used in strengthening metals are often dislocation free. There is evidence that the particles are under great elastic strain. These elastically stressed inclusion particles not only act as barriers to dislocations but also may carry substantial load. However, for alloys dispersed with nearly equiaxed particles, the portion of load which can be carried by the particle is limited. This is because the stresses in the hard particles are transmitted into the matrix through the particle–matrix interface. For interfaces with strong bonding, the transmitted stress tends to deform the matrix metal of which the yield strength is lower than that of the particle. The stress concentration at interfaces may eventually cause the matrix material to flow around the particles. Thus the particles lose their function as effective barriers to dislocations. Another possibility is weak interfacial bondings. In this case the incompatibility in deformation of the particles and the matrix tends to split the interface. As a result, stresses can no longer be transmitted from the matrix to the particles and the dispersed phase fails to effectively carry the loading on the alloy. In view of the interaction between matrix and particles, it is reasonable to expect that the strength of alloys can be improved if the second phase is made to carry part of the load as well to hinder the motion of dislocations. This can be achieved by manufacturing the reinforcing material in the form of long fibres and plates. The increase in interfacial area, through which loads can be transferred, makes the reinforcing materials effective load-carrying members. Furthermore, the reinforcing materials in the form of fibres and plates are also more effective in hindering dislocation movements than dispersed particles. This is actually the principle reason for making fibrous and laminated composites.[22]

Composite materials provide a balanced pursuit of both strength and ductility. The stiff reinforcing materials are responsible for carrying the load. The ductility, however, is due to the tough, but less strong, matrix. Fracture of the reinforcements in a brittle manner cannot be propagated continuously, owing to the fact that it is retarded by a soft matrix. The combination of matrix and reinforcement offers strength as well as ductility, which cannot be separately attained by either component. In conventional structural materials, the strengths are almost identical in all directions, although they are usually only subjected to uniaxial or biaxial loadings. However, in composite materials, the reinforcements can be aligned in the directions of loadings. As a result, the weight of structures employing composite materials is greatly reduced. The excellent structural characteristics as well as the fracture behaviour of the fibre-reinforced composite materials will be discussed in Chapters 6–8.

1.7   PROBLEMS

1.1.   Discuss a naturally occurring composite material. Describe its structure, properties and significance.

1.2.   How is the monoclinic crystal system defined? In this crystal system there are two kinds of Bravais lattices, namely, simple monoclinic and base-centred monoclinic. Why is the body-centred monoclinic lattice not included?

1.3.   Derive the ideal shear strength expression in eqn. (1.6).

1.4.   An edge dislocation line is situated along the z-axis. The extra-half atomic plane and the slip plane associated with the dislocation are indicated by a vertical and a horizontal line segment, respectively, as in the following figure.

(a) Determine the types of shear stress components τ_{xy} in the vicinity of the dislocation by using the expression of eqn. (1.9).

(b) Determine the effects of this stress field on the movements of dislocations sitting at positions A, B, C and D with the same Burgers vector as the one at the origin.

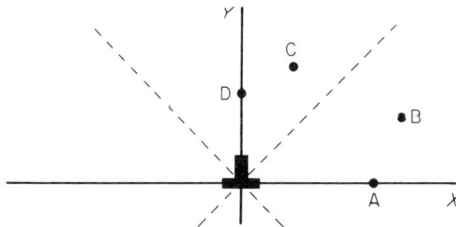

(c) Repeat (a) and (b) for screw dislocations and use the stress expression of eqn. (1.10).

1.5. For a large number of dislocations piling up in a grain of diameter d, the number of dislocations is given by $n = \tau d/2A$. τ is the applied shear stress on the slip plane and A is $Gb(2 - \nu)/4\pi(1 - \nu)$ for mixed dislocations. Derive the Hall–Petch relation between the yield strength and grain size. The yield condition is assumed to be reached when the stress on the leading dislocation is equal to the critical stress necessary to propagate plastic deformation at the grain boundary.

1.6. Use the simplified form of Orowan stress, $\tau = Gb/l$, as given in eqn. (1.11) to plot the theoretical tensile yield strength versus the ratio of barrier spacing l and Burgers vector b. By assuming the theoretical strength of $G/25$ for copper and $G/15$ for aluminium and nickel, determine the necessary l/b values to achieve these strengths. Compare these strength values with the strengths of commercially available alloys. The highest yield strengths obtained at room temperature by precipitation hardening alone are about 100 ksi ($6 \cdot 9 \times 10^8$ N/m^2) in aluminium alloy (containing magnesium, zinc and copper), 160 ksi (11×10^8 N/m^2) in copper alloy (containing beryllium) and about 200 ksi ($13 \cdot 8 \times 10^8$ N/m^2) in nickel base alloy (the nimonics) (see Reference 21).

REFERENCES

1. Needham, J., *Science and Civilization in China*, Vol. 4, Cambridge University Press (1965).
2. *The Holy Scriptures*, according to the Masoretic Text, Jewish Publication Society of America.
3. Kraus, H., 'On the Mechanical Properties and Behavior of Human Compact Bone', *Advances in Biomedical Engineering and Medical Physics*, S. N. Levine, ed., Vol. 2, Interscience Publishers, New York (1968), p. 169.
4. Moffatt, W. G., Pearsall, G. W., and Wulff, J., *The Structure and Properties of Materials*, Vol. 1, John Wiley and Sons, Inc., New York (1966).
5. van Vlack, L. H., *Materials Science for Engineers*, Addison-Wesley Publishing Company, Inc., Reading, Mass. (1970).
6. McClintock, F. A. and Argon, A. S., *Mechanical Behavior of Materials*, Addison-Wesley Publishing Company, Inc., Reading, Mass. (1966).
7. Guy, A. G., *Introduction to Materials Science*, McGraw-Hill Book Company, New York (1972).
8. Hilsch, R., 'Amorphous Layers and Their Physical Properties', *Non-Crystalline Solids*, V. D. Fréchette, ed., John Wiley and Sons, New York (1958).
9. Zachariasen, W. H., 'The Atomic Arrangement in Glass', *J. Am. Chem. Soc.*, **54**, 3841 (1932).
10. Gordon, M., *High Polymers*, Iliffe Books Ltd., London (1963).
11. Mark, H. F., 'The Nature of Polymeric Materials', *Scientific American*, **217**, 148 (1967).
12. Macmillan, N. H., 'Review: The Theoretical Strength of Solids', *J. Mater. Sci.*, **7**, 239 (1972).
13. Nabarro, F. R. N., *Theory of Crystal Dislocations*, Oxford University Press, London (1967).

14. Chou, T. W., 'Elastic Behavior of Disclinations in Nonhomogeneous Media', *J. Appl. Phys.*, **42**, 4931 (1971).
15. Chou, T. W. and Pan, Y. C., 'Elastic Energy of Disclinations in Hexagonal Crystals', *J. Appl. Phys.*, **44**, 63 (1973).
16. Weertman, J. and Weertman, J. R., *Elementary Dislocation Theory*, Macmillan, New York (1964).
17. Hirth, J. P. and Lothe, J., *Theory of Dislocations*, McGraw-Hill Book Co., New York (1968).
18. Swann, P. R., 'Dislocation Arrangements in Face-Centered Cubic Metals and Alloys', *Electron Microscopy and Strength of Crystals*, G. Thomas and J. Washburn, ed., Interscience Publishers, New York (1963), p. 131.
19. Bailey, J. B. and Hirsch, P. B., 'The Dislocation Distribution, Flow Stress and Stored Energy in Cold Worked Polycrystalline Silver', *Phil. Mag.*, **5**, 485 (1960).
20. Kelly, A., 'The Nature of Composite Materials', *Scientific American*, **217**, No. 3 (1967).
21. Zackay, V. F. and Parker, E. R., 'Some Fundamental Considerations in Design of High-Strength Metallic Materials', *High Strength Materials*, V. F. Zackay, ed., John Wiley and Sons, Inc., New York (1965).
22. Kelly, A., 'Theory of Strengthening of Metals', *Composite Materials*, edited by the Institution of Metallurgists, American Elsevier Pub. Co., New York (1966).
23. Bunn, C. W., *Chemical Crystallography*, Oxford University Press, London (1945).

CHAPTER 2

FIBRE-REINFORCED COMPOSITE MATERIALS

2.1 INTRODUCTION

The history of the development of modern composite materials[1] can be dated back to the early 1940s. It was in 1940 when fibreglass was first employed to reinforce plastics. The first successful application was found in its use in nose radar domes, known as radomes, to protect aircraft antennas. Glass–polyester radomes were used to replace the plywood and canvas–urea domes. This type of fibrous composite demonstrated excellent load-bearing capacity, thermostability, and resistance to weathering. But, above all, it is their 'transparency' to electromagnetic waves that made them most desirable for housing electronic equipment. Radomes made from fibreglass-reinforced plastics have since been widely deployed on ships and ground installations, as well as aircraft. The first fibreglass boat was moulded in 1942. In the same year, laminates of fibreglass with polyester resin were produced. Then, in 1946, the United States Government patented the first filament-winding process.

In view of their unique fatigue characteristics, reinforced plastics were applied to aircraft propeller blades around 1950: this year also marked the beginning of an era of rapid development in the application of fibre-reinforced composites in the aeronautic and aerospace industries. Early studies on reinforced phenolics indicated that they possess superior ablation characteristics. The first re-entry nose-cone made from phenolic resin and asbestos fibres was proved successful in 1956. In the following year, a re-entry cone was recovered from the Pacific for the first time. Since then fibre-reinforced composites have also played an important role in the manufacture of large rocket motors.

The savings in both weight and cost gained by the use of composite materials are also impressive. For example, the Boeing 727 jet airliner introduced in 1960 used 5000 pounds (2273 kg) of reinforced plastics. The substitution of metals by composites resulted in a 33 per cent reduction in weight. The recent design of Boeing's SST calls for the use of more than 6000 pounds (2727 kg) of reinforced plastics and another 6000 pounds (2727 kg) of unreinforced plastics. The application of composites to the automobile industry began two decades ago when fibreglass reinforced polyester was used on the Corvette body. It was at nearly the same time that reinforced plastics were applied to hydrospace vehicles on a large scale. Because of their insulation characteristics, some composites also find their applications in electrical laminates. For example, fibreglass–epoxy laminates were first used in 1956 for printed circuit boards.

The development of advanced composite materials was initiated in the past decade with the successful manufacture of several high-strength fibres. S-glass and boron fibres have shown tensile strength of over one-half million psi ($3 \cdot 45 \times 10^9$ N/m^2). The fibre moduli are near 50 million psi ($3 \cdot 45 \times 10^{11}$ N/m^2). The commercial production of continuous graphite filaments was also achieved with moduli over 50 million psi ($3 \cdot 45 \times 10^{11}$ N/m^2). The success that has been made in the developments of boron composites, metal matrix composites, and, more recently, in carbon–carbon composites has demonstrated the ever-increasing potential of composite materials.

The usage of fibre-reinforced composite materials covers an extremely broad spectrum, ranging from bathtubs and fishing rods to helicopter fuselages. The early development of composite materials was hindered by the high raw material costs and the slow and expensive processing methods. However, it is now feasible that the cost of some composites can be drastically reduced. Fully-automated methods have been developed for manufacturing some composite products. These also result in better quality control and lower cost. The impact of fibre-reinforced composite materials is vital to all industries. This fact is demonstrated by the exponential growth rate in the consumption of fibreglass-reinforced polyesters in the United States. The amount used was 100 million pounds ($4 \cdot 5 \times 10^7$ kg) in 1950, and nearly 1200 million pounds ($5 \cdot 45 \times 10^8$ kg) in 1970. In view of the limited possible improvements in metallic materials, the potential of growth of fibrous composites will eventually make them competitive with most other structural materials.

The strength theories for composite materials are discussed in detail in Chapter 8. The major functions of the individual phases in fibrous composites are now briefly outlined. The fibres in a composite are often deliberately aligned with the directions of loadings. Externally-applied forces are transferred to and distributed among the fibres. The transferring and distribution of loadings are achieved through the matrix material and take place at the fibre–matrix interface. The fibres are also responsible for the stiffness of the composite. In metal matrix composites, the fibres may lower the density of the composite. Although fibres are generally more brittle than the matrix the breaking of fibres at localised regions does not trigger catastrophic failure of the composite. This is due to the fact that cracks originating in the fibres are arrested by the ductile matrix. Likewise, cracks extending in the matrix cannot find a continuous path when they encounter the harder fibre material. Besides imparting toughness to the composite, the matrix material also binds the fibres together and protects them from hostile environments.

The fibre–matrix interface plays a rather unique role in the behaviour of composite materials. The nature of the interface determines the wetting and bonding between the constituent phases. This, in turn, controls the effectiveness of load-transference and the strength of composites. Transport of mass between fibres and matrix may take place across the interface. The mechanical coupling of a fibre and a matrix at an interface may result in additional stress fields locally, and may further complicate the mechanical behaviour of the components. It is also realised that imperfect bonding between the fibre and matrix can enhance the ductility of the composite, because cracks originally propagating normal to the fibres may split a weakly bonded interface. Consequently, the crack front is blunted and cracks extending in the direction of loading eventually become inactive.

In this chapter the various types of fibre-reinforced composite materials are examined by first discussing the properties of the component phases. The deformation of a fibrous composite is analysed in Chapters 6 and 7 for plate- and shell-type structures. The behaviour of dispersion-strengthened composites is discussed in Chapter 3.

2.2 PLASTIC MATRIX MATERIALS

Reinforced with various kinds of fibres, plastics are widely used in composites as matrix materials. They are introduced in this section, since their structure and properties are quite different from those of other matrix materials such as metals. Plastic resins used as matrix materials for fibre reinforced products are classified as *thermosetting* and *thermoplastic*. The properties of a thermosetting plastic are developed by chemical reactions which link together small monomer molecules to form long interlinked polymer molecules. Catalysts or curing agents and the application of energy (heat, microwave, etc.) are usually required to accomplish this reaction. As a consequence of the formation of the three dimensional network of covalent bonds, the thermosetting plastics are fairly rigid. They will not soften upon further heating since the polymerisation reaction is irreversible. At high temperatures, however, the covalent bonds may break. This irreversible destruction of the network causes the resin to lose its rigidity and strength at elevated temperatures. Thermoset polymers such as polyester, epoxy, phenolic, and silicone are widely used as matrix materials of fibre-reinforced composites. Thermoplastic materials, on the other hand, do not possess a rigid network structure. The linear molecular chains are formed prior to the development of the matrix and consequently are linked together mainly through the weak van der Waals bonds (Section 1.2). A thermoplastic material can be softened upon heating to a temperature known as the *glass-transition temperature* and the viscosity of the plastic decreases as the temperature rises. Thermoplastic materials softened at temperatures somewhat above the glass-transition temperature regain their strength upon cooling. At elevated temperatures the covalent bonds of the polymer are destroyed and the resin decomposes. Since the glass-transition temperature is considerably lower than the temperature required to destroy covalent bonds, thermoplastic resins are limited to lower temperature applications than thermosetting resins. Thermoplastic materials such as polyethylene, polystyrene, and nylon have been used in composites as moulding compounds. Resins can be obtained from suppliers in various forms such as solutions in organic solvents, powders, and granules.

Composite materials with resin matrices and in particular, thermoplastic matrices, are attractive because of their relatively low fabrication cost. The low density, low electrical and thermal conductivity, corrosion resistance and translucence of the resin matrices further add to their attractions. The percentage of fibre composites based upon thermoplastics is much lower than that based upon thermosetting plastics. However, diligent efforts are being made in the development of fibre-reinforced thermoplastics in order to take advantage of their unusual formability. The structure and properties of some common plastic matrix materials, based upon the reviews made in References 2–6, are discussed in more detail in the following sections.

2.2.1 Polyester Resin

In Section 1.2, the principles of polymerisation of unsaturated organic compounds were discussed. The general reaction in *ester* polymerisation involves the combination of di-basic and di-acidic monomers and can be expressed as

$$n(HO—R—OH) \ + \ n(H—P—H) \ \rightarrow \ -[R—P]_n— \ + \ 2n\,H_2O$$

Di-basic and di-acidic monomers such as ethylene glycol

and fumaric acid

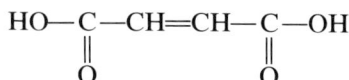

can condense by splitting out water to form the following unsaturated polyester

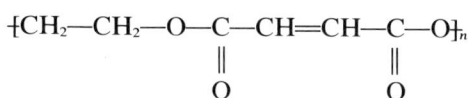

The double bonds between adjacent carbon atoms indicate that the polyester molecules are unsaturated. These unsaturation sites provide cross-links to be established between a molecular chain and an unsaturated material such as a styrene monomer. A strong three-dimensional network thus can be formed. The reaction to form the cross-linked structure is known as *curing*, and the curing agent then becomes part of the network structure. There is a great variety of choices of reactants, which all tend to affect the resin molecular weight and the distance between the sites of unsaturation along the chain. For example, if the unsaturated polyester is reacted with an unsaturated monomer such as styrene, the cross-linking reaction forms a network as illustrated below:

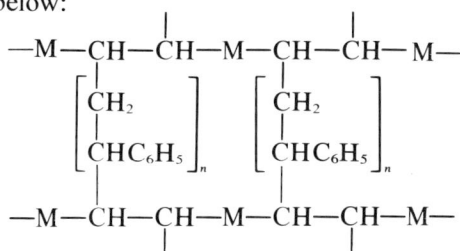

where

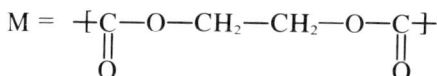

These various choices give rise to a family of polyester resins with each member of the family possessing different sets of properties. Since the strength as well as the thermal and chemical stability depends on the degree of cross-linking, it is important to choose the appropriate combination of reactants and monomers for the desired end product. The degree and the rate of curing is controlled by *catalysts*. The function of a catalyst is to act as an initiator for the polymerisation process through the formation of free radicals (Section 1.2). The rate of their formation and, hence, the cure time is affected by the temperature. By varying the catalyst, curing can be completed in times ranging from several days to less than a minute at temperatures from 70°F to 300°F (21·1°C to 148·9°C). The molecular weight and viscosity of the resulting polyester resin can be controlled by the duration of reaction in this temperature range. In view of the catalyst effect, it is desirable to store thermosetting polymer resins at room temperature and away from sunlight. The high viscosity of polyester resin as received from suppliers can be modified by dissolving it in styrene to obtain the desired viscosity. Most polyesters can be used satisfactorily at temperatures up to 482°F (250°C). The strength of polyester resins deteriorates with increase in service temperature. A major disadvantage of polyester resins is that they shrink when cured. The shrinkage of 4 to 8 per cent by volume is due to the shrinking of the monomer at curing. Consequently, it is difficult to obtain smooth surfaces in the final products.

Polyester resins can be incorporated with reinforcement materials very easily. It is often desirable to add mineral *filler* to polyester resins. In addition to lowering the cost of resins, filler materials also improve the surface appearance, resistance to water, and reduce shrinkage. Some typical properties of polyester resins are given in Table 2.1.

2.2.2 Epoxy Resin

The structure of epoxy resin is characterised by the epoxy group $-CH-CH_2$, also known as epoxide. Epoxy resins of several families are now available ranging from viscous liquids to high-melting solids. Among them, the conventional epoxy resins manufactured from epichlorohydrin and bisphenol A remain the major type used. The

TABLE 2.1

TYPICAL PROPERTIES OF SOME RESIN MATRIX MATERIALS

Properties	Materials		
	Polyester[a]	Epoxy[b]	Polyimide[c]
Specific gravity	1·28	1·11	1·90
Tensile modulus 10^6 psi (10^{10} N/m^2)	0·51 (0·35)	0·5 (0·34)	4·5 (3·1)
Shear modulus 10^6 psi (10^{10} N/m^2)	—	0·2 (0·14)	—
Tensile strength ksi (10^8 N/m^2)	7·97 (0·55)	5 (0·34)	28 (1·93)
Compressive strength ksi (10^8 N/m^2)	20·3 (1·40)	—	41·9 (2·89)
Poisson's ratio	—	0·4	—
Strain at break	2%	10%	—
Thermal expansion in./in.-°F(cm/cm-°C)	$55·6 \times 10^{-6}$ (100×10^{-6})	32×10^{-6} ($57·6 \times 10^{-6}$)	8×10^{-6} ($14·4 \times 10^{-6}$)
Thermal conductivity Btu/h ft °F (kcal/h m °C)	0·116 (0·17)	1·7 (2·5)	3·59 (5·34)

[a] Reference 56.
[b] Reference 57.
[c] Reference 58.

chemistry of manufacturing epoxy resins can be found in Reference 3. Commercially available resins may contain modifiers, or they may be a mixture of resin types. There are a variety of hardeners or curing agents generally used for epoxy resins. The amine type compounds are often used in structural applications. The hardening effect is achieved through the formation of cross-links between the resin polymer chain and the hardener, or by direct linkage among the epoxy groups. The characteristics of the resin-hardener system can be varied through the addition of modifiers. A typical type of modifier is a diluent. Diluents, as the term implies, tend to lower the viscosity of epoxy resins. Some diluents can retard the degradation of resin properties by reacting with the epoxy resins. Epoxy resins are also modified by *plasticisers.* Plasticisers are low-molecular-weight additives and serve to separate molecular chains from one another. The addition of plasticisers thus yields a non-crystalline solid with improved impact and low-temperature properties. Just as for polyester resins, filler materials can be added in epoxy resins as modifiers.

Composite materials using epoxy resins show far better resistance to chemicals and water than those using polyester resins. The shrinkage of epoxy is less than 2% and there is no water or volatile by-product generated during curing. As a result there is less chance for the epoxy matrix to pull away from the reinforcement and the subsequent exposure of fibres at the crack interfaces. Epoxy resins can be cured in the temperature range of 5 to 180 °C (41 to 356 °F). The cured epoxy has demonstrated thermal stability up to 250 °C (482 °F). The most desirable characteristic of epoxy resins lies in the fact that the properties of these resin systems can be 'tailor made' to fit a wide variety of applications through proper choice of resins, curing agents, and modifiers. The cost of epoxy resins is higher than polyesters and phenolics. Care should be exercised in handling certain hazardous resins and hardeners. The properties of different epoxy types are characterised by their viscosity, softening point, and epoxide value. Typical properties of epoxy resins are given in Table 2.1.

The versatility in form and property of epoxy resins has provided a wide spectrum of applications ranging from building and construction to electrical application. Their uses in filament-wound structures are extremely broad. These include pipes and tanks, electrical conduits, transformer and switchgear components, and pressure vessels, as well as aerospace and hydrospace vehicles.

2.2.3 High-Temperature-Resistant Polymers

Phenolics
The phenolic resins are derived from the condensation of phenols
(C_6H_5OH) and aldehydes such as formaldehyde (HCHO). During
curing, phenolics undergo three distinct stages. In the initial stage, the
condensation stage, water is formed as a by-product. Upon heating,
phenolics transform from a soluble, thermoplastic material, to a hard,
insoluble, and infusible thermosetting polymer. The strength of
phenolics is attributed to the cross-link formed by the formaldehyde in
the polymerised materials. Phenolics can be used at high temperature
up to 300 °C (572 °F). The cost of phenolic resins is the lowest among
thermosetting plastic materials. Some forms of phenolic resins may
also generate volatile by-products during curing. Hence, it may be
necessary to apply pressure when laminates are made of fibres
impregnated with phenolic resin.

Silicone
Silicone resin is characterised by the structure in which the polymer
chains consist of alternating oxygen and silicon atoms. By-products are
also generated during condensation. Silicone resins are known for their
thermal and oxidative resistance. They have been used in the temperature range of 500° to 1000 °F (260 to 537·8 °C). Although the price of
silicone resin is relatively high, the excellence in both mechanical and
electrical properties at elevated temperature has made fibre-reinforced
silicone resins ideal for radomes and supersonic vehicle components.

High Performance Polymers
Because of the advantages of polymeric resins discussed previously,
considerable effort has been expended to increase their service temperature range. Several polymeric materials have been developed which
can be used in the temperature range 600 °F to 1000 °F (315·6 °C to
537·8 °C). The properties of these high-performance polymers are attributed to the rigidity of the polymer chain achieved by the incorporation of ring structures. Of the various high performance polymers, the
polyimides are currently the most widely used. A typical polyimide
repeat unit is sketched below:[55]

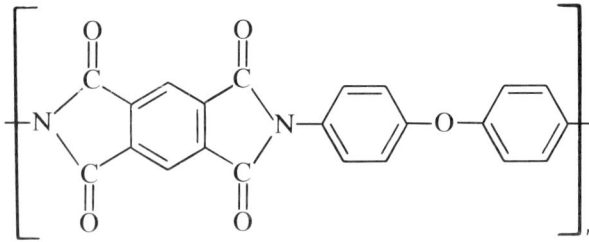

Although the increase in rigidity has enhanced the strength and modulus of the polymer, the brittle material offers less resistance to crack propagation. These polymers are also more susceptible to cure and thermal shrinkage.

To conclude the discussion on resin matrix materials the typical thermal and mechanical properties of epoxy, polyester and polyimides are listed in Table 2.1.[56–58]

2.3 REINFORCEMENT MATERIALS

The principal function of fibres in composite materials is to carry the load along the direction of reinforcements. Besides providing strength, fibres also enhance stiffness of the composites. It is suggested by the American Society for Testing and Materials that the term *fiber* be used for any material in an elongated form such that it has a minimum length to a maximum average transverse dimension of 10:1, a maximum cross-sectional area of 7.9×10^{-5} in.^{2} (5.1×10^{-4} cm^2), and a maximum transverse dimension of 0·010 in. (0·0254 cm). Continuous fibres in composites are usually called *filaments*. The term *wire* is usually reserved for metallic filaments. When a fibre has a rectangular shape with the ratio of width-to-thickness at least 4:1, it is called a *ribbon*.

Fibrous reinforcement materials can be categorised according to their structures in the following.

2.3.1 Amorphous Materials

Amorphous materials are generally characterised by the lack of long-range order. Glass and fused silica fibres are typical of this category. Their composition and mechanical properties are discussed in the following.

Glass Fibres

As depicted in Section 1.2 the structures of glasses in their solid states consist of rigid networks of silica. The network structure is responsible for the superior strength of glasses. The variation of glass properties can be achieved by modifying the structure of silica with various additives. Among all the glass fibres used in composites, *E-glass* is the type most widely used. The average weight percentages of the major components in E-glass are SiO_2 (54·4%), Al_2O_3 (14·4%), CaO (17·5%), B_2O_3 (8.0%) and MgO (4·5%). Oxides such as Na_2O, K_2O, TiO_2 and Fe_2O_3 exist in small quantities. Other types such as *S-glass* and *D-glass* also have been employed as reinforcements in composite materials. The composition of S-glass is SiO_2 (65%), Al_2O_3 (25%) and MgO (10%). S-glass possesses higher tensile strength and modulus of elasticity than E-glass. The strength of D-glass is lower than those of E- and S-glasses. D-glass is suitable for high-performance electronic application due to its low dielectric constant. For more details concerning glass fibre see, for example, Reference 7.

Glass fibres used in composite materials are manufactured as continuous filaments. In this process, raw glass is fed into a furnace and melted. Then fibres are drawn from the molten glass at high speed. Commercially produced filaments have diameters ranging from 0.000 10 to 0.000 75 in. ($2·4 \times 10^{-4}$ to $1·9 \times 10^{-3}$ cm). These fibres, after receiving a protective coating, are gathered in the form of bundles, called *strands*. The treatment of the glass fibres provides a suitable surface condition compatible with that of the resin matrix, so that good adhesion can be achieved. Unsatisfactory bonding between the fibre and matrix can cause interfacial failure, and stress will not be transferred effectively at the weakened spots. When a group of parallel strands are gathered in the form of a ribbon, and wound onto a cylindrical tube, it is called a *continuous roving*. Rovings are designated on the basis of yards per pound. If the continuous strands are looped back and forth upon themselves and held in roving form by a slight twist, it is called a *spun roving*. Plate III[7] shows both continuous and spun rovings. Strands can be cut to lengths of $\frac{1}{8}$ to 2 in. (0·3175 to 5·08 cm). They are known as *chopped strands*. Both continuous and chopped strands can be laid in random orientations to make *reinforcing mats*. It is sometimes desirable to assemble strands by twisting them together. Twisted strands are known as *yarns*. The strands may have either a left-handed or a right-handed spiral. Since a yarn consisting of a single, twisted strand tends to form kinks, it is desirable to make yarns with an equal

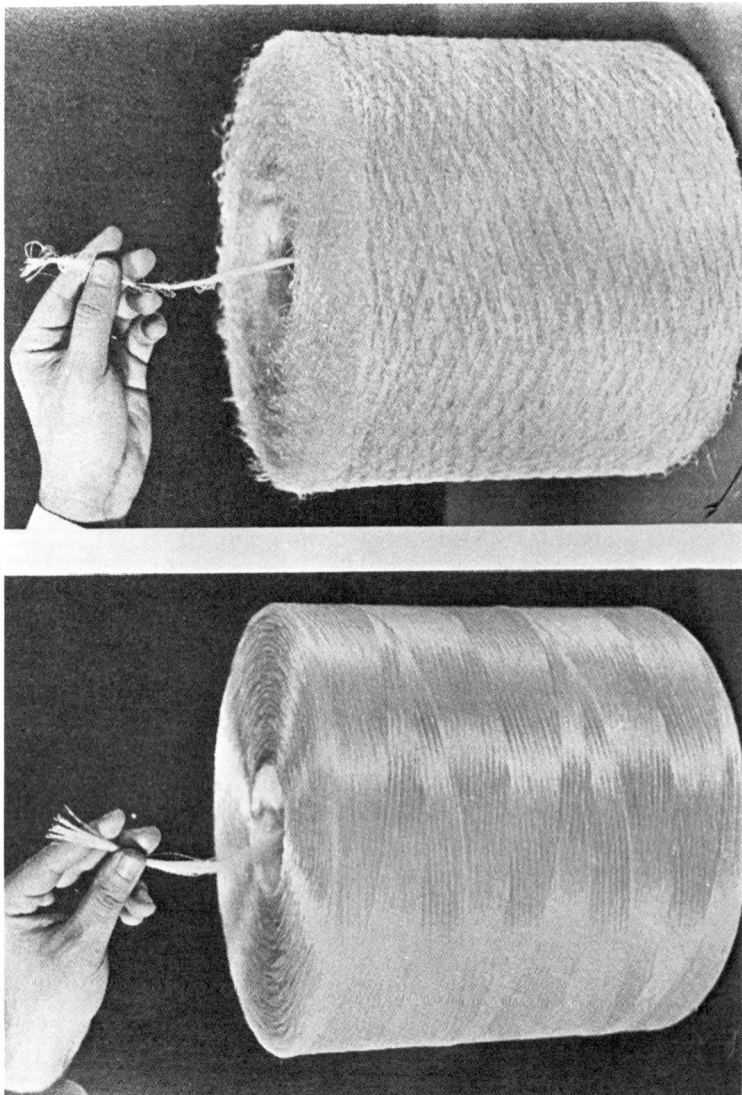

Plate III.   (a) Continuous roving. (b) Spun roving.[7] (Reprinted by permission of Owens-Corning Corp.)

number of strands twisting in different orientations. Both glass fibre strands and yarns can be interlaced to form *woven rovings* and *woven fabrics*, respectively. They are so named because the appearance of the former is much coarser than the latter.

TABLE 2.2

TYPICAL ROOM TEMPERATURE PROPERTIES OF E-, S- AND D-GLASSES
(Reproduced from Reference 7 by permission of Van Nostrand Reinhold Co.)

Property	E-Glass	S-Glass	D-Glass
Specific gravity[a]	2·54	2·49	2·16
Tensile strength[a]	500	665	350
ksi (10^9 N/m^2)	(3·45)	(4·59)	(2·41)
Young's modulus[a]	10·5	12·4	7·5
10^6 psi (10^{10} N/m^2)	(7·24)	(8·55)	(5·17)
Elastic strain (%)[a]	4·8	5·4	4·7
Coefficient of			
thermal expansion[b]	2·8	1·6	1·7
10^{-6} in./in.°F			
(10^{-6} cm/cm°C)	(5·04)	(2·88)	(3·06)
Softening point[b]	1 555	1 778	1 420
°F (°C)	(846)	(970)	(771)
Dielectric constant[b]			
(10^6 Hz)	5·80	4·53	3·56
Index of refraction[b]	1·547	1·523	1·47

[a] Properties measured on glass fibres.
[b] Properties measured on bulk glass.

The typical physical properties of E-, S- and D-glasses are given in Table 2.2.[7] The major advantage of using glass fibres as reinforcement materials is their lower costs compared to any other kind of fibres currently available. They can be easily fabricated in continuous filaments. As can be seen from Table 2.2, glass fibres generally have rather high tensile strengths but relatively low Young's moduli. The low softening point of glass fibres also restricts their uses to resins and low melting-point matrices. Since the strength of glass fibres is affected by the existence of any flaws on their surfaces, great care must be taken in handling them.

High Silica and Quartz Content

High silica content fibres are produced by treating common glass fibres with a hot acid.[8] During this process the impurities can be leached away and fibres with a high silica content will result. High silica content fibres used in composite materials have purities ranging from 95% to 99·4% SiO_2. They are available in forms mentioned for common glass fibres. *Quartz fibres* are produced from high-purity natural quartz crystals and their typical SiO_2 content is 99·95%. Filaments drawn from quartz crystals are extremely thin; they are usually combined into strands and other textile forms.

High silica and quartz content fibres show excellent thermal stability. They can be used at temperatures up to 3000°F (1648·9°C). Hence, these fibres are used in heat shields for re-entry vehicles, as well as other thermal blast shields. The price of quartz fibres is, naturally, higher than that of high silica content fibres, due to the great difference in the cost of the raw materials.

2.3.2 Whiskers

Whiskers are characterised by their fibrous, single-crystal structures, which have almost no crystalline defects. Numerous materials, including metals, oxides, carbides, halides, and organic compounds, have been prepared in the form of whiskers. The strengths of whiskers are comparable to those of glass fibres. The Young's moduli of whiskers are superior to those of glasses and close to those of boron fibres. What is equally impressive is that, regardless of their high strength and modulus, whiskers are tough and can tolerate elastic strains of 3 to 4 per cent. The very remarkable properties of high-strength, high-modulus, and high-melting-point ceramic whiskers are listed in Table 2.3.[9] The performance of whiskers at elevated temperatures is far better than that of any other fibres. The ceramic whiskers are particularly good in resisting mechanical and chemical damages. Whisker-type fibres are generally used to reinforce resin and metallic matrix materials. At the present time the cost of whisker fibres is higher than most of the other fibres. The typical diameters of whisker fibres are in the range of 0·001 in. to 0·000 005 in. (0·0025 cm to 0·000 013 cm). The fine dimensions of whisker fibres make them difficult to handle. Furthermore, they cause problems in mechanical tests due to difficulties in gripping and aligning the fibres, making cross-sectional area measurements, and detecting fibre elongation accurately. The various kinds of whiskers are discussed below according to their methods of production.

TABLE 2.3

PHYSICAL PROPERTIES OF WHISKERS
(Reproduced from Reference 9 by permission of Van Nostrand Reinhold Co.)

Material	Specific gravity	Melting point °F (°C)	Tensile strength 10^6 psi (10^{10} N/m^2)	Young's modulus 10^6 psi (10^{10} N/m^2)
Aluminium oxide	3·9	3 780 (2 082)	2–4 (1·38–2·76)	100–350 (68·9–241·3)
Aluminium nitride	3·3	3 990 (2 199)	2–3 (1·38–2·07)	50 (34·5)
Beryllium oxide	1·8	4 620 (2 549)	2·0–2·8 (1·38–1·93)	100 (68·9)
Boron carbide	2·5	4 440 (2 449)	1 (0·69)	65 (44·8)
Graphite	2·25	6 500 (3 592)	3 (2·07)	142 (97·9)
Magnesium oxide	3·6	5 070 (2 799)	3·5 (2·41)	45 (31·0)
Silicon carbide (alpha)	3·15	4 200 (2 316)	1–5 (0·69–3·45)	70 (48·3)
Silicon carbide (beta)	3·15	4 200 (2 316)	1–5 (0·69–3·45)	100–150 (68·9–103·4)
Silicon nitride	3·2	3 450 (1 899)	0·5–1·5 (0·34–1·03)	55 (37·9)

Metallic whiskers can be grown with reasonable ease through an *evaporation–condensation process.*[10] In this method, metals located at the source of evaporation are heated to a temperature higher than their melting points. A temperature gradient is maintained between the source and the condensation site in the growth chamber. The temperature at the substrate for vapour condensation is kept below the melting point of the metal. The supersaturation and subsequently the condensation of vapours at low temperature enable whiskers to grow. The gradient of vapour pressure is controlled by the temperature gradient in the growth cell. Consequently, the transport of mass through the vapour phase can be maintained. The gradient of vapour pressure can be further manipulated by introducing an inert gas into the growth chamber. Metallic whiskers such as nickel, copper, iron, aluminium, silver, germanium, etc. have been grown by this method in various sizes. The evaporation–condensation process is not suitable for the growth of oxide whiskers in view of their high melting points and low vapour pressures.

Another crystal growth mechanism important to the preparation of whiskers is known as the *vapour–liquid–solid mechanism.*[11] The understanding of this mechanism stemmed from the observation of the growth of silicon whiskers from vapour deposition. It was noted that the addition of a small amount of impurity promoted the growth of filamentary crystals. Impurities such as gold, nickel, copper, and magnesium all have this effect. The growth rate is much faster in length than in thickness. There are other impurities which do not promote filamentary growth and, hence, films or modular deposits are a result. It is also confirmed that the growth of silicon whiskers by this mechanism is generally free of dislocations, and a liquid layer is observed on the surface of the crystal growing from the vapour. The many unique features concerning the growth of silicon crystals have led to the proposition of the vapour–liquid–solid mechanism.

In the controlled growth of silicon whiskers by this mechanism, single crystal silicon wafers are used as substrates. An impurity such as gold is added to the substrate. The temperature of the substrate is then raised to the range of 600 °C (1112 °F) to 800 °C (1472 °F) which is below the melting point of both gold and silicon. In this temperature range a liquid alloy of gold and silicon is formed. This is due to the fact that the alloy phase melts at a temperature below the melting points of the component phases. A discussion of the temperature–alloy phase relation is presented in Section 3.2. A mixture of H_2 and $SiCl_4$ gases is introduced into the deposition chamber. Silicon vapour is generated upon the reduction of $SiCl_4$. The surface of the liquid alloy layer provides preferential sites for the condensation of the vapour phase. The increasing content of silicon in the liquid alloy finally reaches a critical value and silicon starts to precipitate through the solid–liquid surface. The transport of silicon in the liquid phase is a diffusion-controlled process. The growth of the whisker takes place along the direction perpendicular to the solid–liquid interface. Desired diameter and length of a whisker can be achieved through carefully controlled experimentation. It is of great importance to keep the substrate, alloying agent, and the gaseous species free of impurities. A change in the whisker diameter results if there is a change of the deposition temperature. Undesirable temperature gradients in the chamber may cause the formation of kinks and branches during growth. It has also been noted that, at the initial stage of growth, the liquid alloy drop is very sensitive to the supply of silicon vapour. The alloy may break into smaller droplets if there is an over supply of the vapour. When the growth reaches a steady condition, the number of silicon atoms deposited on the surface of the liquid alloy should equal the number of atoms transported to the solid region. A schematic drawing of the vapour–liquid–solid growth process is given in Fig. 2.1.[11] The principle underlying the growth of silicon whiskers can be applied to the growth of any crystal based upon the vapour–liquid–solid mechanism. Crystals

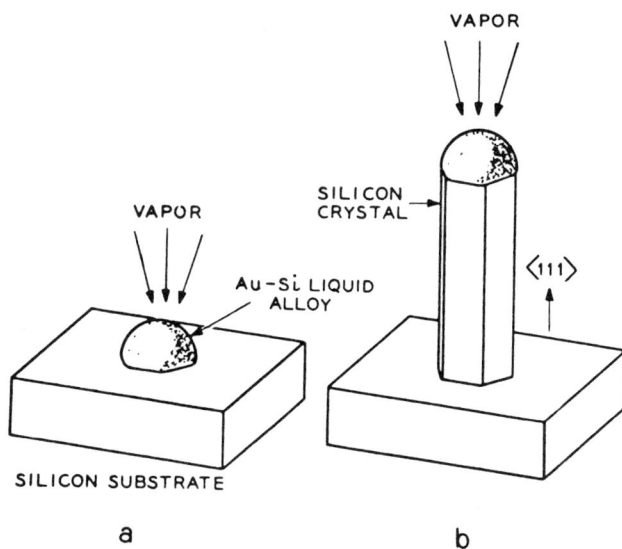

*Fig. 2.1.   VLS mechanism for the growth of a silicon crystal. (Reproduced
from Reference 11 by permission of Wiley-Interscience.)*

including alumina, boron, germanium, magnesium oxide, nickel oxide,
and selenium, as well as silicon, have been grown by this mechanism.

Whiskers grown by the vapour–liquid–solid mechanism are relatively free of dislocations. Definitely, dislocations do not play a
significant role in the growth of whiskers. It has been observed that
dislocations, originated in the substrate, may grow into the whisker
crystal. These dislocations usually terminate at the side faces of the
whisker. Dislocations are found in whiskers only in the region near the
whisker–substrate interface. The low density of dislocations in the
crystal may be attributed to the protection of the growth interface from
contamination by the liquid alloy. This is evident by the fact that in the
regions where silicon vapour condenses directly on to the substrate
and the growing crystal, the concentration of dislocations is much
higher than that in the bulk of the whisker.

Whiskers grown by the vapour–liquid–solid mechanism can be
produced in large quantity. One method of production involves the
usage of a silicon substrate on to which a film of gold is deposited. A
gold–silicon alloy is formed upon heating the substrate and the liquid
film tends to break up, due to surface tension, into small droplets. As a

result, silicon vapour can be deposited, and whiskers can grow at high density. Whiskers have also been produced by *chemical reduction*. Hydrogen reduction of halides of copper, iron, nickel, and cobalt are widely used. The process of *vapour-phase reaction* has been used to produce aluminium oxide and silicon nitride whiskers. The *melt growth* technique can be used to produce continuous whiskers.

2.3.3 Polycrystalline Materials

The fibrous structures in polycrystalline form consist of randomly oriented grains. The common ceramic fibres in this category are alumina, mullite, lithia–alumina spinel, etc. Polycrystalline metallic wires include, among others, tungsten, beryllium, and stainless steel. Carbon filaments are also polycrystalline in structure and usually have a certain degree of crystallinity in the form of graphite.

Among these fibres, the metallic filaments are widely used in composite materials.[12] They are easy to fabricate in large quantity and can be bonded to plastics and metals. Metal filaments are conventionally produced by drawing the wires through a series of diamond dies of successively smaller sizes. A modification of the conventional wire-drawing can be achieved by *bundle drawing*. In this process, filaments of relatively larger diameter are encased in a soft matrix. The entire composite bundle is drawn through a series of dies. The matrix material is then leached away from the bundle. It is desirable to use fine wires to achieve better strength and flexibility. However, the process of making fine wires by drawing is very tedious, and the cost of fabrication increases tremendously as wire diameter gets smaller. *Electrochemical plating* is another method that has been used, but the number of metals suitable for this system is limited, and the time needed for plating seems to impose another restriction on mass production. Metal filaments can be readily produced by *melt drawing*. The Taylor Process of melt drawing is suitable for making micron- and submicron-diameter wires. In this process, glass tubes filled with molten metal are drawn to fine diameter. Since the surfaces of the filaments are well-protected, wires prepared by the Taylor Process have strengths which are much greater than those produced by conventional wire drawing. The method of vapour-deposition, widely used in whisker preparation, can also be used to produce metal filaments.

Metallic wires are characterised by their high elastic moduli. Among them, molybdenum and tungsten are most outstanding. However, the density of metal filaments is higher than that of ceramic whiskers with the probable exception of beryllium. Metallic wires have been used to reinforce ceramic, resin, and metal matrix materials. Metallic wires are generally more ductile than glass and other polycrystalline-type fibres. A comparison of the physical properties of metal filaments is given in Table 2.4.[12,13] The strength of polycrystalline materials in general is greatly affected by their grain sizes (Section 1.5). The effects of recrystallisation and grain growth tend to reduce their strengths at high temperatures. The refractory metal wires and, in particular, their high-temperature performance are discussed in Section 2.5.

Carbon filaments have gained increasing importance in the composite technology. This is due to their low density, chemical inertness, and great strength at elevated temperatures. The graphite whiskers have demonstrated extremely high tensile strength and modulus of elasticity. More recently, considerable progress has also been made in producing continuous graphite filaments. The fabrication of the filaments involves a *precursor conversion* process.[14-16] Textile fibres, usually rayon and polyacrylonitrile (PAN)-based fibres are used as precursor materials. Polyvinyl alcohol, phenolics, and pitches also have been used as precursors. In the manufacturing process the molecular orientation in these fibres is preserved, and the strong carbon–carbon bonds are aligned along the fibre axis. The organic fibres are first pyrolysed by heating at low temperature. Carbonisation of the fibres is then followed at temperatures above 1000°C (1832°F). In the case of PAN-based fibres, tension is applied on the precursor when it undergoes pyrolysis in a protective atmosphere. In rayon-based fibres, the stretching is applied during graphitisation. Higher fibre modulus and strength result from stretching and the alignment of the graphite basal planes. It is well known that the graphite single crystals assume a hexagonal–close-packed structure and have a density of 2·25 g/cm³ (140·5 lb/ft³). The spacing between the adjacent basal planes is 3·354 Å. The carbon formed at low temperature exhibits a random amorphous structure in which the basal planes are not stacked in the regular sequence of a HCP crystal and the interlayer spacing is larger than 3·354 Å. This type of carbon structure, also known as turbostratic, can be transformed into the regular graphite structure by heat treatment of carbon at elevated temperatures. This process is known as *graphitisation* and is performed in the temperature range of 2000°C–3000°C (3632°F–5432°F). Carbon fibres produced after graphitisation contain graphite crystallites. The extent of graphitisation

TABLE 2.4

PHYSICAL PROPERTIES OF METALLIC WIRES

(Reproduced from Reference 12 by permission of Addison-Wesley Co.)

Material	Specific gravity	Melting point °C (°F)	Tensile strength ksi (10^8 N/m^2)	Young's modulus 10^6 psi (10^{10} N/m^2)	Coefficient of thermal expansion 10^{-6} cm/cm °C (10^{-6} in./in. °F)
Aluminium	2·71	660 (1 220)	42 (2·9)	10 (6·89)	23·6 (13·1)
Beryllium	1·85	1 350 (2 462)	160 (11)	45 (31)	11·6 (6·44)
Copper	8·90	1 083 (1 981)	60 (4·14)	18 (12·4)	16·5 (9·17)
Tungsten	19·3	3 410 (6 170)	420 (29)	50 (34·5)	4·6 (2·56)
Austenitic stainless steel	7·9	1 539 (2 802)	347 (24)	29 (20)	8·5 (4·72)
Molybdenum	10·2	2 625 (4 757)	320 (22)	48 (33·1)	

affects the density of fibres. PAN-based fibres are round and are available in untwisted strands. Rayon-based fibres are crenelated and are furnished in yarns. Carbon fibres produced by precursor conversion have achieved a tensile modulus of 4–60 million psi ($2\cdot76\times10^{10}$–$4\cdot14\times10^{11}\,N/m^2$) and a tensile strength of 100–400 ksi. ($6\cdot89\times10^{11}$ – $2\cdot76\times10^{12}\,N/m^2$). The densities of the fibres so far manufactured are in the range of $1\cdot32$–$1\cdot96\,g/cm^3$ ($82\cdot4$–$122\cdot4\,lb/ft^3$). The series of Thornel fibres are rayon-based and are manufactured by the Union Carbide Corporation. These fibres are available in the form of a yarn which consists of two plies twisted together. Each ply has 720 filaments. The nominal yarn diameter is $0\cdot02$ in. ($0\cdot0508$ cm). Some properties of Thornel fibres are listed in Table 2.5.[15]

TABLE 2.5

PROPERTIES OF COMMERCIAL THORNEL FIBRES[a]
(Reproduced from Reference 15 by permission of the AIME.)

Name	Filament diameter (μm)	Specific gravity	Initial tensile modulus $10^6\,psi$ ($10^{10}\,N/m^2$)	Average ultimate tensile strength ksi ($10^9\,N/m^2$)
Thornel 16	8·3	1·33	14 (9·6)	167 (1·15)
Thornel 25	7·1	1·42	27 (18·6)	180 (1·24)
Thornel 40	6·7	1·56	40 (27·6)	250 (1·73)
Thornel 50	6·5	1·67	57 (39·2)	315 (2·18)
Thornel 75	5·6	1·82	79 (54·4)	380 (2·64)

[a] Mechanical properties measured on epoxy-impregnated strands.

2.3.4 Multiphase Materials

Fibres in this category consist of two or more phases, of which one or more phases may be crystalline. Fibres made of boron, boron carbide, silicon carbide, and borides all can take the form of multiphase materials. For the purpose of illustration, the properties and preparation of boron fibres are discussed in this section.[17,18]

The development of boron fibre-reinforced plastics marked the beginning of the era of advanced composite technology. Boron fibres have shown superior average tensile strength of 500 ksi ($3\cdot45\times10^9\,N/m^2$) and elastic modulus of 60 million psi ($4\cdot14\times10^{11}\,N/m^2$). The density of boron fibre is $2\cdot6\,g/cm^3$ ($162\cdot32\,lb/ft^3$) and is close to that of glass. The specific strength (see Problem 2.3) of boron fibre is comparable to those of glass fibres and is far superior to those of metallic filaments. The specific modulus of boron is five times those of glass and metallic filaments with the exception of beryllium. Although the specific moduli of boron and beryllium are comparable, the specific strength of beryllium is only one-third of that of boron. Boron has a high melting point of 2050°C (3722°F) and is hard and brittle. It is not feasible to produce boron fibres by conventional fabrication means. Boron fibres used in advanced composite materials are produced by *chemical vapour plating*. In this process boron vapour can be obtained by either hydrogen reduction of halide systems or thermal decomposition of boron hydrides into boron and hydrogen vapours. Boron vapours generated in both cases are deposited on a substrate. The substrate material must have a high melting point and possess thermal stability at high temperatures: tungsten wires have performed satisfactorily as substrate in boron filaments. In the deposition process, tungsten filaments of 0·0005 in. (0·001 27 cm) in diameter are first heated to about 2200°F (1204°C) in a hydrogen atmosphere to remove surface contamination and oxides. After the treatment, tungsten filaments are fed into the plating chamber together with the reactant gases. For instance, in the hydrogen reduction of halide systems, the mixture of boron trichloride and hydrogen in the plating chamber reacts in the following manner:

$$2BCl_3 + 3H_2 \rightarrow 2B + 6HCl$$

The growth rate of boron is affected by the substrate temperature. It has been observed that an optimum growth rate of 0·0001 in. (0·000 25 cm) in diameter per second is obtained at the substrate temperature of 2000°F (1093°C). Boron filaments used in composites usually have diameters of about 0·004 in. (0·01 cm).

The multiphase structure of boron fibre is rather interesting. A close look at the microstructure can begin at the core structure. At elevated temperatures, there is considerable diffusion of boron atoms into the tungsten core. The diffusion in the reverse direction, namely, tungsten diffusing into boron, is carried out at a much lower rate. Tungsten is observed only in the immediate vicinity of the interface. As a result of the difference in rates at which atoms are transported from one region to another, voids are observed to form in the boron phase. On the other hand, compounds of tungsten–boride form in the tungsten core. Among them, W_2B_5 and WB_4 usually exist in appreciable amounts which induces an increase in volume of the core. Since the core region enclosed in boron is not free to expand, a severe compressive stress near 200 ksi ($1\cdot38 \times 10^9$ N/m^2) is induced in this region. Naturally, the boron phase surrounding the core is then under tensile stress. This stress is usually released by the cracking of boron filament. The cracks extend radially outward from the tungsten–boron interface but do not propagate to the free surface. This is due to the fact that boron near the filament surface is under compressive stress. Internal stress in the boron filament can result from the growth process and the differential of thermal expansion between tungsten and boron.

The structure of deposited boron consists of extremely small crystallites with diameters of the order of 20–30 Å. It is for this reason that the boron structure is usually referred to as 'amorphous'. The boron grown from the nucleation sites on the substrate takes the shape of cones. The growth cones are distinctly visible in the longitudinal section of a boron fibre (Plate IV). These cones are also responsible for the nodular or 'corncob' surface (Plate V). The nucleation and the subsequent growth of boron are affected by many factors. Scratches on substrates produced at wire-drawing and particles of graphite left from the

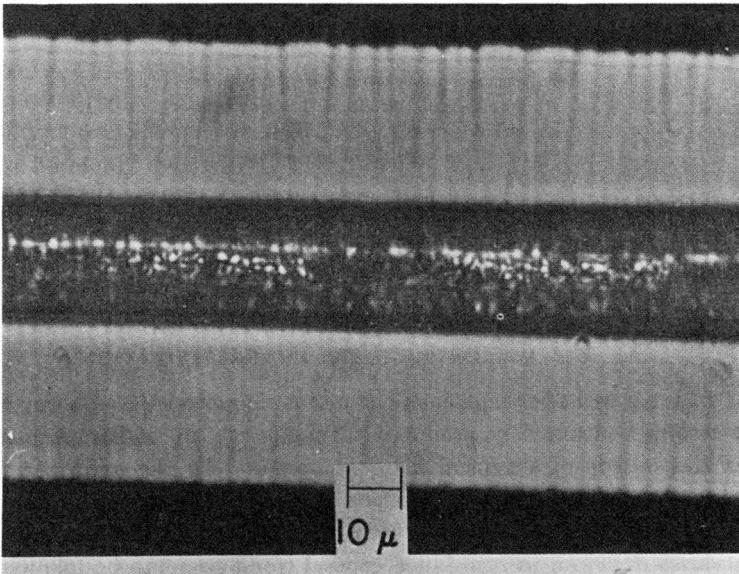

Plate IV. *Longitudinal cross-sectional view showing cone growth.*
(*Reproduced from Reference* 18 *by permission of Addison-Wesley Co.*)

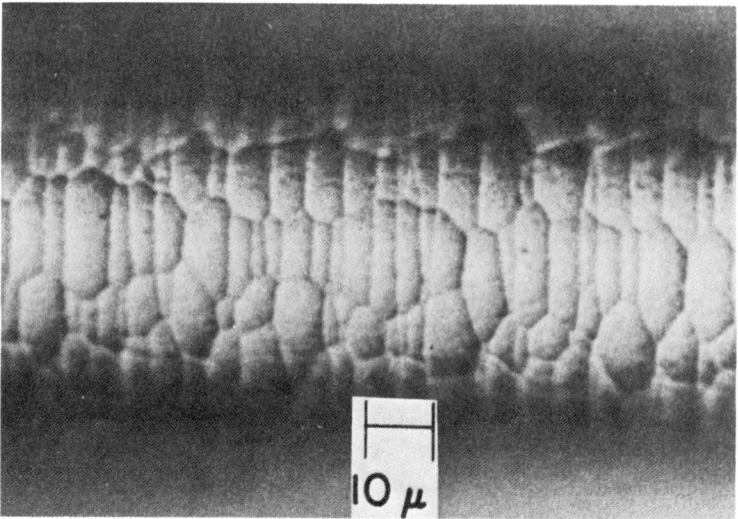

Plate V.   *Nodular structure of a typical boron filament surface.* (*Repro - duced from Reference* 18 *by permission of Addison - Wesley Co.*)

die-lubricant in drawing all provide preferential sites for nucleation and growth. The rapid growth associated with these sites actually is reflected on the filament surface by the large nodules. The overall sizes and distribution of the nodules are further controlled by the rate of deposition of boron vapour.[18]

The strength of boron fibres may deteriorate due to the presence of various kinds of defects. Flaws may be present around an enlarged nodule or on the surface of the tungsten substrate. Inclusion particles, such as graphite, used for lubrication and left on the substrate after drawing, or impurity particles picked up during deposition, are all detrimental to the strength of boron filaments. The major disadvantage of boron fibres is that their weights and dimensions are largely affected by the presence of tungsten cores. Experimental measurements indicate that the boron in boron filaments generally shows tensile strength of more than one million psi ($6 \cdot 89 \times 10^9 \, N/m^2$). However, as indicated earlier, the strength of boron–tungsten filament is about 500 ksi ($3 \cdot 45 \times 10^9 \, N/m^2$). This difference in strength is due mainly to the presence of flaws and impurities. Since it is noted that the surface condition of a substrate can affect the growth process and consequently the surface morphology of deposited boron, improvements may be made by substituting the tungsten core with silica substrates having a smooth surface. The interface bondings between boron fibre and resin matrices, such as epoxides and polyimides, are found to be very satisfactory. Since boron filaments tend to react with metals at elevated temperatures, their use in metal-matrix composites is limited to low-melting metals. The treatment of boron filaments for improving their high-temperature performance is discussed in Section 8.2. It should be pointed out, however, that the tensile modulus of boron filaments is less affected by temperature and barely affected by the process condition.

In addition to boron, filaments including silicon carbide, boron carbide, titanium diboride, titanium carbide, carbon, etc., all can be produced by the vapour-plating process. These filaments also show better stability at high temperature than boron filaments.

Fibres having the various structures described above can also be produced in the forms of *ribbon, flake* and *platelet*. For application in composite materials, glass fibres and whiskers are usually made into the form of *tapes*. Continuous tapes are produced by aligning the fibres and bonding them together in a matrix. Both metal and resin matrices can be used. The internal structure and the degree of surface perfection of fibres are affected by the fabrication methods. Several general statements may be made with respect to the effect of forming processes on the strength of various fibres.[19] It is well known that fibres produced by the same fabrication procedure show higher strengths when their diameters are made smaller. This is apparently due to the statistical nature of defect distribution on the fibre surface. The probability of finding flaws of certain critical length diminishes as the total surface area becomes smaller in thinner fibres. For instance, the strength of steel wires prepared by drawing nearly triples when the wire diameter is reduced from 0·05 in. (0·127 cm) to 0·005 in. (0·0127 cm). The strength of glass fibres drawn from melt doubles when the fibre diameter is reduced from 0·005 in. (0·0127 cm) to 0·0005 in. (0·001 27 cm). Among all the fibre types produced, the strength of whiskers are the highest. Alumina whiskers, for example, have a tensile strength of three-million psi ($2·07 \times 10^{10}$ N/m^2) when prepared by vapour deposition. This strength is nearly ten times that of alumina fibre produced by a drawing procedure. Similarly, graphite whiskers have a strength of about three million psi ($2·07 \times 10^{10}$ N/m^2), which is one hundred times the strength of graphite fibres prepared by extrusion. However, the choice of reinforcement materials for composites calls for the consideration of not only strength, but also other factors such as service conditions, compatibility of fibres with a matrix, and economy.

To summarise the discussion of reinforcing materials, it is interesting to compare their relative sizes in a schematical manner. This is shown in Fig. 2.2.[19] Since the reduction in weight is a major consideration for using composite materials, it is noteworthy to compare fibres on the basis of their specific strengths and specific moduli. A plot of the specific strengths versus specific moduli for the various types of fibres (see Problem 2.3 and References 60 and 61) will indicate that whiskers perform the best while polycrystalline fibres, with the exception of carbon and beryllium, do poorly. Carbon and beryllium fibres are favoured on such a diagram because of their low densities. A further consideration of reinforcing materials is their performance at elevated

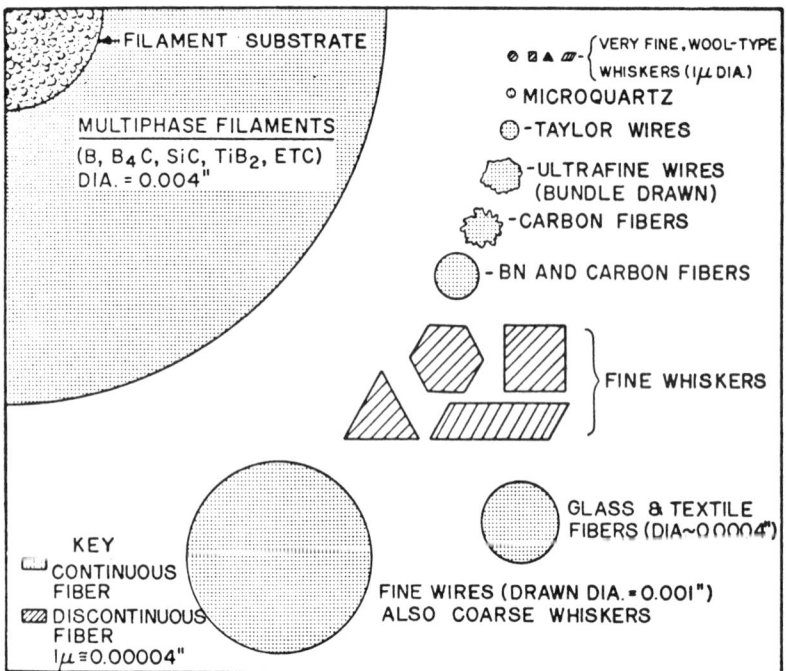

Fig. 2.2.   *Shapes and sizes of several fibres* (1 in. = 2·54 cm). (*Reproduced from Reference* 19 *by permission of Academic Press.*)

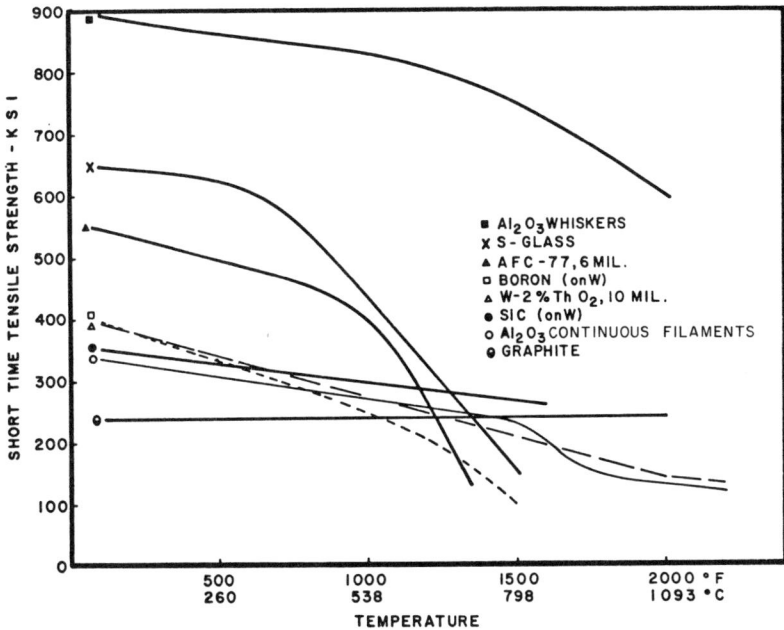

Fig. 2.3.   Short-time elevated temperature strength of reinforcements (1 ksi =
6·89 × 10⁶ N/m²). (*Reproduced from Reference* 13 *by permission of the Chemical
Rubber Co.*)

temperatures. Figure 2.3 indicates the tensile strength of some typical
fibres over a wide range of temperature. At 2000°F (1093°C) the
performance of alumina whiskers clearly stands out. Graphite fibres
show slight increase in strength with temperature. At 2000°F (1093°C)
W–2% ThO₂ and alumina filaments also retain nearly half of their
room-temperature strength.[13]

2.4   RESIN MATRIX COMPOSITE MATERIALS

2.4.1   Fabrication Methods of Resin Matrix Composites

Hand Lay-Up
Fabrication by hand lay-up[20] is the simplest method of making fibre-
reinforced plastics. In this method, a layer of random fibre mat or

woven fabric is first laid on a form. Thermosetting plastics are then brushed on to the reinforcing materials. This process is repeated until the desired thickness of the composite is reached. Air trapped in the resin is squeezed out by a roller. The resin which is catalysed is then allowed to cure at room temperature. Polyester and epoxy are the two most common resins used in hand lay-up fabrications. Very often, organic or inorganic fillers such as wood-flour, sawdust, clay, or sandstone are added to the resin for the purposes of reducing inflammability, providing extra weight, and decoration. The content of reinforcement achieved by this method is relatively low, about 30 per cent by weight.

The moulds used in hand lay-up may be made of a variety of materials such as wax, clay, wood, metal, paper, and plastic sheets. Mould releases, for example, polyvinyl alcohol, silicone, and mineral oils are often used for facilitating release of the final product from the mould. The tooling-up cost for hand lay-up is relatively low. This method is commonly used in making models and prototypes where only limited production is required. It also has been used to fabricate complex products which are not practical to make with matched dies, as well as for extremely large parts such as radomes and boat hulls.

Spray-Up
In the method of spray-up,[20] fibres are first fed through a chopper and cut into desired lengths. A mixture of fibre, resin, and catalyst is sprayed on to a mould. A roller is then used to smooth the surface and remove entrapped air.

Bag-Moulding
Bag-moulding is used to improve the quality of hand lay-up products by further removing the entrapped air. The three basic bag-moulding methods are *vacuum bag, autoclave,* and *pressure bag.*[21] In applying the vacuum-bag method, the lay-up of resin and reinforcing materials is first covered with a perforated parting film and a layer of jute bleeder material. The combination allows the bleeding of air and excess resin. Then the lay-up is covered with a flexible film diaphragm, such as cellophane or nylon, which is sealed to the mould. The vacuum is then drawn upon the whole system with a pressure about 12 psi ($8 \cdot 28 \times 10^4$ N/m^2). The bagging process should immediately follow the lay-up in order to avoid the hardening of resins. The entire bagged system can be cured either in an oven or an autoclave system.

In the case of autoclave curing, a large, metal pressure vessel is used and is pressurised with a gas—typically nitrogen. The pressure applied is in the range from 50 to 100 psi ($3 \cdot 45 \times 10^5$ to $6 \cdot 89 \times 10^5 \, \text{N/m}^2$). The autoclave system is heated and the hot gas is circulated to provide a uniform temperature within the vessel. Sophisticated autoclave systems provide electronic controls which produce programmed temperature/pressure-time cycles. After the laminate has been bagged and subjected to vacuum, it is placed inside the autoclave for cure. The vacuum system continues to function during the cure cycle in order to remove additional air and volatiles emitted during polymerisation of the matrix system. Next, the temperature/pressure-time cycle is initiated and carried out. The laminate is then removed from the autoclave for debagging. The pressure bag concept provides an economical alternative to the autoclave system when a heated platen press or comparable equipment is available. The pressure bag may be pressurized with air and is confined by the platens of the press. The pressure bag concept does not, however, possess the general flexibility of the autoclave system. Schematic diagrams showing these three methods are given in Fig. 2.4.[21]

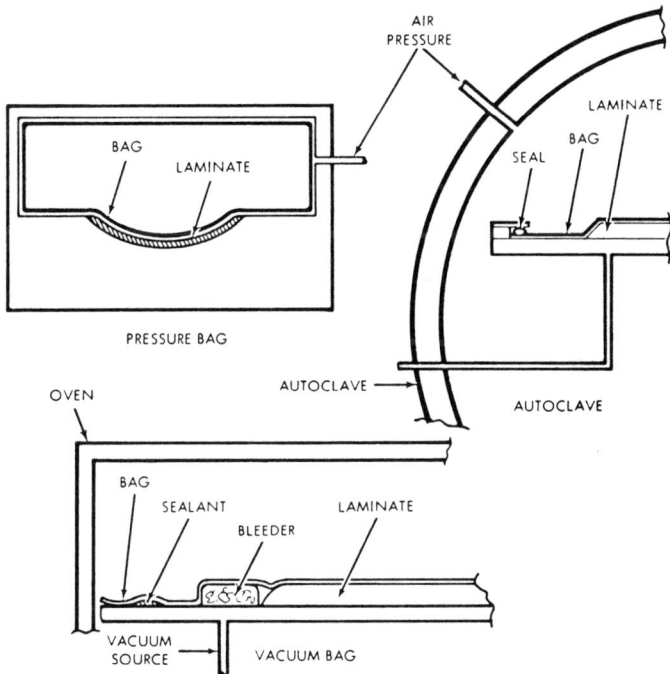

Fig. 2.4.   Schematic views of bag moulding. (Reproduced from Reference 21 by permission of Van Nostrand Reinhold Co.)

Filament Winding

The method of filament winding[22] employs continuous filaments which are wound onto a mandrel in predetermined orientations. Since it is possible to align the reinforcements along the direction of high stress, the strength of filaments can be utilised in an efficient manner. Structural applications of filament winding have been used in the aerospace industry to fabricate rocket-motor cases and radomes, as well as in commercial applications such as storage tanks, pipes, and pressure vessels. The single, most important, reinforcement used in filament winding is fibreglass. The weight percentage of glass fibres attainable in filament winding is the highest among all fabrication methods. At the present time, fibres of boron and graphite have not been used in large quantity for this purpose because of their relatively high cost. Depending upon the temperature range of application, plastic resins such as polyesters, epoxies, silicones and phenolics are used for filament winding. Two types of fabrication, *wet-winding* and *dry-winding*, are often employed. In wet-winding, the fibres are impregnated with resin just before winding. The filaments are pre-impregnated in dry-winding. Continuous fibreglass rovings or strands are usually wound over mandrels in two different patterns. In the *planar winding*, the mandrel is stationary. A layer of fibre is wound on to the mandrel when the fibre feed arm rotates about a longitudinal axis for one complete revolution. The fibres in each layer lie adjacent to but not crossing over one another. The second basic winding type is called *helical winding*. Unlike planar winding, the mandrel rotates while the feed carriage shuttles back and forth. As a result, the filaments are wound in a helical form. Helical winding is characterised by the filament crossovers on the mandrel. The product of winding is then allowed to cure with or without the applications of heat and pressure. For objects with open-ended shape such as cylindrical pipes, the mandrel can be forced out easily. For closed pressure vessels, the mandrels cannot be removed intact. In these cases, the mandrels are usually made of water-soluble hard salt or low melting-point metals: they are removed by solution or melting.

The tension exerted in the filaments can be controlled during winding. This tension can affect the void content, resin content, and, hence, the thickness of the laminae. The laminated composites resulting from filament winding generally have the filaments oriented in the following manners: (1) an angle-ply consisting of two monolayers oriented at $\pm\theta$; (2) a cross-ply consisting of two monolayers at right angles to each other; and (3) an angle-ply and a monolayer at 90 degrees. For example, in a closed pressure vessel, the orientation required as well as the amount of filament used can be determined from given hoop and longitudinal stresses.

Continuous Production Techniques

Continuous production techniques[23] are the truly automated methods used in the fibre composite industry. One of the *continuous pultrusion* methods uses continuous glass fibres which are first impregnated with a thermoset resin. These fibres are then drawn through a die to obtain the desired shape of the end product. The curing is achieved by using an oven, or through external heating of the die. Glass content as high as 60 to 80 per cent by weight can be achieved.

Another typical continuous production method is to be found in the production of continuous laminated sheet. In this method chopped glass fibres, impregnated with polyester resin, are sandwiched between two cellophane webs. The resulting products have been widely used as translucent building panels in the forms of corrugated and flat sheets. Still other continuous fabrication methods have been developed in the production of filament-wound pipes, in facing plywood panels with a reinforced plastic skin, and in manufacturing rigid laminated strips for sporting goods.

Closed Moulding Methods

Matched-die moulding, pre-mix moulding, and *injection moulding* all belong to this category. They are employed when detail is important and when a two-sided finish is desired. Two-piece male and female moulds are used in these methods. In contrast to closed moulding methods, those discussed previously are known as open-moulding methods, which are suitable for the fabrication of large objects and complex shapes.

Matched-die moulding[24] is usually employed in the fabrication of structural parts where the contour is complex and the tolerance is close. Reinforcing materials in the form of glass mats, chopped glass *preforms,* or fabrics are used. Plastic resins are then applied onto the reinforcements, and curing is achieved through heating and pressing the moulds at pressures which are sometimes higher than 1000 psi ($6·89 \times 10^6$ N/m^2). Chopped glass preforms are used when moulding articles which have considerable contours. The preform is manufactured by first depositing chopped glass fibres onto a screen which has the desired contour. A resin binder is sprayed onto the reinforcements. The dried preform when removed from the screen is ready for matched-die moulding. The content of reinforcement material achieved by this method is higher than that of bag-moulding. Pre-mix moulding[25] differs from matched-die moulding in that it employs a ready-for-use moulding compound. This compound is a mixture of chopped glass fibres, resin, catalyst, and filler. The main attraction of this method is the comparatively low cost and the ease with which it can be moulded. It is particularly suitable for moulding objects with variable wall thicknesses and sharp contour changes. Because the reinforced moulding compounds have good electrical properties and resistance to corrosion, heat, and flame, they have been used in a variety of applications to replace wood, metal, and ceramics. Common applications can be found in circuit panels for telephones, air-conditioning partitions, shower floors, etc.

The injection moulding process[25] also utilises a moulding compound which is injected into the cavity of the mould. However, unlike pre-mix moulding, the moulding compound used in injection moulding consists of thermoplastics. Typical resins used include nylon, polyethylene, polypropylene, polycarbonate, polystyrene, and polyvinyl chloride. Glass fibres are the most commonly used reinforcement material. Moulding compounds received from suppliers are usually in the form of pellets which contain fibres, thermoplastics, pigments, and lubricants. Pellets are produced by chopping impregnated glass rovings or by blending chopped glass strands and thermoplastics and feeding them to a compounding extruder. Thermoplastics reinforced with glass have shown marked increases in strength and rigidity as well as decreases in thermal expansion. The commercial applications of reinforced thermoplastics are developing rapidly in recent years in areas of appliances, electronics and household goods.

2.4.2 Properties of Resin Matrix Composites

Glass fibres are the most widely used reinforcing materials in resin-based composites. Reinforced glass fibre composites are available in tape form. One of the commercially available composite tapes is the 'Scotchply' produced by the Minnesota Mining and Manufacturing Company. Both unidirectional and cross-ply fibre orientations are available. Table 2.6[26] lists some properties of the Scotchply tapes. It is seen that plastics reinforced with S-glass fibres have tensile strength of 275 ksi ($1 \cdot 90 \times 10^9$ N/m^2), along the fibre direction. The Young's mod-

TABLE 2.6

SOME PROPERTIES OF NON-WOVEN UNIDIRECTIONAL REINFORCED PLASTICS

(Reproduced from Broutman, L. J. (1967), in *Modern Composite Materials* (Ed. L. J. Broutman and R. H. Krock), by permission of Addison-Wesley Co.)

Material	Glass contents (% by volume)	Specific gravity	Per cent of glass parallel to load	Tensile strength ksi (10^8 N/m^2)	Tensile modulus 10^6 psi (10^{10} N/m^2)	Compression strength ksi (10^8 N/m^2)
E-glass Scotchply 1009	73·3	2·17	100	238 (16·4)	8·1 (5·58)	
E-glass Scotchply 1002	56·5	1·97	100	149 (10·3)	6·2 (4·27)	87 (6)
E-glass Scotchply 1002	56·5	1·97	54·5[a]	75 (5·17)	3·5 (2·41)	71 (4·9)
E-glass Scotchply 1002	56·5	1·97	0[a]	5 (0·35)	1·47 (1·01)	20 (1·38)
S-glass Scotchply	71·6	2·12	100	275 (19)	9·6 (6·62)	200 (13·8)
S-glass Scotchply	71·6	2·12	0[a]	8 (0·55)		20 (1·38)

[a] Cross-ply composite so that remainder of glass is perpendicular to load.

ulus along the same direction is nearly 10 million psi ($6 \cdot 89 \times 10^{10}$ N/m^2). The strength in the direction transverse to the fibres is much lower. The contribution to this strength is mainly due to that of the resin matrix. The stress–strain curves for some non-woven glass filament reinforced plastics are shown in Fig. 2.5.[26] Figure 2.5(a) indicates the strength of unidirectionally-reinforced plastics. The failure strains in the directions parallel and normal to the reinforcement are approximately $0 \cdot 028$ and $0 \cdot 004$, respectively. In Fig. 2.5(b), the stress–strain curves show change of slopes in cross-ply composites. This change is attributed to the breaking of the fibres aligned perpendicular to the loading direction.[26] This is verified by the fact that the knees in the curves occur at a strain approximately equal to the failure strain of the 90-degree fibres in Fig. 2.5(a). Broutman has further shown that the change of slope of the curve is also consistent with the change of composite elastic modulus due to the failure of fibres that are transverse to the loading direction.

The highest glass content in resin-matrix composites is achieved by filament winding. A weight percentage of 60–90 can be achieved by this method. The glass contents, and hence the strengths of glass-fabric composites and glass-mat reinforced plastics, are lower than those of the filament-wound composites. The low strength of glass-fabric composites is in part attributed to the crimped glass fibres in the woven structure. The matrix material is subject to severe stress when the crimped fibres are straightened out under an applied load.[27] The woven process in making glass fabrics may also be detrimental to the strength of glass fibres.[26]

When discontinuous fibres are used to strengthen matrix materials, the strengths of composites rely upon the length of the reinforcing materials. For fibres below the *critical length* l_c, they will be pulled out of the matrix before it can be stressed to its full strength. The average stress in a fibre approaches the fibre ultimate tensile strength if the fibre length is much longer than the critical length. The strength of the composite in this case is close to that of a composite with continuous fibres. The fibre critical length is a function of the strengths of the fibre and the fibre–matrix interface, the fibre diameter, and temperature. Detailed discussions on fibre critical length and the strengthening effects are given in Section 8.2. Since the fibre critical length is an important parameter in designing composite materials reinforced with discontinuous fibres, considerable attention has been given to the experimental determination of this parameter.

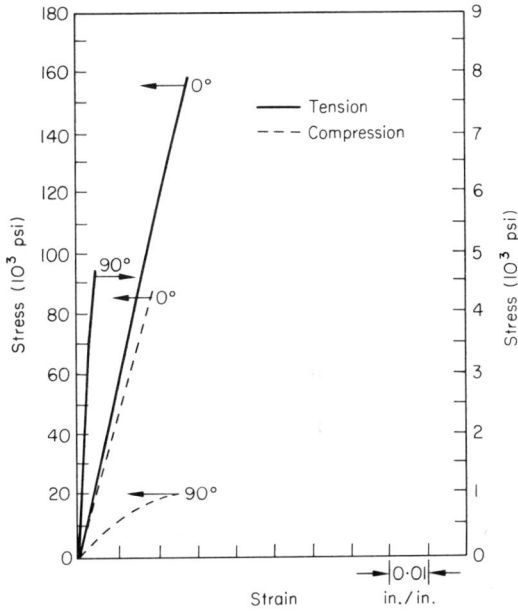

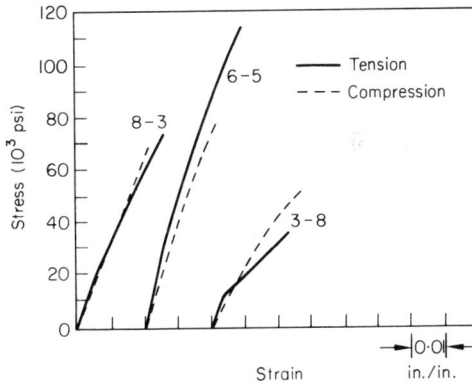

Fig. 2.5. *Stress–strain curves for non-woven, glass-filament reinforced epoxy plastics (Scotchply 1002, 35 per cent resin content by weight).* (a) *Tension and compression curves for unidirectional composites stressed at 0° and 90° to the fibre direction.* (b) *Tension and compression curves for 11-ply cross-ply composites. Three types of composites are shown, with various number of plies in the load direction. For example, 8–3 refers to 8 plies in the load direction and 3 normal to the load* (1 ksi = 6·89 × 10⁸ N/m²). (*Reproduced from Broutman, L. J.* (1967), *in* Modern Composite Materials (*Ed. L. J. Broutman and R. H. Krock*), *by permission of Addison-Wesley Co.*)

The effect of fibre length on the strength of glass-fibre-reinforced epoxy resin has been examined on uniaxially aligned materials.[28] The fibres were in the form of rovings which consisted of 32 ends, each containing approximately 140 fibres of diameter 5×10^{-4} in. (12·7 μm). The resin used was Araldite 219. The rectangular beam specimens prepared were tested using a three-point bending. The strength of the matrix and the fibres were measured and found to be $93 \cdot 25 \times 10^4$ N/m² (135 psi) and 1755×10^4 N/m² (2545 psi), respectively. Based upon these values, the strength of the composite at all fibre lengths can be calculated from eqn. (8.15). Three theoretical curves of composite strength versus fibre length are given in Fig. 2.6 for different l_c values. In the same diagram, the failure stresses of the composite beams are also recorded. These stresses were obtained for beams reinforced with fibres of different lengths. It can be seen that the experimental results agree almost exactly with the theoretical predictions when the critical length is taken to be 0·5 in. (12·7 mm).

The use of fibreglass-reinforced plastics in aircraft structures has been limited mainly to applications such as radomes and aerial

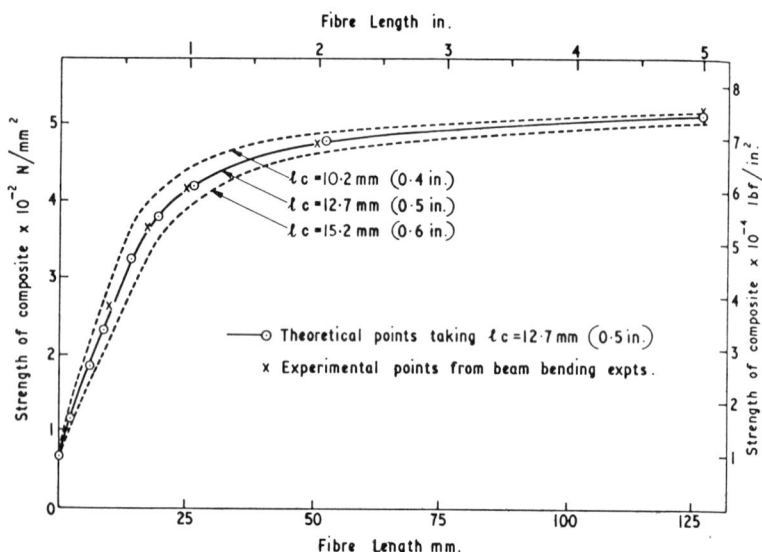

Fig. 2.6.   *Effect of fibre length on the strength of the fibre-reinforced composites. (Reproduced from Reference 28 by permission of Chapman and Hall.)*

windows. Fibreglass-reinforced plastics, although superior in specific strength and dielectric properties, are poor in specific stiffness. However, the development of boron filaments and their incorporation into epoxy resin bring about a bright future for their application in primary aircraft structures. The strength of boron–epoxy composites is comparable to that of glass-fibre composites, while their specific stiffness is far better than glass-fibre composites. It is believed that aircraft structures using boron–epoxy composites may result in weight savings of 25 to 40 per cent.[29] The present application of boron composites has been restricted by the high cost of both filament and fabrication. However, the rapid advance of technology in materials, design, and fabrication will certainly bring the cost down and make their use in the construction of civil aircraft feasible.

The properties of a boron–epoxy composite material, 'Narmco 5505', were examined and summarised by Hadcock.[29] This composite is available in tape form of 100 to 250 ft (30 to 76 m) in length and 3 in. (7·62 cm) wide. The tape has 616 to 624 boron filaments across the width, each with a diameter of 0·004 in. (0·01 cm). The filaments have an average tensile strength of 460 ksi ($3\cdot17 \times 10^9$ N/m^2) and average modulus of elasticity 58 million psi (4×10^{11} N/m^2). The resin matrix used in the tape has a tensile strength about 8 ksi ($5\cdot52 \times 10^7$ N/m^2) and tensile modulus 0·3 million psi ($2\cdot07 \times 10^9$ N/m^2). The layers have an average thickness of about 0·0051 in. (0·0130 cm) and a fibre volume fraction of about 43–54 per cent. The physical and mechanical properties of this tape are given in Table 2.7. It is seen that all the properties are highly directional and the retention of strength at high temperature is very good. The unidirectionally-reinforced tapes can be further consolidated into laminates for structural applications. The tapes in a laminate, depending upon the loading condition, are usually arranged in several orientations. The anisotropic nature of the laminates enables the designer to optimise the laminate load-carrying capacity. The weight of structures can thus be saved through the efficient arrangement of reinforcing materials. Discussions of the deformation and moduli of laminated structures are given in Chapter 5. It needs to be pointed out that the thermal expansion coefficient in the transverse direction is considerably larger than that in the longitudinal direction. This is essentially caused by the different thermal properties of the boron fibre and the resin matrix. As a result, thermal stress is developed in and between the plies when laminated structural parts are cooled down from their curing temperature.

TABLE 2.7

PROPERTIES OF 'NARMCO 5505' BORON–EPOXY COMPOSITE
(Reproduced from Reference 29 by permission of Van Nostrand Reinhold Co.)

Property	Room temperature		$260°F$		$350°F$	
Strength ksi (10^8 N/m^2)						
longitudinal tension	198·0	(13·65)	174·0	(12)	149·0	(10·27)
transverse tension	6·5	(0·45)	5·9	(0·4)	4·9	(0·34)
longitudinal compression	230·0	(15·86)	176·0	(12·13)	159·0	(10·96)
transverse compression	30·9	(2·13)	19·4	(1·38)	14·5	(1·0)
in-plane shear	9·0	(0·62)	5·0	(0·35)	3·0	(0·21)
Modulus 10^6 psi (10^{10} N/m^2)						
longitudinal tension	30.6	(21·1)	30·3	(20·9)	25·6	(17·65)
transverse tension	3·5	(2·41)	2·1	(1·45)	1·0	(0·69)
longitudinal compression	34·0	(23·4)	33·0	(22·75)	32·0	(22·06)
transverse compression	3·7	(2·55)	2·35	(1·62)	1·5	(1·0)
in-plane shear	1·00	(0·69)	0·80	(0·55)	0·22	(0·15)
major Poisson's ratio	0·36		0·35		0·30	
minor Poisson's ratio	0·033		0·025		0·017	
Coefficient of thermal expansion 10^{-6} in./in. °F (10^{-6} cm/cm °C)						
longitudinal	2·5	(4·5)	2·5	(4·5)	2·5	(4·5)
transverse	13·1	(23·6)	17·7	(31·9)	20·2	(36·4)

Finally, the excellent mechanical property of boron–epoxy composite is demonstrated by comparing its specific strength with those of alloys used in aircraft structures. Figure 2.7 shows that the specific strengths of the composite are superior to those of aluminium and titanium alloys in tension, compression, and shear. The strengths of a fibreglass–epoxy composite are also indicated. The aluminium alloy is further limited by its usage at elevated temperatures, while the titanium alloys have lower specific stiffness than the aluminium alloy.

Resin matrices also have been reinforced with strong whiskers. Parratt[30] examined the reinforcing effects of silicon-nitride whiskers in an epoxide resin, AZ 18. The whiskers were 2 μm in diameter and 300 μm or longer in length. The Young's modulus and tensile strength of the whiskers were in the ranges of 50–60 million psi ($3·45 \times 10^{11} - 4·14 \times 10^{11}$ N/m^2) and 5–15 ksi ($3·45 \times 10^7 - 1·03 \times 10^8$ N/m^2), respectively. The resin matrix had a much lower tensile modulus of 0·5 million psi ($3·45 \times 10^9$ N/m^2). When the resin matrix is reinforced with silicon-nitride whiskers in random orientation, significant improve-

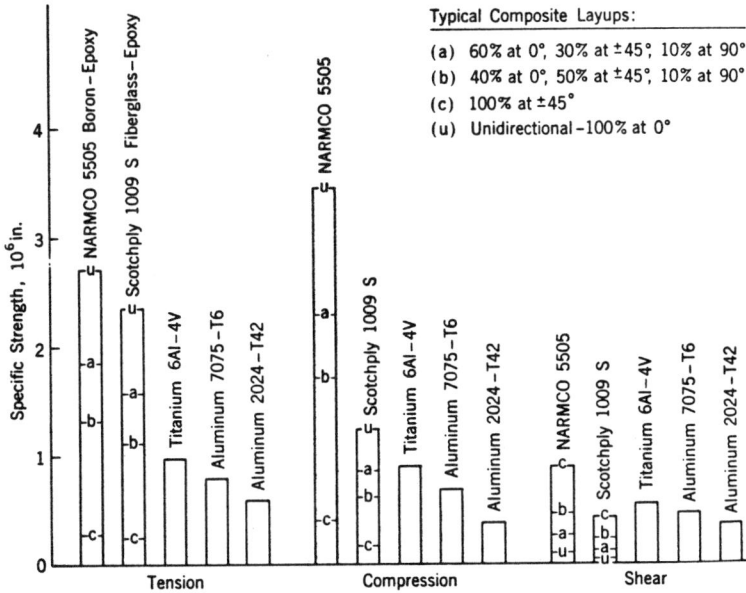

Typical Composite Layups:

(a) 60% at 0°, 30% at ±45°, 10% at 90°
(b) 40% at 0°, 50% at ±45°, 10% at 90°
(c) 100% at ±45°
(u) Unidirectional – 100% at 0°

Fig. 2.7. *Specific strength of composite materials and metals at room temperature* (1 in. = 2·54 cm). (*Reproduced from Reference* 29 *by permission of Van Nostrand Reinhold Co.*)

ments in both tensile modulus and strength were observed. Although the improvement in modulus is well in excess of those observed in similar matrices reinforced with conventional fibres, it was short of the theoretical value predicted by the 'rule of mixtures' (Section 5.6). The discrepancy was partially attributed to the breaking of whiskers at the consolidation pressure and consequently their inability of acting as long continuous fibres. A more significant effect is due to the difference of two orders of magnitude in moduli of the matrix and the reinforcement. The high modulus whiskers thus act as rigid fibres relative to the epoxide resin. As a result, severe gradients of the strain components were developed near the interfacial region. This led to a further departure from the uniform strain condition assumed in the rule of mixtures.

Recently, a fibre with the trade name of 'KEVLAR 49' has been produced by the E. I. duPont de Nemours & Company. This is an organic fibre believed to belong to the nylon family.[54] It is available in the forms of yarn, roving, and fabric. It has a specific gravity of 1·45.

The fibres have tensile strength of 525 ksi (362×10^7 N/m^2) and tensile modulus of 19 million psi (131×10^9 N/m^2). Since the density of this fibre is lower than those of graphite, S-glass, E-glass, and boron fibres, its specific strength and modulus compare very favourably with other existing fibres. The unidirectional composites of 'KEVLAR 49'-reinforced epoxy resin have shown tensile strength of 200 000 psi (1379×10^6 N/m^2) and tensile modulus of 11 million psi ($75 \cdot 8 \times 10^9$ N/m^2) along the fibre direction.

Another system of Al$_2$O$_3$ whiskers in epoxy resin has been examined by Sutton et al.[52] Metal wires also have been used to reinforce resin matrices. The resulting composites generally have low specific stiffness and their usages are limited. The reinforcement of epoxy resins with carbon fibres is discussed in Section 2.6.

2.5 METAL-MATRIX COMPOSITE MATERIALS

The technology of metal-matrix composite materials is being developed very rapidly. Compared to glass-fibre-reinforced plastics, metal-matrix composites are superior for their performance at elevated temperatures. The strength and elastic moduli of metal matrices are higher than those of resin matrices over a wide range of temperature. As to deformation of the composites, metal matrices can greatly enhance the ductility of the composite. The stress concentrations induced by cracked fibres can be relaxed through the plastic deformation of the matrix. As a result there is less chance of a brittle failure of the composite. The majority of the load applied to a metal matrix composite is carried by the reinforcing fibres. Since the metal matrices are strong in shear strength and, in general, well bonded to the fibres, short fibres can be used effectively for the purpose of strengthening. An obvious disadvantage of this type of composites is their relatively high density.

2.5.1 Fabrication Methods for Metal-Matrix Composites

In this Section the methods used for fabricating metal-matrix composites are briefly reviewed.[13,19,31] These methods serve first to combine the fibre and matrix materials and then to consolidate the combination to form the desired shape of the end-product.

Powder Metallurgy Technique

The technique of *powder metallurgy* involves the compacting of solid materials in the form of powders. The powder process has been used for ceramic as well as metallic materials. The product resulting from the powder process is uniform in composition, in contrast to alloys produced by casting. In the latter case, segregation of the component phases often occurs during solidification, and homogenisation of the alloy is needed. Since no melting or casting is involved, the powder process is more economical than many other fabrication techniques. In this process, powders of ceramics or metals are first prepared and then fed into a mould of desired shape. Pressure is then applied to further compact the powder. In order to facilitate the bonding among powder particles, the compact is often heated to a temperature which is below the melting point but high enough to develop significant solid-state diffusion. The use of heat to bond solid particles is known as *sintering* or *firing*. There is no separate bonding phase generated in the sintering process. Through diffusion, the point of contact between two neighbouring particles develops into a surface and the bonding between them is hence strengthened. The driving force for sintering is the elimination of particle surface area.

Metallic materials such as copper, nickel, aluminium, cobalt, and steel are often used in the powder process as matrix materials. The metal matrices in the form of powders are first mixed with whiskers or chopped fibres. The combination is then consolidated by pressing, sintering, hot extrusion, or rolling, in order to enhance the density and strength of the composite. The exposure to high temperature and pressure for long periods may be detrimental to some composite systems.

Liquid Metal Infiltration

Metals such as aluminium, magnesium, silver, and copper have been used as matrix materials in this process because of their relatively lower melting points. The method of liquid metal infiltration is desirable in producing relatively small size composite specimens. Fibres collimated in a mould are infiltrated with liquid metal. Plate VI[31] indicates a cross-sectional view of a boron filament–aluminium matrix specimen produced by this method. By employing the idea of liquid metal infiltration, it is possible to cast composite structures such as rods and beams by passing a bundle of filaments through a liquid-metal bath in a continuous manner. Structures solidified in this manner have uniform cross-sections with uniaxial reinforcement, and need little additional work. The application of the liquid-metal infiltration process is limited by the available choice of matrix and reinforcing materials. The degradation of many fibres at high temperatures rules out their

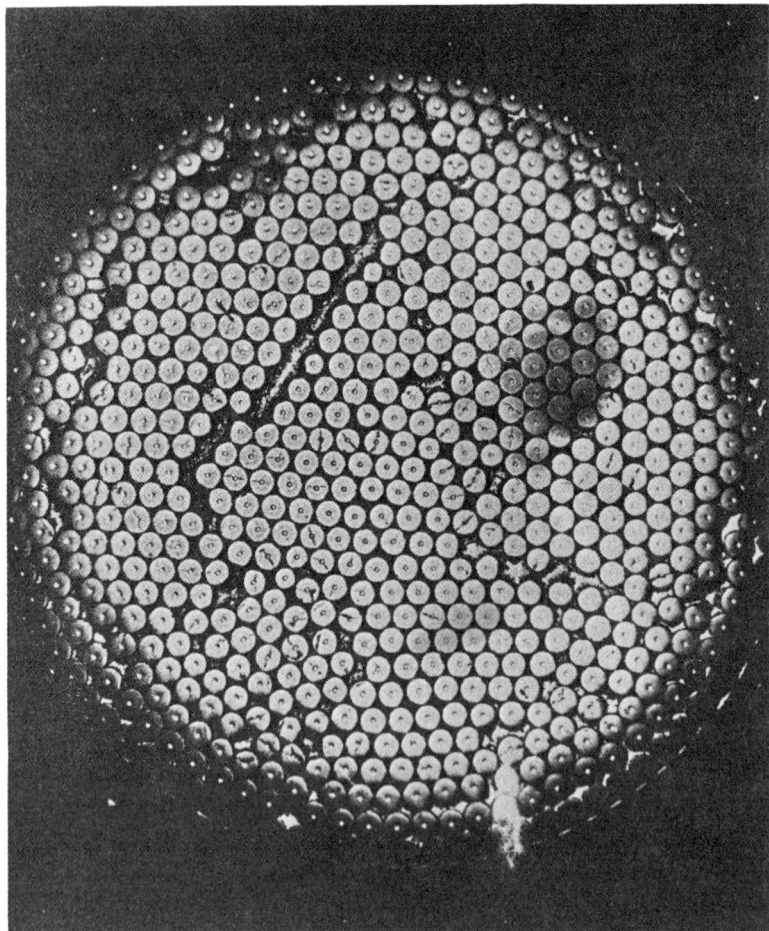

Plate VI.   Cross-section of boron filament–aluminium matrix specimens made by liquid infiltration. (Reproduced from Reference 31 by permission of ASTM.)

use. Another consideration is the wetting of reinforcements by the liquid metal. This problem will be discussed in Section 8.5.

Diffusion Bonding
Just as in the case of sintering, the diffusion bonding process is carried out under high pressure and elevated temperature. Filaments of stainless steel, boron, and silicon carbide have been used with matrices such as aluminium and titanium alloys. Unlike the powder process, the matrix metals used in most commercially available composites are in the form of metal foils. In order to fully develop the bonding strength among the foils and between the foil and the fibre, they all have to be thoroughly cleansed. The fibres are then laid on the metal sheets in predetermined spacing and orientation. Alternate layers of metal foils and reinforcing fibres can be arranged for the desired content of reinforcements. The lay-up is encased in a metal can which is sealed and evacuated. The whole assembly is subsequently heated and pressed to facilitate the development of diffusion bonding. The applied pressure and temperature, as well as their durations for diffusion bonding to develop, vary with the composite systems. For instance, the boron–aluminium composite develops satisfactory bonding at 850°F (454°C) under a pressure of 6 ksi ($4 \cdot 14 \times 10^7$ N/m^2) for 1 hr. A tensile strength of 220 ksi ($1 \cdot 52 \times 10^9$ N/m^2) has been observed for 6061 Al reinforced with 48 volume per cent of boron fibres by diffusion bonding. Prolonged hot pressing may cause reduction in the composite strength. The change of microstructure accompanying the reduction in strength is demonstrated in Plate VII for a stainless steel–aluminium composite material. The formation of an inter-metallic compound in the vicinity of the fibre–matrix interface is responsible for the weakening of the composite strength. The diffusion bonding process may also be used to consolidate tape preforms produced by methods such as plasma spray, hot rolling, and vapour deposition. The tapes are easy to handle and can be arranged in predetermined orientations.[31]

Electroforming
Electroforming has the advantage of combining the fibre and matrix materials at low temperatures, and thus degradation of reinforcing materials can be avoided. The major apparatuses for electroforming consist of a plating bath and a mandrel which serves as the cathode in the deposition process. A continuous filament is wound onto the mandrel while the metal matrix material is being deposited. The

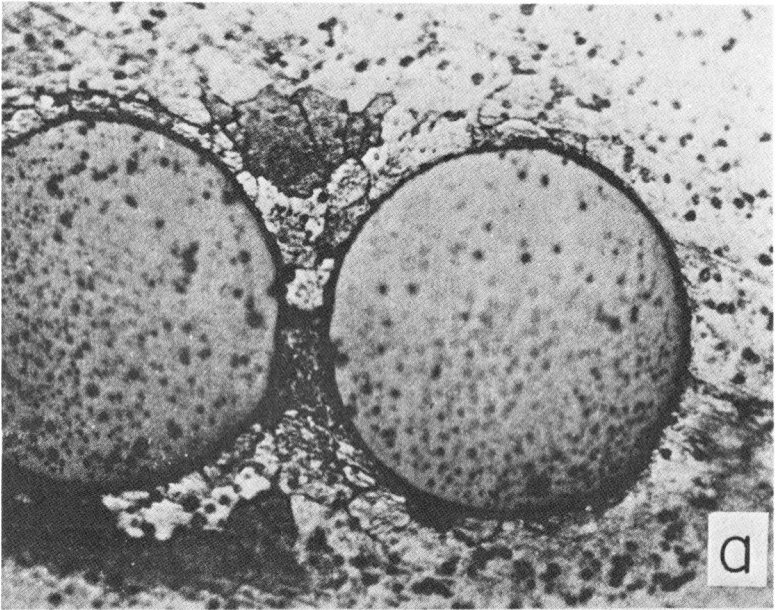

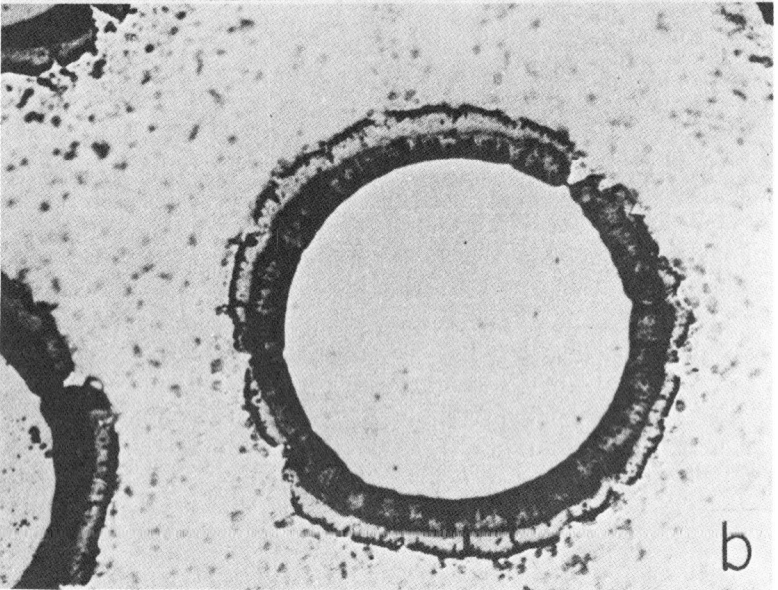

Plate VII. *Boundary microstructure with intermetallic reaction.* (*a*) *Hot pressed* 950°F (510°C), $\frac{1}{2}$ *hr, ultimate tensile strength* 109 300 psi (7·12 × 10^8 N/m²). (*b*) *Hot pressed* 950°F (510°C), $8\frac{1}{2}$ *hr, ultimate tensile strength* 80 940 psi (5·58 × 10^8 N/m²). (*Reproduced from Reference* 31 *by permission of ASTM.*)

spacing of filaments can be closely controlled in the winding process, and high-volume fractions of fibre content can be achieved. Monolayer tapes formed by this process can be further consolidated into composite structure members by diffusion bonding. For multilayer composites formed in this manner, voids tend to form between fibres and between successively deposited layers. Filaments of boron, silicon carbide, and tungsten have been successively incorporated into a nickel matrix by electroforming. Other matrix metals, as well as alumina whiskers, also have been employed in this method of fabrication.

Vapour Deposition
The process of vapour deposition is carried out by decomposing a compound of the metal matrix material and its subsequent deposition on the reinforcing materials. The reinforcements can be in the forms of continuous filaments or random whisker mats. A main advantage of this technique is that the chemical decomposition process can be accomplished at a relatively low temperature and the degradation of fibres can be minimised. High-volume fraction of fibre can be attained by this process. However, the slow and costly process of vapour deposition is its major disadvantage. Metals such as aluminium and nickel have been used for deposition.

Rolling
Both hot and cold rolling can be employed to incorporate filaments and metal strips into continuous tapes. In these processes, the fibres and metal strips are fed through rollers under applied pressure. The rollers are heated to a high temperature in the case of hot-rolling. Both pressure and temperature serve to accelerate diffusion bonding, although the contact time of the composite assembly with the applied pressure and temperature is relatively short. The metal strips can be grooved in order to provide precise alignment of the filaments. The sandwich construction of continuous tapes by the rolling process is restricted to a few layers in thickness. However, tapes fabricated in this manner can be layed up and further consolidated by diffusion bonding. The rolling process can also be used to consolidate continuous fibres coated with a metallic matrix material.

Extrusion
One method of extrusion is known as *co-extrusion*, which does not need the application of high temperature. Figure 2.8 indicates a design of the extrusion tooling for making composite wires. The wire,

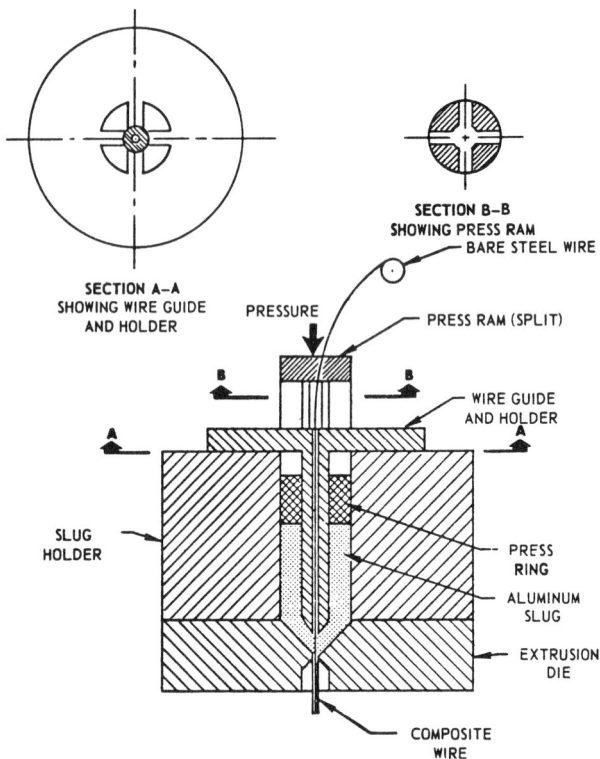

Fig. 2.8.   Design of extrusion tooling. (Reproduced from Reference 31 *by permission of ASTM.)*

consisting of a steel core surrounded by an aluminium alloy sheath, is produced by simultaneous feeding of the reinforcing filament and extruding of the matrix metal. Composite wires formed in this manner can be further rolled into tapes and plates through diffusion bonding. There are other methods of extrusion in which fibres are first aligned in matrix powders, and this assembly is pressed into the form of bars. A single preform, or several of them sealed in a can, are then extruded to the desired dimension.[31]

Other Methods
Methods including *plasma spray, pneumatic impaction,* and the simultaneous growth of the reinforcing and matrix materials from a melt also have been used for producing metal-matrix composites. The method of

plasma spray is suitable for low melting point metals. It employs a plasma torch which sprays matrix materials in the form of liquid droplets onto a rotating mandrel covered with aligned fibres. The composite formed is then removed from the mandrel and hot-pressed to eliminate voids. As one example, aluminium has been successively sprayed on silicon carbide-coated boron fibre. Composite tapes formed in this manner can be further consolidated into structural parts by diffusion bonding. Pneumatic impaction has been used to consolidate the mixture of metal powders and reinforcing materials by applying high-pressure impact. In view of its basic difference in the fabrication process relative to all the other methods, the method of growth from a melt is discussed in Chapter 3.

An examination of the fabrication methods reviewed above indicates that the matrix metals used in the consolidation processes may assume different forms. These include the metallic powders used in pneumatic impaction and the powder metallurgy technique; the liquid form used in liquid metal infiltration and plasma spray; the molecular form of matrix metals appearing in the electroforming and vapour deposition processes; and the metal foils employed in diffusion bonding and the rolling process. It is also noted that one form of matrix metal incorporated with reinforcements may be subjected to more than one fabrication process. For instance, composite materials produced by sintering are subsequently extruded and rolled. Monolayer tapes fabricated by various methods may also be further consolidated by rolling and diffusion bonding.

2.5.2 Properties of Metal-Matrix Composites

Metallic materials have been reinforced with all the kinds of fibre materials discussed in Section 2.3. Some representative composite systems in each category are now briefly reviewed.

When amorphous fibres are used as reinforcements, the matrices are restricted to low melting point metals such as aluminium, lead, and zinc. The composite of glass fibre–aluminium can be used as a model system.[32] Glass fibres tend to deteriorate in strength when the composites are fabricated by hot pressing or liquid metal infiltration. A protective coating is often applied to the filaments before fabrication. A film of aluminium on glass filament can prevent the degradation caused by molten aluminium. Both E-glass and fused silica fibre-reinforced aluminium have shown good strength-retention at elevated temperatures. For instance, the composite with 40 per cent in volume of glass fibres has tensile strength of 24 ksi (1.65×10^8 N/m^2) at 540 °C (1004 °F). The composite with 50 per cent in volume of fused silica fibres has shown much higher strength than composites reinforced with glass fibres of the same volume fraction. It has a room-temperature strength of about 115 ksi (7.93×10^8 N/m^2) and is able to maintain 40 per cent of the room-temperature strength at 500 °C (932 °F). For the volume percentages of the second-phase materials examined by Sutton and Chorńe, both glass fibres and composites have shown higher strengths than the particle-strengthened aluminium. This is also true with respect to strength-retention at elevated temperatures.

In view of their high specific strength and modulus, whiskers provide an excellent potential for application in metal matrices, especially at elevated temperatures. Composite systems using alumina, boron carbide, silicon nitride, and silicon carbide whiskers in the reinforcement of aluminium, nickel, silver, and copper alloys have been explored. Until now, the relatively high cost of whiskers and the difficulty of aligning them during composite fabrication have restricted their use.

The model system to be introduced here is the sapphire–silver composite investigated by Sutton and Chorńe.[32] This system has limited practical significance. Nevertheless, it demonstrates the feasibility of whisker-reinforced metals. The fine whiskers used have diameters ranging from 1 to 15 μm and fibre length-to-diameter ratios (aspect ratio) of from 1000 to 15 000. A thin, adherent metallic film was coated onto the oxide fibres by vapour deposition. Nickel and platinum can be used to promote the wetting of whiskers by the silver matrix. The coated fibres were then packed into capillary tubes of fused silica. The composite was fabricated by infiltrating molten silver into the tubes at about 1000 °C (1832 °F). The composite specimens so prepared were 0·01 to 0·02 in. (0·025 to 0·051 cm) in diameter. The composite tensile strength seemed to vary linearly with the whisker content. The increase in tensile strength with respect to the unreinforced matrix is very significant. The improvement in high-temperature strength is even more impressive. Figure 2.9 shows the variation of tensile strength with temperature when the specimens were heated for 15 to 30 minutes at the test temperature. The pure silver has very little retention of strength at elevated temperature. However, when it is reinforced with whiskers, nearly half of the room-temperature strength is maintained at temperatures close to the melting point of silver. The strength of the composite also increases with the fibre aspect ratio for reasons to be discussed in Section 8.2.

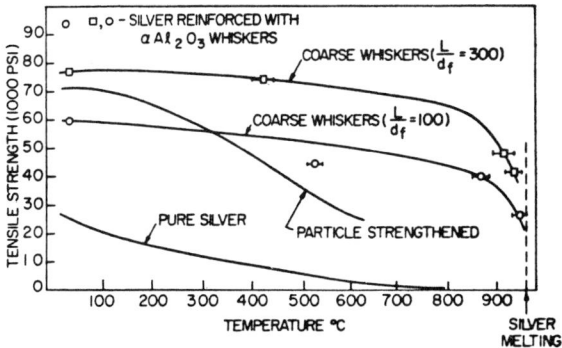

Fig. 2.9. Tensile strength of silver and silver strengthened with Al₂O₃ parti-cles and Al₂O₃ whiskers at various temperatures. L/d_f = aspect ratio.
(1 ksi = 6·89 × 10⁶ N/m²). (Reproduced from Reference 32 by permission of the ASM.)

When sapphire whiskers were incorporated into an aluminium matrix, a duplex coating of nickel and titanium proved to be desirable.[33] The whiskers produced were the wool type with cross-sectional area ranging from 50 to 0·1 μm^2. The coating was applied by a sputtering process, and the fibres were carefully aligned. Liquid aluminium was then infiltrated rapidly so as not to dissolve the coating. For a fibre content of 27 per cent in volume, composite tensile strength of 54·9 ksi (3·78 × 10⁸ N/m²) was observed at 500°C (932°F). This represents nearly a hundredfold increase in strength over pure aluminium. On the specific strength basis, the performance of the composite is compar-able to titanium at elevated temperature. Improvements in fibre alignment and increase in fibre length certainly will further enhance the strength.

Multiphase boron fibres have been used to reinforce metal matrices including aluminium and nickel. The continuous form of the reinforc-ing material is a great advantage in design and fabrication. Boron filament-reinforced aluminium has demonstrated marked improvement in its creep resistance and fatigue life. The composite also shows good strength-retention at elevated temperature. The boron fibres frac-ture in a brittle manner, while the aluminium matrix shows extensive plastic deformation before fracture. A typical stress–strain curve for a unidirectionally reinforced boron–aluminium composite has three dif-ferent stages as was originally pointed out by McDanels, et al.[34] These stages include the elastic behaviour of both components, the plastic deformation of the matrix material, and finally the region of fibre breakage. The composite stress–strain curve thus consists of a linear portion for the first stage and non-linear portions in the subsequent stages.[35] Boron–aluminium composites have been fabricated by diffusion bonding, liquid metal infiltration, plasma spray, hot pressing, and electroforming. Plates VIII and IX are micrographs showing the fibre and matrix structures in an aluminium–boron composite material.[36] This composite was fabricated from aluminium foils of 0·003 and 0·005 in. (0·0076 and 0·0127 cm) thickness and boron filaments of 0·004 in. (0·010 cm) of diameter. The diffusion bonding was conducted under a pressure of 10 ksi (6·89 × 10^7 N/m^2) at 975 °F (524 °C) for 1½ hr. The property of the composite is very much affected by the quality of the interfacial bonding between the fibre and the matrix as well as between the matrix foils. It can be seen in Plate VIII that the aluminium–aluminium interface was eliminated and the bonding is complete. Plate IX is used to demonstrate an incomplete aluminium–aluminium bond. The deformed foil interfaces which appeared in this micrograph were attributed to the insufficiency of the metal flow needed to rupture the oxide skin on the aluminium foil. The advantage of bonding aluminium and boron through diffusion is evidenced by the very limited reaction near the interfacial region. Boron–aluminium composite specimens prepared by melt infiltration are subjected to severe residual stresses because the linear thermal expansion coefficient of aluminium, 13 × 10^{-6} in./in. °F (23·4 × 10^{-6} cm/cm °C), is nearly three times that of boron. The effect is reflected by the poor tensile strength of the composite fabricated by melt infiltration.[37]

The major aluminium alloy used as matrix is the alloy 6061 (Al + 1% Mg + 0·5% Si). The degradation of boron filaments at the high processing temperature is detrimental to the composite strength. Boron fibres tend to degrade and lose their strength at temperatures above 500 °C (932 °F). To improve the high-temperature stability, and to minimise the reaction with matrices, boron fibres can be coated with silicon carbide (Borsic). The properties of Borsic–aluminium composites are given in Table 2.8. These composites can be used satisfactorily at temperatures up to 600 °F (315·6 °C). Experimental results also indicated that thermal and mechanical treatment of the 6061 aluminium alloy tend to enhance the tensile strength of the composite.[38] Significant improvements in the transverse tensile strength and interlaminar shear strength of boron–aluminium composite materials can be

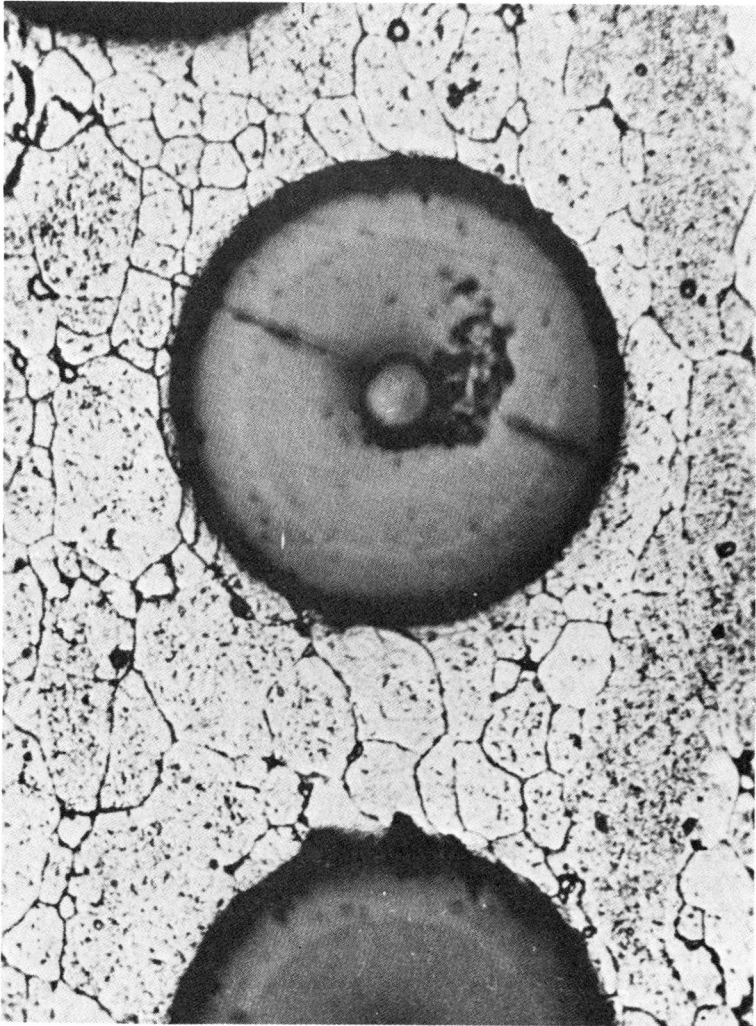

Plate VIII. *Aluminium–boron composite showing complete aluminium–aluminium bond, etched. (Reproduced from Reference 36 by permission of the ASTM.)*

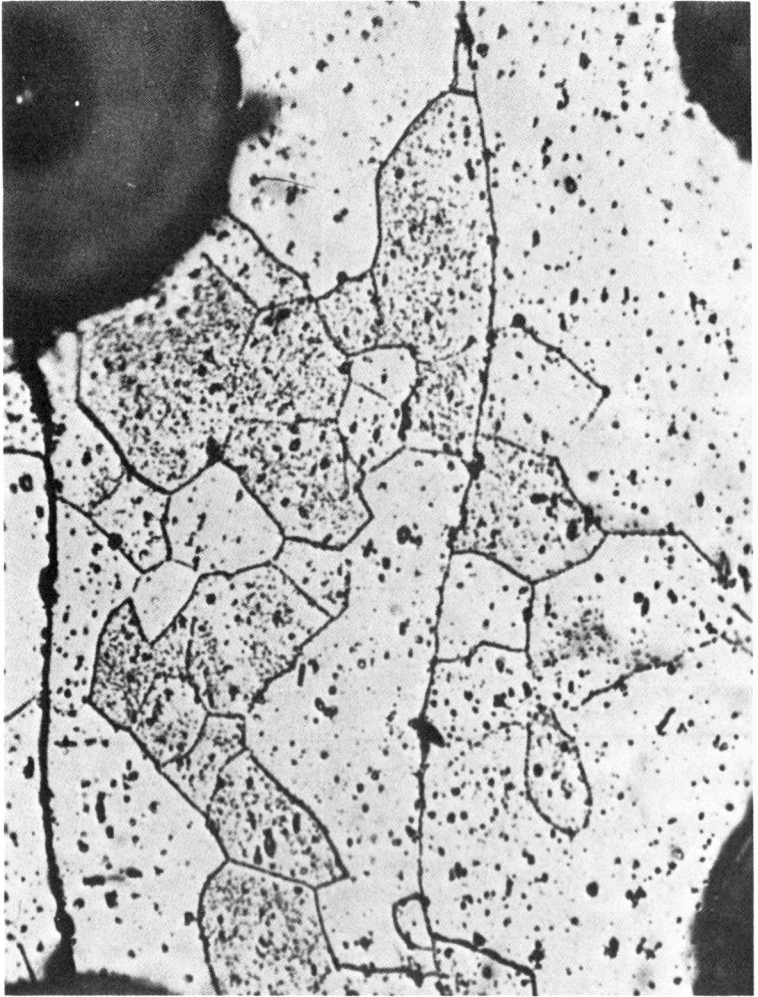

Plate IX.   *Aluminium–boron composite showing incomplete aluminium–aluminium bond, etched* (× 400). (*Reproduced from Reference* 36 *by permission of the ASTM.*)

TABLE 2.8

PROPERTIES OF BORSIC–ALUMINIUM COMPOSITES[a]
(Reproduced from Reference 16 by permission of Gordon and Breach.)

Density	
g/cm^3 (lb/in.^{3})	2·7 (0·097)
Elastic Properties	
Young's modulus (fibre direction)	
10^6 psi (10^{10} N/m^2)	30 (20·7)
Young's modulus (transverse direction)	
10^6 psi (10^{10} N/m^2)	12 (8·27)
Shear modulus	
10^6 psi (10^{10} N/m^2)	7 (4·83)
Major Poisson's ratio	0·22 ± 0·01
Strength	
Ultimate tensile strength (fibre direction)	
ksi (10^8 N/m^2)	140–190 (9·65–13·1)
Ultimate tensile strength (transverse direction)	
ksi (10^8 N/m^2)	12–15 (0·83–1·03)
Interlaminar shear strength	
ksi (10^8 N/m^2)	up to 13 (up to 0·90)

[a] Data for composites made from Hamilton Standard, Division of United Aircraft Corporation, tape. The composite has 50 per cent by volume of fibre and 6061 aluminium alloy matrix.

achieved by the addition of a small amount of high-strength stainless steel wires perpendicular to the boron fibres.[51] For applications at temperatures as high as 700°F (371°C), composites of Borsic and silicon-carbide fibres in titanium and nickel matrices need to be considered. The tensile strength and modulus of Borsic–titanium are 175 ksi (1·21 × 10^9 N/m^2) and 34 million psi (2·34 × 10^{11} N/m^2), respectively. The density of this composite is 3·6 g/cm^3 (0·13 lb/in.^{3}). Both Borsic–aluminium and Borsic–titanium composites possess good erosion resistance.[16]

Refractory metal wires also have been successfully used to reinforce metal matrices. The systems of copper–tungsten and copper–molybdenum have been well examined by Kelly and Tyson[39,40] and are introduced as model systems. The combination of copper matrix with tungsten and molybdenum wires were chosen because there is no chemical reaction between components. Molybdenum wires are ductile at room temperature. On the other hand, tungsten is brittle at room temperature and exhibits a brittle–ductile transition at 200°C (392°F). These systems provide an excellent chance for the study of elastic and plastic behaviour of fibre-reinforced composite materials. The wires used were 0·2 mm and 0·5 mm in diameter. Composite specimens were fabricated by molten-metal infiltration. Both continuous and discontinuous fibres were used.

For the copper–tungsten system with continuous wires, some typical tensile stress–elongation curves obtained at room temperature are shown in Fig. 2.10(a). At high, fibre-volume fraction, the composites fail in a brittle manner. However, composites with very low fibre content show considerable ductility. The serrated tensile stress–elongation curve results from the continuous breaking of fibres into smaller and smaller segments. The variation of composite tensile strength with the fibre volume fraction V_f is given in Fig. 2.10(b). The scattering of data at high-volume fraction of fibre is attributed to the inability of the thin matrix film to release the local stress concentration. Consequently, the tensile strength of the brittle tungsten fibres are not fully utilised. When the copper–tungsten system was tested at 250 °C (480 °F), both the fibres and matrix behaved in a ductile manner. Figure 2.11(a) provides their tensile stress–elongation curves. Several interesting points can be observed from this diagram. First, the ductile fibres in a composite can be stretched to an ultimate strain much larger than that observed in pure fibres. Secondly, the ultimate tensile strains of the composites lie in between the ultimate tensile strains of the pure fibre and matrix materials. Thirdly, at very low fibre contents, the composites show the effect of strain hardening. Results of composite tensile strength versus fibre volume fraction at 250 °C (482 °F) are given in Fig. 2.11(b). Severe necking of the specimens was observed in all fibre contents. However, the failures of the composites were controlled by the different components at different ranges of fibre contents. The failure of composites at high fibre content is caused by the failure of fibres and the inability of the matrix to take over the force originally carried by the reinforcements. On the other hand, at very low V_f values, the fibres are ineffective in strengthening the composite, and the failures of specimens occur due to the necking of the matrix. The behaviour of copper–molybdenum composites is similar to that of the copper–tungsten system with ductile wires. A detailed discussion of the strength theories of fibrous composite materials is given in Section 8.2.

Another area of development which draws considerable interest is in the application of refractory metals and super-alloys to composite materials. The combination of refractory metal wires and super-alloys

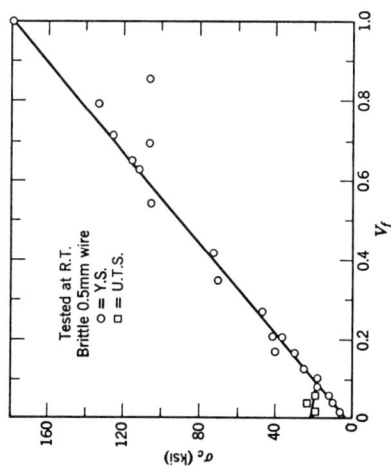

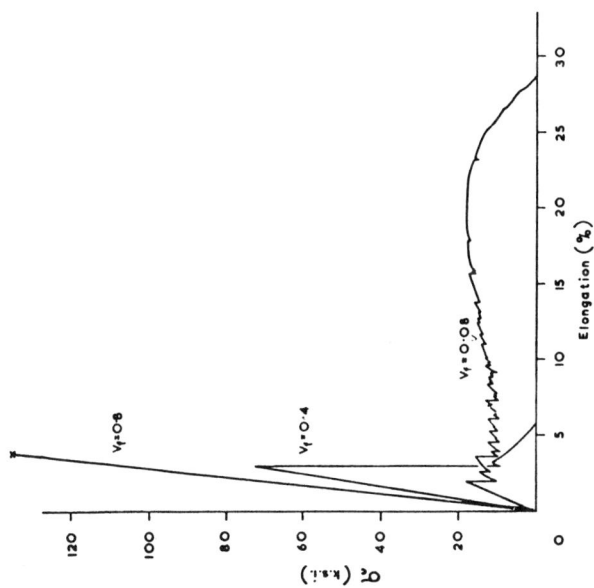

Fig. 2.10. (a) Tensile stress–elongation curves at room temperature for copper–tungsten specimens. (Reprinted with permission of Microforms International Marketing Corp. exclusive copyright licensee of Pergamon Press journal backfiles.) (b) Tensile strength versus volume fraction of continuous, 0·5 mm diameter, tungsten wires in copper, tested at room temperature. (Reproduced from Reference 39 by permission of Wiley and Sons.)

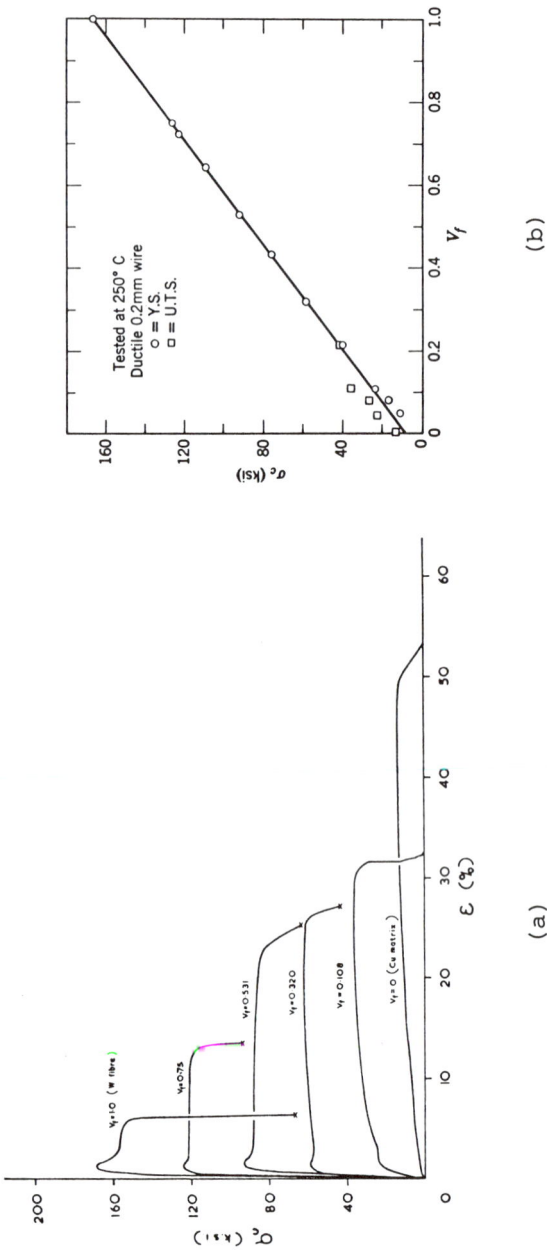

Fig. 2.11. (a) Nominal stress–strain curves for copper–tungsten composites at 250°C (482°F). (Reprinted with permission of Microforms International Marketing Corp. exclusive copyright licensee of Pergamon Press journal backfiles.) (b) Tensile strength versus volume fraction of continuous, 0·2 mm diameter tungsten wires in copper, tested at 250°C (482°F). (Reproduced from Reference 39 by permission of Wiley and Sons.)

offers the great potential of heat-resistant composite materials. As an example, the application of refractory metal wires-reinforced super-alloys to gas-turbine blades has been explored by researchers.[41,42] The refractory metal wires used in this study consist of two tungsten alloys (W–Re–Hf–C and W–Hf–C), a tantalum alloy (ASTAR 811C), and a columbium alloy (B-88). The wires were prepared by drawing with or without annealing during the drawing process. Wire properties were tested in the as-drawn condition. The ultimate tensile strength of these wires tested in vacuum at room temperature: 2000 °F (1093 °C) and 2200 °F (1204 °C), are listed in Table 2.9. The wire diameters are all 0·015 in. (0·038 cm). The hard-drawn W–Re–Hf–C wires are the strongest among all the kinds studied. They show strengths of 458 ksi ($3·16 \times 10^9$ N/m^2) at room temperature and 281 ksi ($1·94 \times 10^9$ N/m^2) at 2200 °F (1204 °C). A comparison of the data in Table 2.9 with Fig. 2.3 indicates that wires of tungsten-base alloys perform better than most of the other fibres, with the only exception being alumina whiskers. The strengths of wires at elevated temperature under prolonged loading are equally impressive. This is demonstrated by their rupture strengths. Figure 2.12 (a) and (b) summarises the time to rupture as a function of stress for the various refractory metal wires. Again, at both 2000 °F (1093 °C) and 2200 °F (1204 °C), the tungsten-base alloy wires have higher rupture strength. The columbium-based alloy behaves relatively poorly. Another way of examining the high-temperature performance of refractory wires is to take into consideration their densities and compare them on the basis of specific rupture strength. Data for this comparison are shown in Table 2.10 for the 100-hour rupture strength. Although the tungsten-base alloys are still outstanding, the columbium wires have improved their positions. On the basis of specific 100-hour rupture strength, the columbium alloy is comparable to the tantalum alloy and has nearly half the specific strength of tungsten alloys. The specific ultimate tensile strength of the refractory wires can also be calculated from the data of Table 2.9. At room temperature, the specific tensile strength of columbium wires is comparable to that of the tungsten wires and is better than that of the tantalum wires. However, at elevated temperatures, the specific tensile strengths of both tantalum and columbium are less than half of those of tungsten wires.

When refractory metal wires are incorporated into super-alloys, the most serious problem encountered is the reaction of the component phases and the recrystallisation of metal wires at elevated tempera-

TABLE 2.9

THE ULTIMATE TENSILE STRENGTHS OF REFRACTORY METAL WIRES
(Reproduced from Reference 41 by permission of NASA.)

Wire materials

Test temperature	W–Hf–C[a]	W–Hf–C[b]	W–Re–Hf–C[b]	ASTAR 811C[a]	B–88[a]
Room temperature	392 ksi $(27 \times 10^8$ N/m$^2)$	326 ksi $(22 \cdot 5 \times 10^8$ N/m$^2)$	458 ksi $(31 \cdot 6 \times 10^8$ N/m$^2)$	253 ksi $(17 \cdot 4 \times 10^8$ N/m$^2)$	235 ksi $(16 \cdot 2 \times 10^8$ N/m$^2)$
2 000°F (1 093°C)	207 $(14 \cdot 3)$	253 $(17 \cdot 4)$	314 $(21 \cdot 6)$	113 $(7 \cdot 79)$	71 $(4 \cdot 96)$
2 200°F (1 204°C)	201 $(13 \cdot 9)$	224 $(15 \cdot 4)$	281 $(19 \cdot 4)$	80 $(5 \cdot 52)$	45 $(3 \cdot 10)$

[a] In-process annealed.   [b] Hard-drawn.

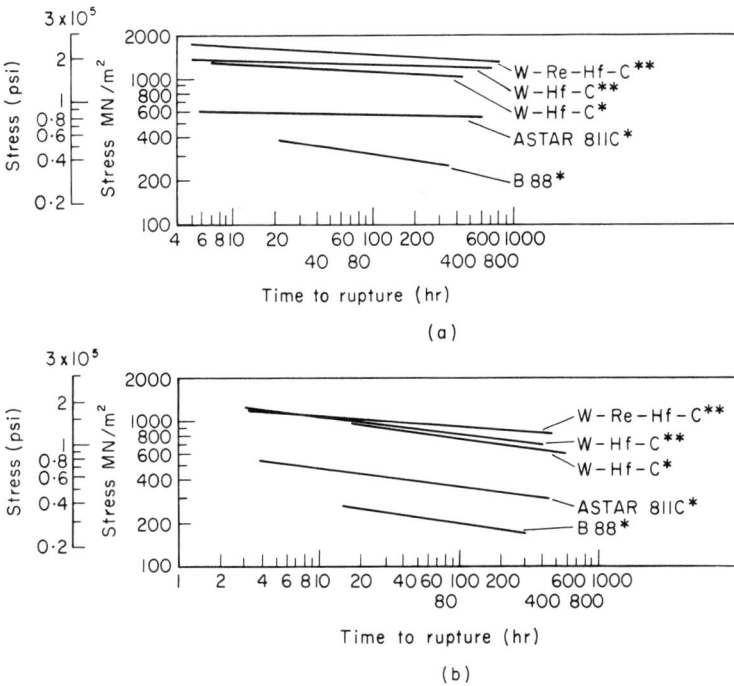

Fig. 2.12. (a) Comparison of time to rupture as functions of stress for wires tested at 1093°C (2000°F). (b) Comparison of time to rupture as functions of stress for wires tested at 1204°C (2200°F). *in-process annealed. **hard drawn. (Reproduced from Reference 41 by permission of NASA.)

tures. In order to limit reactions, composites have been fabricated by processes such as powder metallurgy, rolling, and hot pressing. Signorelli[42] investigated a nickel-base super-alloy Ni–25W–16Cr–2Al–2Ti. Refractory metals are added to the matrix to reduce the diffusion of matrix element into the wire and the subsequent recrystallisation of the metallic wire near the interface. The addition of aluminium and titanium also tend to reduce the diffusion of nickel by the formation of intermetallic compounds. The performance of refractory-metal-wire-reinforced super-alloys has demonstrated superior stability at elevated temperatures over that of the bulk super-alloys. For super-alloys reinforced with 70 volume per cent of W–Hf–C wires, the 1000-hr rupture strength of 60 ksi ($4 \cdot 14 \times 10^8$ N/m²) at 2000°F (1093°C) has been observed. This is eight times the rupture strength of bulk super-alloy. It

TABLE 2.10

THE 100-HOUR RUPTURE STRENGTH AND THE 100-HOUR
SPECIFIC RUPTURE STRENGTH OF REFRACTORY METAL WIRES
(Reproduced from Reference 41 by permission of NASA.)

Wire materials	Density lb/in³ (g/cm³)		2 000°F(1 093°C)		2 200°F(1 204°C)	
		Rupture strength	Specific rupture strength (inches)	Rupture strength	Specific rupture strength (inches)	
W–Hf–C[a]	0·7 (19·4)	180 ksi (12·41 × 10⁸ N/m²)	257 000 (652 780 cm)	120 ksi (8·27 × 10⁸ N/m²)	171 000 (434 340 cm)	
W–Re–Hf–C[a]	0·7 (19·4)	205 (14·13)	293 000 (744 220)	132 (9·10)	189 000 (480 060)	
W–Hf–C[b]	0·7 (19·4)	161 (11·1)	230 000 (584 200)	111 (7·65)	159 000 (403 860)	
ASTAR 811C[b]	0·61 (16·9)	84 (5·79)	138 000 (350 520)	51·5 (3·55)	84 000 (213 360)	
B–88[b]	0·373 (10·32)	44 (3·03)	118 000 (297 720)	29 (2·0)	78 000 (198 120)	

[a] Hard-drawn. [b] In-process annealed.

is anticipated that further improvement in composite rupture strength can be obtained by coating the refractory metal wires to create a diffusion barrier. The excellent mechanical properties so far obtained have shown the high potential of refractory-metal-wire-reinforced super-alloys for use in turbine blades in advanced aircraft and land-based turbine engines.

2.6 CARBON-FIBRE-REINFORCED COMPOSITES

Carbon-fibre-reinforced composites are being developed with great enthusiasm by various industries. The major attractions of the composites are their high degree of stiffness and strength, low density, and good fatigue and damping capacity. Carbon-fibre-reinforced epoxy resins and carbon matrices are introduced in this section.

2.6.1 Carbon–Epoxy Composites

The mechanical properties of unidirectional, carbon-fibre-reinforced, epoxy matrix composites have been studied by, among others, Rothman and Molter.[43] The reinforcement used was Thornel-50, supplied by the Union Carbide Corporation. Thornel-50 is a graphite yarn composed of two plies twisted together. Each ply consists of 720 filaments. The nominal yarn diameter is 0·02 in. (0·051 cm). Some properties of Thornel fibres are given in Section 2.3. The epoxy resin system used was ERL-2256, also supplied by the Union Carbide Corporation. The composite was fabricated in an aluminium vacuum-injection-type matched die. The specimens were prepared in the form of unidirectionally-reinforced composite tape. The specimen thickness was 0·125 in. (0·3175 cm), and the nominal fibre volume was 58·2 per cent. Some mechanical properties of the carbon–epoxy composites are[43]

tensile strength	109×10^3 psi (752×10^6 N/m^2)
tensile modulus	24×10^6 psi (165×10^9 N/m^2)
flexural strength	103×10^3 psi (710×10^6 N/m^2)
flexural modulus	$24·7 \times 10^6$ psi ($170·3 \times 10^9$ N/m^2)
shear strength	$3·5 \times 10^3$ psi ($24·1 \times 10^6$ N/m^2)
shear modulus	$0·71 \times 10^6$ psi ($4·90 \times 10^9$ N/m^2)

The flexural modulus was measured by the four-point beam method. The inter-laminar shear strength was determined by a short-beam shear test, which is sensitive to the selected span-to-depth ratio and may not produce pure shear. The above shear strength and shear modulus data were obtained from a nominal span-to-depth ratio of 5. The carbon composites usually suffer from low compressive and interlaminar shear strengths. Graphite fibre reinforced epoxy is available from manufacturers in prepreg tapes. These tapes can be cured and consolidated into laminates by using autoclave or press moulding.

The effect of fibre volume on the tensile strength of a carbon–epoxy composite has been reported by Elkin et al.[44] The Thornel-50 fibres were impregnated with the USP's 798 (U.S. Polymeric, Inc.) resin system. It was observed that the longitudinal tensile strength tends to peak between 60 and 65 per cent fibre and the transverse tensile strength dropped sharply at fibre volumes greater than 55 per cent. The shear strength also showed a tendency to drop at above 50 to 55 per cent fibre.

2.6.2 Carbon–Carbon Composites

Composite materials with carbon fibres reinforcing carbon matrices are being developed very rapidly. The applications of carbon–carbon composites have been made in aircraft structures, re-entry vehicle heatshields, rocket-engine nozzles, and turbine blades. Other applications including disc-brakes, hot-pressing dies, hot seals, and bearing materials are also being considered. The excellent high-temperature performance of the composites for these applications is due to their low density, improved strength at elevated temperature, thermal shock resistance, and chemical inertness. Reviews of the status of carbon–carbon composites have been made by Stroller et al.[14,15] The content of this Section is largely based upon these reviews.

Two basic approaches have been used to incorporate carbon matrices with carbon fibres. They are known as the *chemical vapour deposition (CVD) process* and the *carbonised organic technique*. In the chemical vapour deposition process, the carbon matrix is formed by decomposing a hydrocarbon gas and the subsequent deposition of carbon on a fibrous substrate. Vapour depositions are usually achieved under isothermal conditions or with a thermal gradient, although the techniques of pressure gradient and pressure pulsation also have been used. Figure 2.13(a) is a schematic view of the isothermal CVD technique. Hydrocarbon gas mixed with an inert carrier gas is introduced into the deposition chamber. An induction coil is used to heat the gas and fibrous substrate to a uniform temperature. Infiltration of

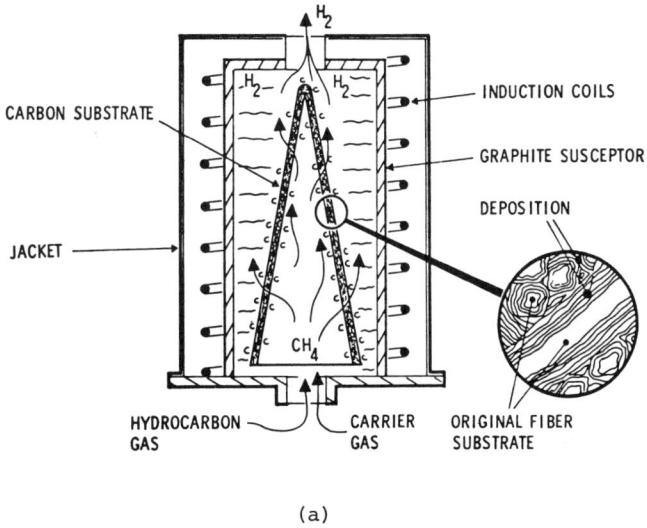

(a)

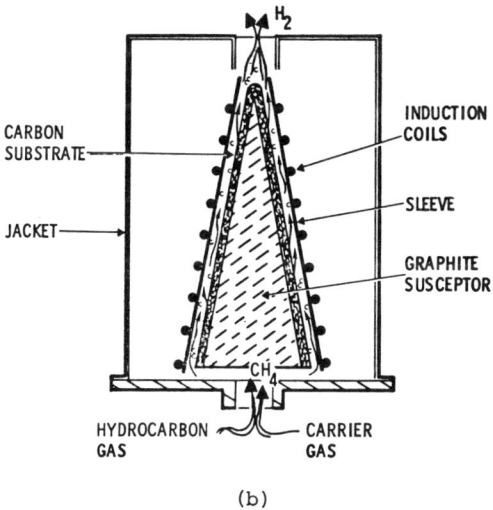

(b)

Fig. 2.13. (a) Isothermal CVD technique. (b) Thermal gradient CVD technique. (Reproduced from Reference 14 by permission of Sandia Laboratories.)

the carbon into the fibres is achieved at 1100°C (2012°F) and at a pressure less than 50 torr. The properties of the pyrolytic carbon matrix formed in this manner are affected by the deposition temperature, pressure, gas composition, and flow rate. A relatively uniform deposition with high density can be achieved by this method. The thermal gradient method is different from the isothermal method in that the induction coils are placed at different locations. As seen from Fig. 2.13(a), the graphite susceptor is heated by the induction coils. Heat radiated from the susceptor thus gives a uniform temperature distribution in the deposition chamber. In the thermal gradient deposition technique (Fig. 2.13(b)), however, the fibrous substrate is mounted on a mandrel. The induction coils are separated from the substrate by a non-conducting sleeve. As a result of this set-up, the substrate is heated on the surface attached to the mandrel while the outer surface is relatively cool. The temperature gradient established through the thickness of substrate enables the carbon to deposit near the inner surface first and then progress outward. The carbon deposition formed by the thermal gradient technique, although less uniform, can be achieved in a continuous manner. However, in the isothermal process, the crust formed on the outer surface of the substrate must be machined away before multiple infiltration can be carried out.

The effects of deposition temperature are reflected by the microstructure and optical anisotropy of the pyrolytic carbon. This is illustrated in Plate X for the thermal infiltration technique. It is seen that as the deposition temperature increases, the optical activity under polarised light diminishes. The two extreme structures at high and low temperatures are commonly designated as isotropic and laminar, respectively. The properties of the as-deposited carbon in the composite have been estimated by extrapolating the corresponding composite properties to pure matrix materials. The densities of the pyrolytic carbon are in the range of 1·4 to 1·7 g/cm³ and the apparent graphite crystallite sizes are 40–100 Å. The Young's modulus of the matrix in the basal plane is comparable to that of the fibre. However, the tensile strengths in the basal plane are relatively low· in the range of 12 to 25 ksi (82·8 × 10⁶ to 172·4 × 10⁶ N/m²). Also, for this estimated range of data, the isotropic carbon usually has a lower value than the laminar carbon.

It is also pertinent at this point to mention the properties of the commercially available bulk pyrolytic graphites which possess very unique thermal, electrical, and mechanical properties. Pyrolytic

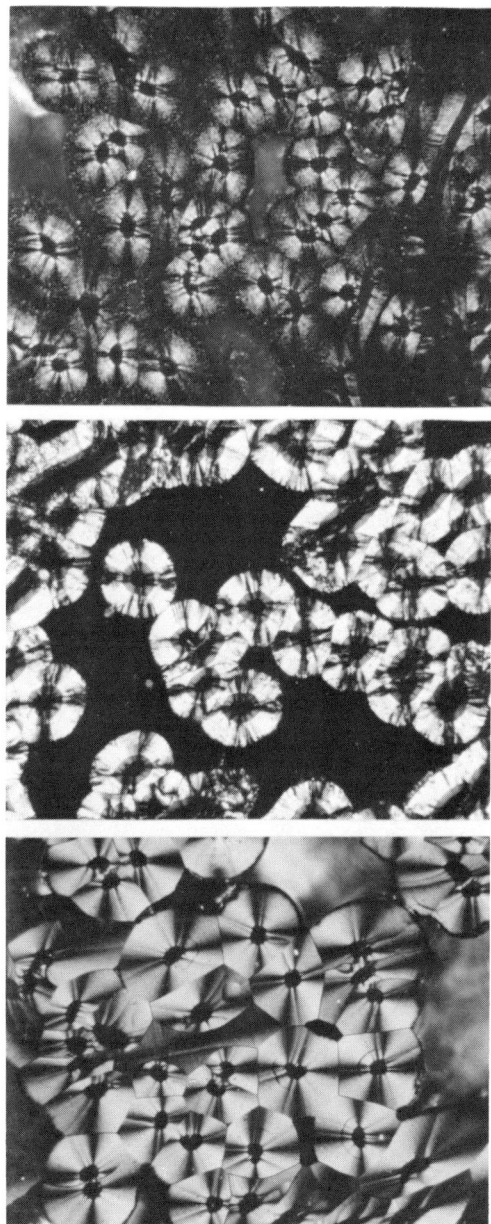

Plate X. Microstructure of CVD/felts prepared by means of temperature gradient infiltration at (a) 1100°C–1300°C, (b) 1350°C–1400°C, and (c) 1500°C polarised light (×400). (Reproduced from Reference 14 by permission of Sandia Laboratories.)

graphite is pure polycrystalline graphite deposited from a carbon-bearing vapour at temperatures in excess of 2000 °C. The crystallites tend to have their basal planes aligned parallel to the surface of deposition. The directionality of the atomic structure is responsible for its many highly anisotropic properties. The spacing between the basal planes in pyrolytic graphite is slightly higher than that in graphite single crystals, and the stacking order is not maintained. The high degree of orientation of pyrolytic graphite also renders a density fairly close to the density of perfect graphite of $2 \cdot 25$ g/cm^3. This is much higher than the densities of the ordinary commercial graphites ($1 \cdot 5$–$1 \cdot 8$ g/cm^3) which have porous structures. The directionality in properties is demonstrated by the data in Table 2.11.[45] It is interesting to note the difference in order of magnitude for the mechanical and physical properties along different directions. The negative Poisson's ratio in the basal plane further compounds the unusual properties of pyrolytic graphite. The marked differences in the thermal expansion coefficients within and normal to the basal plane can result in high thermal stresses during the cooling of structural parts from the deposition temperature. The elastic constants of pyrolytic graphite which has been ordered by

TABLE 2.11

PROPERTIES OF PYROLYTIC GRAPHITE

(Reproduced from Reference 45 by permission of Space Age Material Corp.)

Properties	Temperature	'$a - b$'[a]	'c'[a]
Tensile strength	Room temperature	15 (1·03)	1·5 (0·103)
ksi (10^8 N/m^2)	5 000 °F (2 760 °C)	60 (4·14)	0·5 (0·034)
Compressive strength	Room temperature	15 (1·03)	60 (4·14)
ksi (10^8 N/m^2)			
Young's modulus	Room temperature	5 (3·45)	1·5 (1·03)
10^6 psi (10^{10} N/m^2)	4 000 °F (2 204 °C)	2 (1·38)	0·5 (0·34)
Poisson's ratio	Room temperature	− 0·2	1·00
Thermal expansion	Room temperature	0·1 (0·18)	13 (23·4)
10^{-6} in./in.°F			
(10^{-6} cm/cm°C)	5 000 °F (2 760 °C)	3·0 (5·4)	13 (23·4)
Thermal conductivity	Room temperature	280–300 (4 167–4 465)	1·2 (17·9)
BTU/hr ft°F		80–120	
(cal/hr cm°C)	2 000 °F (1 093 °C)	(1 191–1 786)	0·6 (8·9)
Electrical resistivity	Room temperature	5×10^{-4}	0·5
ohm cm	2 000 °F (1 093 °C)	4×10^{-4}	0·2

[a] a and b axes are in the basal plane; c axis is normal to the basal plane.

annealing under compressive stress are available in Reference (53). Applications of pyrolytic graphite have been made in thermal protection systems, rocket nozzles, and re-entry-vehicles.

The carbonised organic technique of manufacturing carbon–carbon composites is quite different from any of the fabrication techniques previously reviewed. The basic idea of this approach is to impregnate the fibres with an organic material and subsequently convert it to carbon in an inert atmosphere. A primary consideration of the choice of organic precursor is the carbon yield. Phenolics, epoxy resins, and coal-tar pitches, having carbon yields of 40–70 weight per cent, have been used as organic precursors. The carbonised matrix material such as coal-tar pitch can be further graphitised upon heat treatment because of its platelet-like structure. On the other hand, the cross-linked structures of phenolic and epoxy prevented them from being graphitised. The glass-like carbon so formed has lower tensile strength and Young's modulus than that formed by the CVD technique.

The fibre substrates used in fabricating composite materials consist of either continuous or discontinuous fibres. Composites with carbonised rayon felt in a pyrolytic carbon matrix and short chopped fibres in a pitch-based matrix are commercially available. The rayon felt commonly carbonised at 1200°C contains about 98 per cent of carbon and has a density of 0.1 g/cm^3. Since the carbonisation process is accompanied by tremendous weight loss and shrinkage of the rayon felt, it is desirable to compact the felt by heat and pressure before carbonisation in order to achieve a higher fibre content. The application of PAN-based felt has the potential of attaining a higher fibre content without compaction. Unlike discontinuous fibre substrates, substrates with continuous fibres result in highly anisotropic composites. It is also not uncommon to use both continuous and discontinuous fibres in a substrate and to incorporate the substrate with the matrix by using both carbonised organic and pyrolytic carbon approaches.

Carbon–carbon composites have consistently demonstrated good strength at elevated temperature. For instance, the tensile strength of CVD/felt composite (7–10 v/o fibre) increases from 7.5 ksi ($51.7 \times 10^6 \text{ N/m}^2$) at room temperature to nearly 13 ksi ($89.6 \times 10^6 \text{ N/m}^2$) at 2500°C. Carbonised pitch reinforced with short, chopped fibres has shown a similar increase from 5 ksi ($34.5 \times 10^6 \text{ N/m}^2$) at room temperature to 8 ksi ($55.2 \times 10^6 \text{ N/m}^2$) at 2500°C. However, there are other composites that have shown some decreases in tensile strengths at extremely high temperature. The Young's moduli of carbon–carbon composites generally decrease at elevated temperature. Tests over a wide temperature range also found reduction in both tensile strengths and Young's moduli when the composites were previously treated at elevated temperature. It is also interesting to note that unlike bulk polycrystalline graphite, carbon–carbon composites are less brittle. The fracture toughness of the composite is higher than that of the bulk graphite. The design of carbon–carbon composites can be optimised when the constituent properties and the behaviour of the composites are better understood.

Recently, Granoff et al.[46] examined the effect of fibre volume per cent on mechanical properties of carbon-felt–carbon-matrix composites. The carbonisation of viscose-rayon fibre was carried out after the densification or compaction treatment of the felt. The composite was then fabricated by the CVD infiltration techniques. Densities of the matrix and fibres were found to be in the ranges of $1 \cdot 95$–$2 \cdot 00$ g/cm^3 and $1 \cdot 55$–$1 \cdot 59$ g/cm^3, respectively. An examination of the composite micro-structure indicated that the compaction treatment had a significant effect. The pyrolytic carbon matrix deposited from the vapour formed a sheath surrounding each fibre. The size of the pyrolytic carbon sheath was smaller in compacted felt than in uncompressed felt. There was also a marked decrease in total porosity in the compacted samples. The bulk density of the composite decreased with increasing fibre volume. However, this is consistent with the density values of the fibre and the matrix. The compaction process also tended to align the fibres in planes parallel to the felt plate. The implication of fibre alignment to composite strength can be understood through the orientation of the matrix carbon. During the infiltration process, the pyrolytic carbon matrix deposited on the fibres tends to have its basal planes aligned parallel to the axes of the felt fibres. As a result, the majority of the basal planes of the carbon matrix are aligned parallel to the plane of the felt. Obviously, the compaction process enhances the strength in the plane of the felt.

It was observed that both the composite in-plane flexural strength and flexural modulus increased with fibre volume. The in-plane tensile strength increased from 5 to $11 \cdot 3$ ksi ($34 \cdot 5 \times 10^6$ to $77 \cdot 9 \times 10^6$ N/m^2) for the fibre volume increase from 9 to 36 per cent. For the same fibre volume change, the trend of the tensile moduli was not very clear. It should be noted that heat treatment of the composites, as pointed out earlier, always results in a decrease in strength and moduli. Tensile tests on fibres from the carbonised felt indicated that the tensile strengths of the individual fibres were considerably higher than that of the infiltrated felt composite. On the other hand, the elastic modulus was only slightly higher for the fibres relative to the composite. In order to account for the increases in composite tensile and flexural strengths with increasing fibre volume, the following contributing effects were suggested: (1) re-orientation of the stronger crystallographic directions; namely, the basal planes of the matrix reoriented themselves into the direction of applied load due to compaction; (2) the decrease in the pyrolytic carbon sheath thickness with increasing fibre volume, and, hence, the refinement of the microstructure; and, (3) decrease in total porosity and the elimination of large voids due to the densification process.

The application of filament-wound carbon–carbon composites to tubes, rings, and bottles, and their mechanical properties can be found in References 47 and 48.

2.6.3 Three-Dimensionally-Reinforced Graphite Composites

Commercially available carbon–carbon composites generally contain random, unidirectional, or bidirectional reinforcements. These materials usually suffer from either low strength or extreme anisotropic behaviour. To overcome these shortcomings, Avco Corporation has developed a series of three-dimensionally fibre-reinforced graphites (3D graphite).[49] The original three-dimensional, reinforcement concept involves the orienting of straight, non-interlaced yarns in three mutually perpendicular directions. 3D graphite blocks up to $8 \times 8 \times 8$ in.^{3} ($20 \cdot 3 \times 20 \cdot 3 \times 20 \cdot 3$ cm^3) and cylinders up to 9 in. (22·9 cm) diameter, 1 in. (2·54 cm) thickness and 15 in. (38·1 cm) long have been processed. The matrix in the composite was provided by liquid precursors such as coal-tar pitch and phenolic resin. In order to achieve high density, repeated impregnation, carbonisation, and graphitisation cycles were used. The chemical vapour deposition procedure also has been used for matrix processing. Graphitisation was carried out at temperatures up to 4900 °F (2704 °C).

The organic precursors were observed to graphitise partially. The coal-tar pitch, among all other precursors, showed the highest degree of graphitisation as well as the highest char yield. The long graphitisation cycles at high temperature also tended to increase the degree of graphitisation of the Thornel fibres of the woven structure. Bulk density of 1·7 g/cm^3 has been reached for the 3D graphite composite. The corresponding porosity content is 10 volume per cent.

A 3D carbon–carbon composite which modifies the original orthogonal construction also has been developed by Avco Corporation and is designated as Mod 3.[59] In this construction the yarns parallel to one of the three orthogonal planes in the composite are replaced by woven fabrics. The final construction is produced by piercing multilayers of fabric with yarns. Typical properties of Mod 3 carbon–carbon composites are given in Table 2.12.

Three-dimensional carbon composites have demonstrated a unique property regarding their resistance to fracture and behave in a 'pseudo-elastic-plastic' manner. Because of the gradual development of fibre debonding and matrix microcracking, and the inability of cracks to propagate through the three-dimensional fibre array, the composites do

TABLE 2.12

PROPERTIES OF MOD 3 CARBON–CARBON COMPOSITE
(Reproduced from Reference 59 by permission of SAMPE.)

Specific gravity	1·65
Tensile strength, ksi (10^8 N/m^2)	
Room temperature	15 (1·03)
4 500 °F (2 482 °C)	10 (0·69) (shear failure)
Tensile modulus, 10^6 psi (10^{10} N/m^2)	
Room temperature	6 (4·14)
4 500 °F (2 482 °C)	1·5 (1·03)
Compressive strength, ksi (10^8 N/m^2)	
Room temperature	10 (0·69)
4 500 °F (2 482 °C)	23 (1·59)
Compressive modulus, 10^6 psi (10^{10} N/m^2)	
Room temperature	3·3 (2·3)
4 500 °F (2 482 °C)	1·5 (1·0)
Flexural strength, ksi (10^8 N/m^2)	
Room temperature	14 (0·97)
3 500 °F (1 927 °C)	15 (1·03)
Flexural modulus, 10^6 psi (10^{10} N/m^2)	
Room temperature	4 (2·76)
3 500 °F (1 927 °C)	2·5 (1·72)
Coefficient of thermal expansion	
10^{-6} in./in. °F (10^{-6} cm/cm °C)	
Room temperature to 2 000 °F (1 093 °C)	0·3 (0·5)
Thermal conductivity	
BTU/hr ft °F (kcal/hr m °C)	
500 °F (260 °C)	32 (48)
4 500 °F (2 482 °C)	14 (21)

not exhibit brittle failure.   The non-brittle nature of the composites, combined with the low coefficient of thermal expansion, makes them desirable for applications where good thermal stress resistance and low sensitivity to flaws are needed. Tests also have shown that the ablative property of the three-dimensional composites is much better than that of the carbon fibre-reinforced phenolic.[59]

Three-dimensional composites using quartz fibres in carbon and phenolic matrices also have been developed and applied as ablative materials.[50]

2.7   PROBLEMS

2.1. Highly oriented polymeric fibres generally consist of a mixture of 'crystalline' and 'amorphous' regions. Two models have been proposed to represent the microstructure of fibres in terms of crystalline–amorphous regions. These models assume that the crystalline and amorphous regions are equally stressed or equally strained from the application of a tensile force in the direction of the fibre axis. Theoretical considerations based on molecular and crystal structures suggest that Young's modulus of the crystalline region is $E_x \cong 10 \times 10^6$ psi (6.89×10^{10} N/m^2) and Young's modulus of the amorphous region is $E_a \cong 0.1 \times 10^6$ psi (6.89×10^8 N/m^2).

A given fibre is observed to consist of 60 volume per cent of crystalline material and has a Young's modulus of $E_f = 0.25 \times 10^6$ psi (17.23×10^8 N/m^2). Which model appears to be appropriate for this fibre?

2.2. Tabulate and compare the following typical properties of the various fibre materials: density, diameter, tensile strength, elastic modulus, applicable temperature range, price, advantages and disadvantages.

2.3. Gather the data of fibre tensile strength and Young's modulus. Plot them on a diagram of specific strength versus specific modulus and discuss the result. The specific strength is defined as the ratio of tensile strength to the fibre weight density. Similarly, the specific modulus is the ratio of Young's modulus to the fibre weight density.

2.4. Tabulate and compare in a qualitative manner, the following properties of the various plastic matrix materials: density, by-products of cure, mechanical property, electrical property, moulding pressure, applicable temperature range, price, advantages and disadvantages.

2.5. It has been suggested that surface effects may modify the properties of plastic resins used in fibre-reinforced composites. Such surface effects should be proportional to the surface-to-volume ratio of the resin phase. Derive an equation relating the surface-to-volume ratio, γ, of the resin phase to the volume fraction of fibre and the surface-to-volume ratio of the fibre. Make a plot of γ versus volume fraction of fibre for whiskers (diameter, $d = 0.000\,04$ in.), glass fibres ($d = 0.0004$ in.) and boron fibres ($d = 0.004$ in.). Which composite system would be most sensitive to surface modification of the resin phase? (See Reference 57.)

2.6. The specific modulus and specific strengths are useful performance indices for materials subject to tensile loads. Derive an appropriate performance index for materials subject to flexural loads. (See Reference 62.)

REFERENCES

1. Rosato, D. V., 'History of Composites', *Handbook of Fiberglass and Advanced Plastics Composites*, G. Lubin, ed., Van Nostrand Reinhold Co., New York (1968), p. 1.
2. Rubin, M., 'Polyester Resins', *Handbook of Fiberglass and Advanced Plastics Composites*, G. Lubin, ed., Van Nostrand Reinhold Co., New York (1969), p. 23.
3. Dorman, E. N., 'Epoxy Resins', *Handbook of Fiberglass and Advanced Plastics Composites*, G. Lubin, ed., Van Nostrand Reinhold Co., New York (1969), p. 46.
4. Doyle, H. J. and Harrier, S. C., 'Phenolics and Silicones', *Handbook of Fiberglass and Advanced Plastics Composites*, G. Lubin, ed., Van Nostrand Reinhold Co., New York (1969), p. 85.
5. Levine, H., 'High Temperature Resistant Polymers', *Handbook of Fiberglass and Advanced Plastics Composites*, G. Lubin, ed., Van Nostrand Reinhold Co., New York (1969), p. 114.
6. Dietz, A. G. H. and Heyser, A. S., *Fiberglass Reinforced Plastics*, R. H. Sonneborn, ed., Reinhold Pub. Corp., New York (1954).
7. Mettes, D. G., 'Glass Fibers', *Handbook of Fiberglass and Advanced Plastics Composites*, G. Lubin, ed., Van Nostrand Reinhold Co., New York (1969), p. 143.
8. Shulock, H., 'High Silica and Quartz', *Handbook of Fiberglass and Advanced Plastics Composites*, G. Lubin, ed., Van Nostrand Reinhold Co., New York (1959), p. 191.
9. Milewski, J, V., Shyne, J. J., and Shaver, R. C., 'Whiskers and Their Composites', *Handbook of Fiberglass and Advanced Plastics Composites*, G. Lubin, ed., Van Nostrand Reinhold Co., New York (1969), p. 255.
10. Campbell, W. B., 'Growth of Whiskers by Vapor-Phase Reactions', *Whisker Technology*, A. P. Levitt, ed., Wiley–Interscience, New York (1970), p. 15.
11. Wagner, R. S., 'VLS Mechanism of Crystal Growth', *Whisker Technology*, A. P. Levitt, ed., Wiley–Interscience, New York (1970), p. 47.
12. Roberts, J. A., 'Metal Filaments', *Modern Composite Materials*, L. J. Broutman and R. II. Krock, ed., Addison-Wesley Pub. Co., Reading, Mass. (1967), p. 228.
13. Lynch, C. T. and Kershaw, J. P., *Metal Matrix Composites*, The Chemical Rubber Company Press, Cleveland, Ohio (1972).
14. Stoller, H. M. and Frye, E. R., 'Processing of Carbon/Carbon Composites—An Overview', SC-DC-71 3653, Sandia Laboratories (1971).
15. Stoller, H. M., Butler, B. L., Theis, J. D., and Lieberman, M. L., 'Carbon Fiber Reinforced–Carbon Matrix Composites', SC-DC-72 1474, Sandia Laboratories (1972). Presented at the 1971 Fall Meeting of the Metallurgical Society of AIME, Detroit, Michigan, October 1971.

16. Galasso, F. S., *High Modulus Fibers and Composites*, Gordon and Breach, Science Publishers, New York (1969).
17. Line, L. E. Jr, and Henderson, U. V. Jr, 'Boron Filament and Other Reinforcements Produced by Chemical Vapor Plating', *Handbook of Fiberglass and Advanced Plastics Composites*, G. Lubin, ed., Van Nostrand Reinhold Co., New York (1969), p. 201.
18. Wawner, F. E. Jr, 'Boron Filaments', *Modern Composite Materials*, L. J. Broutman and R. H. Krock, eds., Addison-Wesley Pub. Co., Reading, Mass. (1967), p. 244.
19. Rauch, H. W. Sr, Sutton, W. H., and McCreight, L. R., *Ceramic Fibers and Fibrous Composite Materials*, Academic Press, New York (1968).
20. Winfield, A. G., 'Hand Lay-up Techniques', *Handbook of Fiberglass and Advanced Plastics Composites*, G. Lubin, ed., Van Nostrand Reinhold Co., New York (1969), p. 285.
21. Monroe, S. F. and Chitwood, B. E., 'Structural Laminate Bag Molding Process', *Handbook of Fiberglass and Advanced Plastics Composites*, G. Lubin, ed., Van Nostrand Reinhold Co., New York (1969), p. 306.
22. Shibley, A. M., 'Filament Winding', *Handbook of Fiberglass and Advanced Plastics Composites*, G. Lubin, ed., Van Nostrand Reinhold Co., New York (1969), p. 438.
23. Goldsworthy, W. B., 'Continuous Production Methods', *Handbook of Fiberglass and Advanced Plastics Composites*, G. Lubin, ed., Van Nostrand Reinhold Co., New York (1969), p. 485.
24. Lubin, G., 'Matched Die Molding—Fabric, Mat and Preform', *Handbook of Fiberglass and Advanced Plastics Composites*, G. Lubin, ed., Van Nostrand Reinhold Co., New York (1969), p. 335.
25. Young, P. R., 'Reinforced Molding Compounds', *Handbook of Fiberglass and Advanced Plastics Composites*, G. Lubin, ed., Van Nostrand Reinhold Co., New York (1969), p. 369.
26. Broutman, L. J., 'Fibre-Reinforced Plastics', *Modern Composite Materials*, L. J. Broutman and R. H. Krocks, ed., Addison-Wesley, New York (1967), p. 337.
27. McGarry, F. J. and Desai, M. B., 'Failure Mechanisms in Fiberglass Reinforced Plastics', Proc. 14th SPI Reinforced Plastics Division, Society of the Plastics Industry, New York (1959), Section 16-E.
28. Hancock, P. and Cuthbertson, R. C., 'The Effect of Fiber Length and Interfacial Bond in Glass Fiber-Epoxy Resin Composites', *J. Mater. Sci.*, 5, 762 (1970).
29. Hadcock, R. N., 'Boron-Epoxy Aircraft Structures', *Handbook of Fiberglass and Advanced Plastics Composites*, G. Lubin, ed., Van Nostrand Reinhold Co., New York (1969), p. 628.
30. Parratt, N. J., 'Reinforcing Effects of Silicon-Nitride Whiskers in Silver and Resin Matrices', *Powder Met.*, 7, 152 (1964).
31. Davis, L. W., 'How Metal Matrix Composites Are Made', *Fiber-Strengthened Metallic Composites*, ASTM STP 427, American Society for Testing and Materials (1967), p. 69.
32. Sutton, W. H. and Chorńe, J., 'Potential of Oxide-Fiber Reinforced Metals,' *Fiber Composite Materials*, S. H. Bush ed., American Society for Metals, Metals Park, Ohio (1965), p. 173.
33. Mehan, R. L., 'Fabrication and Evaluation of Sapphire Whisker Reinforced Aluminum Composites', *Metal Matrix Composites*, ASTM STP 438, American Society for Testing and Materials (1968), p. 29.
34. McDaniels, D. L., Jech, R. W., and Weeton, J. W., 'Stress Strain Behavior of

Tungsten Fiber Reinforced Copper', NASA TN D-1881, National Aeronautics and Space Administration, October 1963.

35. Kreider, K. G., 'Mechanical Testing of Metal Matrix Composites', *Composite Materials: Testing and Design*, ASTM STP 460, American Society for Testing and Materials (1969), p. 203.

36. Stuhrke, W. F., 'The Mechanical Behavior of Aluminum Boron Composite Materials', *Metal Matrix Composites*, ASTM STP 438, American Society for Testing and Materials (1968), p. 108.

37. Lenoe, E. M., 'Micromechanics of Boron Filament Reinforced Aluminum Composites', *Metal Matrix Composites*, ASTM, STP 438, American Society for Testing and Materials (1968), p. 150.

38. Shimizu, H. and Dolowy, J. E. Jr, 'Fatigue Testing and Thermal Mechanical Treatment Effects on Aluminum–Boron Composites', *Composite Materials: Testing and Design*, ASTM STP 460, American Society for Testing and Materials (1968), p. 192.

39. Kelly, A. and Tyson, W. R., 'Fiber-Strengthened Materials', *High Strength Materials*, V. F. Zackay, ed., J. Wiley and Sons, Inc. New York (1965).

40. Kelly, A. and Tyson, W. R., *J. Mech. Phys. Solids*, 13, 329 (1965).

41. Petrasek, D. W., 'High Temperature Strength of Refractory-Metal Wires and Consideration for Composite Application', NASA TN D-6881, National Aeronautics and Space Administration, August 1972.

42. Signorelli, R. A., 'Review of Status and Potential of Tungsten-Wire–Superalloy Composites for Advanced Gas Turbine Engine Blades', NASA TM X-2599, National Aeronautics and Space Administration, September 1972.

43. Rothman, E. A. and Molter, G. E., 'Characterization of the Mechanical Properties of a Unidirectional Carbon Fiber Reinforced Epoxy Matrix Composite', *Composite Materials: Testing and Design*, ASTM STP 460, American Society for Testing and Materials (1969), p. 72.

44. Elkin, R. A., Fust, G., and Hanley, D. P., 'Characterization of Graphite Fiber/Resin Matrix Composites', *Composite Materials: Testing and Design*, ASTM STP 460, American Society for Testing and Materials (1969), p. 321.

45. Data Sheet, Space Age Material Corp., New York (1963).

46. Granoff, B., Pierson, H. O., and Schuster, D. M., 'Carbon–Felt Carbon–Matrix Composites: Dependence of Thermal and Mechanical Properties on Fiber Volume Percent', *J. Comp. Mater.*, 7, 36 (1973).

47. Bert, C. W. and Guess, T. R., 'Mechanical Behavior of Carbon/Carbon Filamentary Composites', *Composite Materials: Testing and Design (Second Conference)*, ASTM STP 497, American Society for Testing and Materials (1972), p. 89.

48. Hamstad, M. A. and Chiao, T. T., 'Acoustic Emission Produced During Burst Tests of Filament-Wound Bottles', *J. Comp. Mater.*, 7, 320 (1973).

49. Taverna, A. R. and McAllister, L. E., 'The Development of High Strength Three Dimensionally Reinforced Graphite Composites', Presented at the American Ceramic Society Annual Meeting, April 1971, Chicago.

50. Adsit, N. R., Carnahan, K. R., and Green, J. E., 'Mechanical Behavior of Three-Dimensional Composite Ablative Materials', *Composite Materials: Testing and Design (Second Conference)*, ASTM STP 497, American Society for Testing and Materials (1972), p. 107.

51. Jones, R. C. and Christian, J. L., 'Analysis of an Improved Boron/Aluminum

Composite', *Composite Materials: Testing and Design (Second Conference)*, ASTM STP 497, American Society for Testing and Materials (1972), p. 439.

52. Sutton, W. H., Rosen, B. W., and Flom, D. G., 'Whisker-Reinforced Plastics for Space Applications', *Soc. Plastics Engrs.*, **20**, 1203 (1964).
53. Blakslee, O. L., Proctor, D. G., Seldin, E. J., Spence, G. B., and Weng, T., 'Elastic Constants of Compression-Annealed Pyrolytic Graphite', *J. Appl. Phys.*, **41**, 3373 (1970).
54. *Kevlar 49 DP-01 Data Manual*, E. I. duPont de Nemours & Company (1973).
55. McGrum, N. G., *A Review of the Science of Fibre Reinforced Plastics*, H. M. Stationery Office, London (1971).
56. *Crystic Monograph No. 2, Polyester Handbook*, Crystic Research Centre, Scott Bader Company Limited, England (1971).
57. McCullough, R. L., *Fundamental Concepts of Composite Materials*, Boeing Scientific Research Laboratories, D1-82-0970, April (1970).
58. *Data Sheet, Polyimide Materials Operation*, General Electric Company, Pittsfield, Massachusetts.
59. McAllister, L. E. and Taverna, A. R., 'Development and Evaluation of Mod 3 Carbon/Carbon Composites', presented at the 17th National Symposium of the Society of Aerospace Materials and Process Engineers, Los Angeles, California, April (1972).
60. 'Composites: Designers Wait and Contemplate', *Industrial Research*, October (1969).
61. Galasso, F. S., 'Research Spurs Fast-Expanding Fibers Market', *Ceramic Age*, May (1972).
62. Williams, M. L., 'Cost-Property Index for Comparing Load Bearing Materials', *J. Comp. Mater.*, **4**, 172 (1970).

CHAPTER 3

DISPERSION-STRENGTHENED AND DIRECTIONALLY SOLIDIFIED EUTECTIC COMPOSITES

3.1 INTRODUCTION

In this chapter, two types of composite materials are discussed. Both of these are distinctly different from the fibre-reinforced composites. The dispersion-strengthened composites are reinforced with precipitates or particles. The mechanisms of strengthening can be interpreted in the light of dislocation theory. In directionally solidified eutectic composites, the reinforcing materials generally assume the forms of rods or plates. A eutectic composite possesses the advantage of having its matrix and reinforcing materials fabricated and incorporated in one single process. The volume fraction of the second phase material has a unique value for each eutectic system. Consequently, it is not possible to manipulate the mechanical and physical properties of a certain eutectic system as in fibrous composites. Since the structural applications discussed in the later chapters are centred upon fibrous composites, the fabrication and properties of dispersion-strengthened alloys and eutectic composites are briefly introduced in this chapter. Interested readers are referred to the references for more detailed discussions.

3.2 DISPERSION-STRENGTHENED METALS

The strengthening of metals by dispersed particles has been mentioned briefly in Chapter 1. Metals are conventionally strengthened by alloying them with second-phase materials dispersed in the matrices. The sizes of the dispersed materials may vary from those of atoms in solution-strengthened alloys to a few microns (μm = 10^{-4} cm) such as in cermets. In this section the dispersion of reinforcing particles by precipitating them from a supersaturated solid solution or by synthetic means is reviewed. The mechanisms of strengthening are then discussed according to the various interactions of the dispersed phase when imperfections are present in the matrices. Finally, comparisons are made with respect to particle- and fibre-reinforced metals.

3.2.1 Precipitation-Hardened Metals

Precipitation hardening in an alloy system is induced by quenching its homogeneous solid solution at a high temperature and subsequent ageing at a low temperature. In this *age-hardening* reaction, a second-phase material precipitates from the supersaturated solid solution. During the precipitation process, aggregates of solute atoms are first formed at the lattice points of the random solid solution. As the precipitation process goes on, the aggregates will grow in size. The precipitates so formed have chemical compositions different from the supersaturated solid solutions, and their crystal structures may also be quite different from the materials from which they are precipitated. The following discussion on the nature of precipitates is mainly based upon the works of Kelly and Nicholson,[1] Fine,[87] and Mehl *et al.*[88]

The crystal structure of the precipitates in an alloy affects the nature of the interface between the precipitate particles and the matrix. This in turn has an effect on the shape of the precipitate. The interfaces, from the viewpoint of crystal structures, can be categorised as coherent, semi-coherent and non-coherent. A *coherent interface* is formed when the crystal lattices of the two component phases are in complete registry at the boundary and the crystal planes are continuous across the interface (Fig. 3.1(a)). There is also a definite crystal orientation relationship between the component phases when a coherent interface exists. For instance, the γ' phase in the aluminium–silver system has an HCP structure with its basal plane coherent with the (111) plane of the aluminium matrix. A plate-shaped precipitate with its plane of the plate parallel to a certain crystallographic plane in the matrix gives rise to the geometric pattern known as *Widmanstätten structure*. Due to the difference in lattice parameters, local strains are usually developed in the interfacial region, as seen in Fig. 3.1(a). When the lattice misfit builds up, it then becomes increasingly difficult to establish full registry of the crystal planes. A release of the distortion at the phase boundary can be achieved through a semi-coherent boundary as shown in Fig. 3.1(b). The atomic planes that terminate at the boundary can be viewed as forming edge dislocations. These

(a)

(b)

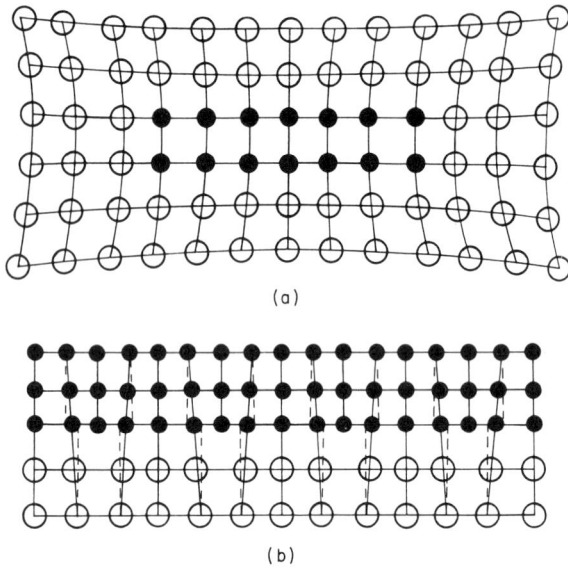

Fig. 3.1. (a) Coherent interface. (b) Misfit dislocations.

dislocations are also known as *misfit dislocations*. The spacing of these dislocations at a boundary depends upon the difference in lattice parameters of the precipitates and the matrix. When these dislocations are closely spaced, the interface can be viewed as virtually *incoherent*.

The Guinier–Preston zones, also known as GP zones, are formed in the early stages of ageing. They can be considered as distorted solute-rich domains. The shape of a zone depends upon the particular alloy system. X-ray results indicate that the GP zones in Al–Cu alloy at room temperature are formed by segregations of Cu atoms along {100} planes in the α matrix, and are only one or two atoms thick and about 25 atoms in diameter.

The Guinier–Preston zones of aluminium alloys containing copper, zinc and silver have crystal structures identical to those of the matrices and all the interfaces are coherent. Other precipitates, such as the θ' phase in aluminium–copper and the γ' phase in aluminium–silver alloys, have only part of the interfaces coherent with the matrix. Interface dislocations in aluminium–copper and aluminium–silver alloys have been examined by Laird and Aaronson.[2,3] They found that the interface dislocations, after long ageing time, may assume the arrangements of parallel lines, grids or hexagonal arrays.

It is seen from Fig. 3.1(a) that due to the difference in the atomic volumes of an atom in the matrix and in the precipitate, elastic strains are developed around the precipitates. The elastic energy associated with the precipitates is proportional to their sizes. Reduction in the distortion energy can be achieved through changes in shape of the precipitates. For instance, precipitates in FCC metals tend to form plates on {100} planes. This type of behaviour is attributed to the fact that the value of Young's modulus of FCC metals has a minimum along the ⟨100⟩ direction. Consequently, less distortion and smaller elastic strain energy is involved at precipitation. Furthermore, since the strain energy associated with a precipitate is in proportion to its size, precipitates originally coherent with the matrix may become non-coherent as they grow. The transition size of a precipitate depends upon many factors such as elastic properties, lattice misfit and the interface energy. Estimates of the transition size have been made by Nabarro[4] and Friedel.[5]

The nucleation of precipitates may take place at defects within the grains or at the grain boundaries. In highly supersaturated alloys the precipitate nucleation occurs rather homogeneously. At low super-saturation, precipitation occurs predominantly at defects. The reason that precipitates tend to form at defects such as dislocations in supersaturated alloys is because the activation energy for the non-homogeneous nucleation process is smaller than that in the homogene-ous nucleation case. The lattice misfit of the precipitate can be partly or wholly accommodated by the dislocations. Cahn[6] has shown that the activation energy for nucleation at dislocations decreases when the Burgers vector of dislocations and the degree of supersaturation increase. The activation energy is less for nucleation at an edge dislocation than at a screw dislocation. Furthermore, Cahn found that the rate of precipitation on a dislocation is very much larger than the rate of homogeneous precipitation. As a result, the precipitates formed on dislocations grow rapidly and their sizes are larger than the matrix precipitates. This phenomenon is obvious at low supersaturation and is evident by the presence of precipitate-free regions in the matrix. However, at high supersaturation, the homogeneous nucleation is also active and precipitates at dislocations do not have a chance to grow to sizes larger than average.

Precipitates which are semicoherent or incoherent with the matrices tend to form on lattice defects such as dislocations and grain bound-aries. The nucleation of precipitates on grain boundaries at low ageing temperature is attributed to the reduction in both surface energy and strain energy. The activation energies for the nucleation processes are lower at grain boundaries than within the grains. Besides the precipitates formed on the grain boundaries, it also has been observed that regions near grain boundaries may be free of precipitates. These regions were initially described as 'denuded zones' and the causes of their formation are twofold. First, the grain boundary precipitates which frequently nucleate before the intragranular precipitates may deplete the solute near the grain boundary. Plate XI demonstrates the formation of massive γ phase on a grain boundary in an aluminium–silver alloy. The Guinier–Preston zones which form in the matrix do not require vacancies to accommodate the lattice misfit. Secondly, the formation of a denuded zone can be caused by the lack of nucleation sites, such as vacancies. Obviously, this effect is important when the precipitated phase has a larger atomic volume than the matrix: the η' phase in aluminium–zinc–magnesium alloy is an example of this kind (Plate XII).[86] Unwin et al.[7] concluded that all the grain boundaries except those having misorientations of less than $2°$ acted as ideal vacancy sinks in quenched, age-hardened alloys. The occurrence of a denuded zone in a rapidly quenched alloy can be entirely explained in terms of the matrix diffusion controlled loss of excess vacancies to the grain boundary. For this latter reason, researchers now prefer to use the term 'precipitation-free zone' as a better description of this type of region. When the precipitated phase has a smaller atomic volume than the matrix, it would be formed more easily in the vacancy-free region than within the vacancy-supersaturated grains. The formation of precipitates as well as precipitate-free regions at a grain boundary may be detrimental to the strength of alloys. It has been found that the precipitate-free zone can be eliminated by plastic deformation prior to ageing. The addition of trace elements can stabilise the vacancies by reducing their mobility near the grain boundaries and also modify the structure of precipitates so that a coherent phase may be formed.

The yield strengths of precipitation-hardened alloys are much higher than those of the pure metal matrices. For instance, the alloying of copper with 12 per cent of beryllium and 0·2 per cent of nickel produces coherent precipitates which give rise to a yield strength of $10^{-2} G$, where G is the matrix shear modulus. This strength is an order of magnitude higher than that of pure copper. Also, at low temperature, alloys containing coherent precipitates work-harden at rates close to

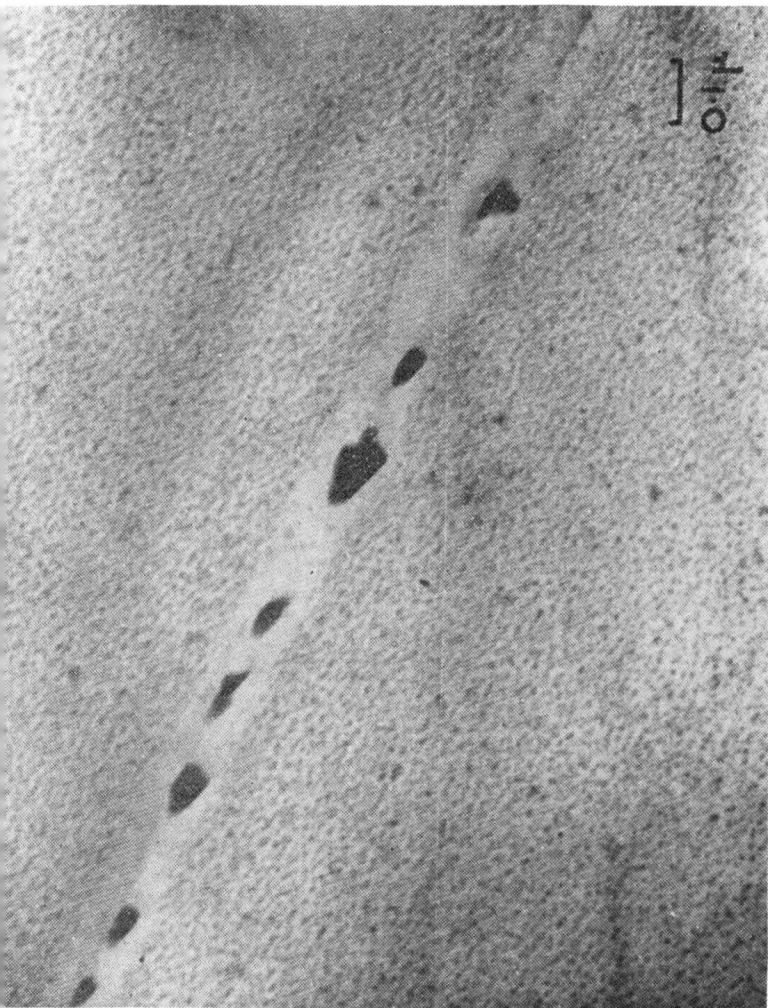

Plate XI. *Electron micrograph of a thin foil of Al-4·4% Ag alloy, water quenched from 525°C, and aged 5 days at 160°C showing precipitation of G.P. zones in the grains and γ in the grain boundaries. The precipitate-free region is ∼300 Å wide. (Reproduced from Reference 1 by permission of Macmillan and Co.)*

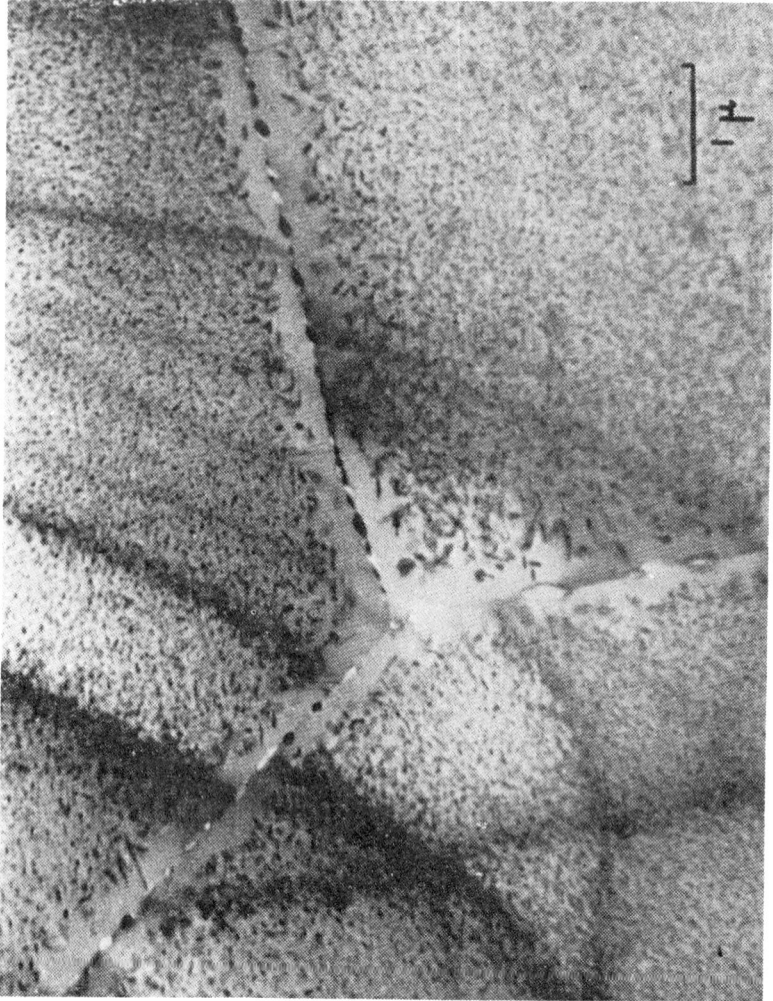

Plate XII. Electron micrograph of a thin foil of Al–3·2% Zn–2·5% Mg alloy, water quenched from 465°C, and aged 20 days at 160°C, showing wide precipitate-free zones adjacent to the grain boundaries. (Reproduced from Reference 86 by permission of the Institute of Metals.)

those of the pure matrix materials. However, in many alloys containing non-coherent precipitates, the work-hardening rates at small strains are much larger than those of the matrices. At larger strains, the rates of work hardening of those alloys decline and the stress–strain curves are nearly parallel to those of the matrices. The precipitates in some matrices are also observed to have the effect of enhancing the retention of strength at high temperatures.

3.2.2 Synthetic Dispersion-Hardened Materials

Alloys hardened by precipitation tend to lose their strength as the temperature increases. At high temperatures, the growth of precipitates is active and larger precipitates are formed at the expense of the smaller ones. As a result, the inter-particle spacing is too large to effectively block the movements of dislocations. When the solid solution temperature of an alloy is approached, the precipitates begin to dissolve back into the matrix. Thus precipitation-hardened alloys are unsuitable for applications at elevated temperatures.

Alloys for high-temperature applications can be achieved by dispersing second-phase particles of oxides, nitrides, carbides and borides. When a metal matrix is reinforced with hard ceramic particles, the composite is known as a *cermet*. These particles are usually insoluble in the matrix metal. The structure of the dispersion-strengthened alloy is achieved by synthetic means, with the capability of obtaining particle sizes as small as 50 Å. The properties of the alloy depend upon the control of volume fraction, size, spacing and uniformity of the particles. As to the strength of the alloys so produced, the dispersed particles behave as very effective barriers to dislocation movements; the structure of the particle and the matrix are quite different and the particles are not likely to be penetrated and cut by dislocations. The various techniques of fabrication have been reviewed by Smith.[8] Some of the usual methods, based upon this review, are briefly introduced below.

Internal oxidation has been used to produce very fine and uniform particles which are extremely stable. The oxide in a composite is formed by diffusing oxygen into the solvent matrix, and the subsequent oxidation of the solute metal. This technique is suitable for solvent metals in which oxygen can diffuse rapidly. The rate of oxidation tends to fall off at locations further away from the surface. This can be remedied by internally oxidising fine powders and then treating them with sintering and cold working. Internal oxidation has been used for silver, copper and nickel base materials, producing particles with spacings less than 500 Å.

The sintered aluminium powder technique also employs oxidation of the reinforcing particles. It has been used for the production of aluminium, titanium, magnesium, lead and tin containing their respective oxides as the dispersed phases. The dispersed phase of a metal is produced from the oxide layer on the surface of the powder particles. The oxide films can be broken and dispersed into the matrix by compaction and cold-working. The content of oxides can be controlled by the particle size and the degree of oxidation. Metals with high solubility of oxygen are not suitable for this method of production.

Direct mixing of oxide powders and a matrix metal, followed by compaction and sintering has been used for many combinations of dispersed phases and matrix metals. The mixing needs to be uniform to avoid agglomerates. It is also possible to mix two oxide powders with the matrix. In view of the high reactivity of very fine oxide powders, mixing is usually carried out in a controlled atmosphere.

Other methods of production include mixing particles with liquid metal, electrodeposition, and the precipitation of a matrix metal into powders. Decker[89] has provided a comprehensive review of these subjects.

The strength of dispersion-strengthened materials is affected by the interfacial energy. For a fixed volume fraction, particles tend to grow larger in size. This is because the total interfacial area is smaller for a lesser number of precipitated particles. During their growth, elastic strain and interfacial energy are built up around the particles. This is especially important when the particles are coherent with the matrix but have structures different from that of the matrix. It has been observed that as the particle size increases, the coherency may be gradually lost.

The low-temperature strengths of some dispersion-strengthened alloys are given in Table 3.1. It is noted that alloys with dispersed particles generally show lower strengths than those containing precipitates. The only exception is found in nickel base alloys. However, alloys reinforced with dispersed particles have shown better strength retention at elevated temperatures than precipitation-hardened materials. This is attributed to the low solubility of the particles at elevated temperatures. Hence, unlike precipitates these particles still can act as effective barriers in hindering the movement of dislocations.

TABLE 3.1

STRENGTHS OF SOME DISPERSION-HARDENED MATERIALS
(Reproduced from Reference 8 by permission of Institution of Metallurgists.)

Base metal	Alloy	Yield strength ksi (10^8 N/m^2)	Ultimate tensile strength ksi (10^8 N/m^2)
Al	Conventional: Al–Zn–Mg (wrought)	78 (5·38)	84 (5·79)
	Dispersed phase: S.A.P. (10% oxide)	28 (1·93)	50 (3·45)
Cu	Conventional: Cu–2·5% Be (fully heat-treated)	130 (8·96)	170 (11·72)
	Dispersed phase: Cu–3·5% Al_2O_3	58 (4·0)	68 (4·69)
Ni	Conventional: Nimonic 90 (fully heat-treated)	100 (6·89)	160 (11·03)
	Dispersed phase: Ni–7% ThO_2	134 (9·24)	138 (9·51)
	Dispersed phase: Ni–12% Mo–7% ThO_2	178 (12·27)	186 (12·82)
Fe	Conventional: Ni–Cr steel (quenched and tempered)	170 (11·72)	200 (13·79)
	Dispersed phase: Fe–16% Al_2O_3	58 (4·0)	66 (4·55)

3.3 THE MECHANISMS OF DISPERSION HARDENING

Several mechanisms have been suggested for the strengths of metals reinforced with dispersed particles. These theories are concerned with the initial yield strength of the composites as well as the behaviour of work-hardening. In the theories of yield strength, criteria were developed upon the stresses necessary to move a single dislocation over a distance comparable to the particle spacing. The dislocation movement in the alloys can be hindered for various reasons upon which the different yield strength theories were established. It is mentioned in Section 3.2 that at low temperatures, alloys containing coherent precipitates generally show work-hardening behaviour similar to those of the pure matrices. On the other hand, at low strains, the work-hardening rate of alloys is greatly enhanced by non-coherent particles. The behaviour of alloy work-hardening can be understood through the interaction among dislocations. In the following, some of the strengthening theories are introduced. Unless pointed out specifically, these theories can be applied equally well to precipitation-hardened alloys and synthetic dispersion-hardened alloys.

3.3.1 Theory of Mott and Nabarro
This is the earliest theory developed to explain the hardening of alloys. The basis of this theory is the difference in atomic volumes between the dispersed phase and the matrix material.[10,11] It is discussed in Section 1.4 that local stresses are induced around a spherical misfit particle and that the magnitude of the stresses are proportional to the amount of misfit ϵ. Mott and Nabarro found that the shear strain developed in the matrix at a distance r from the centre of the particle is $\epsilon r_0^3/r^3$ (Problem 3.1). r_0 denotes the radius of a dispersed particle. For an alloy with N particles per unit volume, the average distance of a point in the matrix to the nearest particle is then $\frac{1}{2}N^{-1/3}$. Hence the average shear strain ($\gamma = 2\epsilon_{ij}$, Section 4.1) in the matrix is taken as

$$\gamma = 8\epsilon r_0^3 N \qquad (3.1)$$

Expressed in terms of f, the volume fraction of the particles, the average strain is approximately equal to

$$\gamma = 2\epsilon f \qquad (3.2)$$

The corresponding shear stress in the matrix material with a shear modulus G is

$$\tau = 2G\epsilon f \qquad (3.3)$$

A more refined calculation on the critical resolved shear stress of solid solutions containing coherent precipitates has been carried out by Gerold and Haberkorn.[90] They found that the increment in alloy strength is proportional to $\epsilon^{3/2}$. High values of coherency strains in some superalloys have resulted in increases in strength as high as 100 ksi $(6 \cdot 89 \times 10^8 \text{ N/m}^2)$.[89]

The misfit between the particle and the matrix may result from various causes. These include the incoherency between the matrix and the particle, the difference in thermal expansion coefficients of the component materials and phase changes in the particle and the matrix. Equation (3.3) gives the self-stress in the matrix of an alloy. This stress can contribute to the prevention of the macroscopic slip of the composite. However, the effect may not be very large. Consider, for instance, $f = 1/100$ and $\epsilon = 5 \times 10^{-3}$,[12] $\tau = 10^{-4} G$. The effect of this self-stress can also be interpreted in terms of the spacing between dispersed particles. In Section 1.4, it is shown that the shear stress necessary to bulge a dislocation through two pinning points separated at a distance l is equal to

$$\tau = Gb/l \qquad (3.4)$$

Combining eqns. (3.3) and (3.4) it is seen that the internal stress can effectively enhance the strength of the alloy when the dispersed particles are separated at distances

$$l \geq \frac{b}{2\epsilon f} \qquad (3.5)$$

If l is much less than this value, a significant contribution to the flow stress arises from that indicated by eqn. (3.4).[1] Elasticity solutions of stresses due to misfitting particles of various shapes in isotropic and anisotropic media and their interaction with dislocations can be found in References 62–5, 83.

3.3.2 Theory of Orowan

The theory of Orowan[13] is concerned with the yield strength of alloys consisting of ductile matrices reinforced with hard particles. The dispersed particles are considered to be non-deformable. In Orowan's theory, the critical condition for a dislocation to bypass the particles in its glide plane is to bend the dislocation to a semicircular arc between the particles. The dislocation with its dipoles annihilated can move forward while dislocation loops are left behind surrounding each particle (Fig. 3.2). The local stress necessary to bulge the dislocation between two particles is given by Orowan as (Problem 3.2)

$$\tau = \frac{2T}{bD} \qquad (3.6)$$

where T is the line tension of the curved dislocation line, and D denotes the mean planar interparticle spacing in the slip plane. The form of eqn. (3.6) and the idea of loop generation in Orowan's theory closely resembles the concept of the Frank–Read source proposed a few years later. The Orowan's criterion indicates that the critical shear stress increases with increasing particle volume fraction and decreasing particle size.

Theoretical refinements and experimental verifications of Orowan's theory of alloy yield-strength have been carried out by many

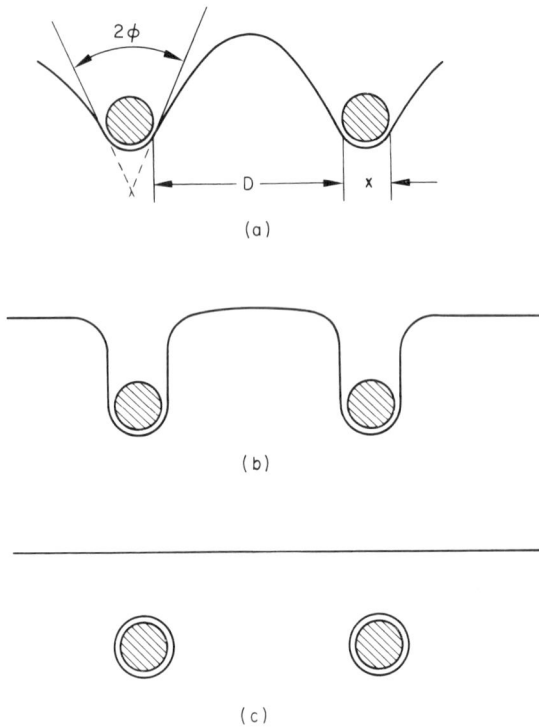

Fig. 3.2.   The Orowan hardening mechanism.

researchers. Ashby,[12] for example, considered a general dislocation configuration such as the one shown in Fig. 3.2(a). The two arms surrounding a dispersed particle are assumed to make an angle 2ϕ. The elastic energy as well as the line tension of the dislocation depends upon the interaction between the adjacent bowed-out loops. This interaction is taken into consideration through the outer cut-off radius in the elastic energy expression. Ashby estimated the elastic energy of a screw dislocation segment bending around a dispersed particle of diameter x to be

$$E(\phi) = \frac{Gb^2}{4\pi} \ln \left\{ \frac{x}{r_0} \left[1 + \left(\frac{D}{x} - 1 \right) \sin \phi \right] \right\} \qquad (3.7)$$

The elastic energy for an edge dislocation segment can be obtained by dividing the above expression by the factor $(1 - \nu)$. The resultant force, F, acting on a dispersed particle due to the line tension T, is

$$F = 2T \cos \phi \qquad (3.8)$$

The line tension along a dislocation depends upon the character of the individual dislocation segments, namely, their orientations relative to the Burgers vector of the dislocation line. By neglecting the dependency on orientation, and assuming $T = E$, the force on a dispersed particle can be obtained from eqn. (3.8). The local stress, τ', acting on the dislocation, that causes the dislocation segments to bow-out between the particles, can be obtained by considering the balance of forces. These forces include the force due to the externally applied stress and the reaction force from the particles. (See Problem 3.2 for the concept of forces on dislocations.) The result for an *edge* dislocation is

$$\tau' = \frac{1}{2\pi} \frac{Gb}{D} \cos \phi \ln \left\{ \frac{x}{r_0} \left[1 + \left(\frac{D}{x} - 1 \right) \sin \phi \right] \right\} \qquad (3.9)$$

The expression of local stress necessary for deforming a screw dislocation against the barrier is equal to $1/(1 - \nu)$ times the stress in eqn. (3.9). It should be noted that the segments of a screw dislocation bulging out between the particles change their characters to those of edge dislocations. The inverse is true for an edge dislocation moving on its slip plane.

The critical values of local stress and angle ϕ can be obtained from eqn. (3.9) by differentiating it with respect to ϕ. The results indicated that the maximum of τ' takes place at angles of ϕ in the range of 0 to 30 degrees. It was also concluded that the critical value of ϕ approaches zero, namely, the arms of the dislocation on both sides of a particle are parallel, when D becomes large. The yield stress of an alloy can be defined as the macroscopic stress necessary for general bypassing to take place. In an alloy, the planar interparticle spacing may vary from one location to another. Hence, it is necessary to take into consideration the distribution of particle spacing. Kocks,[14] and Foreman and Makin[15] have suggested a statistical factor of about 0·85 to relate the macroscopic shear stress τ and the microscopic, local stress, τ'

$$\tau = 0·85 \, \tau'$$

for the limiting case of $\phi = 0°$. This statistical factor is smaller for ϕ greater than zero. The average interparticle spacing D is assumed as

$$D = N_s^{-1/2} \qquad (3.10)$$

where N_s is the number of particles per unit area of the slip plane. Correction should be made for the value of D when the particle size x is large. The significance of Ashby's work is that it re-examined the critical configuration for unstable dislocation expansion. Approximations exist in regard to the line tension and the outer cut-off radius of a dislocation bulging between dispersed particles. A comparison has been made between the theory and the experimental data obtained by Ebeling and Ashby,[16] and Jones and Kelly[17] for copper single crystals reinforced with SiO_2 and BeO particles, respectively. In view of the simplifications employed in the development of this theory, its agreement with experimental data is satisfactory.

An important consequence of the interaction between dislocation lines and alloying particles is the bypassing of the barriers due to dislocation cross-slip, which affects the yield strength as well as the strain-hardening of alloys. The resultant structure of defects is strongly dependent on the type of dislocations interacting. Figure 3.3(a) indicates the cross-slip mechanism for edge dislocations. The arms ab and cd surrounding an impurity particle are screw components. The cross-slip of these segments results in jogs at a, b, c and d. Under the applied shear stress along the direction of Burgers vector, the two screw dislocations, one left-handed and the other right-handed, tend to bow out and attract each other. When they are in contact, the segments ab and cd annihilate each other and a prismatic dislocation results from the interaction. The situation is different when a screw dislocation bypasses a barrier by cross-slip. It is seen from Fig. 3.3(b) that two prismatic loops, one on each side, are created by the cross-slip process. The stress necessary for cross-slip to take place may arise from various long-range and short-range stress sources. It is likely that the Orowan stress has to be exceeded so that the screw segments can be formed around a particle and then cross-slip.[18] If the stress for cross-slip is not large enough, an Orowan loop may be formed first. Then under the influence of a second dislocation, the first loop can cross-slip (Fig. 3.3(c)). A detailed discussion of the cross-slip of Orowan loops at incoherent particles is given in Reference 93. Plate XIII shows a typical structure of a single crystal of copper reinforced with alumina particles. The rows of dislocation loops are clearly displayed, and were observed to take place predominantly by the movement of edge dislocations. Hirsch and Humphreys[19] found that the microstructures resulted from dislocation cross-slip in copper and that copper alloys containing small particles could be explained in

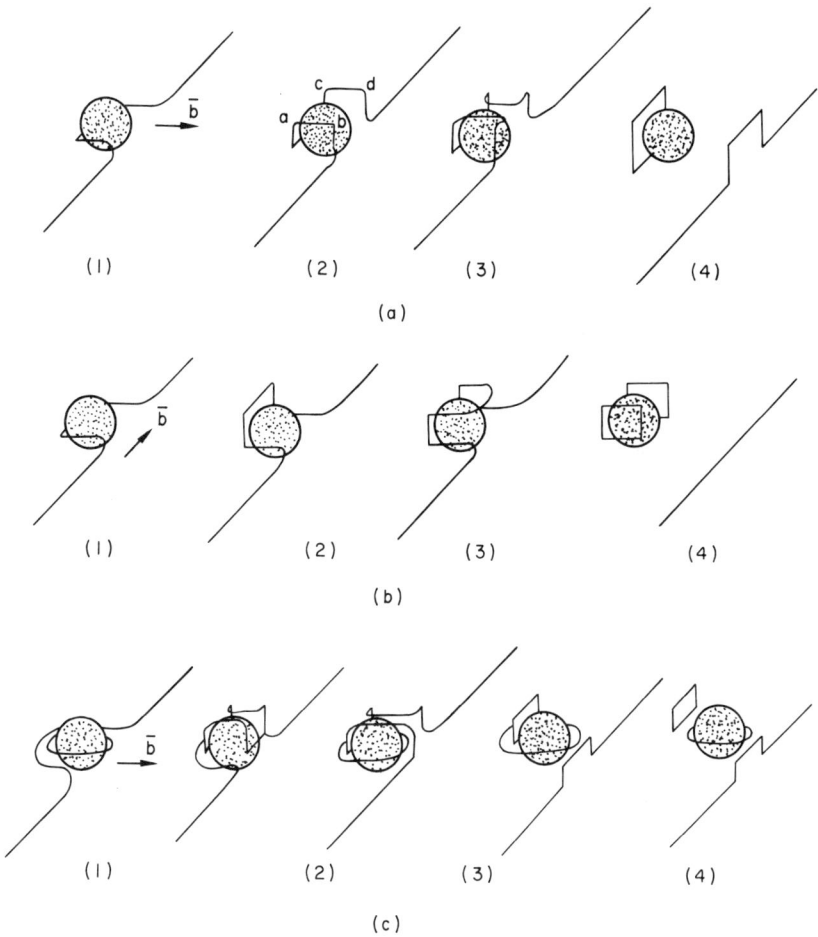

Fig. 3.3.   (a) An edge dislocation bypassing a particle. (b) A screw dislocation bypassing a particle.   (c) A dislocation bypassing a particle by the Orowan mechanism and cross-slip. (Reproduced from Argon, A. S., (Ed.) (1969), Physics of Strength and Plasticity, by permission of the MIT Press.)

terms of the mechanisms introduced above. They also concluded that cross-slip could occur with or without leaving a small number of Orowan loops.

The cross-slip of dislocations also has an effect on the rate of strain-hardening of alloys. For alloys containing coherent precipitates,

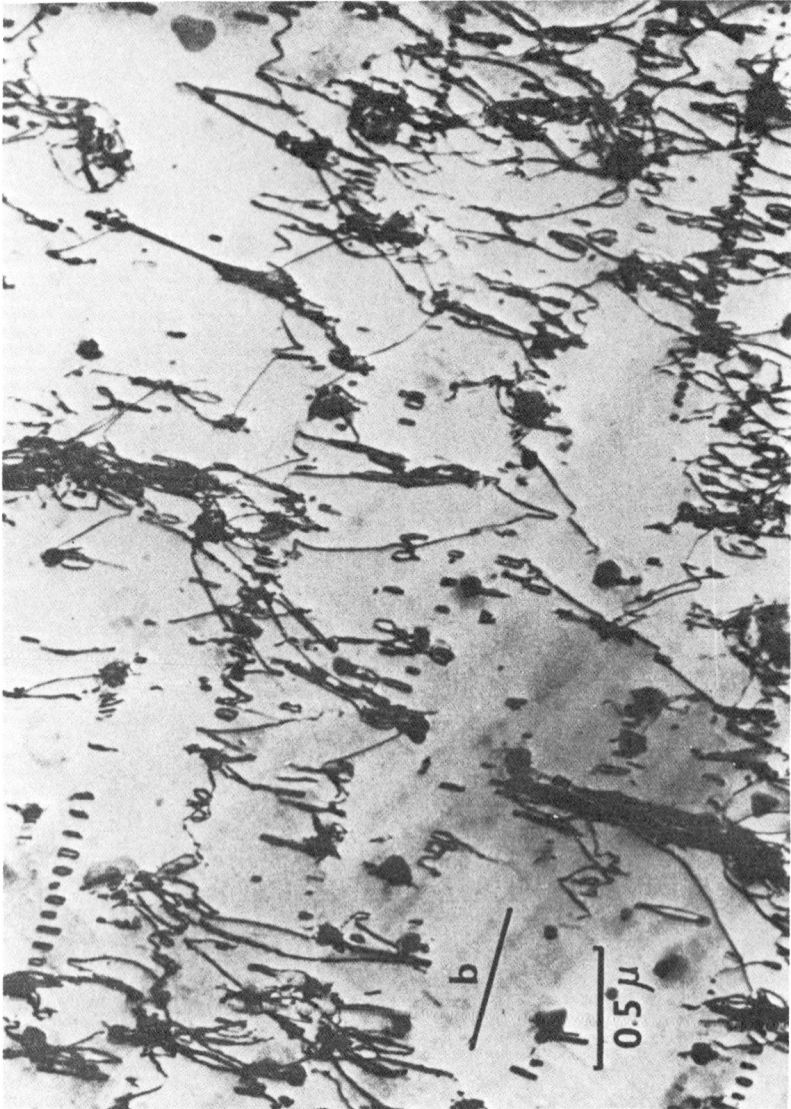

Plate XIII.   Section parallel to the primary slip plane of a single crystal of copper containing alumina particles ($V_f = 2.2 \times 10^{-3}$) deformed at room temperature to a shear strain of 0.15. (Reproduced from Reference 19 by permission of the MIT Press.)

the strain-hardening rate is not greatly enhanced by the second-phase particles, because the dislocations can move through the precipitates without accumulation. On the other hand, in alloys with dispersed hard particles, the rate of work-hardening is always very rapid, because of the high dislocation-loop density built up by cross-slip. The rate of work-hardening was found to be greater for plate-shaped particles than spheres. Also an increase in volume fraction or a decrease in spacing between particles tends to enhance the rate of hardening.[1]

Hirsch and Humphreys[19,91,92] have also suggested a theory of work-hardening. They proposed that work-hardening was due to the interaction between screw dislocations and the prismatic dislocation loops (Fig. 3.4) forming helices. The consequence of the reaction is that the particles are converted into linear obstacles whose lengths increase with increase in strain. Hence plastic deformation is induced by propagating dislocations through an array of parallel linear barriers. The distances between the obstacles are naturally smaller than those

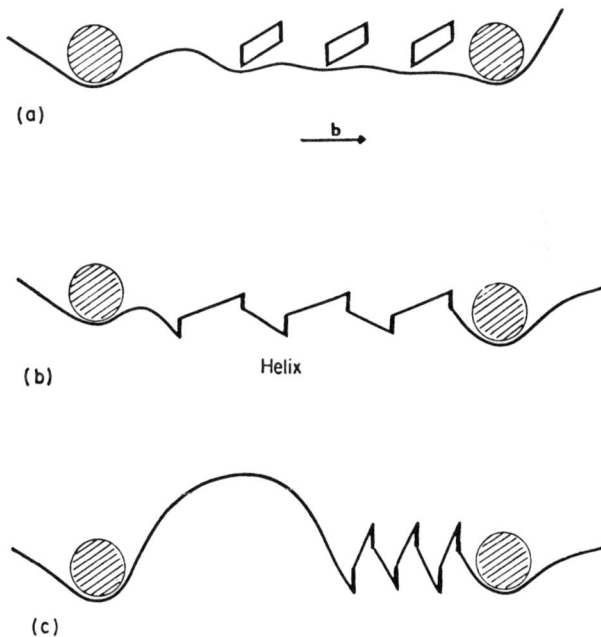

Fig. 3.4.   A screw dislocation and a row of prismatic loops interacting to form a helix. (Reproduced from Reference 19 by permission of the MIT Press.)

between the particles. Based upon this reasoning Hirsch and Humphreys developed a linear work hardening theory. The rate of hardening is proportional to the volume fraction of the alloying element. The theory agrees with experimental results at low strains.

3.3.3   Hardening Due to Modulus Mismatch

It has been mentioned in Section 1.5 that dispersed second phase materials with elastic moduli higher than those of the matrix materials tend to retard the motion of dislocations. Consequently, a stress higher than that in a homogeneous matrix material needs to be applied in order to induce general plastic yielding, for example, 10 per cent of the strength of a super-alloy has been due to this effect.[89] In this Section, the effects of reinforcing materials on dislocations arising from the inhomogeneity in elastic properties are discussed in a quantitative manner.

The interaction between a screw dislocation and the interface of a bimaterial system has been examined by Head,[20] Hsieh and Dundurs.[94] Head employed the close analogy between screw dislocations and electrostatic line charges. In particular, the interaction of a screw dislocation with changes in elastic constants of the medium is analogous to the electrostatic problem of a line charge in a medium with non-homogeneous dielectric constant. Consequently, the elastic stress field of a screw dislocation near a bimaterial interface can be constructed by the *method of images* just as in the solution of electrostatic problems. Consider a bimaterial system with a planar interface which is taken as the plane $x = 0$. The shear moduli G_1 and G_2 are used to designate the phases at $x > 0$ and $x < 0$, respectively. Assuming a perfect bonding at the interface, the complete stress field due to a right-handed screw dislocation in phase 1 situated at $(t, 0)$, parallel to the boundary, is given as

$$\tau_{yz} = \frac{G_1 b}{2\pi} \left[\frac{x - t}{(x - t)^2 + y^2} + K \frac{x + t}{(x + t)^2 + y^2} \right] \qquad x > 0$$

$$\tau_{yz} = \frac{G_1 b}{2\pi} \frac{(1 + K)(x - t)}{(x - t)^2 + y^2} \qquad x < 0$$

$$\tau_{xz} = \frac{-G_1 b}{2\pi} \left[\frac{y}{(x - t)^2 + y^2} + K \frac{y}{(x + t)^2 + y^2} \right] \qquad x > 0$$

$$\tau_{xz} = \frac{-G_1 b}{2\pi} \frac{(1 + K)y}{(x - t)^2 + y^2} \qquad x < 0$$

(3.11)

where $K = (G_2 - G_1)/(G_2 + G_1)$ and b is the Burgers vector. By comparing the stress expressions of eqns. (3.11) and (1.10), the above stress expressions can be interpreted in terms of the real and *image dislocations*. For $x > 0$, the stress field is the same as that due to the original dislocation plus a dislocation of strength Kb at the image point $(-t, 0)$. For $x < 0$, the stress field is the same as that due to a dislocation of strength $(1 + K)b$ at $(t, 0)$ in an infinite medium. Using this arrangement of real and image dislocations, it can be easily shown that the governing equations for elastic deformation as well as the continuity of the relevant stress and displacement components at the interface are all obeyed. Using the definition of forces on dislocations (see Problem 3.2), it can be shown that a dislocation will be attracted toward the boundary if $G_2 < G_1$, and repelled from it if $G_2 > G_1$. This is a long-range force which varies with $1/t$.

The interaction between a planar phase boundary and an edge dislocation is more complicated. The complete elastic field associated with a single edge dislocation cannot be obtained using a simple image construction. This problem has been examined by Head,[21] and Dundurs and Sendeckyj.[22] Depending upon the combination of elastic constants, there are three possible types of interactions. These have been categorised as follows: (a) $\alpha - \beta^2 > 0$: attraction in phase 2, and repulsion in phase 1, (b) $\alpha - \beta^2 < 0$, $\alpha + \beta^2 > 0$: repulsion by interface on both sides, and (c) $\alpha + \beta^2 < 0$: repulsion by interface in phase 1 and attraction by it in phase 2. The elastic constant factors associated with the phases 1 and 2 are defined as

$$\alpha = [\Gamma(k_1 + 1) - (k_2 + 1)]/[\Gamma(k_1 + 1) + k_2 + 1]$$
$$\beta = [\Gamma(k_1 - 1) - (k_2 - 1)]/[\Gamma(k_1 + 1) + k_2 + 1]$$
$$\Gamma = G_2/G_1, \qquad k = 3 - 4\nu$$

The behaviour of a dislocation in front of an interface is independent of the orientation of its Burgers vector relative to the interface.

The interaction between phase boundaries and dislocation loops is also a problem of practical interest as can be seen in the discussion in Section 3.3.2. The long-range interaction of dislocation loops and phase boundaries was discussed by Vagera.[23,24] More recently, a complete solution of the elastic field of a dislocation loop in a two-phase material has been obtained by Salamon and Dundurs.[25-27] It was concluded that for a prismatic loop parallel to an interface, the interaction between them is not necessarily restricted to simple repulsion and attraction as in the case of straight dislocation lines. They found that for certain combinations of elastic constants, stable or unstable equilibrium positions were possible. The equilibrium behaviour of dislocations near a second-phase material also has been found when a straight dislocation line approaches a surface layer. Solutions of this kind can be found in the works of Head,[21] Conners,[28] Weeks, et al.[29] and Kurihara.[95]

In order to simulate the interaction between dislocations and precipitates, it is necessary to take into consideration the finite sizes of the precipitates. Fleischer[30] first attempted to examine the interaction between a screw dislocation and a plate-like precipitate. The plate cross-section has a height d and thickness a. It was assumed that the dislocation in the matrix strained the precipitate in the same manner as would be the case were the precipitate not present. The stress acting on the screw dislocation due to the presence of the precipitate was estimated to be

$$\tau = \frac{\Delta Gb}{4\pi^2} \left\{ \frac{a/x}{x+a} \tan^{-1} \frac{d/2}{x+a} + \log\left(1 + a/x\right) \left[\frac{d/2}{(d/2)^2 + (x+a)^2} \right] \right\} \quad (3.12)$$

where ΔG is obtained by subtracting the shear modulus of the matrix from that of the precipitate. Also, the dislocation was assumed to be situated at halfway of the plate height and at a distance x from the plate surface. For the case that $d/2 \gg x + a$, eqn. (3.12) can be approximated as

$$\tau = \frac{\Delta Gb}{8\pi x} \left(\frac{a}{x+a} \right) \quad (3.13)$$

Fleischer, using eqn. (3.13), estimated the strengthening effect of Guinier–Preston zones in copper alloy. A range of stress on dislocations between 11 and 50 kg/mm^2 was obtained, depending upon the distance x.

The exact solution of the elastic field of a screw dislocation in and near a lamellar inclusion was obtained by Chou[31] using the image method. The model consists of an elastic layer of width a sandwiched between two elastic half-planes. Depending upon the combination of elastic properties, it is also possible for a screw dislocation to assume an equilibrium position near the precipitate. For the case that the side phases are identical, the arrangement simulates a plate-like precipitate in a homogeneous matrix material. The image force acting on a screw dislocation in the matrix at a distance x from the interface is given by

$$F = \frac{G_1 b^2}{4\pi} \frac{K}{x} \left[1 - (1 - K^2) \sum_{n=1}^{\infty} K^{2n-2} \frac{x/a}{x/a + n} \right] \qquad (3.14)$$

where G_1 and G_2 are the shear moduli of the matrix and the inclusion, respectively. A comparison of eqns. (3.13) and (3.14) is asked for in Problem 3.4.

The interactions of phase boundaries and dislocations in anisotropic crystals also have been studied. Solutions for image forces on both screw and edge dislocations were obtained for crystals with cubic and orthotropic symmetries.[32-5,95]

Besides a plate-like shape, precipitates may assume other morphologies such as circular, spherical and ellipsoidal. The non-homogeneity of elastic properties in these cases also tends to interact with dislocations. The closed form solutions of the interaction energy and interaction force between a circular cylindrical inclusion and a parallel screw dislocation has been obtained by Dundurs.[36] A hard inclusion expels screw dislocations that are in the inclusion and repels dislocations in the matrix. In this case, the centre of the inclusion is an unstable equilibrium position. The interaction of edge dislocations with circular inclusions is more intricate than that of screw dislocations. Besides simple repulsion and attraction, edge dislocations in the matrix may also exist in stable or unstable equilibrium positions, depending upon the relative magnitude of elastic constants.[37-9] Other works concerning interactions of dislocations with a single inclusion, multiple inclusions, inclusions with imperfect bondings and elliptically cylindrical inclusions can be found in References (21) and (40-44).

The interaction of dislocations with spherical particles such as impurity atoms and voids plays a rather unique role in strengthening and in controlling the yield strength of alloys. The interaction has been discussed qualitatively in Section 1.5 from the viewpoint of elastic misfit. The following discussions examine the interaction arising from the difference in elastic properties. A quantitative evaluation of this interaction is a relatively difficult task. It is simply because of the three-dimensional nature of the elasticity problem concerning a point defect and a straight dislocation line. Weeks *et al.*[45] first attempted to find the solution of the long-range interaction between a straight screw dislocation and a spherical void, and were able to derive the expression for the interaction energy. By further using an estimate for the short-range interaction, they discussed the implications with regard to mechanical strength at low and high temperatures. The exact solution of screw dislocation and void interaction was later obtained by Willis *et al.*[46] in closed form. The induced dilatation field was also obtained and used to discuss the resulting drift of interstitials into the void. Comninou and Dundurs[47] derived the long-range interaction force between a screw dislocation and spherical inclusions with finite rigidity. When the dislocation is distant from the inclusion, the stress around the inclusion due to the dislocation can be assumed to be uniform. The long-range interaction is then obtained by considering the disturbance of the uniform stress field due to the presence of the inclusion. An inclusion tends to repel the dislocation if its shear modulus is higher than that of the matrix. The distribution of the interaction force on the dislocation is non-uniform and is highly affected by Poisson's ratio of the matrix. Figure 3.5 indicates the variation of the image force. It is interesting to note that for ν above $\frac{1}{4}$ the force distribution ceases to be bell-shaped and the maximum occurs at some distance from the point nearest the inclusion. A similar approach has been used by Lin and Mura[44] to study the long-range elastic interaction between a dislocation and an ellipsoidal inclusion in cubic crystals. The Peach–Koehler force on the dislocation was calculated for the limiting cases, such as spherical, disc-shaped and rod-shaped inclusions. Numerical solutions that also take into consideration the short-range interactions can be found in the works of Yoo and Ohr,[48] and Sines *et al.*[49–51]

In conclusion, the following pertinent remarks need to be made. First, the elasticity solutions concerning the interactions between dislocations and second-phase materials in alloys have provided us with some fundamental understandings of the nature of the interactions. The most important information obtained from these solutions is the interaction force on the dislocations arising from the presence of the second-phase materials. Depending upon the size and distribution of the dispersed phase, the interaction force may be long- or short-range. The local retarding or attraction forces on dislocations can be related to the macroscopic strength of the alloy and a strengthening effect is thus achieved. Secondly, dislocations may pile up against dispersed particles if they cannot bypass these barriers. The formation of dislocation pile-ups can further affect the flow stress and strain hardening behaviour of the alloys. For comprehensive reviews of the physical and mathematical bases of the dislocation pile-up theory, the reader is referred to References 52 and 53. Thirdly, because of the discontinuity in elastic properties, the elasticity theory of dislocations always predicts an interaction force infinite in magnitude whenever a dislocation approaches a phase boundary. This type of singular

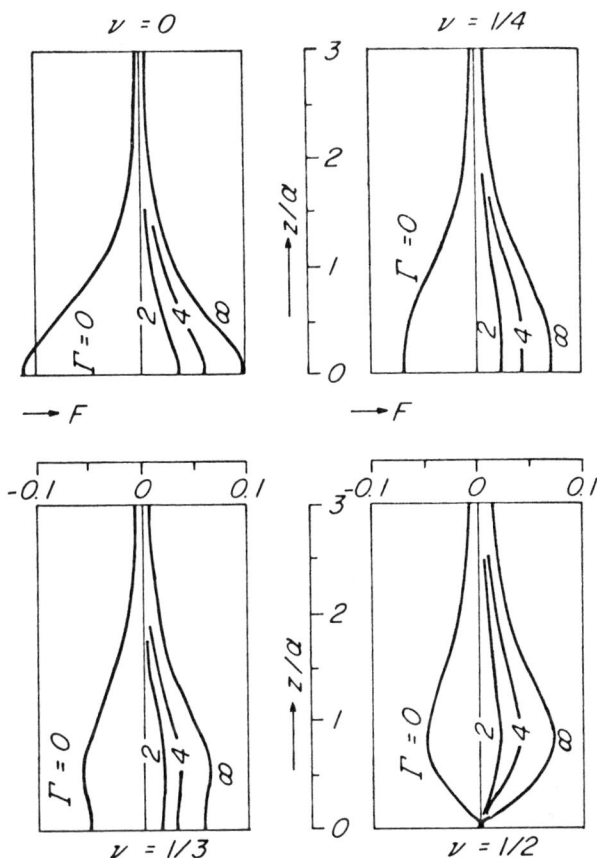

Fig. 3.5. *Distribution of force acting on dislocation.* (*Reproduced from Reference 47 by permission of the American Institute of Physics.*) *The force F is measured along the dislocation line as a function of the distance z from the point nearest to the inclusion. The unit of F is c^4/b^2Ga^3, where c, b, G, a and ν are the shortest distance between the dislocation and inclusion centre, Burgers vector, matrix shear modulus, inclusion radius and matrix Poisson's ratio, respectively.*

behaviour has no real physical significance and a dislocation can penetrate into the dispersed second phase if it is crystallographically allowed to do so. Attempts have been made to eliminate such a barrier which is actually a mathematical rather than a physical one. Pacheco and Mura[54] considered the dislocation to be continuously distributed in accordance with the model of Peierls. This solution yields a finite interaction force at the interface and is consistent with the solution of a singular dislocation when the dislocation is at about three atomic spacings from the interface. Nakahara *et al.*[55] approached this problem by finding the elastic field of an interface dislocation in a planar phase boundary. They considered that a dislocation is in the interface when it is within a distance equal to the core radius from the interface. The interaction energy can thus be defined at and near the interface. Similar attempts also have been applied to defect loops[56,57] and to interface dislocations in anisotropic materials.[34,58] In these cases, it is concluded that the displacement and stress components for a dislocation vary continuously with positions in and near a boundary. This should be the direct consequence of the limiting process employed. Finally, the interactions of dislocations and inclusions discussed above are all caused by the effects of the inhomogeneities on the dislocation elastic fields. In actual crystal lattices, these interactions could be complicated by other considerations such as externally applied stresses. These applied stresses can be modified near the inclusions, and result in local stress concentrations. These stress concentrations near inclusions cause additional interactions between dislocations and dispersed second-phase materials. Discussions on the distribution of local stress due to inhomogeneity and its effect on dislocations can be found in References 63, 66, 85, and 96.

3.3.4 Short-Range Interactions

The interaction between a dislocation and the misfit stress produced around a coherent precipitate is short-range in nature, as has been discussed by Mott and Nabarro. There are other short-range interactions in alloys. In this section the discussions are centred on the short-range interactions induced by passing dislocations through an ordered precipitate and by the formation of misfit dislocations. The precipitates may be ordered internally, such as those in the Guinier–Preston zones. If the Burgers vector of a moving dislocation is not equal to the lattice parameter of the ordered structure, work must be done to create a disordered interface across the slip plane. In the more general cases, the precipitates may not have an ordered structure. Then, the passage of dislocations through these precipitates may alter the number of nearest neighbours of a particular type at the surface of the precipitates.[1]

Kelly and Fine[59] estimated the stress necessary to force a dislocation through a Guinier–Preston zone in an aluminium alloy containing 2 atomic per cent copper and in an aluminium alloy containing 13 per cent silver. The Guinier–Preston zones in the aluminium–copper alloy examined were estimated to be in the form of discs about 100 Å in diameter and consisting of a single layer of copper atoms on the {100} planes of the matrix. The stress σ necessary to force each atomic length of a dislocation through a Guinier–Preston zone is given by the following equation

$$\sigma b^2 d = \frac{\Delta E}{2} \tag{3.15}$$

d is the average distance between the centres of the zones in the slip-plane, and is estimated to be 1.5×10^{-6} cm. ΔE is the energy difference between a copper atom in a Guinier–Preston zone, and in a solution of the aluminium matrix, and it can be estimated from the heat of reversion of the alloy, namely, the thermal energy necessary to redissolve the zones. For a 2 atomic per cent copper alloy, $E = 0.077$ eV per atom of copper. Using these data, eqn. (3.15) gives the value of σ around 0.5 kg/mm^2. An estimate of the flow stress based upon the bulging of dislocations in between the Guinier–Preston zones gives a considerably higher stress value.

The Guinier–Preston zones in the aluminium–silver alloy assume the shape of spheres, with an average distance between zones of about 10^{-6} cm. Because of the similarity in size of aluminium and silver atoms, there is virtually no strain caused by elastic misfit. The flow stress computed from the heat of reversion data is 6.2 kg/mm^2. This value is again lower than that calculated from the Orowan's criterion. These results led Kelly and Fine to suggest that the process of shearing the zones determined the initial flow stress in the age-hardened alloys.

Gleiter and Hornbogen[97,98] examined the precipitation of the ordered $Ni_3Al(\gamma')$ particles in a Ni–Cr–Al superalloy. The resulting mixture of ordered and disordered phases is free of coherency stress. The ordered particles can be sheared off by dislocation pairs. The critical shear stress depends on the particle size and distribution, and the anti-phase-boundary energy.

The possibility of penetration of dislocations into precipitates versus that of bypassing also has been discussed by Friedel.[60] Fleischer[61] suggested that misfit dislocations may be formed at precipitate surfaces when slip occurs across a change in lattice spacing. The formation of misfit dislocations may be another form of short-range interaction.

In concluding the discussion on dispersion hardening, it can be said that the strengths of alloys are contributed to by many factors. The deformability of the dispersed phase and its coherency with the matrix; the size and volume fraction of the dispersed phase; the inter-particle spacings; the extent of age-hardening treatment and the strength of interfaces between dispersed particles and matrices. The strength of alloy systems is most likely affected by more than one strengthening mechanism.

The effectiveness of the hardening mechanisms varies with the stages of ageing.[89] It is believed that if the dispersed particles are soft or at short ageing time when the precipitates are small, a mobile dislocation can cut through the particles lying in its slip plane. As a result, the hardening mechanisms in this case may include: the formation of anti-phase boundaries in ordered precipitates; the coherency strains due to lattice mismatch; the elastic interaction due to modulus mismatch; and drag to dislocations created by mismatch in flow stress. The deformation mode in this case is characterised by the planar and coarse appearance of slip lines. The strength of an alloy increases as the particle size or ageing time increases because the effects of the above mentioned hardening mechanisms are more pronounced for larger particles. However, beyond a certain transition size or ageing time, alloy strength tends to decrease. This is because instead of cutting the particles, a dislocation can by-pass them by bulging through the openings between the barriers at a lower stress level. The stress necessary for dislocations to by-pass the barriers is further reduced when the particles cluster together at long ageing time. During this stage, it is believed that the hardening of alloys is mainly contributed through the Orowan mechanism and mismatches in lattices and moduli. Furthermore, additional dislocations are generated from the necessity of retaining continuity between non-deformable particles and the matrix.[99] The dislocation cell structures formed on the particles tend to further inhibit dislocation movement. The contribution to alloy strength may obey the Hall–Petch type relation given in eqn. (1.12). Depending upon the alloy systems, the parameter d may be taken as particle spacing, cell size or grain size.

In order to attain the maximum degree of strengthening, it is desirable, at least in principle, to make the physical properties of precipitates and dispersed hard particles as different from the matrix as possible. This then implies that coherency stresses, interface dislocations and modulus effects all tend to increase the friction stress to dislocations. However, it should be borne in mind that the yield strength of an alloy is limited by the shear modulus of the metal matrix.

This is because the alloy shear yield strength can never exceed its theoretical strength which is related to the shear modulus.

Finally, it needs to be mentioned that cold work of alloys with dispersed hard particles and precipitates can enhance their stability and resistance to recovery and recrystallisation. This is also due to the fact that dislocations generated by cold-work form networks which prohibit the movement of dislocations. The resistance to recovery and recrystallisation for some alloys was also observed to be inversely proportional to the particle spacing. The experimental results of Preston and Grant[67] on copper alloys containing aluminium oxide particles indicated that these alloys maintained their hardness when tested after 1 hr of annealing at temperatures close to the matrix melting point. Precipitation-hardened copper alloys containing beryllium and cobalt also showed retention of hardness at temperatures up to one-half of the melting temperature of the copper matrix.[68] Unlike alloys with hard particles, precipitation-hardened alloys lose their resistance to creep when the precipitates coarsen and dissolve back into the matrices at elevated temperatures.

3.3.5 Comparison With Fibrous Composites

Composites reinforced with dispersed phases, such as precipitates and hard particles are different from fibrous composites in many aspects. In fibrous composites, externally applied loads are transferred through the matrix to the fibres which are the principal load carrying elements. However, the matrix of a dispersion-strengthened alloy carries the majority of the load. The function of the dispersed phases in this case is to hinder the movements of dislocations. But the effectiveness of these barriers is limited, especially with equiaxial particles. This is because great stresses are induced in the matrix near the hard particles when the composite is deformed. Flow of the matrix around the particles can take place through interface failure, dislocation cross-slip and climb, and attainment of the matrix theoretical strength. As a result, the hard particles are not likely to be loaded to their fracture strength. On the other hand, dislocations impeded by fibres cannot easily bypass the barriers. Some deformation of the matrix is necessary in order to load the fibres and to make full use of their strengths.

Alloys hardened with precipitates and dispersed particles generally have volume fractions of the second-phase materials in the range of 1 to 15 per cent. There is a problem of dispersing the second-phase materials uniformly if its volume fraction is higher. The fibres in composites can be accurately aligned and thus very high volume fraction can be achieved. The strengths of fibrous composites are further affected by the fibre aspect ratio, interfacial bonding and the chemical activity between the fibre and matrix, especially at elevated temperatures. The effectiveness of strengthening in composites with dispersed hard particles relies very much on the inter-particle spacings. The inter-facial bonding between the particles and the matrix may be important but certainly not as critical to composite strength as in fibrous composites. Furthermore, because of their highly directional properties, the analysis of structures made of fibrous composites is more complicated than the isotropic alloys.

It also needs to be noted that the behaviour of cermets is somewhat different from the other types of dispersion-strengthened composites. This is due to the facts that the ceramic particles are larger than 1 μm in diameter and their volume fractions range between 25 and 70 per cent. Consequently, in this type of particulate composites, both the matrix and the particles share the load-bearing function. Cermets, such as tungsten carbide and tungsten thoria have long been used as cutting tools and lamp filaments, respectively. Discussions on their mechanical behaviours can be found in Reference 9. Comprehensive comparisons of fibrous composites with dispersion-strengthened composites are given in References 68, 69, and 82.

3.4 DIRECTIONALLY SOLIDIFIED EUTECTIC COMPOSITES

3.4.1 Introduction of the Phase Diagram

Phase diagrams provide useful information for various material systems. To facilitate the understanding of the basic principles governing the construction and analysis of phase diagrams, the diagram of the Fe–Fe$_3$C system in Fig. 3.6 is chosen as a convenient example. The metallurgical system under consideration is known as a *binary system* since it is composed of two *components*. The raw materials forming the components of a system can be pure elements, multiphase mixtures or compounds such as Fe$_3$C. The abscissa designates the weight percentage of the components while the ordinate indicates the temperature. Then the regions in this phase diagram represent certain equilibrium thermodynamic *phases* as determined by various combinations of temperature and composition. More specifically, a phase represents a physically distinct region of matter which has characteristic atomic

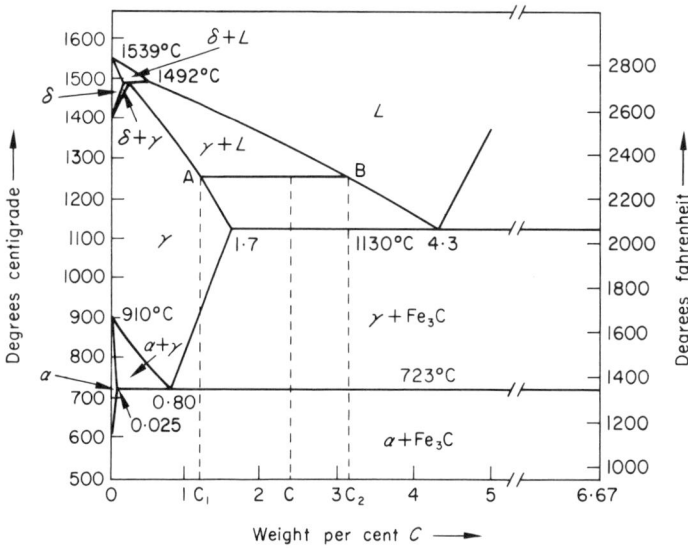

Fig. 3.6. The Fe–Fe₃C *phase diagram.*

structure and properties. The phases of a system are, in principle, mechanically separable.[70] In Fig. 3.6, there are several distinct phases. At elevated temperatures, the binary system exists in the liquid state (L). There are also four distinct solid phases designated as α, γ, δ and Fe₃C. As pointed out in Chapter 1, the α and δ irons, assume BCC structures and the γ phase has a FCC structure. The iron carbide crystal is known to be orthorhombic. The single-phase region of Fe₃C is denoted by a line since its composition is fixed by the carbon content of 6·67 per cent. The single-phase regions in a phase diagram are always separated by multiphase regions which are mixtures of the neighbouring single-phase materials. The various two-phase regions of the Fe–Fe₃C system are also shown in Fig. 3.6.

Each region in a phase diagram is represented by certain admissable combinations of temperature and composition. These combinations are governed by the *Gibb's phase rule*. The phase rule states that in a system having a thermodynamic equilibrium, the number of phases, P, the number of components, C, and the number of independent variables, V, of the system obey the following relation

$$P + V = C + 2 \tag{3.16}$$

The independent variables include temperature, pressure and composition. Since phase diagrams are usually determined under one atmosphere pressure, only temperature and composition are left as independent variables. In a binary system, $C = 2$. For a single phase region in Fig. 3.6, eqn. (3.16) indicates that all the temperature and composition combinations within the regions are allowed under one atmosphere pressure. In the two-phase regions, one has the freedom to specify either temperature or composition. Consider, for instance, an alloy composed of a mixture of γ and L. At an arbitrary temperature T, the composition is fixed by the horizontal line AB drawn at this temperature. The compositions of the γ and L phases in the two-phase mixture are given by C_1 and C_2, respectively. The line AB is also known as a *tie line*.

Although the compositions of the individual phases in the two-phase mixture is fixed, the weight ratio of the phases in the mixture varies with the composition of the alloy. Consider the alloy with composition C at temperature T, the weight percentages of γ and L phases in the alloy can be derived as

$$x_\gamma = \frac{C_2 - C}{C_2 - C_1}$$

$$x_L = \frac{C - C_1}{C_2 - C_1}$$

(3.17)

The relation given by eqn. (3.17) is known as the *lever rule*. It needs to be emphasised that at temperature T, alloys may have different weight fractions of the individual phases. However, the compositions of the individual phases are fixed once the temperature is chosen.

Another important feature of the phase diagram in Fig. 3.6 is the phase transformation associated with an alloy of composition 4·3 per cent carbon at 1130°C. Upon cooling, the high-temperature liquid phase solidified into two solid phases, γ and Fe_3C. The alloy with this particular composition solidifies at a unique temperature just as pure elements. This reaction is known as a *eutectic reaction*.

In ternary alloys, three-dimensional representations are needed for the phase relations. For multicomponent systems, it is hardly possible to handle them diagrammatically.

3.4.2 Eutectic Composites

These types of composites are produced by solidifying an alloy of eutectic composition. The solid material so formed can be considered as a composite with two distinct phases. In order to achieve great strength it is most desirable to align the second-phase material in one direction to produce a unidirectionally solidified composite. The state of the art of eutectic composites has been reviewed by Salkind, et al.[71] Part of the present discussion is based upon this review.

Active research work in eutectic composites was initiated by Kraft[72] and his colleagues at the United Aircraft Research Laboratories; now a large number of eutectics have been studied. Data for simple eutectic systems as well as eutectics of complex oxides, complex halides, organics and complex intermetallics have been compiled by Galasso.[73] The unidirectional solidification process can be carried out by moving a heating coil and, hence, a molten zone along an ingot. Alternatively, solidification can be accomplished by keeping the heating coil stationary and drawing the molten specimen slowly. Two types of microstructures are typical among all the eutectics. These are the rod-like whiskers embedded in a matrix and the lamellar structures. Other structures in the forms of dendrites, platelets, blades and spirals also have been observed. A representative eutectic having rod-like whiskers can be found in the Al–Ni system. The eutectic of this system occurs at 640°C and is composed of an inter-metallic compound Al_3Ni at 25 atomic per cent of nickel and the α phase which is a solid solution of 0·02 atomic per cent of nickel in aluminium. The eutectic alloy occurs at 2·8 atomic per cent of nickel. Based upon these data, the volume percentages of the individual phases in the eutectic can be calculated. The microstructure of unidirectionally solidified Al–Al_3Ni eutectic is shown in Plates XIV and XV.[74,75] The microstructure of unidirectionally solidified Al–$CuAl_2$ eutectic is lamellar as shown in Plates XVI and XVII. The eutectic of the Al–Cu system occurs at 548°C and has 54 volume per cent of Al. The microstructures of eutectics are affected by the volume fraction of the second-phase material. The lamellar and rod-like structures often prevail at volume fractions of second-phase materials above and below 30 per cent, respectively.[71]

The unidirectionally solidified Al–Al_3Ni eutectic composite has a tensile strength of 48 ksi ($3·31 \times 10^8 \, N/m^2$), which is considerably higher than the strengths of conventionally cast Al–Al_3Ni eutectic (11 ksi, $7·58 \times 10^7 \, N/m^2$) and pure aluminium matrix (4·2 ksi, $2·9 \times 10^7 \, N/m^2$). The strength of the Al_3Ni whiskers is around 400 ksi ($2·76 \times 10^9 \, N/m^2$). The high-temperature performance of the Al–Al_3Ni eutectic is superior to precipitation-hardened aluminium alloys, and is

Plate XIV. Microstructure of unidirectionally solidified Al–Al₃Ni eutectic exhibiting rod-like whiskers—transverse section. (*Reproduced from References 74 and 75 by permission of the AIME.*)

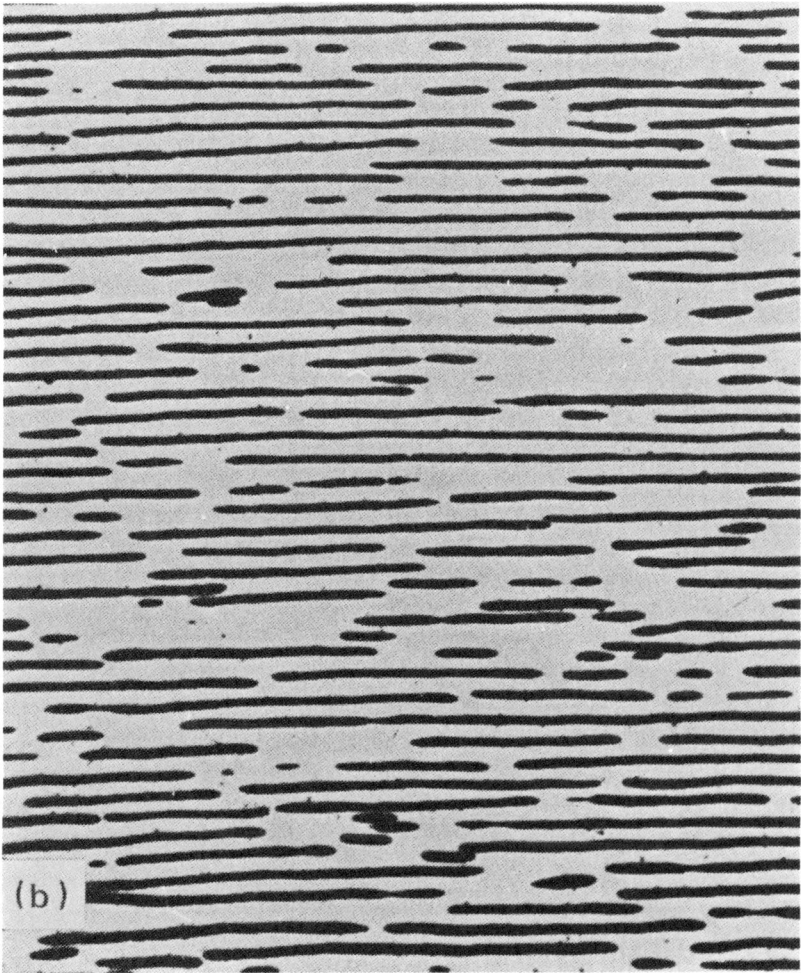

Plate XV. *Microstructure of unidirectionally solidified* Al–Al₃Ni *eutectic exhibiting rod-like whiskers—longitudinal section. (Reproduced from References 74 and 75 by permission of the AIME.)*

Plate XVI. *Microstructure of unidirectionally solidified Al–CuAl$_2$ eutectic exhibiting lamellar structure—longitudinal section. (Reproduced from Reference 71 by permission of Wiley-Interscience Publishers.)*

comparable to sintered aluminium powder. It retains nearly half of the room temperature strength at 900°F (482°C). On the other hand, the strength of aluminium alloy 7075-T6 deteriorates from 81 ksi (5.58×10^8 N/m^2) at room temperature to nearly 7 ksi (4.83×10^7 N/m^2) at 700°F (371°C).[76]

The Al–CuAl$_2$ eutectic can be used as a model system for studying lamellar eutectics. At room temperature this composite has a tensile

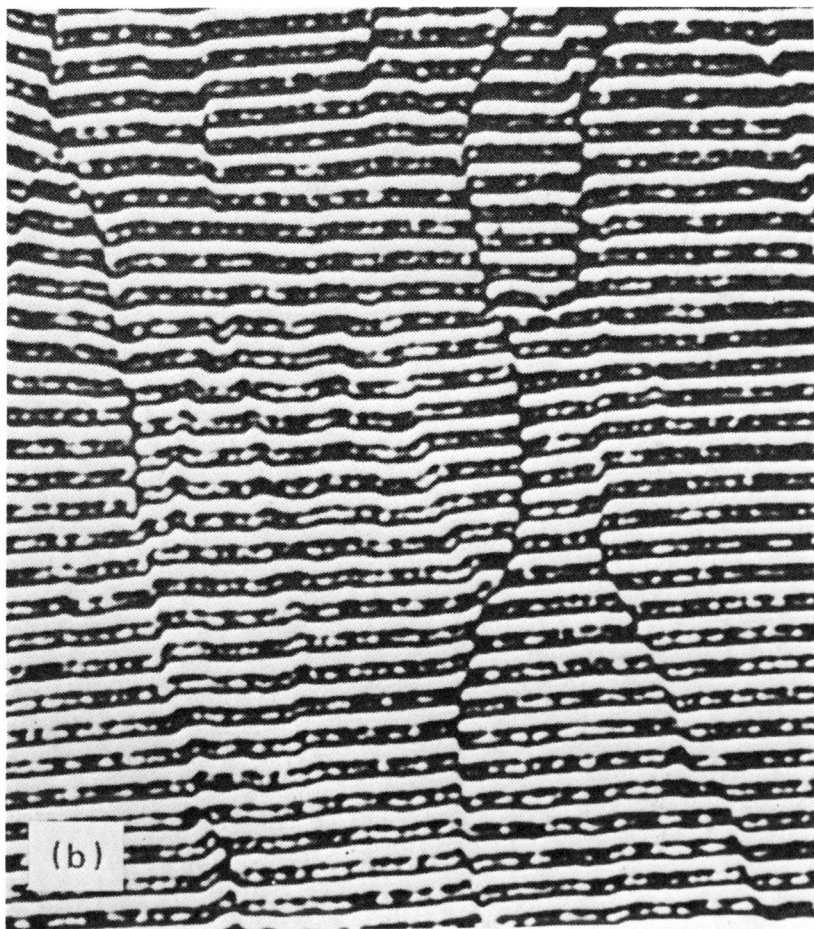

Plate XVII. *Microstructure of unidirectionally solidified* Al–CuAl₂ *eutectic exhibiting lamellar structure—transverse section.* (*Reproduced from Reference 71 by permission of Wiley-Interscience Publishers.*)

strength of 38 ksi ($2 \cdot 62 \times 10^8$ N/m²). Although the volume fraction of the $CuAl_2$ lamellae is high, it does not appear to reinforce the matrix very effectively. This composite also has demonstrated high strength retention at elevated temperatures.

The strength of eutectic composites reinforced with rod-like whiskers seems to obey the prediction of the rule of mixtures. In lamellar eutectics the effect of the interfaces between the lamellae plays a rather significant role in strengthening the composites. Cline and Stein[77] examined this effect in the directionally solidified eutectic of silver–copper alloy. This system was chosen because the eutectic has tensile strength substantially exceeding the strength of either of the metals in the composite. The eutectic of this alloy has a fine lamellar structure of silver-rich and copper-rich phases with an interlamellar spacing ranging from 0·1 to 3·7 μm in the small ingots. The two phases had the same orientation and the habit plane of the interface was a (100) plane. In view of the large difference in lattice parameters, the interphase boundaries were partially coherent. The flow stress of the small ingots grown at different rates was plotted as a function of the inverse square root of the lamellar spacing as seen in Fig. 3.7. The variation of the strength with lamellar spacing seems to follow a Hall–Petch type of relation. At the finest spacing tested, the flow stress is comparable to the strongest commercial silver–copper alloy.

Based upon the expression of eqn. (3.11) Cline and Stein suggested that the maximum value of the image stress acting on a screw

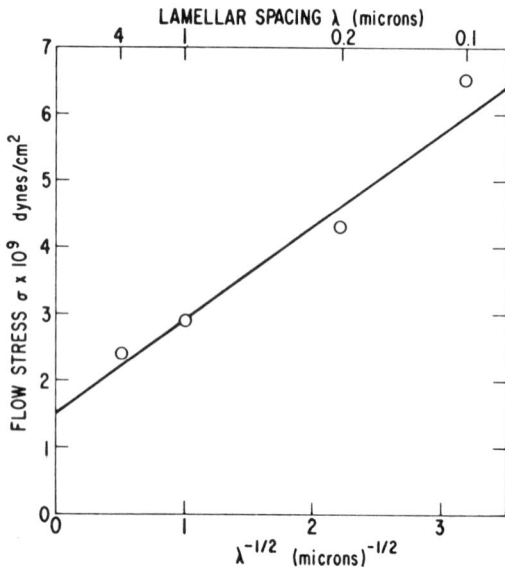

Fig. 3.7.   *The flow stress at room temperature as a function of the inverse square root of the lamellar spacing. (Reproduced from Reference 77 by permission of the AIME.)*

dislocation at a distance from the interface smaller than a certain cut-off distance Δ is

$$\tau_{max} = \left(\frac{G_2 - G_1}{G_2 + G_1}\right) \frac{G_1 b_1}{4\pi\Delta} \qquad (3.18)$$

G_1 is the shear modulus of the phase where the dislocation is situated. When n dislocations are forced by the applied tensile stress σ against the interface, the stress concentration on the leading dislocation is n times the resolved applied stress (see Section 1.5). In order for the material to flow, the stress on the leading dislocation must exceed the maximum image stress given by eqn. (3.18). By taking into consideration relative orientations of the tensile axis and the slip plane, Cline and Stein found that the number of dislocations in the pile-up, image stress, and the applied tensile stress can be related as

$$n\sigma = 2\tau_{max}$$

or

$$n\Delta = \left(\frac{G_2 - G_1}{G_2 + G_1}\right) \frac{G_1 b_1}{2\pi\sigma} \qquad (3.19)$$

Experimental results indicated that the flow stresses of the lamellar composites increased as the lamellar spacing decreased. Then, from eqn. (3.19), the quantity $n\Delta$ also decreased. Cline and Stein suggested that the variation in the number of dislocations in a pile-up with the lamellar spacing might explain the effect of lamellar spacing on the flow strength of the eutectic composite. The effect of hardening due to solid solution was also examined. The x-ray data indicated that there was less than 1 atomic per cent of copper in silver and less than 4 atomic per cent of silver in copper. The strengthening effect due to solid solution was estimated to be less than 1×10^7 N/m^2. This value is negligible in comparison to the hardening due to the image force.

Another strengthening mechanism due to interface dislocation has been suggested by Walter, et al.[78] In their study of the NiAl–Cr eutectic, regular arrays of interface dislocations were observed at the boundary between the chromium-rich rods and the NiAl-rich matrix. Both phases are cubic and the lattice mismatch is small. The {100} planes and ⟨001⟩ directions in the rods and in the matrix are parallel. The dislocation network reduces the internal stresses caused by the lattice mismatch at the interface. The apparent lattice mismatch calculated from the measured spacing of dislocations in the network was 0·35 per cent. This represents a difference of 0·01 Å between the lattice parameters of NiAl and the chromium rod and one dislocation for every 270 atomic planes. Based upon these data, the strain energy per unit area of the dislocation network was estimated at 140 erg/cm^2. The strengthening due to the dislocation network was calculated for the stress necessary to push a dislocation segment through two adjacent pinning points. The magnitude of this stress is about 10 ksi (6·89 × 10^7 N/m^2). It also has been suggested that additional strengthening should result from the formation of imperfect dislocations as slip dislocations cross the network into the chromium rods. Furthermore, the differential thermal expansion can result in changes in interface dislocation spacing on heating and cooling the eutectic composites.[79]

From the above discussion, it is obvious that directionally solidified eutectic composites have certain advantages over fibrous composite materials discussed in Chapter 2. Since the reinforcing materials are incorporated into the matrix in a one-step operation, eutectic composites eliminate the need for a separate process for fibre production. As a result, the many problems encountered in handling and in wetting the fibres discussed before can also be avoided. Another desirable property of eutectic composites is their excellent thermal stability at elevated temperatures. Several eutectic systems have demonstrated remarkable strength at high temperatures. Among them, the Ta–Ta$_2$C eutectic has a room temperature strength of 150 ksi (1·03 × 10^9 N/m^2) and is superior to one of the best tantalum alloys, Ta–10W, at temperatures above 2000 °F (1093 °C).[80] The Nb–Nb$_2$C eutectic has strength higher than any commercial Nb-base alloy and is able to maintain its strength at 2500 °F (1371 °C) after 250 hours.[80] The Ni–NiMo eutectic has a lamellar structure. It has room temperature strength of 160 ksi (1·10 × 10^9 N/m^2) and Young's modulus of 40 × 10^6 psi (2·76 × 10^{11} N/m^2). It is stronger than any Ni-base super-alloys at elevated temperatures.[81] In view of their excellent properties eutectic composites have great potential of being applied at elevated temperatures.

An obvious disadvantage of using eutectic composites is the inability of varying the volume fractions of the second-phase materials. However, in view of the numerous existing eutectics, it is not too difficult to find the particular eutectic with the desired material property and volume fraction of the second phase. The applications of eutectics are frequently restricted to simple geometric shapes such as plates and rods and high purity of the initial metals is required for faster solidification and satisfactory end products.

3.5 PROBLEMS

3.1 The elastic field in an isotropic material due to the presence of a misfit particle is spherically symmetrical. The shear strain at a distance r from the centre of a particle with radius r_0 is $\epsilon r_0^3/r^3$ where ϵ denotes the misfit. Derive this expression based upon the non-vanishing displacement component in the radial direction $U_r = \epsilon r_0^3/r^2$.

3.2. Derive eqn. (3.6). In the derivation of this equation, it is necessary to understand the concept of the Peach–Koehler force on a dislocation. The source of this force may be due to stresses arising from external loading or internal stress field. These stresses can be regarded as producing a force that moves the dislocation through a crystal. The magnitude of this force is obtained by equating the work done by the Peach–Koehler force and that due to the stresses. The force per unit length on a dislocation due to applied shear stress τ is proportional to the product τb (see Reference 84 for detailed discussions).

3.3. Figure 3.6 indicates the formation of a helix through the interaction of a screw dislocation and a row of prismatic loops. Prove that, in general, a non-equilibrium concentration of point defects can react with a screw dislocation to form a helix. By considering the total force on a dislocation, show that the helix is an equilibrium form for a dislocation line (see Reference 84 for detailed discussions).

3.4. By assuming various combinations of G_1 and G_2 and plate thickness, compare the magnitude of image forces as given by eqns. (3.12) and (3.13).

3.5. The hardening effects due to misfit particles and plastic work can be estimated by considering the interactions between crystalline defects. Find the maximum shear stress on the plane at a distance h from a misfit particle. Also find the maximum shear stress an edge dislocation exerts on another parallel dislocation with identical Burgers vector. The separation distance between the dislocations is h.

3.6. Calculate the volume percentages of precipitated particles necessary to attain the theoretical alloy strengths given in Problem 1.6.

3.7. The following questions are related to the design of a high temperature material that is hardened with second-phase particles.

(a) What fundamental factors would you consider in trying to obtain a material for long time high temperature, high strength stability?
(b) What classes of materials would you look for?
(c) What would you expect to be the stability of a unidirectionally solidified lamellar eutectic? How would you expect this material to change in structure during long exposure at temperatures close to the melting point? What would you do to develop a finer structure?

REFERENCES

1. Kelly, A. and Nicholson, R. B., *Precipitation Hardening*, Progress in Materials Science, Vol. 10, B. Chalmers, ed., Macmillan, New York (1963).
2. Laird, C. and Aaronson, H. I., 'The Dislocation Structure of the Broad Faces of Widmanstätten γ Plates in an Al–15% Ag Alloy', *Acta Met.*, **15**, 73 (1967).
3. Laird, C. and Aaronson, H. I., 'Structures and Migration Kinetics of Alpha: Theta Prime Boundaries in Al–4 Pct Cu: Part I—Interfacial Structures', *Met. Trans. AIME*, **242**, 1393 (1968).
4. Nabarro, F. R. N., 'The Influence of Elastic Strain on the Shape of Particles Segregating in an Alloy', *Proc. Phys. Soc.*, **52**, 90 (1940).
5. Friedel, J., *Dislocations*, Addison-Wesley Pub. Co., Reading, Mass. (1964).
6. Cahn, J. W., 'Nucleation on Dislocations', *Acta Met.*, **5**, 169 (1957).
7. Unwin, P. N. T., Lorimer, G. W., and Nicholson, R. B., 'The Origin of the Grain Boundary Precipitate Free Zone', *Acta Met.*, **17**, 1363 (1969).
8. Smith, G. C., 'Dispersion Strengthened Materials', *Composite Materials*, Edited by the Institution of Metallurgists, American Elsevier Pub. Co., New York (1966), p. 27.
9. Tetelman, A. S. and McEvily, A. J. Jr, *Fracture of Structural Materials*, John Wiley and Sons, Inc. New York, 1967.
10. Mott, N. F. and Nabarro, F. R. N., 'An Attempt to Estimate the Degree of Precipitation Hardening with a Simple Model', *Proc. Phys. Soc.*, **52**, 86 (1940).
11. Nabarro, F. R. N., 'The Mechanical Properties of Metallic Solid Solutions', *Proc. Phys. Soc.*, **58**, 669 (1946).
12. Ashby, M. F., 'On the Orowan Stress', *Physics of Strength and Plasticity*, A. S. Argon, ed., The MIT Press, Cambridge, Mass. (1969), p. 113.
13. Orowan, E., *Symposium on Internal Stresses in Metals and Alloys*, Institute of Metals (1948), p. 451.
14. Kocks, U. F., 'A Statistical Theory of Flow Stress and Work Hardening', *Phil. Mag.*, **13**, 541 (1966).
15. Foreman, A. J. E. and Makin, M. J., 'Dislocation Movement Through Random Arrays of Obstacles', *Phil. Mag.*, **14**, 911 (1966).
16. Ebeling, R. and Ashby, M. F., 'Dispersion Hardening of Copper Single Crystals', *Phil. Mag.*, **13**, 805 (1966).
17. Jones, R. L. and Kelly, A., *Proc. Second Bolton Landing Conference on Oxide Dispersion Strengthening*, Gordon and Breach, New York (1968).
18. Humphreys, F. J. and Martin, J. W., 'Effect of Dispersed Phases Upon Dislocation Distributions in Plastically Deformed Copper Crystals', *Phil. Mag.*, **16**, 927 (1967).
19. Hirsch, P. B. and Humphreys, F. J., 'Plastic Deformation of Two-Phase Alloys Containing Small Nondeformable Particles', *Physics of Strength and Plasticity*, A. S. Argon, ed., The MIT Press, Cambridge, Mass. (1969), p. 189
20. Head, A. K., 'The Interaction of Dislocations and Boundaries', *Phil. Mag.*, **44**, 92 (1953).
21. Head, A. K., 'Edge Dislocations in Inhomogeneous Media', *Proc. Phys. Soc. London*, B**66**, 793 (1953).
22. Dundurs, J. and Sendeckyj, G. P., 'Behavior of an Edge Dislocation near a Bimetallic Interface', *J. Appl. Phys.*, **36**, 3353 (1965).
23. Vagera, I., 'Long Range Elastic Interaction of Prismatic Dislocation Loop with a Grain Boundary', *Czech. J. Phys.*, B**20**, 702 (1970).

24. Vagera, I., 'Long Range Force on a Dislocation Loop due to Grain Boundary', *Czech. J. Phys.*, **B20**, 1278 (1970).

25. Salamon, N. J. and Dundurs, J., 'Elastic Fields of a Dislocation Loop in a Two-Phase Material', *J. Elasticity*, **1**, 153 (1971).

26. Dundurs, J. and Salamon, N. J., 'Circular Prismatic Dislocation Loop in a Two-Phase Material', *Phys. stat. sol.*, **50**, 125 (1972).

27. Salamon, N. J., 'A Dislocation Loop in an Inhomogeneous Material', Ph.D. Dissertation, Northwestern University, 1971.

28. Conners, G. H., 'The Interaction of a Dislocation with a Coated Plane Boundary', *Int. J. Engng Sci.*, **5**, 25 (1967).

29. Weeks, R., Dundurs, J., and Stippes, M., 'Exact Analysis of an Edge Dislocation near a Surface Layer', *Int. J. Engng Sci.*, **6**, 365 (1968).

30. Fleischer, R. L., 'Theory of Hardening by Thin Zones', *Electron Microscopy and Strength of Crystals*, G. Thomas and J. Washburn, eds., Interscience Publishers, New York (1963), p. 973.

31. Chou, Y. T., 'Screw Dislocations in and near Lamellar Inclusions', *Phys. stat. sol.*, **17**, 509 (1966).

32. Head, A. K., 'The Dislocation Image Force in Cubic Polycrystals', *Phys. stat. sol.*, **10**, 481 (1965).

33. Chou, Y. T., 'On Dislocation-Boundary Interaction in an Anisotropic Aggregate', *Phys. stat. sol.*, **15**, 123 (1966).

34. Braekhus, J. and Lothe, J., 'Dislocations at and near Planar Interfaces', *Phys. stat. sol.*, **43**, 651 (1971).

35. Pande, C. S. and Chou, Y. T., 'Edge Dislocations in Anisotropic Inhomogeneous Media', *J. Appl. Phys.*, **43**, 840 (1972).

36. Dundurs, J., 'On the Interaction of a Screw Dislocation with Inhomogeneities', *Recent Advances in Engineering Science*, A. C. Eringen, ed., Gordon and Breach, New York (1967), p. 223.

37. Dundurs, J. and Mura, T., 'Interaction Between an Edge Dislocation and a Circular Inclusion', *J. Mech. Phys. Solids*, **12**, 177 (1964).

38. Dundurs, J. and Sendeckyj, G. P., 'Edge Dislocation Inside a Circular Inclusion', *J. Mech. Phys. Solids*, **13**, 141 (1965).

39. Sendeckyj, G. P., 'On the Behavior of Edge Dislocations near Inhomogeneities', *Scripta Metallurgica*, **3**, 763 (1969).

40. Smith, E., 'The Interaction Between Dislocations and Inhomogeneity', *Int. J. Engng Sci.*, **6**, 129 (1968).

41. Sendeckyj, G. P., 'Screw Dislocations in Inhomogeneous Solids', *Fundamental Aspects of Dislocation Theory*, J. A. Simmons, R. de Wit and R. Bullough, eds., Nat. Bur. Stand., Spec. Publ., 317, Washington, D.C. (1970), p. 57.

42. Dundurs, J. and Gangadharan, 'Edge Dislocation near an Inclusion with a Slipping Interface', *J. Mech. Phys. Solids*, **17**, 459 (1969).

43. Knesl, Z. and Kroupa, F., 'Interaction Between Screw Dislocations with Hollow Cores', *Czech. J. Phys.*, **B20**, 1054 (1970).

44. Lin, S. C. and Mura, T., 'Long-Range Elastic Interaction Between a Dislocation and an Ellipsoidal Inclusion in Cubic Crystals', *J. Appl. Phys.*, **44**, 1508 (1973).

45. Weeks, R. W., Pati, S. R., Ashby, M. F., and Barrand, P., 'The Elastic Interaction Between a Straight Dislocation and a Bubble or a Particle', *Acta Met.*, **17**, 1403 (1969).

46. Willis, J. R., Hayns, M. R. and Bullough, R., 'The Dislocation Void Interaction', *Proc. Roy. Soc. Lond.*, A329, 121 (1972).

47. Comninou, M. and Dundurs, J., 'Long-Range Interaction Between a Screw Dislocation and a Spherical Inclusion', *J. Appl. Phys.*, 43, 2461 (1972).

48. Yoo, M. H. and Ohr, S. M., 'Elastic Interaction of a Point Defect with a Prismatic Dislocation Loop in Hexagonal Crystals', *J. Appl. Phys.*, 43, 4477 (1972).

49. Goodman, J. W. and Sines, G., 'Mixed Dislocations in Anisotropic Cubic Crystals: Interaction with Point-Defect Stress Fields', *J. Appl. Phys.*, 43, 2086 (1972).

50. Masumura, R. A., Grupen, W. B., and Sines, G. H., 'The Elastic Interaction Between Point Defects and a Disk Inhomogeneity in a Hexagonal Crystal', *Crystal Lattice Defects*, Gordon and Breach, Vol. 3 (1972), p. 103.

51. Navi, P., Goodman, J. W., and Sines, G., 'Finite Size Lattice Defect in Anisotropic Copper: Interaction with an Extended Edge Dislocation', *Scripta Met.*, 6, 71 (1972).

52. Li, J. C. M. and Chou, Y. T., 'The Role of Dislocations in the Flow Stress Grain Size Relationships', *Met. Trans.* 1, 1145 (1970).

53. Chou, Y. T. and Li, J. C. M., 'Theory of Dislocation Pile-ups', *Mathematical Theory of Dislocations*, T. Mura, ed., American Society of Mechanical Engineers, New York (1969), p. 116.

54. Pacheco, E. S. and Mura, T., 'Interaction Between a Screw Dislocation and a Bimetallic Interface', *J. Mech. Phys. Solids*, 17, 163 (1969).

55. Nakahara, S., Wu, J. B. C., and Li, J. C. M., 'Dislocations in a Welded Interface Between Two Isotropic Media', *Mat. Sci. Engng*, 10, 291 (1972).

56. Wu, J. B. C. and Li, J. C. M., 'Stress Field and Energy of Interface Dislocation Loop', Presented at The Metallurgical Society of AIME, Spring Meeting, 1973, Philadelphia, Pa.

57. Chou, T. W., 'Elastic Behaviours of Disclinations in Nonhomogeneous Media', *J. Appl. Phys.*, 42, 4931 (1971).

58. Chou, Y. T. and Pande, C. S., 'Interfacial Screw Dislocations in Anisotropic Two-Phase Media', *J. Appl. Phys.*, 44, 3355 (1973).

59. Kelly, A. and Fine, M. E., 'The Strength of an Alloy Containing Zones', *Acta Met.*, 5, 365 (1957).

60. Friedel, J., 'On the Chemical Hardening by Coherent Precipitates', *Physics of Strength and Plasticity*, A. S. Argon, ed., The MIT Press (1969), p. 181.

61. Fleischer, R. L., 'Effects of Non-uniformities on the Hardening of Crystals', *Acta Met.*, 8, 598 (1960).

62. Eshelby, J. D., 'The Determination of the Elastic Field of an Ellipsoidal Inclusion and Related Problems', *Proc. Roy. Soc.*, A241, 376 (1957).

63. Chang, C. S. and Conway, H. D., 'Stress Analysis of an Infinite Plate Containing an Elastic Rectangular Inclusion', *Acta Mechanica*, 8, 160 (1969).

64. Mori, T. and Tanaka, K., 'Average Stress in Matrix and Average Elastic Energy of Materials with Misfitting Inclusions', *Acta Met.*, 21, 571 (1973)

65. Ashby, M. F. and Johnson, L., 'On the Generation of Dislocations at Misfitting Particles in a Ductile Matrix', *Phil. Mag.*, 20, 1009 (1969).

66. Dundurs, J., 'Elastic Interaction of Dislocations with Inhomogeneities', *Mathematical Theory of Dislocations*, T. Mura, ed., The American Society of Mechanical Engineers, New York (1969), p. 70.

67. Preston, O. and Grant, N. J., 'Dispersion Strengthening of Copper by Internal Oxidation', *Trans. AIME*, 221, 164 (1961).

68. Kelly, A., 'The Strengthening of Metals by Dispersed Particles', *Proc. Roy. Soc. Lond.*, A282, 63 (1964).
69. Sutton, W. H. and Chorné, J., 'Potential of Oxide–Fiber Reinforced Metals', *Fiber Composite Materials*, American Society for Metals, Metals Park, Ohio (1965), p. 173.
70. Brophy, J. H., Rose, R. M., and Wulff, J., *The Structure and Properties of Materials*, Vol. 2, John Wiley and Sons, Inc., New York (1964).
71. Salkind, M. J., Lemkey, F. D., and George, F. D., 'Whisker Composites by Eutectic Solidification', *Whisker Technology*, Wiley–Interscience, New York (1970), p. 343.
72. Kraft, R. W. and Albright, D., 'Microstructure of Unidirectionally Solidified Al–CuAl$_2$ Eutectic', *Trans. AIME*, 221, 95 (1961).
73. Galasso, F. S., *High Modulus Fibers and Composites*, Gordon and Breach, New York (1969).
74. Lemkey, F. D., Hertzberg, R. W., and Ford, J. A., 'The Microstructure, Crystallography, and Mechanical Behavior of Unidirectional Solidified Al–Al$_3$Ni Eutectic', *Trans. Met. Soc. of AIME*, 233, 335 (1965).
75. Hertzberg, R. W., Lemkey, F. D., and Ford, J. A., 'Mechanical Behavior of Lamellar (Al–CuAl$_2$) and Whisker Type (Al–Al$_3$Ni) Unidirectional-Solidified Eutectic Alloys', *Trans. Met. Soc. of AIME*, 233, 342 (1965).
76. Salkind, M. J., Bayles, B. J., George, F. D., and Tice, W. K., 'Investigation of Fracture Mechanisms, Thermal Stability and Hot-Strength Properties of Controlled Polyphase Alloys', United Aircraft Research Laboratories Report D910239-4, 1965.
77. Cline, H. E. and Stein, D. F., 'Strengthening by Interfaces in the Ag–Cu Directionally Solidified Eutectic', *Trans. Met. Soc. of AIME*, 245, 841 (1969).
78. Walter, J. L., Cline, H. E., and Koch, E. F., 'Interface Dislocation in Directionally Solidified NiAl–Cr Eutectic', *Trans. Met. Soc. of AIME*, 245, 2073 (1969).
79. Cline, H. E., Walter, J. L., Koch, E. F., and Osika, L. M., 'The Variation of Interface Dislocation Networks with Lattice Mismatch in Eutectic Alloys', *Acta Met.*, 19, 405 (1971).
80. Lemkey, F. and Salkind, M., *Crystal Growth*, Suppl. to *J. Phys. Chem. Solids*, Pergamon Press, New York (1967), p. 171.
81. Thompson, E., 'The Structure and Properties of Ni–NiMo Eutectic Prepared by Unidirectional Solidification', *Proceedings of the Twelfth Meeting of the Refractory Composites Working Group*, AFML TR67-228, 1967, p. 214.
82. Clauser, H. R., 'Advanced Composite Materials', *Scientific American* (July 1973), p. 36.
83. Barnett, D. M. and Nix, W. D., 'The Interaction Force Between Tetragonal Defects and Screw Dislocations in Cubic Crystals', *Acta Met.*, 21, 1157 (1973).
84. Weertman, J. and Weertman, J. R., *Elementary Dislocation Theory*, Macmillan, New York (1964).
85. Dundurs, J. and Sendeckyj, G. P., 'Interaction of Dislocations with Inhomogeneities in Presence of Applied Stresses', *Composite Materials Workshop*, S. W. Tsai, J. C. Halpin and N. J. Pagano, eds., Technomic Pub. Co. Inc., Stamford, Conn. (1968), p. 44.
86. Thomas, G. and Nutting, J., 'The Ageing Characteristics of Aluminum Alloys', *J. Inst. Metals*, 88, 81 (1959–60).
87. Fine, M. E., *Introduction to Phase Transformations in Condensed Systems*, Macmillan, New York (1964).

88. Mehl, R. F. *et al.*, *Precipitation from Solid Solution*, American Society for Metals, Cleveland, Ohio (1959).
89. Decker, R. F., 'Alloy Design, Using Second Phases', *Met. Trans.*, **4**, 2495 (1973).
90. Gerold, V. and Haberkorn, H., 'On the Critical Resolved Shear Stress of Solid Solutions Containing Coherent Precipitates', *Phys. stat. sol.*, **16**, 675 (1966).
91. Hirsch, P. B. and Humphreys, F. J., 'The Deformation of Single Crystals of Copper and Copper–Zinc Alloys Containing Alumina Particles, I. Macroscopic Properties and Workhardening Theory', *Proc. Roy. Soc. Lond.*, A**318**, 45 (1970).
92. Humphreys, F. J. and Hirsch, P. B., 'The Deformation of Single Crystals of Copper and Copper–Zinc Alloys Containing Alumina Particles, II. Microstructure and Dislocation-Particle Interactions', *Proc. Roy. Soc. Lond.*, A**318**, 73 (1970).
93. MacEwen, S. R., Hirsch, P. B. and Vitek, V., 'Cross-Slip of Orowan Loops at Incoherent Particles', *Phil. Mag.*, **28**, 703 (1973).
94. Hsieh, C. F. and Dundurs, J., 'Continuous Distributions of Dislocations in Bonded Half Spaces', *Int. J. Engng Sci.*, **11**, 933 (1973).
95. Kurihara, T., 'Edge Dislocations in an Anisotropic Material with a Surface Layer', *Int. J. Engng Sci.*, **11**, 891 (1973).
96. Dundurs, J., 'Effect of Elastic Constants on Stress in a Composite Under Plane Deformation', *J. Comp. Mater.*, **1**, 310 (1967).
97. Gleiter, H. and Hornbogen, E., 'Beobachtung der Wechselwirkung von Versetzungen mit kohärenten geordneten Zonen (II)', *Phys. stat. sol.*, **12**, 251 (1965).
98. Gleiter, H. and Hornbogen, E., 'Precipitation Hardening by Coherent Particles', *Mat. Sci. Engng*, **2**, 285 (1967–68).
99. Ashby, M. F., *Proc. Second Bolton Landing Conf. on Oxide Dispersion Strengthening*, Gordon and Breach, New York (1968), p. 143.

CHAPTER 4[a]

INTRODUCTION TO PLATE AND SHELL THEORY

4.1 BASIC EQUATIONS OF ELASTICITY

Sokolnikoff[1] derives in detail the formulation of the governing differential equations of elasticity in the first three chapters. This will not be repeated here, but rather the equations are presented and then utilised in the process of deriving the governing equations for a rectangular plate and cylindrical shell.

Consider an elastic body of any general shape and a typical *material point* in the interior of the elastic body. If one assigns a Cartesian reference frame with axes x, y, and z, as shown in Fig. 4.1, it is then convenient to assign a rectangular parallelepiped shape to the material point, and label it a *control element* of dimensions dx, dy, and dz, as shown in Fig. 4.1. The control element is defined to be infinitesimally small compared with the size of the elastic body, yet infinitely large compared to elements of the molecular structure, in order that the material can be considered a continuum.

On the surfaces of the control element there can exist both normal stresses (those perpendicular to the plane of the face) and shear stresses (those parallel to the plane of the face). On any one face these three stress components comprise a vector, called a *surface traction*.

It is important to note the sign convention and the subscript meaning of these surface stresses. For a stress component on a positive face, that is, a face whose outer normal is in the direction of the positive axis, that stress component is positive when it is directed in the direction of the positive axis. Conversely, when a stress is on a negative face of the control element, it is positive when it is directed in the negative axis direction. This procedure is followed in Fig. 4.1.

[a] Excerpted from Vinson, J. R., *Structural Mechanics: the Behaviour of Plates and Shells* (1974), reprinted by permission of John Wiley and Sons, Inc.

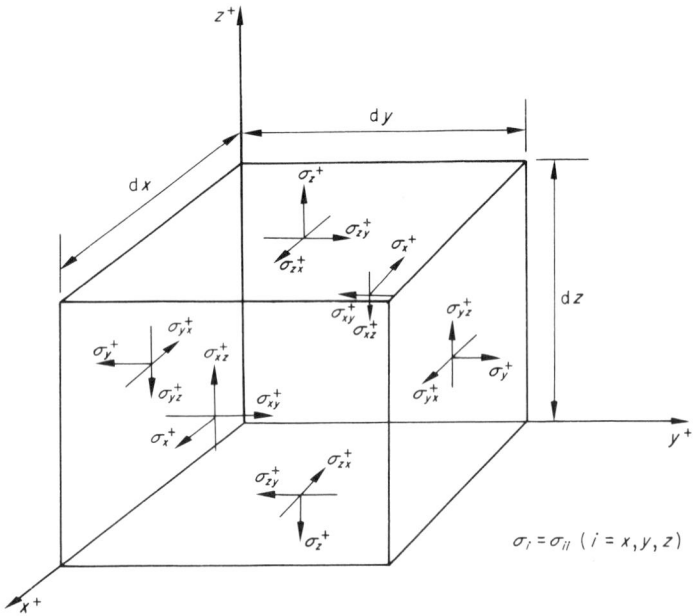

Fig. 4.1.   *Control element in an elastic body.*

Likewise, the first subscript of any stress component signifies the axis parallel to the outer normal of the face on which the stress component acts. The second subscript refers to the axis which is parallel to the stress component.

The u, v, and w displacements are parallel to the x, y, and z-axes respectively, and are positive when in the positive axis direction.

Strains in an elastic body are of two types, extensional and shear. *Extensional strains*, ϵ_{ii}, where $i = x$, y, or z, are directed parallel to each of the axes respectively and are a measure of the change in dimension of the control volume in the subscripted direction due to the normal stresses acting on all surfaces of the control volume. Looking at Fig. 4.2, one can define shear strains. The shear strain ϵ_{ij} (where i, $j = x$, y or z, and $i \neq j$) is a change of angle. As an example, in the x–y plane, defining

$$\gamma_{xy} \equiv \frac{\pi}{2} - \phi \qquad \text{(in radians)}$$

then

$$\epsilon_{xy} = \tfrac{1}{2}\gamma_{xy}$$

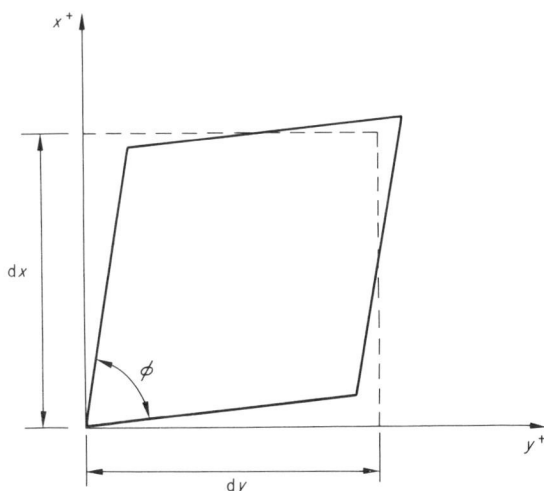

Fig. 4.2.   Shearing of a control element.

It is important to define the shear strain ϵ_{xy} to be one-half the angle γ_{xy} in order to use tensor notation. However, in some texts and papers the shear strain is defined as γ_{xy}. Care must be taken to ensure awareness of which definition is used when reading or utilising a text or a research paper.

The *modulus of elasticity* of an isotropic material, *E*, and the *shear modulus, G*, are defined and discussed in great detail in any strength of materials text.

One final quantity must be defined in some detail due to its use later in anisotropic materials—Poisson's ratio—denoted by ν_{ij}. It is defined as the ratio of the negative of the strain in the *j* direction to the strain in the *i* direction caused by a stress in the *i* direction, σ_{ii}. With this definition it is a positive quantity of magnitude $0 \le \nu_{ij} = \nu \le 0 \cdot 5$, for all existing isotropic materials (i.e. *isotropic* refers to a material whose properties do not vary with orientation), although theoretically Poisson's ratio could have a range of $-1 \le \nu \le 0 \cdot 5$.

The basic equations of elasticity for a control element of an isotropic elastic body in a Cartesian reference frame can now be written. They are written in full, but the compact Einsteinian notation of tensor calculus is also provided in parenthesis.

Equilibrium equations: $(\sigma_{ki,k} + F_i = 0)$
If there are no body forces (F_i) acting, the equilibrium equations are written as

$$\frac{\partial \sigma_x}{\partial x} + \frac{\partial \sigma_{yx}}{\partial y} + \frac{\partial \sigma_{zx}}{\partial z} = 0 \tag{4.1}$$

$$\frac{\partial \sigma_{xy}}{\partial x} + \frac{\partial \sigma_y}{\partial y} + \frac{\partial \sigma_{zy}}{\partial z} = 0 \tag{4.2}$$

$$\frac{\partial \sigma_{xz}}{\partial x} + \frac{\partial \sigma_{yz}}{\partial y} + \frac{\partial \sigma_z}{\partial z} = 0 \tag{4.3}$$

Stress–strain relations: $(\sigma_{ij} = c_{ijkl}\epsilon_{kl})$

$$\epsilon_x = \frac{1}{E}[\sigma_x - \nu(\sigma_y + \sigma_z)] \tag{4.4}$$

$$\epsilon_y = \frac{1}{E}[\sigma_y - \nu(\sigma_x + \sigma_z)] \tag{4.5}$$

$$\epsilon_z = \frac{1}{E}[\sigma_z - \nu(\sigma_x + \sigma_y)] \tag{4.6}$$

$$\epsilon_{xy} = \frac{1}{2G}\sigma_{xy} \quad \text{where} \quad G = \frac{E}{2(1+\nu)} \tag{4.7}$$

$$\epsilon_{yz} = \frac{1}{2G}\sigma_{yz} \tag{4.8}$$

$$\epsilon_{zx} = \frac{1}{2G}\sigma_{zx} \tag{4.9}$$

Linear strain–displacement relations: $(\epsilon_{ij} = \frac{1}{2}(u_{i,j} + u_{j,i}))$

$$\epsilon_x = \frac{\partial u}{\partial x} \tag{4.10}$$

$$\epsilon_y = \frac{\partial v}{\partial y} \tag{4.11}$$

$$\epsilon_z = \frac{\partial w}{\partial z} \tag{4.12}$$

$$\epsilon_{xy} = \frac{1}{2}\left(\frac{\partial u}{\partial y} + \frac{\partial v}{\partial x}\right) \tag{4.13}$$

$$\epsilon_{xz} = \frac{1}{2}\left(\frac{\partial u}{\partial z} + \frac{\partial w}{\partial x}\right) \tag{4.14}$$

$$\epsilon_{yz} = \frac{1}{2}\left(\frac{\partial v}{\partial z} + \frac{\partial w}{\partial y}\right) \tag{4.15}$$

Compatibility equations: $\epsilon_{ij,kl} + \epsilon_{kl,ij} - \epsilon_{ik,jl} - \epsilon_{jl,ik} = 0$

$$\frac{\partial^2 \epsilon_{xx}}{\partial y \partial z} = \frac{\partial}{\partial x} \left(-\frac{\partial \epsilon_{yz}}{\partial x} + \frac{\partial \epsilon_{xx}}{\partial y} + \frac{\partial \epsilon_{xy}}{\partial z} \right) \tag{4.16}$$

$$\frac{\partial^2 \epsilon_{yy}}{\partial z \partial x} = \frac{\partial}{\partial y} \left(-\frac{\partial \epsilon_{zx}}{\partial y} + \frac{\partial \epsilon_{xy}}{\partial z} + \frac{\partial \epsilon_{yz}}{\partial x} \right) \tag{4.17}$$

$$\frac{\partial^2 \epsilon_{zz}}{\partial x \partial y} = \frac{\partial}{\partial z} \left(-\frac{\partial \epsilon_{xy}}{\partial z} + \frac{\partial \epsilon_{yz}}{\partial x} + \frac{\partial \epsilon_{zx}}{\partial y} \right) \tag{4.18}$$

$$2 \frac{\partial^2 \epsilon_{xy}}{\partial x \partial y} = \frac{\partial^2 \epsilon_{xx}}{\partial y^2} + \frac{\partial^2 \epsilon_{yy}}{\partial x^2} \tag{4.19}$$

$$2 \frac{\partial^2 \epsilon_{yz}}{\partial y \partial z} = \frac{\partial^2 \epsilon_{yy}}{\partial z^2} + \frac{\partial^2 \epsilon_{zz}}{\partial y^2} \tag{4.20}$$

$$2 \frac{\partial^2 \epsilon_{zx}}{\partial z \partial x} = \frac{\partial^2 \epsilon_{zz}}{\partial x^2} + \frac{\partial^2 \epsilon_{xx}}{\partial z^2} \tag{4.21}$$

Because of the symmetry of the stress-and-strain tensor

$$\sigma_{ij} = \sigma_{ji} \quad \text{and} \quad \epsilon_{ij} = \epsilon_{ji} \quad (i, j = x, y, z)$$

the unknowns in the above equations are: six stress components, six strain components and three displacements.

When a form for the displacements or strains is assumed such that the displacements are guaranteed to be continuous and single-valued everywhere in the region under consideration, the compatibility equations are automatically satisfied, and need not be used. Under that condition there are fifteen equations, namely (4.1) to (4.15), and mathematically the elasticity equations are well posed.

4.2 ASSUMPTIONS OF PLATE THEORY

In classical, linear, thin-plate theory, there are a number of assumptions that are necessary in order to reduce the three-dimensional equations of elasticity to a two-dimensional set that can be solved. Consider an elastic body shown as in Fig. 4.3, comprising the region $0 \leqslant x \leqslant a$, $0 \leqslant y \leqslant b$, and $-h/2 \leqslant z \leqslant h/2$, such that $h \ll a$ and $h \ll b$. This is called a *plate*.

The following assumptions are made.

1. A lineal element of the plate extending through the plate thickness, normal to the mid surface, x–y-plane, in the unstressed state, upon the application of load: (a) undergoes at most a

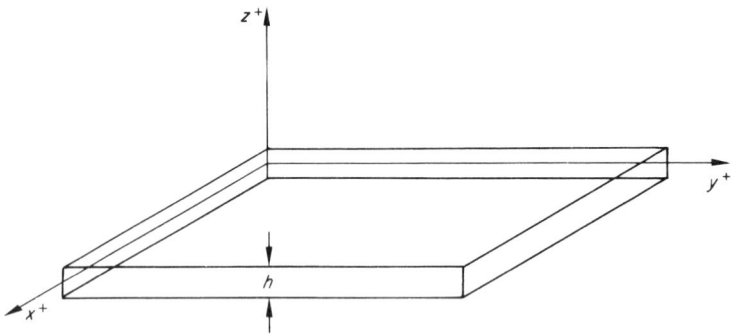

Fig. 4.3.   Rectangular plate.

translation and a rotation with respect to the original co-ordinate system; (b) remains normal to the deformed middle surface.
2. A plate resists lateral and in-plane loads by bending, transverse shear stresses, and in-plane action, not through block-like compression or tension in the plate in the thickness direction.

From 1(a) the following is implied:

3. The lineal element does not elongate or contract.
4. The lineal element remains straight upon load application.

In addition,

5. St Venant's Principle applies.

It is seen from 1(a) that the most general form for the two in-plane displacements is

$$u(x, y, z) = u_0(x, y) + z\alpha(x, y)$$
$$v(x, y, z) = v_0(x, y) + z\beta(x, y)$$

where u_0 and v_0 are the in-plane middle surface displacements ($z = 0$), and α and β are rotations undefined as yet. Assumption 3 requires that $\epsilon_z = 0$, which in turn means that the lateral deflection w is at most (from eqn. 4.12).

$$w = w(x, y)$$

We also ignore eqn. (4.6).

Assumption 4 requires that for any z, $\epsilon_{xz} = $ constant and $\epsilon_{yz} = $ constant at any specific location (x, y) on the plate middle surface for all z. Assumption 1(b) requires that the constant be zero, hence

$$\epsilon_{xz} = \epsilon_{yz} = 0$$

Assumption 2 means that $\sigma_z = 0$ in the stress–strain relations.

Incidentally, the assumptions above are identical to those of thin classical beam, ring, and shell theory.

4.3 DERIVATION OF THE EQUILIBRIUM EQUATIONS FOR A PLATE

Figure 4.4 shows the positive directions of stress quantities to be defined when the plate is subjected to lateral and in-plane loads.

We shall now define the following

$$\begin{Bmatrix} M_x \\ M_y \\ M_{xy} \end{Bmatrix} = \int_{-h/2}^{h/2} \begin{Bmatrix} \sigma_x \\ \sigma_y \\ \sigma_{xy} \end{Bmatrix} z \, dz \qquad (4.22)$$

$$\begin{Bmatrix} Q_x \\ Q_y \end{Bmatrix} = \int_{-h/2}^{h/2} \begin{Bmatrix} \sigma_{xz} \\ \sigma_{yz} \end{Bmatrix} dz \qquad (4.23)$$

$$\begin{Bmatrix} N_x \\ N_y \\ N_{xy} \end{Bmatrix} = \int_{-h/2}^{h/2} \begin{Bmatrix} \sigma_x \\ \sigma_y \\ \sigma_{xy} \end{Bmatrix} dz \qquad (4.24)$$

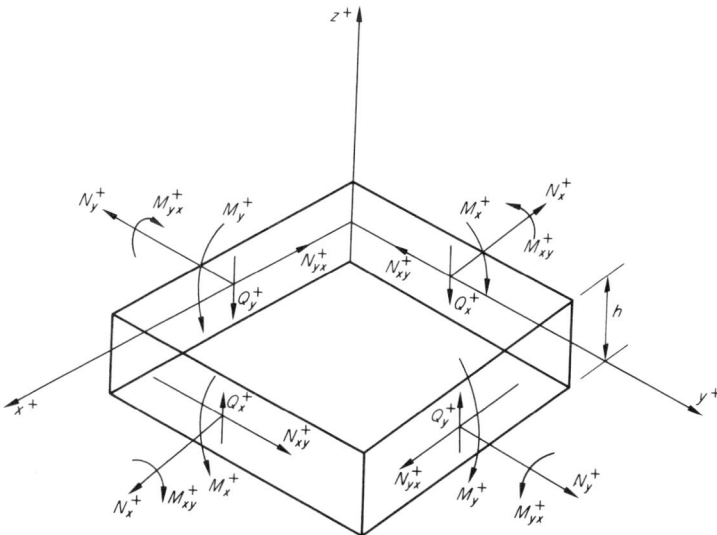

Fig. 4.4. *Positive directions of stress resultants and couples.*

In plate theory, the details of the control element under consideration are disregarded, and one integrates across the thickness h, so instead of dealing with stresses on a control element, one really utilises stress couples, (4.22), shear resultants, (4.23), and stress resultants, (4.24).

Proceeding, multiply eqn. (4.1) by $z \, dz$ and integrate from $-h/2$ to $+h/2$.

$$\int_{-h/2}^{h/2} \left(z \frac{\partial \sigma_x}{\partial x} + z \frac{\partial \sigma_{xy}}{\partial y} + z \frac{\partial \sigma_{xz}}{\partial z} \right) dz = 0$$

$$\frac{\partial}{\partial x} \int_{-h/2}^{h/2} \sigma_x z \, dz + \frac{\partial}{\partial y} \int_{-h/2}^{h/2} \sigma_{xy} z \, dz + \int_{-h/2}^{h/2} z \frac{\partial \sigma_{xz}}{\partial z} \, dz = 0$$

$$\left[\frac{\partial M_x}{\partial x} + \frac{\partial M_{xy}}{\partial y} + z\sigma_{xz} \right]_{-h/2}^{h/2} - \int_{-h/2}^{h/2} \sigma_{xz} \, dz = 0$$

Looking at the third term, $\sigma_{xz} = \sigma_{zx} = 0$ when there are no shear loads on the surface. This is not true for laminated plates. There, defining $\tau_{1x} = \sigma_{xz}(+h/2)$ and $\tau_{2x} = \sigma_{xz}(-h/2)$ the results are shown in eqn. (4.25). It should be noted that, for plates supported on the edge by knife edges, σ_{xz} does not go to zero at $\pm h/2$, and so the theory does not give good results at that edge, but due to St Venant's Principle, the solutions are accurate away from the knife edges.

$$\frac{\partial M_x}{\partial x} + \frac{\partial M_{xy}}{\partial y} + \frac{h}{2}(\tau_{1x} + \tau_{2x}) - Q_x = 0 \qquad (4.25)$$

Likewise eqn. (4.2) becomes

$$\frac{\partial M_{xy}}{\partial x} + \frac{\partial M_y}{\partial y} + \frac{h}{2}(\tau_{1y} + \tau_{2y}) - Q_y = 0 \qquad (4.26)$$

where $\tau_{1y} = \sigma_{yz}(+h/2)$ and $\tau_{2y} = \sigma_{yz}(-h/2)$. Looking now at eqn. (4.3) multiplying it by dz and integrating between $\pm h/2$, results in the following

$$\int_{-h/2}^{h/2} \left(\frac{\partial \sigma_{zx}}{\partial x} + \frac{\partial \sigma_{yz}}{\partial y} + \frac{\partial \sigma_z}{\partial z} \right) dz = 0$$

$$\left[\frac{\partial Q_x}{\partial x} + \frac{\partial Q_y}{\partial y} + \sigma_z \right]_{-h/2}^{h/2} = 0 \qquad (4.27)$$

$$\frac{\partial Q_x}{\partial x} + \frac{\partial Q_y}{\partial y} + p_1(x, y) - p_2(x, y) = 0$$

where $\sigma_z(+h/2) = p_1(x,y)$ and $\sigma_z(-h/2) = p_2(x,y)$.

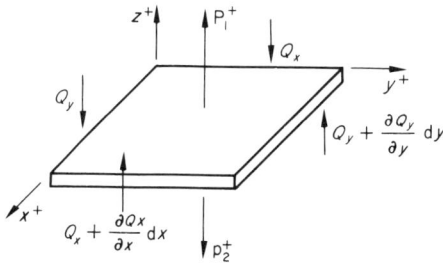

One could also derive this by considering vertical equilibrium of a plate element as shown above. One may ask why we make use of σ_z in this equation and not in the stress–strain relations? The foregoing is not really inconsistent, because σ_z does not appear explicitly in eqn. (4.27), and once away from the surface the pressure is absorbed by shear rather than by σ_z in the plate interior.

Similarly, multiplying eqns. (4.1) and (4.2) by dz, and integrating across the plate thickness results in

$$\frac{\partial N_x}{\partial x} + \frac{\partial N_{xy}}{\partial y} + (\tau_{1x} - \tau_{2x}) = 0 \tag{4.28}$$

$$\frac{\partial N_{xy}}{\partial x} + \frac{\partial N_y}{\partial y} + (\tau_{1y} - \tau_{2y}) = 0 \tag{4.29}$$

Thus eqns. (4.25) to (4.29) comprise the five equilibrium equations for a rectangular plate loaded statically by lateral distributed loads and surface shear stresses.

4.4 DERIVATION OF PLATE MOMENT–CURVATURE RELATIONS AND INTEGRATED STRESS RESULTANT–DISPLACEMENT RELATIONS

We now obtain the plate equations corresponding to the stress–strain relations. We shall not deal with ϵ_x, ϵ_y and ϵ_{xy} since we have averaged the stresses by integrating through the surface, but we shall deal with the displacements. Thus, combining eqns. (4.4) to (4.15) gives the following, remembering that σ_z has been assumed zero, and excluding eqn. (4.6) as stated previously

$$\frac{\partial u}{\partial x} = \frac{1}{E} [\sigma_x - \nu\sigma_y] \tag{4.30}$$

$$\frac{\partial v}{\partial y} = \frac{1}{E} [\sigma_y - \nu\sigma_x] \qquad (4.31)$$

$$\frac{1}{2} \left(\frac{\partial u}{\partial y} + \frac{\partial v}{\partial x} \right) = \frac{1}{2G} \sigma_{xy} \qquad (4.32)$$

$$\frac{1}{2} \left(\frac{\partial v}{\partial z} + \frac{\partial w}{\partial y} \right) = \frac{1}{2G} \sigma_{yz} \qquad (4.33)$$

$$\frac{1}{2} \left(\frac{\partial w}{\partial x} + \frac{\partial u}{\partial z} \right) = \frac{1}{2G} \sigma_{xz} \qquad (4.34)$$

First recall the form of the admissible displacements resulting from the plate theory assumptions given previously:

$$u = u_0(x, y) + z\alpha(x, y) \qquad (4.35)$$

$$v = v_0(x, y) + z\beta(x, y) \qquad (4.36)$$

$$w = w(x, y) \qquad (4.37)$$

From the assumption of plate theory, remember, a linear element through the plate will experience translations, rotations, but no extension or contractions.

Remember, too, that the assumptions of classical plate theory require that transverse shear deformation is zero. So if $\epsilon_{xz} = 0$ then from (4.33) and (4.34)

$$\frac{1}{2} \left(\frac{\partial u}{\partial z} + \frac{\partial w}{\partial x} \right) = 0 \quad \text{or} \quad \frac{\partial u}{\partial z} = -\frac{\partial w}{\partial x}; \quad \text{likewise,} \quad \frac{\partial v}{\partial z} = -\frac{\partial w}{\partial y}$$

Hence, from eqns. (4.35) and (4.36), it is seen that

$$\alpha = -\frac{\partial w}{\partial x} \qquad (4.38)$$

$$\beta = -\frac{\partial w}{\partial y} \qquad (4.39)$$

Thus, the rotations are defined for classical plate theory. Substituting (4.35) into (4.30), then multiplying it by $z \, dz$ and integrating from $-h/2$ to $+h/2$, one obtains

$$\int_{-h/2}^{h/2} \frac{\partial u_0}{\partial x} z \, dz + \int_{-h/2}^{h/2} z^2 \frac{\partial \alpha}{\partial x} \, dz = \int_{-h/2}^{h/2} \frac{1}{E} [\sigma_x - \nu\sigma_y] z \, dz \qquad (4.40)$$

Likewise (4.36) and (4.31) result in

$$\int_{-h/2}^{h/2} \frac{\partial v_0}{\partial y} z \, dz + \int_{-h/2}^{h/2} z^2 \frac{\partial \beta}{\partial y} \, dz = \int_{-h/2}^{h/2} \frac{1}{E} [\sigma_y - \nu\sigma_x] z \, dz \qquad (4.41)$$

and eqns. (4.35), (4.36) and (4.32) give

$$\int_{-h/2}^{h/2} \left[\frac{\partial u_0}{\partial y} + \frac{\partial v_0}{\partial x} \right] z \, dz + \int_{-h/2}^{h/2} \left(z^2 \frac{\partial \alpha}{\partial y} + z^2 \frac{\partial \beta}{\partial x} \right) dz = \int_{-h/2}^{h/2} \frac{1}{G} \sigma_{xy} z \, dz$$

(4.42)

Integrating (4.40), (4.41) and (4.42) gives, using (4.38) and (4.39)

$$\frac{h^3}{12} \frac{\partial \alpha}{\partial x} = \frac{1}{E} [M_x - \nu M_y] = -\frac{h^3}{12} \frac{\partial^2 w}{\partial x^2}$$

(4.43)

$$\frac{h^3}{12} \frac{\partial \beta}{\partial y} = \frac{1}{E} [M_y - \nu M_x] = -\frac{h^3}{12} \frac{\partial^2 w}{\partial y^2}$$

(4.44)

$$\frac{h^3}{12} \left(\frac{\partial \alpha}{\partial y} + \frac{\partial \beta}{\partial x} \right) = \frac{1}{G} M_{xy} = -\frac{h^3}{6} \frac{\partial^2 w}{\partial x \partial y}$$

(4.45)

Since

$$G = \frac{E}{2(1+\nu)}$$

(4.46)

$$M_{xy} = -(1-\nu)D \frac{\partial^2 w}{\partial x \partial y} \quad \text{where} \quad D = \frac{Eh^3}{12(1-\nu^2)}$$

(4.47)

D is the *flexural stiffness* for the plate (note its similarity to $EI = Ebh^3/12$ for a beam of rectangular cross section of width b and height h). Solving (4.43) and (4.44) for M_x and M_y results in,

$$M_x = -D \left[\frac{\partial^2 w}{\partial x^2} + \nu \frac{\partial^2 w}{\partial y^2} \right]$$

(4.48)

$$M_y = -D \left[\frac{\partial^2 w}{\partial y^2} + \nu \frac{\partial^2 w}{\partial x^2} \right]$$

(4.49)

Likewise, substituting (4.48) and (4.49) into eqns. (4.25) and (4.26) results in

$$Q_x = -D \frac{\partial}{\partial x} (\nabla^2 w) + \frac{h}{2} (\tau_{1x} + \tau_{2x})$$

(4.50)

$$Q_y = -D \frac{\partial}{\partial y} (\nabla^2 w) + \frac{h}{2} (\tau_{1y} + \tau_{2y})$$

(4.51)

Substituting eqns. (4.35) and (4.36) into eqns. (4.30) to (4.32), then multiplying the latter three equations by dz, and integrating across the thickness, results in the following integrated stress–strain relations

$$N_x = K \left[\frac{\partial u_0}{\partial x} + \nu \frac{\partial v_0}{\partial y} \right]$$

(4.52)

$$N_y = K \left[\frac{\partial v_0}{\partial y} + \nu \frac{\partial u_0}{\partial x} \right]$$

(4.53)

$$N_{xy} = N_{yx} = Gh \left[\frac{\partial u_0}{\partial y} + \frac{\partial v_0}{\partial x} \right] \tag{4.54}$$

where $Eh/(1 - \nu^2) = K$, the extensional stiffness. (Note the similarity of K to $EA = Ebh$ for a rectangular beam.)

4.5   DERIVATION OF THE GOVERNING EQUATIONS FOR A PLATE

From the foregoing it is seen that the equations governing the lateral deflections, bending and shearing action of a plate are given by eqns. (4.25), (4.26), (4.27), (4.47), (4.48), and (4.49). The equations governing the in-plane stress resultants and in-plane midsurface displacements are given by eqns. (4.28), (4.29), (4.52), (4.53) and (4.54). It should be noted that in this classical thin plate theory the first six equations, related to bending and shear, are completely uncoupled from the latter five equations that deal with in-plane loads and displacements. [*Note*: in Section 4.10, we shall see that when the in-plane loads are highly compressive, the in-plane loads do indeed cause lateral displacements (buckling), but a more sophisticated theory will be evolved at that time.]

From eqn. (4.27) it should be noted that the plate can only tell the difference between tractions on the upper and lower surfaces. Hence, we define

$$p_1(x, y) - p_2(x, y) \equiv p(x, y) \tag{4.55}$$

Substituting the suitable derivatives of (4.25) and (4.26) into (4.27) results in the following for the case of no shear stresses on the plate upper and lower surfaces

$$\frac{\partial^2 M_x}{\partial x^2} + 2\frac{\partial^2 M_{xy}}{\partial x \partial y} + \frac{\partial^2 M_y}{\partial y^2} + p(x, y) = 0$$

Substituting (4.47), (4.48) and (4.49) into this results in

$$-D\left[\frac{\partial^4 w}{\partial x^4} + \nu\frac{\partial^4 w}{\partial x^2 \partial y^2}\right] - 2(1 - \nu)D\frac{\partial^4 w}{\partial x^2 \partial y^2} - D\left[\frac{\partial^4 w}{\partial y^4} + \nu\frac{\partial^4 w}{\partial x^2 \partial y^2}\right] +$$

$$+ p(x, y) = 0$$

$$D\left[\frac{\partial^4 w}{\partial x^4} + 2\frac{\partial^4 w}{\partial x^2 \partial y^2} + \frac{\partial^4 w}{\partial y^4}\right] = p(x, y) \tag{4.56}$$

$$D\nabla^4 w = p(x, y)$$

where

$$\nabla^2(\ ) = \frac{\partial^2(\ )}{\partial x^2} + \frac{\partial^2(\ )}{\partial y^2} \quad \text{and} \quad \nabla^4(\ ) = \nabla^2(\nabla^2(\ ))$$

Next, treating the in-plane displacements and forces, by substituting eqns. (4.52) to (4.54) into the two equilibrium equations (4.28) and (4.29) they become, after considerable manipulation

$$\nabla^4 u_0 = 0 \tag{4.57}$$

$$\nabla^4 v_0 = 0 \tag{4.58}$$

Equation (4.56) can now be used in discussing some similar equations. By merely saying $\partial(\ )/\partial y = 0$, letting $\nu = 0$, and multiplying by b, eqn. (4.56) becomes the governing differential equation for a beam

$$EI\frac{d^4 w}{dx^4} = q(x) = bp(x) \tag{4.59}$$

where b is the width of the beam and $I = bh^3/12$ for a beam of rectangular cross-section and $q(x)$ the load in units of load per unit length of the beam.

For a vibrating plate, an inertial load per unit planform area is added as an equivalent force per unit area, resulting in

$$D\nabla^4 w(x, y, t) = p(x, y, t) - \rho h \frac{\partial^2 w}{\partial t^2} \tag{4.60}$$

where ρ is the mass density of the plate material, and t is the co-ordinate of time.

4.6   BOUNDARY CONDITIONS

Since we have a fourth order partial differential equation in x and y, then we need four boundary conditions in x and four in y, that is to say two at each edge. For the clamped and simply supported edges, our knowledge of beam theory dictates the following:

For a clamped edge   For a simply supported edge

$$\left.\begin{array}{c} w = 0 \\ \dfrac{\partial w}{\partial n} = 0 \end{array}\right\} \qquad \left.\begin{array}{c} w = 0 \\ M_n = 0 \end{array}\right\} \tag{4.61}$$

where n is the direction normal to the edge.

For a free edge

At a free edge such as $x = 0$, M_x, M_{xy} and Q_x all equal 0. We therefore have one too many boundary conditions. If we included transverse shear deformation the equation system would be sixth-order and all of these boundary conditions could be satisfied (Fig. 4.5). It is now necessary to combine M_{xy} and Q_x to form an effective shear resultant.

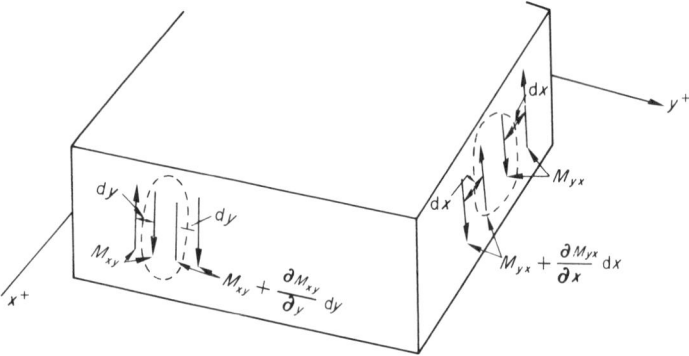

Fig. 4.5

It is clear that the effect of the twisting moment M_{xy} on an element of length dy is the same as that produced by opposite vertical forces of magnitude M_{xy} with moment arm between them. As we move along the edge from one element to the next, it is clear that there exists at each point an unbalanced vertical force of magnitude Q_x dy.

Thus the effective shear at the free boundary is

$$- M_{xy} + M_{xy} + \frac{\partial M_{xy}}{\partial y}\, \mathrm{d}y + Q_x\, \mathrm{d}y = 0$$

or

$$V_x \equiv Q_x + \frac{\partial M_{xy}}{\partial y} = 0 \tag{4.62}$$

Likewise, consideration of the other face results in

$$V_y = Q_y + \frac{\partial M_{yx}}{\partial x} = 0 \tag{4.63}$$

The second boundary condition for a free edge is

$$M_n = 0 \tag{4.64}$$

where n refers to the direction normal to the edge. These are known as the *Kirchoff Boundary Conditions.*

4.7 STRESS DISTRIBUTION WITHIN A PLATE

In plate theory because all equations are integrated across the thickness only integrated stress quantities are known. For stresses at a control element or material point within a plate, one must *assume* a stress distribution. This is done by means of an analogy to beam theory. Thus

$$\begin{Bmatrix} \sigma_x \\ \sigma_y \\ \sigma_{xy} \end{Bmatrix} = \frac{1}{h} \begin{Bmatrix} N_x \\ N_y \\ N_{xy} \end{Bmatrix} + \frac{z}{(h^3/12)} \begin{Bmatrix} M_x \\ M_y \\ M_{xy} \end{Bmatrix} \tag{4.65}$$

$$\begin{Bmatrix} \sigma_{xz} \\ \sigma_{yz} \end{Bmatrix} = \frac{3}{2h} \left[1 - \left(\frac{z}{h/2}\right)^2 \right] \begin{Bmatrix} Q_x \\ Q_y \end{Bmatrix} - \frac{1}{4} \begin{Bmatrix} S_x \\ S_y \end{Bmatrix} \tag{4.66}$$

where

$$\begin{Bmatrix} S_x \\ S_y \end{Bmatrix} = \begin{Bmatrix} \tau_{1x} \\ \tau_{1y} \end{Bmatrix} \left[1 - 2\left(\frac{z}{h/2}\right) - 3\left(\frac{z}{h/2}\right)^2 \right] +$$
$$+ \begin{Bmatrix} \tau_{2x} \\ \tau_{2y} \end{Bmatrix} \left[1 + 2\left(\frac{z}{h/2}\right) - 3\left(\frac{z}{h/2}\right)^2 \right] \tag{4.67}$$

S_x and S_y are functions which account for the effect of shear stresses acting on the upper and lower surfaces of the plate.

4.8 DOUBLE SERIES SOLUTION (NAVIER SOLUTION)

Having derived the governing equations for a rectangular isotropic plate, eqns. (4.56) to (4.58), we now turn to their solution.

In rectangular plate problems one can usually obtain solutions using a double series such as

$$w(x, y) = \sum_{m=1}^{\infty} \sum_{n=1}^{\infty} A_{mn} f_m(x) g_n(y)$$

Such solutions are usually lengthy to compute with, due to very slow convergence. One usually tries to obtain a solution in which the function of one spatial variable is summed such that

$$w(x, y) = \sum_{n=1}^{\infty} \phi_n(y) f_n(x)$$

This trick is particularly useful when two of the opposite edges are simply supported.

However, first treating the double series solution, consider a rectangular plate simply supported on all four edges in the region $0 \leqslant x \leqslant a$, $0 \leqslant y \leqslant b$, $-h/2 \leqslant z \leqslant h/2$. The governing equation is

$$\nabla^4 w = p(x, y)/D$$

The solution can be written as

$$w(x, y) = \sum_{m=1}^{\infty} \sum_{n=1}^{\infty} A_{mn} \sin \frac{m\pi x}{a} \sin \frac{n\pi y}{b} \tag{4.68}$$

The lateral load must be expanded in the same series solution

$$p(x, y) = \sum_{m=1}^{\infty} \sum_{n=1}^{\infty} B_{mn} \sin \frac{m\pi x}{a} \sin \frac{n\pi y}{b} \tag{4.69}$$

where

$$B_{mn} = \frac{4}{ab} \int_0^a \int_0^b p(x, y) \sin \frac{m\pi x}{a} \sin \frac{n\pi y}{b} \, \mathrm{d}y \, \mathrm{d}x$$

From the above

$$\sum_{m=1}^{\infty} \sum_{n=1}^{\infty} A_{mn} \pi^4 \left\{ \frac{m^4}{a^4} + 2\frac{m^2 n^2}{a^2 b^2} + \frac{n^4}{b^4} \right\} \sin \frac{m\pi x}{a} \sin \frac{n\pi y}{b}$$

$$= \frac{1}{D} \sum_{m=1}^{\infty} \sum_{n=1}^{\infty} B_{mn} \sin \frac{m\pi x}{a} \sin \frac{n\pi y}{b}$$

$$A_{mn} = \frac{B_{mn}}{D\pi^4 \{(m^2/a^2) + (n^2/b^2)\}^2}$$

Thus, the solution is easily found for this case.

4.9   SINGLE SERIES SOLUTION (METHOD OF M. LEVY)

Consider a plate with opposite edges simply supported, as shown in Fig. 4.6. Again

$$\nabla^4 w = \frac{p(x, y)}{D} \tag{4.70}$$

Boundary conditions on the y edges are

$$w(x, 0) = w(x, b) = 0$$
$$M_y(x, 0) = M_y(x, b) = 0 \tag{4.71}$$

From (4.49)

$$M_y = -D\left[\frac{\partial^2 w}{\partial y^2} + \nu \frac{\partial^2 w}{\partial x^2}\right]$$

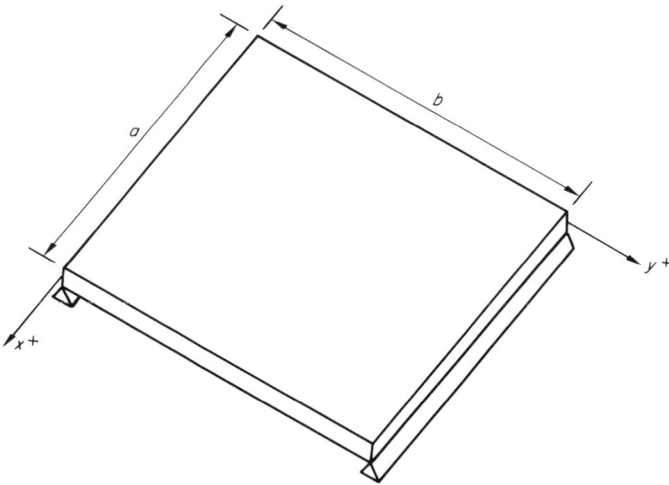

Fig. 4.6

Hence

$$\frac{\partial^2 w}{\partial y^2}\left(x, \frac{0}{b}\right) + \nu \frac{\partial^2 w}{\partial x^2}\left(x, \frac{0}{b}\right) = 0$$

but $\dfrac{\partial^2 w}{\partial x^2}\left(x, \dfrac{0}{b}\right) = 0$ because the curvature is zero parallel to the edge.

Therefore

$$\frac{\partial^2 w}{\partial y^2}\left(x, \frac{0}{b}\right) = 0$$

Assume the form of the solution to be as follows, which satisfies the boundary condition on the y edges given by (4.71)

$$w(x, y) = \sum_{n=1}^{\infty} \phi_n(x) \sin \frac{n\pi y}{b} \qquad (4.72)$$

For this example, the lateral pressure is given by the following

$$p(x, y) = g(x) h(y) \qquad (4.73)$$

where $g(x)$ and $h(y)$ are known. It is therefore necessary to expand $h(y)$ in a series solution corresponding to (4.72). Hence

$$h(y) = \sum_{n=1}^{\infty} A_n \sin \frac{n\pi y}{b}$$

where

$$A_n = \frac{2}{b} \int_0^b h(y) \sin \frac{n\pi y}{b} \, dy \qquad (4.74)$$

Substituting (4.72) and (4.73) into (4.70) gives

$$\sum_{n=1}^{\infty} (\phi_n^{\text{IV}} - 2\lambda_n^2\phi_n'' + \lambda_n^4\phi_n) \sin\frac{n\pi y}{b} = \frac{1}{D} \sum_{n=1}^{\infty} A_n g_n(x) \sin\frac{n\pi y}{b} \quad (4.75)$$

where

$$\lambda_n = n\pi/b$$

For this to be true, the series must be equated term by term

$$\phi_n^{\text{IV}}(x) - 2\lambda_n^2\phi_n''(x) + \lambda_n^4\phi_n(x) = \frac{1}{D} A_n g_n(x) \quad (4.76)$$

Note: We have arrived at this point without specifying any boundary conditions on the other two edges. Hence, any time a problem has two opposite edges simply supported, one can arrive at (4.76) without any other information.

Now, as an example, let $x = 0$, a be simply supported edges, and let $p(x, y)$ be a function of y only, hence $g(x) = 1$.

Boundary conditions

$$w(0, y) = 0, \qquad w(a, y) = 0$$
$$M_x(0, y) = 0, \qquad M_x(a, y) = 0$$

So

$$\frac{\partial^2 w}{\partial x^2}\left(\genfrac{}{}{0pt}{}{0}{a}, y\right) = 0$$

Because of the form assumed for the lateral deflection (4.72) these boundary conditions require that

$$\left.\begin{array}{l}\phi_n(0) = \phi_n(a) = 0\\ \phi_n''(0) = \phi_n''(a) = 0\end{array}\right\} \quad (4.77)$$

and (4.76) becomes

$$\phi_n^{\text{IV}} - 2\lambda_n^2\phi_n'' + \lambda_n^4\phi_n = \frac{A_n}{D} \quad (4.78)$$

From (4.78) we must now solve for ϕ_n. Proceeding in the customary way, for the complementary solution we let $\phi_n = e^{sx}$, such that

$$s^4 - 2\lambda_n^2 s^2 + \lambda_n^4 = 0$$
$$(s^2 - \lambda_n^2)(s^2 - \lambda_n^2) = 0$$

where $\lambda_n^2 > 0$

$$s = \pm\lambda_n, \pm\lambda_n$$

So, the complementary solution is

$$\phi_n(x) = (C_1 + C_2 x) \cosh \lambda_n x + (C_3 + C_4 x) \sinh \lambda_n x$$

The particular solution is easily seen to be $\phi_n = A_n / D\lambda_n^4$. Therefore the complete solution is

$$\phi_n(x) = (C_1 + C_2 x) \cosh \lambda_n x + (C_3 + C_4 x) \sinh \lambda_n x + \frac{A_n}{D\lambda_n^4} \quad (4.79)$$

Now all that is left to do is to substitute (4.79) into the boundary conditions on the x edges (4.77) to obtain the values of the constants C_i ($i = 1, 2, 3, 4$). For this example the results are

$$\left. \begin{array}{l}
C_1 = \dfrac{-A_n}{D\lambda_n^4} \\[2ex]
C_2 = \dfrac{A_n}{2D\lambda_n^3} \dfrac{(1 - \cosh \lambda_n a)}{\sinh \lambda_n a} \\[2ex]
C_3 = \dfrac{A_n a}{2D\lambda_n^3 (1 + \cosh \lambda_n a)} \left[\dfrac{2}{\lambda_n a} \sinh \lambda_n a - 1 \right] \\[2ex]
C_4 = \dfrac{A_n}{2D\lambda_n^3}
\end{array} \right\} \quad (4.80)$$

Thus, the complete solution for the lateral deflection is

$$w(x, y) = \sum_{n=1}^{\infty} \left[(C_1 + C_2 x) \cosh \lambda_n x + (C_3 + C_4 x) \sinh \lambda_n x + \frac{A_n}{D\lambda_n^4} \right] \sin \lambda_n y$$

$$(4.81)$$

where C_1 to C_4 are given by (4.80).

It should be noted that we could have solved this problem by assuming the deflection to be any one of these following ways

$$w(x, y) = \sum_{m=1}^{\infty} \sum_{n=1}^{\infty} A_{mn} \sin \frac{m\pi x}{a} \sin \frac{n\pi y}{b}$$

$$w(x, y) = \sum_{n=1}^{\infty} \phi_n(x) \sin \frac{n\pi y}{b}$$

$$w(x, y) = \sum_{m=1}^{\infty} \psi_m(y) \sin \frac{m\pi x}{a}$$

The first of these methods converges very slowly, while the second and third converge more rapidly, as stated previously.

For the case of the x edges being clamped or free, and with the same loading, $p = p(y)$ only, eqn. (4.79) may be used, with the appropriate boundary conditions to obtain the solution.

For rectangular plates, the Navier method and the Levy method comprise the two most often used methods of obtaining solutions for rectangular plates. If at least two opposite edges of a plate are not simply supported, solutions become more difficult and frequently energy methods are used to obtain approximate solutions in those cases. These are discussed in Section 4.11.

More detailed treatment of rectangular plates, as well as circular plates, and thermoelastic effects are dealt with in References 2–4.

4.10   ELASTIC STABILITY OF PLATES

We have previously derived the governing equations for a thin plate subjected to both in-plane and lateral loads. In them there was one governing equation describing the relationship between the lateral deflection and the laterally distributed loading

$$D\nabla^4 w = p(x, y) \tag{4.82}$$

and other equations dealing with in-plane displacements, related to in-plane loads

$$\nabla^4 u_0 = \nabla^4 v_0 = 0 \tag{4.83}$$

As discussed previously, the equation involving lateral displacements and lateral loads is completely independent (uncoupled) from the in-plane loadings and in-plane displacements and vice versa.

However, it is found that when in-plane loads are compressive, upon attaining certain discrete values, these compressive loads do produce lateral displacements. Thus, there does occur a coupling between in-plane loads and lateral displacement, w. As a result, a more inclusive theory must be developed to account for this phenomenon, which is called *buckling* or *elastic instability*. You have noted this phenomena previously in mechanics of materials in the study of the buckling of columns.

Unlike in developing the governing plate equations earlier wherein we began with the three-dimensional equations of elasticity, we shall begin the following with looking at the in-plane forces acting on a plate element, and it is seen that these stress resultants are functions of x and y (Fig. 4.7).

Viewing the plate from the midsurface in the position y direction (Fig. 4.8), the relationship between forces and displacements is seen, when the plate is subjected to both lateral and in-plane forces.

Hence, as can be seen from Fig. 4.8, the z component of the N_x

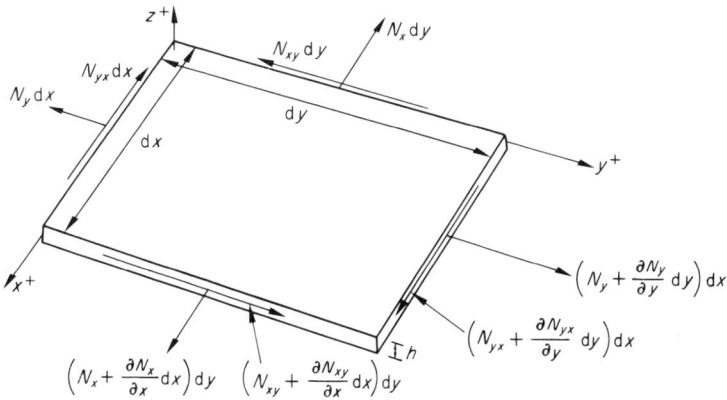

Fig. 4.7

loading per unit area is, for small slopes

$$\frac{1}{dx\,dy}\left[\left(N_x+\frac{\partial N_x}{\partial x}dx\right)dy\left(\frac{\partial w}{\partial x}+\frac{\partial^2 w}{\partial x^2}dx\right)-N_x\,dy\frac{\partial w}{\partial x}\right].$$

Neglecting terms of higher order, the force per unit planform area in the z direction is seen to be

$$N_x\frac{\partial^2 w}{\partial x^2}+\frac{\partial N_x}{\partial x}\cdot\frac{\partial w}{\partial x}$$

Similarly the z component of the N_y force per unit planform area is seen to be

$$N_y\frac{\partial^2 w}{\partial y^2}+\frac{\partial N_y}{\partial y}\cdot\frac{\partial w}{\partial y}$$

Now, to investigate the z component of the in-plane shear resultants, see Fig. 4.9.

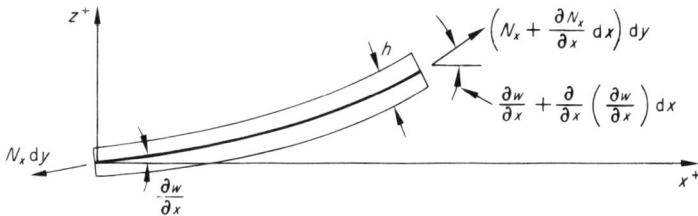

Fig. 4.8

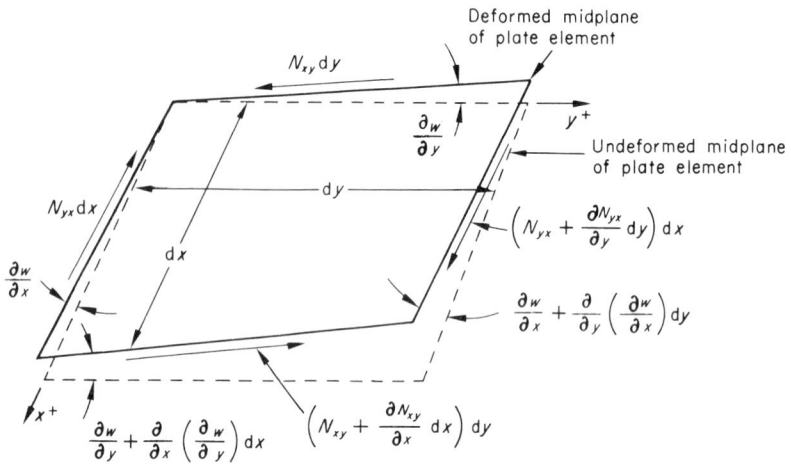

Fig. 4.9

Hence, the z component per unit area of the in-plane shear resultants is

$$\frac{1}{dx\,dy}\left\{\left(N_{xy}+\frac{\partial N_{xy}}{\partial x}\,dx\right)\left(\frac{\partial w}{\partial y}+\frac{\partial^2 w}{\partial x\partial y}\,dx\right)dy+\right.$$

$$+\left(N_{yx}+\frac{\partial N_{yx}}{\partial y}\,dy\right)\left(\frac{\partial w}{\partial x}+\frac{\partial^2 w}{\partial x\partial y}\,dy\right)dx-$$

$$\left.-N_{xy}\frac{\partial w}{\partial y}\,dy-N_{yx}\frac{\partial w}{\partial x}\,dx\right\}$$

Neglecting higher order terms, this results in

$$N_{xy}\frac{\partial^2 w}{\partial x\partial y}+\frac{\partial N_{xy}}{\partial x}\frac{\partial w}{\partial y}+N_{yx}\frac{\partial^2 w}{\partial x\partial y}+\frac{\partial N_{yx}}{\partial y}\frac{\partial w}{\partial x}$$

With all the above z components of forces per unit area, the governing plate equation (4.82) is modified to include the z components of the in-plane forces

$$D\nabla^4 w = p(x,y)+N_x\frac{\partial^2 w}{\partial x^2}+N_y\frac{\partial^2 w}{\partial y^2}+2N_{xy}\frac{\partial^2 w}{\partial x\partial y}+$$

$$+\frac{\partial N_x}{\partial x}\frac{\partial w}{\partial x}+\frac{\partial N_y}{\partial y}\frac{\partial w}{\partial y}+$$

$$+\frac{\partial N_{xy}}{\partial x}\frac{\partial w}{\partial y}+\frac{\partial N_{yx}}{\partial y}\frac{\partial w}{\partial x}$$

However from in-plane force equilibrium (see eqns. 4.28 and 4.29), it is remembered that

$$\frac{\partial N_x}{\partial x} + \frac{\partial N_{yx}}{\partial y} = 0$$

$$\frac{\partial N_{xy}}{\partial x} + \frac{\partial N_y}{\partial y} = 0$$

for the case of a plate with no surface shear stresses. Hence, the final form of the equation is seen to be

$$D\nabla^4 w = p(x, y) + N_x \frac{\partial^2 w}{\partial x^2} + N_y \frac{\partial^2 w}{\partial y^2} + 2N_{xy} \frac{\partial^2 w}{\partial x \partial y} \tag{4.84}$$

This equation is analogous to the beam–column equation which can be obtained by multiplying the above by b (the width of the beam) and letting $\partial(\ )/\partial y = 0$, $\nu = 0$, $\bar{P} = -bN_x$ and $q(x) = bp(x)$, to provide

$$\frac{d^4 w}{dx^4} + k^2 \frac{d^2 w}{dx^2} = \frac{q(x)}{EI} \tag{4.85}$$

where $k^2 = \bar{P}/EI$.

Solving eqn. (4.85) the solution can be written as

$$w(x) = A \cos kx + B \sin kx + C + Dx + w_p(x)$$

where $w_p(x)$ is the particular solution for the loading $q(x)$. Consider the case wherein $q(x) = 0$, and the column is simply supported at each end. The boundary conditions are then, at $x = 0$, L,

$$w(0) = w(L) = 0 \tag{4.86}$$

$$M_x\left(\begin{matrix} L \\ 0 \end{matrix}\right) = -EI\frac{d^2 w}{dx^2} = 0 \quad \text{or} \quad \frac{d^2 w(0)}{dx^2} = \frac{d^2 w(L)}{dx^2} = 0$$

From the first boundary condition, $A + C = 0$, from the third $A = 0$; hence, $C = 0$ also. From the second boundary condition $B \sin kL + DL = 0$, and from the fourth boundary condition

$$Bk^2 \sin kL = 0 = \frac{B\bar{P}}{EI} \sin kL = 0 \tag{4.87}$$

Note that in eqn. (4.87) when $kL \neq n\pi$, then $B = D = 0$; when $kl = n\pi$, then $D = 0$, $B \neq 0$ but indeterminate and

$$\bar{P} = n^2\pi^2 \frac{EI}{L^2} \tag{4.88}$$

It is thus seen that for most values of $\bar{P}$ (the axial compressive loading) the lateral deflection w is zero, and the in-plane and lateral forces and responses are uncoupled. However, for a countable infinity of discrete values of $\bar{P}$, there is a lateral deflection of an indeterminate magnitude. Mathematically, this is referred to as an eigenvalue problem, and the discrete values given in (4.88) are called *eigenvalues*. The resulting deflections, in this case are

$$w(x) = B \sin kx$$

and are called *eigenfunctions*.

The natural vibrations of elastic bodies are also eigenvalue problems, where in that case the natural frequencies are the eigenvalues and the vibration modes are the eigenfunctions.

As to buckling, looking at eqn. (4.88), where $\bar{P}$ increases, it is clear that the lowest buckling load occurs when $n = 1$, and at that particular load, the column will either inelastically deform and strain harden, or the column will fracture. Hence, $n > 1$ has no physical significance. The load

$$\bar{P} = \frac{\pi^2 EI}{L^2} \tag{4.89}$$

is therefore the critical buckling load for this column with these boundary conditions. In this particular case the buckling load is called the Euler buckling load, after the Swiss scientist who first successfully studied the problem.

Another way to phrase the buckling problem is exemplified by solving eqn. (4.85) letting $q(x) = q_0 =$ constant. The resulting particular solution is $q_0 x^2 / 2\bar{P}$. If the column is simply supported, solving the boundary value problem for the lateral deflection, the result is

$$w(x) = \frac{q_0}{\bar{P}k^2 \sin kL} \left[\begin{array}{c} \cos kx \sin kL + \sin kx - \cos kL - \sin kL \\ - Lx \sin kL + k^2 x^2 \sin kL \end{array} \right] \tag{4.90}$$

In eqn. (4.90), the solution of a boundary value problem, when the axial load $\bar{P}$ has values given in (4.88), $\sin kL = 0$, $w(x)$ goes to infinity; or more properly, since we have a small deflection linear mathematical model, $w(x)$ becomes indefinitely large.

Hence, whether we solve for the homogeneous solution of eqn. (4.85), resulting in an eigenvalue problem, or we solve the non-homogeneous equation (4.85), resulting in a boundary value problem, the results are identical, when $\bar{P}$ has values given by (4.88), or physically where $\bar{P}$ attains the value given by (4.89), the column 'buckles'.

Note also that the buckling load, eqn. (4.89), is not affected by any load $q(x)$.

These elastic stability considerations are very important in analysing or designing any structure in which compressive stresses result from the loading.

Plate buckling is qualitatively analogous to column buckling, except that the mathematics is more complicated, and the conditions that result in the lowest eigenvalue (the real buckling load) are not so clear in many cases.

Whenever the in-plane forces are compressive, and are more than a few per cent of the plate buckling loads (to be defined later), eqn. (4.84) must be used rather than eqn. (4.82).

For the plate, just as in the case of the beam-column, since the in-plane load that causes an elastic stability is not dependent upon a lateral load, to investigate the elastic stability we shall assume $p(x, y) = 0$ in eqn. (4.84).

Consider, as an example, a simply supported plate, subjected to an in-plane load N_x. Hence, we shall assume the solution of eqn. (4.84) to be of the Navier form

$$w(x, y) = \sum_{m=1}^{\infty} \sum_{n=1}^{\infty} A_{mn} \sin \frac{m\pi x}{a} \sin \frac{n\pi y}{b} \qquad (4.91)$$

Substituting eqn. (4.91) into eqn. (4.84) we obtain, where $N_{xy} = N_y = p(x, y) = 0$,

$$(N_x)_{cr} = -\frac{D\pi^2 a^2}{m^2} \left[\frac{m^2}{a^2} + \frac{n^2}{b^2} \right]^2. \qquad (4.92)$$

The loaded plate is shown below

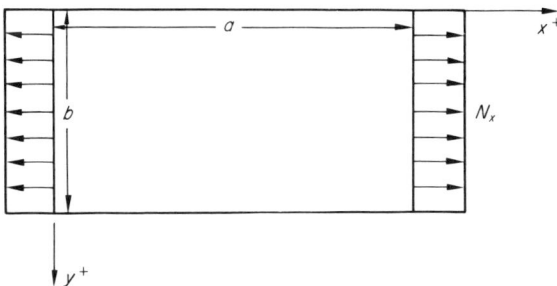

It is obvious from eqn. (4.92) that to obtain the minimum value of $N_{x_{cr}}$, $n = 1$, since it only appears in the numerator. Thus for an isotropic plate, simply supported on all four edges, subjected only to a uniaxial in-plane load the buckling mode will always be one-half sine wave across the span, regardless of the length or width of the plate.

Thus, since $n = 1$, eqn. (4.92) can be written as

$$(N_x)_{cr} = -\frac{D\pi^2}{b^2}\left(\frac{m}{r}+\frac{r}{m}\right)^2 \quad \text{(where } r = a/b) \quad (4.93)$$

Note that the first term represents the Euler buckling load for a strip of unit width and length a. The second term indicates what proportion the stability of the plate has greater than the stability of an isolated strip.

Note if $a < b$ (the plate wider than it is long), the second term is always less than the first, hence $(N_x)_{min}$ is always obtained by letting $m = 1$. Hence for $a < b$, the buckling mode for the simply supported plate is always

$$w(x, y) = a_{11} \sin\frac{\pi x}{a} \sin\frac{\pi y}{b}$$

In that case

$$(N_x)_{cr} = -\frac{D\pi^2}{b^2}\left(\frac{1}{r}+r\right)^2 \quad (4.94)$$

To find out at what aspect ratio r, that N_x is truly a minimum, let

$$\frac{d(N_x)_{cr}}{dr} = 0 = -\frac{2D\pi^2}{b^2}\left(\frac{1}{r}+r\right)\left(-\frac{1}{r^2}+1\right)$$

Therefore $r = 1$.

Hence for $m = 1$, $(N_x)_{cr}$ is minimum where $a = b$. Under that condition, from (4.94)

$$(N_x)_{cr_{a=b}} = -\frac{4D\pi^2}{b^2} = -\frac{4D\pi^2}{a^2} \quad (4.95)$$

Comparing this with the Euler buckling load for a simply supported column, it is seen that the continuity of a plate and the support along the sides of the plate provide a factor of at least 4 over the buckling of a series of strips (columns) that are neither continuous nor supported along the unloaded edges.

Now as the length-to-width ratio increases, as a/b increases, the buckling load (4.94) will increase, and one can ask, will $m = 1$ always result in a minimum buckling load, or is there another value of m which will provide a lower buckling load as r increases (i.e. $(N_x)_{m=2} \leqslant (N_x)_{m=1}$ for some r?).

Mathematically, this can be phrased as

$$\left(\frac{m}{r}+\frac{r}{m}\right)^2 \overset{?}{\leqslant} \left(\frac{m-1}{r}+\frac{r}{m-1}\right)^2 \qquad \text{(for some } r\text{)}$$

This states the condition under which the plate of aspect ratio r will buckle in m half sine waves in the loaded direction rather than $m-1$ since waves. Manipulating the above results in

$$m(m-1) \leqslant r^2 \qquad (4.96)$$

Equation (4.96) states that the plate will buckle in two half sine waves in the axial direction rather than one when $r \geqslant \sqrt{2}$. The plate will buckle in three half sine waves in the axial direction rather than two, when $r \geqslant \sqrt{6}$, etc.

Again one can ask that when the plate buckles into $m = 2$ configuration, does a minimum buckling load occur, if so at what r and what is the corresponding minimum of $N_{x_{cr}}$?

From eqn. (4.93), for $m = 2$

$$\frac{\mathrm{d}(N_x)_{cr}}{\mathrm{d}r} = 0 = -\frac{D\pi^2}{b^2} 2 \left(\frac{2}{r}+\frac{r}{2}\right)\left(-\frac{2}{r^2}+\frac{1}{2}\right) = 0$$

or $r^2 = 4$, $r = 2$

$$(N_x)_{cr,\min} = -\frac{4\pi^2 D}{b^2} \qquad (4.97)$$

This is the same value as given in eqn. (4.95) for $m = 1$. Proceeding with all values of r and m, the graph shown in Fig. 4.10 can be drawn, which clearly shows the results.

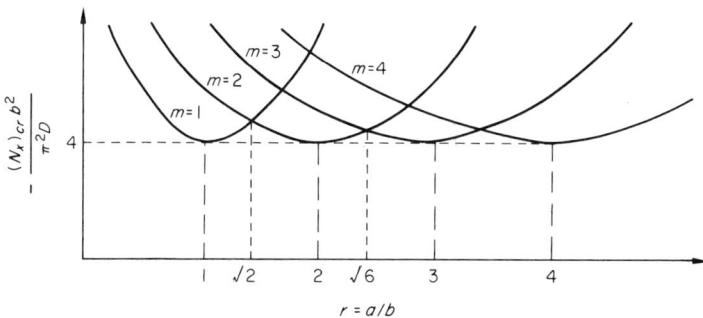

Fig. 4.10

Hence knowing the value of r, the figure provides one with the actual value of $(N_x)_{cr}$ and what m is. However, in practice, for $r > 1$, universally one simply uses eqn. (4.95) or (4.97) for the buckling load and there is little interest really in the value of m. However, looking more closely at eqn. (4.96), as m increases we see that

$$m(m-1) \to m^2 = r^2 \quad \text{or} \quad m = r = a/b$$

This says that for long plates, the number of half sine waves of the buckles have lengths approximately equal to the plate width. Another way of saying it is that the plate attempts to buckle into a number of square cells.

Remembering that $\sigma_x = N_x/h$, eqns. (4.95) and (4.97) can be written as for $a/b \geqslant 1$:

$$\sigma_{cr} = -\frac{\pi^2 E}{3(1 - \nu^2)} \left(\frac{h}{b}\right)^2 \tag{4.98}$$

All the preceding discussion has been in regard to an isotropic plate simply supported on all four edges. Solving the governing differential equations for other boundary conditions often results in eigenvalues that are lengthy, complicated, transcendental equations. They are difficult to use, and lack the clarity that has been shown in the case of a simply supported plate. Therefore, in most cases, for plates with other boundary conditions, the use of energy methods is employed to obtain approximate values to the buckling loads, and these are discussed in the next section.

4.11   ENERGY METHODS

As an alternative to developing the governing differential equations for a material point in an elastic body and specifying certain boundary conditions, as has been done in the earlier sections of this chapter, one can develop an expression involving the strain energy and the work done by forces acting on the body. In the area of solid mechanics, there are three primary energy methods: Potential energy, Complementary energy, and Reissner's variational theorem. The first of these methods will be dealt with here, and the last method is treated in Section 6.3.

4.11.1   Minimum Potential Energy
For an elastic body, the potential energy can be written as

$$V = \int_R W \, dR - \int_{S_t} T_i u_i \, dS - \int_R F_i u_i \, dR \qquad (4.99)$$

where

W = Strain energy density function.
R = Volume of the elastic body.
T_i = ith component of the surface traction.
u_i = ith component of the deformation.
F_i = ith component of the body force.
S_t = Portion of the surface over which tractions are prescribed.

The first term on the right-hand side is the strain energy of the body. The second and third are the work done by the surface tractions and the body forces, respectively.

The theorem of minimum potential energy can be stated as: 'Of all displacements satisfying compatability and the given boundary conditions, those which satisfy the equilibrium equations make the potential energy a minimum.' This theorem is discussed in detail in Reference 1.

Mathematically, the operation can be written as

$$\delta V = 0 \qquad (4.100)$$

Here the lower-case delta is a mathematical operation called a variation. Operationally, it is analogous to partial differentiation, and to solve problems in structural mechanics only the following two operations are needed

$$\frac{d(\delta y)}{dx} = \delta\left(\frac{dy}{dx}\right), \qquad \delta(y^2) = 2y\delta y \qquad (4.101)$$

For the case of elastic stability, and neglecting body forces, substituting eqn. (4.99) into (4.100), one sees that

$$\delta V = \delta\left[\int_R W \, dR - \int_{S_t} T_i u_i \, dS\right] \qquad (4.102)$$

One way to interpret this is that the amount of potential energy in the body does not change (i.e. the variation in potential energy is zero) when the elastic body passes from a configuration of equilibrium (prior to buckling) to an infinitesimally near adjacent configuration (after buckling). One then seeks the forces T_i that will cause such a situation to occur. When stated in this manner the theorem is sometimes referred to as the theorem of stationary potential energy.

The strain energy density function W is defined as

$$W \equiv \tfrac{1}{2}\sigma_{ij}\epsilon_{ij} \tag{4.103}$$

For a body described in a Cartesian co-ordinate frame

$$W = \tfrac{1}{2}\sigma_x\epsilon_x + \tfrac{1}{2}\sigma_y\epsilon_y + \tfrac{1}{2}\sigma_z\epsilon_z +$$

$$+ \sigma_{xy}\epsilon_{xy} + \sigma_{xz}\epsilon_{xz} + \sigma_{yz}\epsilon_{yz} \tag{4.104}$$

To utilise the theorem of minimum potential energy it is necessary to express the strain energy density function in terms of displacements.

Furthermore, if for example, one wishes to obtain the strain energy for a plate, then one makes use of all of the plate theory assumptions in developing the strain energy relationship, namely

$$\left. \begin{array}{l} \sigma_z = \epsilon_z = \epsilon_{xz} = \epsilon_{yz} = 0 \\[2mm] u = u_0(x, y) + z\alpha(x, y) \quad \text{where} \quad \alpha = -\dfrac{\partial w}{\partial x} \\[2mm] v = v_0(x, y) + z\beta(x, y) \quad \text{where} \quad \beta = -\dfrac{\partial w}{\partial y} \end{array} \right\} \tag{4.105}$$

From eqns. (4.4), (4.5) and (4.7) and using eqn. (4.105) above, one obtains

$$\left. \begin{array}{l} \sigma_x = \dfrac{E}{(1 - \nu^2)} \left[\epsilon_x + \nu\epsilon_y \right] \\[3mm] \sigma_y = \dfrac{E}{(1 - \nu^2)} \left[\epsilon_y + \nu\epsilon_x \right] \\[3mm] \sigma_{xy} = \dfrac{E}{1 + \nu}\, \epsilon_{xy} \end{array} \right\} \tag{4.106}$$

Substituting eqns. (4.106) and (4.105) into (4.104) results in

$$W = \frac{E}{2(1 - \nu^2)}\, \epsilon_x \left[\epsilon_x + \nu\epsilon_y \right] + \frac{E}{2(1 - \nu^2)}\, \epsilon_y \left[\epsilon_y + \nu\epsilon_x \right] + \frac{E}{(1 + \nu)}\, \epsilon_{xy}^2 \tag{4.107}$$

Now for a plate subjected to bending only $u = -z(\partial w/\partial x)$, $v = -z(\partial w/\partial y)$, $w = w(x, y)$, the strain energy density function becomes

$$W = \frac{Ez^2}{2(1 - \nu^2)} \left[\left(\frac{\partial^2 w}{\partial x^2}\right)^2 + \left(\frac{\partial^2 w}{\partial y^2}\right)^2 + 2\nu \left(\frac{\partial^2 w}{\partial x^2}\right)\left(\frac{\partial^2 w}{\partial y^2}\right) \right] +$$

$$+ \frac{E(1 - \nu)z^2}{(1 - \nu^2)} \left(\frac{\partial^2 w}{\partial x\partial y}\right)^2 \tag{4.108}$$

The strain energy, $U\ (= \int_R W\, dR)$ can then be found, where it is seen that in (4.108) the only function of z is the z^2 found throughout, hence one can easily integrate with respect to z to obtain

$$U = \frac{D}{2} \int_0^a \int_0^b \left\{ \left(\frac{\partial^2 w}{\partial x^2} + \frac{\partial^2 w}{\partial y^2} \right)^2 - 2(1 - \nu) \left[\left(\frac{\partial^2 w}{\partial x^2} \right) \left(\frac{\partial^2 w}{\partial y^2} \right) - \left(\frac{\partial^2 w}{\partial x \partial y} \right)^2 \right] \right\} \, dx \, dy$$

(4.109)

It is seen that the first term on the right-hand side is proportional to the square of the average plate curvature. The remaining term is known as the *Gaussian curvature*.

Returning to eqn. (4.103), by using other appropriate assumptions on stresses, strains, and form of displacements one can obtain the strain energy expression for a beam, ring, shell, etc.

Returning to eqn. (4.99), consider a plate subjected to a lateral loading $p(x, y)$, and in-plane loads N_x, N_y and N_{xy}.

For the lateral loading $p(x, y)$ it is clear that the work done is

$$\int_{s_t} T_i u_i \, dS = \int_0^a \int_0^b p(x, y) w(x, y) \, dx \, dy$$

(4.110)

To obtain the work done by the in-plane forces to account for deformation prior to buckling as well as buckling deformations it is necessary to employ the first two terms of the expansion of strains in terms of displacements, namely

$$\epsilon_x = \frac{\partial u_0}{\partial x} + \frac{1}{2} \left(\frac{\partial w}{\partial x} \right)^2$$

$$\epsilon_y = \frac{\partial v_0}{\partial y} + \frac{1}{2} \left(\frac{\partial w}{\partial y} \right)^2$$

(4.111)

$$\epsilon_{xy} = \frac{1}{2} \left(\frac{\partial u_0}{\partial y} + \frac{\partial v_0}{\partial x} \right) + \frac{1}{2} \frac{\partial w}{\partial x} \frac{\partial w}{\partial y}$$

rather than the simple relationships of eqns. (4.10), (4.11) and (4.13).

These, of course, include non-linear terms, but the complete explanation of the reason for using (4.111) will not be dealt with here. In eqn. (4.111) the first terms on the right-hand side reflect the deformations caused by stretching (or compressing) the midsurface of the plate under the in-plane loads, which is the minimum energy mode of deformation up to the value of the buckling load. The second term reflects the lateral deformations caused when the in-plane loads reach the buckling load, wherein lateral deformations, $w(x, y)$, occur, and where there is no stretching or shortening of the plate middle surface ($u_0 = v_0 = 0$). Hence, under the action of in-plane loads the complete expression for the work done is

$$
\int_{s_t} T_i u_i \, ds = -\int_0^a \int_0^b \left\{ N_x \left[\frac{\partial u_0}{\partial x} + \frac{1}{2}\left(\frac{\partial w}{\partial x}\right)^2 \right] + N_y \left[\frac{\partial v_0}{\partial y} + \frac{1}{2}\left(\frac{\partial w}{\partial y}\right)^2 \right] + \right.
$$

$$
\left. + N_{xy} \left[\left(\frac{\partial u_0}{\partial y} + \frac{\partial v_0}{\partial x}\right) + \left(\frac{\partial w}{\partial x}\right)\left(\frac{\partial w}{\partial y}\right) \right] \right\} dx \, dy \qquad (4.112)
$$

If one is seeking only buckling loads, let $u_0 = v_0 = 0$ in eqn. (4.112). Furthermore, the minus sign in (4.112) occurs solely because N_x, N_y are tensile stress resultants, and N_{xy} is defined consistent with the others.

To summarise, for investigating the buckling of a plate in the presence of lateral loads also the potential energy is written as (remember that the buckling load is independent of the lateral loads, however)

$$
V = \frac{D}{2} \int_0^a \int_0^b \left\{ \left(\frac{\partial^2 w}{\partial x^2} + \frac{\partial^2 w}{\partial y^2}\right)^2 - 2(1-\nu)\left[\left(\frac{\partial^2 w}{\partial x^2}\right)\left(\frac{\partial^2 w}{\partial y^2}\right) - \left(\frac{\partial^2 w}{\partial x \partial y}\right)^2 \right] \right\} dx \, dy
$$

$$
- \int_0^a \int_0^b p(x,y)w(x,y)\, dx \, dy + \frac{1}{2}\int_0^a \int_0^b \left[N_x\left(\frac{\partial w}{\partial x}\right)^2 + \right.
$$

$$
\left. + N_y\left(\frac{\partial w}{\partial y}\right)^2 + 2N_{xy}\left(\frac{\partial w}{\partial x}\right)\left(\frac{\partial w}{\partial y}\right) \right] dx \, dy \qquad (4.113)
$$

One can use the theorem of minimum potential energy in three ways. First, it can be used to obtain a governing equation and the natural boundary conditions for a problem that is consistent with the assumptions made in obtaining the strain energy expression. Secondly, if one knows what the deformation pattern is under the applied loads, except for the amplitude, then by substituting the displacement function into eqn. (4.113) one can obtain the amplitude of the deformation under $p(x,y)$, or the values of the critical loading $(N_x)_{cr}$, $(N_y)_{cr}$ or $(N_{xy})_{cr}$. Thirdly, if one does not know the form of the deformation exactly but can make a good estimate, then one can use the theorem of minimum potential energy to obtain approximate deformations due to $p(x,y)$ or approximate values of buckling loads. By far the most useful of these to engineers is the third method.

Each of these methods will be demonstrated, and to maintain simplicity and lucidity, the first two usages will be illustrated with a beam in this section.

4.11.2 The Bending of a Beam due to a Lateral Load

Consider a beam of length a and width b, where as before the displacement and load are functions of x only. Without in-plane loads, eqn. (4.113) can be written as

$$V = \frac{EI}{2} \int_0^a \left(\frac{d^2 w}{dx^2}\right)^2 dx - \int_0^a q(x)w(x)\,dx \qquad (4.114)$$

where $q(x) = bp(x)$, and

$$I = \frac{bh^3}{12} \qquad \text{(Remember for a beam } \nu = 0\text{)}$$

We will now use minimum potential energy to obtain the governing differential equation for a beam under a lateral loading and the natural boundary conditions for the problem. From eqn. (4.110)

$$\delta V = 0 = EI \int_0^a \delta\left(\frac{d^2 w}{dx^2}\right)^2 dx - \int_0^a q(x)\delta w\,dx \qquad (4.115)$$

To solve for the governing equation, the first term on the right-hand side must be integrated by parts several times. The details of the process are

$$\frac{EI}{2}\int_0^a \delta\left(\frac{d^2 w}{dx^2}\right)^2 dx = EI \int_0^a \left(\frac{d^2 w}{dx^2}\right)\delta\left(\frac{d^2 w}{dx^2}\right) dx$$

$$= EI \int_0^a \frac{d^2 w}{dx^2}\frac{d^2}{dx^2}(\delta w)\,dx$$

$$= \left[\left(EI\frac{d^2 w}{dx^2}\right)\delta\left(\frac{dw}{dx}\right)\right]_0^a - EI\int_0^a \frac{d^3 w}{dx^3}\frac{d}{dx}(\delta w)\,dx$$

$$= \left[\left(EI\frac{d^2 w}{dx^2}\right)\delta\left(\frac{dw}{dx}\right)\right]_0^a - \left[\left(EI\frac{d^3 w}{dx^3}\right)\delta w\right]_0^a +$$

$$+ EI \int_0^a \frac{d^4 w}{dx^4}\delta w\,dx \qquad (4.116)$$

Substituting (4.116) into (4.115) and rearranging, one obtains

$$\delta V = 0 = \left[\left(EI\frac{d^2 w}{dx^2}\right)\delta\left(\frac{dw}{dx}\right)\right]_0^a - \left[\left(EI\frac{d^3 w}{dx^3}\right)\delta w\right]_0^a +$$

$$+ \int_0^a \left[EI\frac{d^4 w}{dx^4} - q(x)\right]\delta w\,dx = 0 \qquad (4.117)$$

For this to be true, it is seen that the following equation must be true

$$EI\frac{d^4w}{dx^4} = q(x) \qquad (4.118)$$

It is seen that this is the governing differential equation for a beam. Obviously, there are other approaches that could have been used to obtain this well-known equation. However, when analysing a structure or elastic body which does not fall into the category of a classical shape, then by making use of physical intuition, experience, or by performing experiments, an engineer can make certain assumptions on displacements, strains and stresses, then utilise minimum potential energy to develop a governing equation for the problem.

When one uses a variational principle in this way, the resulting differential equation is called the Euler–Lagrange equation.

Look also at eqn. (4.117), and one sees that there are several natural boundary conditions specified. Looking at the first term on the right-hand side, it states that at each end of the beam, $x = 0$, a either

$$M_x = -EI\frac{d^2w}{dx^2} = 0 \quad \text{or} \quad \frac{dw}{dx} \qquad (4.119)$$

must be specified. (i.e. the variation of $dw/dx = 0$).

Likewise, from the second term, at the ends of the beams $x = 0$, a, either

$$V_x = -EI\frac{d^3w}{dx^3} = 0 \quad \text{or} \quad w \qquad (4.120)$$

must be specified.

It is easily seen that these natural boundary conditions include for the beam all the classical boundary conditions: simple support, clamped, free, and spring supported.

Again, when analysing the behaviour of elastic bodies through the use of minimum potential energy, one not only obtains Euler–Lagrange equations that are consistent with the assumptions adopted, but one also obtains sets of natural boundary conditions which are also consistent with the assumptions made. These can be of great utility in 'real-life' problems.

4.11.3 The Buckling of a Simply-Supported Column Due to an Axial Load

To illustrate the second use of the theorem of minimum potential energy, consider a beam of length a, width b, wherein displacements vary only in the x directions, and are subjected to an axial load N_x. Suppose one assumes that the column will respond to the load N_x by assuming the following deflection shape

$$w(x) = A \sin \frac{\pi x}{a} \qquad (4.121)$$

From eqn. (4.113) the potential energy expression is seen to be

$$V = \frac{EI}{2} \int_0^a \left(\frac{d^2 w}{dx^2}\right)^2 dx + \frac{1}{2} \int_0^a P\left(\frac{dw}{dx}\right)^2 dx \qquad (4.122)$$

where $P = N_x b$.
Substituting eqn. (4.121) into (4.122) results in

$$V = \frac{EI}{2} A^2 \frac{\pi^4}{a^4} \int_0^a \sin^2 \left(\frac{\pi x}{a}\right) dx + \frac{1}{2} PA^2 \frac{\pi^2}{a^2} \int_0^a \cos^2 \left(\frac{\pi x}{a}\right) dx$$

$$= \frac{A^2}{2} \frac{\pi^2}{a^2} \frac{a}{2} \left[\frac{EI\pi^2}{a^2} + P\right] \qquad (4.123)$$

Now using eqn. (4.100) it is seen that from (4.121) and (4.123) the function of x was prescribed, and the only symbol which can have a variation is the amplitude A. Hence

$$\delta V = 0 = \frac{a\pi^2}{4L^2} 2A\delta A \left[\frac{\pi^2 EI}{a^2} + P\right] = 0$$

For this to be true, it requires that

$$P = -\frac{\pi^2 EI}{a^2} \qquad (4.124)$$

This then is the value of the load P which will cause the assumed deformation (4.121). Since, however, eqn. (4.121) is the exact buckling mode for a simply-supported column, then the load is also exactly the Euler buckling load obtained previously.

Hence, in using minimum potential energy for any elastic body, if one assumes the exact form of the deformation by design or accident, one will obtain the exact value of the buckling load, or the exact amplitude of the deformation in a boundary value problem.

However, as stated previously, the most useful application of energy theorems is when one can make a reasonable assumption as to the displacement, and then solve for an approximate solution. This will be demonstrated in Chapter 6 using Reissner's Variational Theorem. Many solutions, design data tables, and figures for the buckling of both plates and shells are found in References 5 and 6.

4.12   CIRCULAR CYLINDRICAL SHELLS UNDER AXIALLY SYMMETRIC LOADS

To systematically develop the following equations from the three-dimensional equations of elasticity, analogous to the development earlier in this chapter for classical rectangular plates, requires a knowledge of curvilinear tensor techniques, curvilinear co-ordinate systems, and much time. Hence as an introduction to classical shell theory, the governing differential equations will be presented below without derivation.

Before proceeding, a sketch of the shell showing the positive directions of displacements, and loads is in order, as presented in Fig. 4.11.

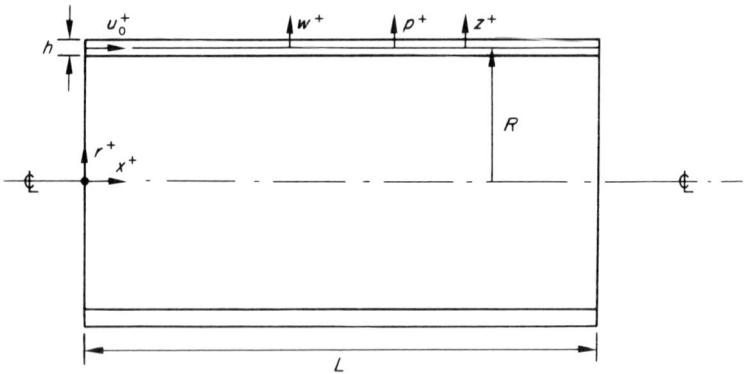

Fig. 4.11

The positive directions of stress resultants and couples are given in Fig. 4.12. Not shown are N_θ and M_θ which are in the circumferential direction.

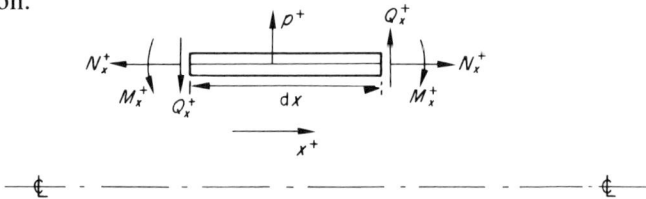

Fig. 4.12

In the above, N_x, M_x and Q_x are defined in eqns. (4.22) to (4.24). The governing equations for the case of axially symmetric loads are

$$\frac{dN_x}{dx} = 0 \qquad (4.125)$$

$$\frac{dQ_x}{dx} - \frac{N_\theta}{R} + p(x) = 0 \qquad (4.126)$$

$$\frac{dM_x}{dx} - Q_x = 0 \qquad (4.127)$$

$$\beta_x + \frac{dw}{dx} = 0 \qquad (4.128)$$

$$N_x = K\left[\frac{du_0}{dx} + \frac{\nu}{R} w\right] \qquad (4.129)$$

$$N_\theta = K\left[\nu \frac{du_0}{dx} + \frac{w}{R}\right] = \nu N_x + \frac{Ehw}{R} \qquad (4.130)$$

$$M_x = -D\frac{d^2 w}{dx^2} \qquad (4.131)$$

$$M_\theta = -D\nu\frac{d^2 w}{dx^2} = \nu M_x \qquad (4.132)$$

$$Q_x = -D\frac{d^3 w}{dx^3} = V_x \qquad (4.133)$$

It is seen from (4.125) that N_x is a constant everywhere in the shell and is uniquely determined by its value at the boundaries.

The nine governing equations, (4.125) to (4.133), can be reduced to two for this axially symmetric case, namely,

$$\frac{d^2 u_0}{dx^2} + \frac{\nu}{R}\frac{dw}{dx} = 0 \qquad (4.134)$$

$$\frac{d^4 w}{dx^4} + \frac{1}{k^2 R^4}w + \frac{\nu}{k^2 R^3}\frac{du_0}{dx} = \frac{p(x)}{D} \quad \left(\text{where } k^2 = \frac{h^2}{12R^2}\right) \quad (4.135)$$

Solving (4.129) for du_0/dx, (4.135) can be written as

$$\frac{d^4 w}{dx^4} + \frac{(1-\nu^2)}{k^2 R^4}w = \frac{1}{D}\left[p - \frac{\nu N_x}{R}\right] \qquad (4.136)$$

Substituting in the value for k^2, (4.136) can be written finally as

$$\frac{d^4 w}{dx^4} + 4\epsilon^4 w = \frac{1}{D}\left[p - \frac{\nu N_x}{R}\right] \qquad (4.137)$$

where

$$\epsilon^4 = \frac{3(1 - \nu^2)}{h^2 R^2} \tag{4.138}$$

The form of the governing equation (4.137) is desirable since it is uncoupled from the other governing equation (4.134). N_x is a constant determined by boundary conditions. In fact it is seen lucidly that the presence of an axial in-plane force is that of an equivalent lateral pressure as far as the lateral displacement w is concerned.

It is also noted that the governing differential equation for the lateral deflection of a circular cylindrical shell has the same form as the governing differential equation for the lateral deflection of a beam on an elastic foundation; it would be identical if D were replaced by EI, and $4\epsilon^4 D$ replaced by k, the foundation modulus. Thus, one may use the physical intuition, as well as the known solutions for beams or elastic foundation in considering these shells.

By standard methods, the roots of the fourth-order equation (4.137) are determined to be $\pm \epsilon(1 \pm i)$. Thus the general solution can be written in the form

$$w(x) = A\, e^{-\epsilon x} \cos \epsilon x + B\, e^{-\epsilon x} \sin \epsilon x + C\, e^{\epsilon x} \cos \epsilon x +$$
$$+ E\, e^{\epsilon x} \sin \epsilon x + w_p(x) \tag{4.139}$$

where A, B, C, and E are constants of integration determined by the boundary conditions, and $w_p(x)$ is the particular integral. The in-plane displacement u_0 can be determined by the first integral of (4.134), and is seen to be

$$u_0(x) = \frac{N_x x}{K} - \frac{\nu}{R} \int w\, dx + F \tag{4.140}$$

where F is a constant of integration.

It is seen that for the case of circular cylindrical shells under axially symmetric loads there are six boundary conditions: four dealing with specifications of the lateral deflection, slope, stress couple, or shear resultant (w or its derivatives); the fifth is N_x, the in-plane stress resultant which is determined at the outset by external equilibrium, and the sixth (F) is determined by the specification of the in-plane displacement at some axial location.

4.12.1   Edge Load Solutions

In the following, only the solutions for the lateral deflections are explicitly determined. The in-plane displacement u_0 can be determined easily from (4.140).

A Semi-Infinite Shell $(0 \le x \le \infty)$ *Subjected to an Edge Moment* $M_x = M_0$ *at* $x = 0$

Since in this case $p(x) = N_x = 0$, only the homogeneous portion of the general solution, (4.139), is needed. The boundary conditions at $x = 0$ are

$$M_x(0) = M_0 = -D \frac{d^2 w(0)}{dx^2} \tag{4.141}$$

$$Q_x(0) = 0 = -D \frac{d^3 w(0)}{dx^3} \tag{4.142}$$

Since we are dealing with small displacements, using linear theory, it is seen that for w to remain finite as $x \to \infty$, it is required that $C = E = 0$. Hence, we have

$$w(x) = A\, e^{-\epsilon x} \cos \epsilon x + B\, e^{-\epsilon x} \sin \epsilon x \tag{4.143}$$

Substituting (4.143) into (4.141) and (4.142), we obtain

$$w''(0) = -2\epsilon^2 B = -\frac{M_0}{D}$$

$$w'''(0) = 2\epsilon^3 (A + B) = 0$$

where primes denote differentiation with respect to x. Thus, $B = -A = M_0/2\epsilon^2 D$, and the solution is

$$w(x) = \frac{M_0}{2\epsilon^2 D}\, e^{-\epsilon x} (\sin \epsilon x - \cos \epsilon x) \tag{4.144}$$

A Semi-Infinite Shell $(0 \le x \le \infty)$ *Subjected to an Edge Shear* $Q_x = Q_0$ *at* $x = 0$

Here the boundary conditions are

$$M_x(0) = 0 = -Dw'' \tag{4.145}$$

$$Q_x(0) = Q_0 = -Dw''' \tag{4.146}$$

Again it is required that $C = E = 0$, and the solution is given by (4.143). Substituting (4.143) into (4.145) and (4.146) results in

$$2\epsilon^2 B = 0$$

$$2\epsilon^3 (A + B) = -\frac{Q_0}{D}$$

The solution is therefore

$$w(x) = -\frac{Q_0}{2\epsilon^3 D}\, e^{-\epsilon x} \cos \epsilon x \tag{4.147}$$

Edge Load Solutions as the Homogeneous Solution

It is seen that the solutions of (4.144) and (4.147) exhibit the same form: a constant times an oscillating (trigonometric) factor and a factor which exhibits an exponential decay away from the edge of the shell. This decay in the lateral deflection due to a stress couple or a transverse shear resultant is one of the characteristics of shells in general, and is one of the most important features of shell behaviour. Since the slope, bending moment and shear resultant away from the edge are all proportional to the derivatives of the lateral deflection, each of these also decays away from the edge where the edge load is acting. This characteristic is called the 'bending boundary layer', and it is seen that bending and shear stresses due to the edge load occur only in this bending boundary layer.

Now the lateral deflection, slope, bending moment, and transverse shear resultant all decay as $e^{-\epsilon x}$ where $\epsilon = [3(1 - \nu^2)]^{1/4}/\sqrt{(Rh)}$. Hence $x/\sqrt{(Rh)}$ is a fundamentally important parameter with regard to shell behaviour. It is seen that for $x/\sqrt{(Rh)} \geq 4$, $\epsilon x \to 5\cdot15$, where $\nu = 0\cdot3$ for example. This then means that $e^{-\epsilon x} \leq 0\cdot006$, and therefore the lateral deflection, slope, moment, and shear are negligibly small. Therefore the length L_B of the bending boundary layer is taken to be $L_B = 4\sqrt{(Rh)}$.[†]

This suggests a very useful solution technique for shell problems. Consider a finite length shell subjected to some axially symmetric loading and some set of stated boundary conditions. Instead of satisfying the boundary conditions directly through obtaining values of A, B, C and E in (4.139), determine values of unknown edge loads, used as a form of the homogeneous solution, to satisfy the stated boundary conditions. The advantage of this method is that the effects of the particular boundary conditions on w and its derivatives vanishes past $x = L_B$ from the edge. Further from the edge only the particular solution will contribute to the lateral deflection, slope, bending moments, and transverse shear. Hence, if the length of the shell L is $L \geq L_B$, the boundary conditions involving w at one end of the shell are uncoupled from those at the other end. Mathematically, this means that for a shell of $L \geq L_B$, instead of solving a 4×4 matrix[†] to obtain the boundary conditions, one solves two 2×2 matrices.

†It can be shown that this can be generalised to any shell of revolution under axially symmetric edge load by stating that $L_B = \sqrt{(R_\theta h)}$, where R_θ is the circumferential radius of curvature at the edge.

‡Actually a 6×6 matrix for a general shell of revolution under axially symmetric load,

Using the edge load form of the solution for the homogeneous solution for a shell of length L, it is written as

$$w(x) = \frac{M_0}{2\epsilon^2 D} e^{-\epsilon x} (\sin \epsilon x - \cos \epsilon x) - \frac{Q_0}{2\epsilon^3 D} e^{-\epsilon x} \cos \epsilon x +$$

$$+ \frac{M_L}{2\epsilon^2 D} e^{-\epsilon(L-x)} [\sin \epsilon (L-x) - \cos \epsilon (L-x)] +$$

$$+ \frac{Q_L}{2\epsilon^3 D} e^{-\epsilon(L-x)} \cos \epsilon (L-x) \qquad (4.148)$$

where the edge loads are considered positive as shown in Fig. 4.13.

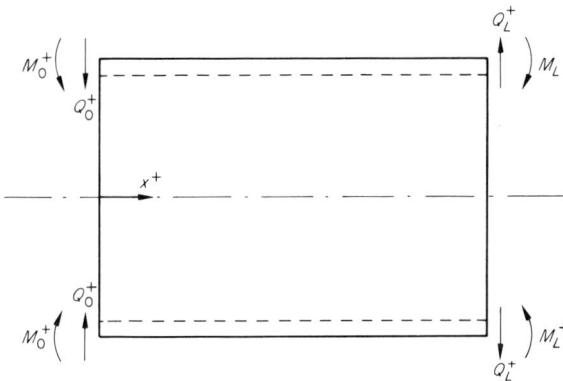

Fig. 4.13

It should be noted from (4.140) that only some of the terms in the $u_0(x)$ equation decay away from the edges, namely the second term as written. The others involve one that increases monotonically in x, while F is a constant. At any rate, away from the bending boundary layer at each edge, the expression for u_0 is simplified.

Even when the length of the shell is so short that $L < L_B$, in which case there is no separation of the boundary conditions, there is no practical advantage in using the form of the homogeneous solution given by (4.139) compared with the form of (4.148). In this case M_0, Q_0, M_L and Q_L would not be the values at edge loads but merely constants to satisfy boundary conditions.

---

but for cylindrical shells, N_x is found by a trivial force balance and the boundary constant F in the $u_0(x)$ equation is always found subsequent to finding the four boundary constants involving w and its derivatives.

4.12.2 A General Solution for Cylindrical Shells Under Axially Symmetric Loads

For reference in solving problems of this kind, all equations needed are catalogued below (see Fig. 4.13). It is restricted to cases where $d^4 p(x)/dx^4 = 0$. Since, from a practical point of view this would almost always be true, the solutions presented are fairly general.

$$w(x) = \frac{M_0}{2\epsilon^2 D} e^{-\epsilon x}(\sin \epsilon x - \cos \epsilon x) - \frac{Q_0}{2\epsilon^3 D} e^{-\epsilon x} \cos \epsilon x$$

$$+ \frac{M_L}{2\epsilon^2 D} e^{-\epsilon(L-x)}[\sin \epsilon(L-x) - \cos \epsilon(L-x)]$$

$$+ \frac{Q_L}{2\epsilon^3 D} e^{-\epsilon(L-x)} \cos \epsilon(L-x) + \frac{1}{4\epsilon^4 D}\left[p(x) - \frac{\nu N_x}{R}\right] \quad (4.149)$$

where $\epsilon = [3(1 - \nu^2)]^{1/4}/\sqrt{(Rh)}$.

$$w'(x) = \frac{dw}{dx} = \frac{M_0}{\epsilon D} e^{-\epsilon x} \cos \epsilon x + \frac{Q_0}{2\epsilon^2 D} e^{-\epsilon x}(\sin \epsilon x + \cos \epsilon x)$$

$$- \frac{M_L}{\epsilon D} e^{-\epsilon(L-x)} \cos \epsilon(L-x)$$

$$+ \frac{Q_L}{2\epsilon^2 D} e^{-\epsilon(L-x)}[\sin \epsilon(L-x) + \cos \epsilon(L-x)] + \frac{1}{4\epsilon^4 D} p'(x)$$

$$(4.150)$$

$$M_x(x) = -Dw''(x) = M_0 e^{-\epsilon x}(\sin \epsilon x + \cos \epsilon x) + \frac{Q_0}{\epsilon} e^{-\epsilon x} \sin \epsilon x$$

$$+ M_L e^{-\epsilon(L-x)}[\sin \epsilon(L-x) + \cos \epsilon(L-x)]$$

$$- \frac{Q_L}{\epsilon} e^{-\epsilon(L-x)} \sin \epsilon(L-x) - \frac{1}{4\epsilon^4} p''(x) \quad (4.151)$$

$$Q_x = -Dw'''(x) = -2M_0\epsilon\, e^{-\epsilon x} \sin \epsilon x + Q_0 e^{-\epsilon x}(\cos \epsilon x - \sin \epsilon x)$$

$$+ 2M_L\epsilon\, e^{-\epsilon(L-x)} \sin \epsilon(L-x)$$

$$- Q_L e^{-\epsilon(L-x)}[-\cos \epsilon(L-x) + \sin \epsilon(L-x)] - \frac{1}{4\epsilon^4} p'''(x)$$

$$(4.152)$$

$$u_0(x) = \left[\frac{1}{K} + \frac{\nu^2}{4R^2\epsilon^4 D}\right] N_x x - \frac{\nu}{R}\left\{-\frac{M_0}{2\epsilon^3 D} e^{-\epsilon x} \sin \epsilon x\right.$$

$$+ \frac{Q_0}{4\epsilon^4 D} e^{-\epsilon x}(\cos \epsilon x - \sin \epsilon x) + \frac{M_L}{2\epsilon^3 D} e^{-\epsilon(L-x)} \sin \epsilon(L-x)$$

$$+ \frac{Q_L}{4\epsilon^4 D} e^{-\epsilon(L-x)}[\cos \epsilon(L-x) - \sin \epsilon(L-x)]\bigg\}$$

$$- \frac{\nu}{4R\epsilon^4 D} \int p(x) \, dx + F \tag{4.153}$$

$$N_x = \text{constant} \tag{4.154}$$

$$N_\theta(x) = \nu N_x + \frac{Ehw(x)}{R} \tag{4.155}$$

$$M_\theta(x) = \nu M_x(x) \tag{4.156}$$

$$\sigma_x = \frac{N_x}{h} + \frac{M_x z}{h^3/12} \tag{4.157}$$

$$\sigma_\theta = \frac{N_\theta}{h} + \frac{M_\theta z}{h^3/12} = \nu \sigma_x + \frac{Ew(x)}{R} \tag{4.158}$$

$$\sigma_{xz} = \frac{3Q_x}{2h}\left[1 - \left(\frac{z}{h/2}\right)^2\right] \tag{4.159}$$

where E is the modulus of elasticity in (4.155) and (4.158).

4.12.3 Sample Solutions

Effects of Simple and Clamped Supports
Consider the circular cylindrical shell shown in Fig. 4.14. The end of the shell $x = 0$ is simply supported, the end $x = L$ is clamped. The plate ends of the shell are assumed rigid. The internal pressure is p_0, and $\nu = 0.3$. From external axial force equilibrium, $N_x = p_0 R/2$. The boundary conditions at $x = 0$ are $w(0) = 0$, and $M_x(0) = 0$. From (4.149)

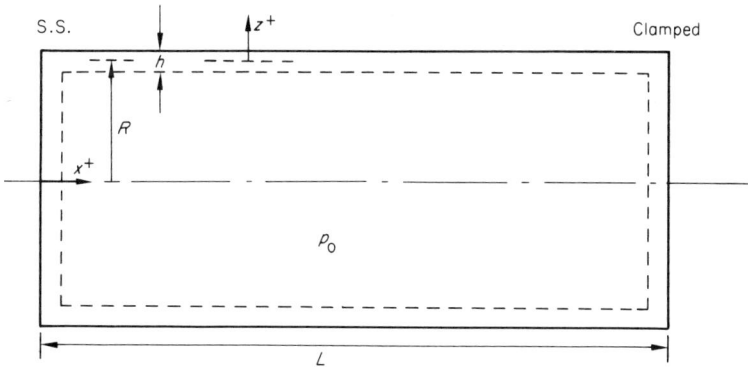

Fig. 4.14

and (4.151), the boundary conditions are

$$w(0) = -\frac{M_0}{2\epsilon^2 D} - \frac{Q_0}{2\epsilon^3 D} + \frac{1}{4\epsilon^4 D} p_0 \left(1 - \frac{\nu}{2}\right)$$

$$M_x(0) = M_0 = 0 \tag{4.160}$$

Hence,

$$Q_0 = \frac{p_0}{2\epsilon}(1 - \nu/2) \tag{4.161}$$

To determine the location of the maximum value of σ_x, it is required to determine the locations of the maximum value of $M(x)$, since N_x is a constant. Since M_x is zero at $x = 0$, and again for $x \geqslant L_B$, an extremum value lies somewhere in the bending boundary layer. It occurs at a value of x when $dM_x/dx = Q_x = 0$ hence, since

$$Q_x = \frac{p_0}{2\epsilon}(1 - \nu/2) \, e^{-\epsilon x} (\cos \epsilon x - \sin \epsilon x) = 0$$

for this condition to exist, $\cos \epsilon x = \sin \epsilon x$, or $\epsilon x = \pi/4,\ 5\pi/4$, etc.; and, because of the exponential decay, the maximum stress occurs at

$$x(Q_x = 0) = \frac{\pi}{4\epsilon}$$

and

$$(M_x)_{\max} = M_x\left(\frac{\pi}{4\epsilon}\right) = p_0(1 - \nu/2)\frac{\sqrt{2}}{4\epsilon^2} e^{-\pi/4}$$

From (4.157)

$$(\sigma_x)_{\max} = \frac{N_x}{h} \pm \frac{6(M_x)_{\max}}{h^2}$$

$$= \frac{p_0 R}{2h} \pm 0{\cdot}498\frac{p_0 R}{h} \qquad (\nu = 0{\cdot}3)$$

$$(\sigma_x)_{\max} = \sigma_x\left(\frac{\pi}{4\epsilon}, +h/2\right) = 0{\cdot}998\frac{p_0 R}{h} \tag{4.162}$$

From (4.160), (4.161), (4.149), and (4.151)

$$w(x) = \frac{p_0(1 - \nu/2)}{4\epsilon^4 D}[1 - e^{-\epsilon x} \cos \epsilon x] \qquad \text{for } x \leqslant L_B \tag{4.163}$$

$$M(x) = \frac{p_0(1 - \nu/2)}{2\epsilon^2} e^{-\epsilon x} \sin \epsilon x \qquad \text{for } x \leqslant L_B \tag{4.164}$$

Therefore from (4.158), (4.163), and (4.164).

$$\sigma_\theta = \frac{p_0 R}{h} \pm \frac{3\nu(1 - \nu/2)Rp_0\, e^{-\epsilon x}\, \sin \epsilon x}{h[3(1 - \nu^2)]^{1/2}} -$$

$$- \frac{p_0(1 - \nu/2)R\, e^{-\epsilon x}\, \cos \epsilon x}{h} \qquad \text{for } z = \pm h/2 \qquad (4.165)$$

For $\nu = 0\cdot3$, this reduces to

$$\sigma_\theta = \frac{p_0 R}{h}(1 \pm 0\cdot464\, e^{-\epsilon x}\, \sin \epsilon x - 0\cdot85\, e^{-\epsilon x}\, \cos \epsilon x)$$

$$\text{for } x \leq L_B, z = \pm h/2$$

Extreme values occur for the condition $\partial\sigma_\theta/\partial x = 0$, which results in the requirement that

$$\pm 0\cdot464(-\sin \epsilon x + \cos \epsilon x) + 0\cdot85(\sin \epsilon x + \cos \epsilon x) = 0$$

The positive requirement occurs at $\epsilon x = 1\cdot875$; the negative when $\epsilon x = 2\cdot845$. Of these two σ_θ is maximum for the former. The maximum value of σ_θ in the large $x \leq L_B$, is

$$(\sigma_\theta)_{max} = \sigma_\theta\left(\frac{1\cdot875}{\epsilon}, + h/2\right) = 1\cdot1072\frac{p_0 R}{h} \qquad \text{for } \nu = 0\cdot3 \quad (4.166)$$

At the clamped end the boundary conditions are $w(L) = w'(L) = 0$. From (4.149), (4.150).

$$w(L) = 0 = -\frac{M_L}{2\epsilon^2 D} + \frac{Q_L}{2\epsilon^3 D} + \frac{p_0}{4\epsilon^4 D}(1 - \nu/2) = 0$$

$$w'(L) = 0 = -\frac{M_L}{\epsilon D} + \frac{Q_L}{2\epsilon^2 D} = 0$$

Hence

$$M_L = -\frac{p_0(1 - \nu/2)}{2\epsilon^2} \qquad (4.167)$$

$$Q_L = -\frac{p_0(1 - \nu/2)}{\epsilon} \qquad (4.168)$$

It can be shown and is physically obvious that at the clamped end, $(M_x)_{max} = M(L) = M_L$.

Hence at $x = L$,

$$\sigma_x = \frac{N_x}{h} + \frac{M_L z}{h^3/12} = \frac{p_0 R}{2h} - \frac{p_0(1 - \nu/2)z}{2\epsilon^2 h^3/12}$$

$$(\sigma_x)_{max} = \sigma_x(L, -h/2) = 2\cdot04\frac{p_0 R}{h} \qquad \text{for } \nu = 0\cdot3 \qquad (4.169)$$

However, $(\sigma_\theta)_{max}$ occurs away from the end of the shell. Analogous to the procedures used at the other end, it is found that

$$(\sigma_\theta)_{max} = \sigma_\theta\left(L - \frac{2\cdot65}{\epsilon}, +h/2\right) = 1\cdot069\frac{p_0R}{h} \qquad (4.170)$$

At $x = L/2$, which is outside the bending boundary layer, it is seen from (4.151) that $M_x = 0$, hence

$$\sigma_x = \frac{N_x}{h} = \frac{p_0R}{2h} \qquad (4.171)$$

which is the membrane solution. Likewise, from (4.158)

$$\sigma_\theta = \nu\sigma_x + \frac{Ew}{R} = \frac{p_0R}{h} \qquad (4.172)$$

which is the membrane solution.

It is seen that the maximum principal stress occurring in the boundary layer at the simply supported end is, from (4.166) $1\cdot1072(p_0R/h)$; correspondingly, at the clamped end it is $2\cdot04(p_0R/h)$ (from (4.169)). The maximum stress predicted by membrane theory is $1\cdot0(p_0R/h)$. Hence, stresses greater than membrane stresses occur in both boundary layers; 10 per cent higher in the simply supported area, and 104 per cent higher in the clamped edge. Thus this shell if designed on a basis of membrane shell theory would be woefully inadequate.

REFERENCES

1. Sokolnikoff, I. S., *Mathematical Theory of Elasticity*, McGraw-Hill Book Company, Inc., New York (1956).
2. Timoshenko, S. and Woinowsky-Krieger, A., *Theory of Plates and Shells*, McGraw-Hill Book Company, Inc., New York (1959).
3. Marguerre, K. and Woernle, H. T., *Elastic Plates*, Blaisdell Publishing Company (1969).
4. Vinson, J. R., *Structural Mechanics: The Behavior of Plates and Shells*, Wiley–Interscience, John Wiley and Sons, Inc., New York (1974).
5. Timoshenko, S. and Gere, J., *Theory of Elastic Stability*, McGraw Hill Book Co., Inc., New York (1961).
6. Bleich, F., *Buckling Strength of Metal Structures*, McGraw-Hill Book Co., Inc., New York (1952).

Note: For problems and answers covering this chapter, see Reference 4 above.

CHAPTER 5

ANISOTROPIC ELASTICITY

In the previous chapter all discussion applied to isotropic materials only. Isotropic materials are those which have identical physical, thermal, and electrical properties in any direction. Other materials are termed anisotropic. In the real world, to assume any material is isotropic is at best an approximation. For instance in polycrystalline metals, the structure is usually made up of a large number of anisotropic grains, whereas isotropy exists only because the grains are randomly oriented. However, these materials can also become macroscopically anisotropic due to cold-working, forging, or spinning during fabrication. Other materials such as wood are naturally anisotropic, the properties being quite different in the direction of the grain compared to those normal to the grain.

Today, composite materials are coming into prominence. Fibrereinforced composite materials are uniquely useful because they can be constructed in such a manner that they have higher strength-to-density ratios or stiffness-to-density ratios of any materials at moderate temperatures, and the fibres can be oriented to provide minimum structural weight for a given structural geometry and given system of loads. The materials science aspects of these materials are discussed in Chapters 2 and 3.

In the structural analysis of plates and shells of composite materials, the equilibrium equations and the strain–displacement relations do not change from those for plates and shells of isotropic materials discussed in Chapter 4. However, it is necessary to drastically alter the stress–strain relations to account for the anisotropy of the materials system. This is the purpose of this chapter.

201

5.1 DERIVATION OF THE ORTHOTROPIC ELASTICITY TENSOR

The following derivation of the stress–strain relations of an orthotropic material has been developed by Sokolnikoff[1] and others, but, for clarification, is included here in some detail.

The general elasticity equation can be written as (see page 158)

$$\sigma_{ij} = c_{ijkl}\epsilon_{kl} \qquad (i, j, k, l = 1, 2, 3) \qquad (5.1)$$

where σ_{ij} is the second-order stress tensor, ϵ_{kl} is the second-order strain tensor, c_{ijkl} is the fourth-order elasticity tensor, and the 1, 2 and 3 directions form a right-handed orthogonal co-ordinate system. The stress and strain tensors are each symmetric, that is to say $\sigma_{ij} = \sigma_{ji}$ and $\epsilon_{kl} = \epsilon_{lk}$ (see Chapters 1 and 2 of Reference 1). The following short-hand notation may be used for the stress and strain components

$$
\begin{aligned}
\sigma_{11} &= \sigma_1; & \sigma_{22} &= \sigma_2; & \sigma_{33} &= \sigma_3 \\
\epsilon_{11} &= \epsilon_1; & \epsilon_{22} &= \epsilon_2; & \epsilon_{33} &= \epsilon_3 \\
\sigma_{23} &= \sigma_4; & \sigma_{31} &= \sigma_5; & \sigma_{12} &= \sigma_6 \\
2\epsilon_{23} &= \epsilon_4; & 2\epsilon_{31} &= \epsilon_5; & 2\epsilon_{12} &= \epsilon_6
\end{aligned}
\qquad (5.2)
$$

The general equations (5.1) may then be written as

$$
\left.
\begin{aligned}
\sigma_1 &= c_{11}\epsilon_1 + c_{12}\epsilon_2 + c_{13}\epsilon_3 + c_{14}\epsilon_4 + c_{15}\epsilon_5 + c_{16}\epsilon_6 \\
&\ \vdots \\
\sigma_6 &= c_{61}\epsilon_1 + c_{62}\epsilon_2 + c_{63}\epsilon_3 + c_{64}\epsilon_4 + c_{65}\epsilon_5 + c_{66}\epsilon_6
\end{aligned}
\right\}
\qquad (5.3)
$$

It is then seen that the symmetry of the stress and strain tensor reduces the number of components of the general elasticity tensor in three-dimensional space from 81 to 36 components.

If the strain energy density function, W, exists, where (see eqn. 4.103)

$$W = \tfrac{1}{2}\sigma_{ij}\epsilon_{ij}$$

such that

$$\frac{\partial W}{\partial \epsilon_{ij}} = c_{ijkl}\epsilon_{kl} = \sigma_{ij} \qquad (5.4)$$

then the independent components of the elasticity tensor are reduced to 21, since eqn. (5.4) requires that $c_{ijkl} = c_{klij}$, or in the shorthand notation, $c_{ij} = c_{ji}$.

In the following, a Cartesian co-ordinate system will be employed for simplicity. However, the results are applicable to any orthogonal curvilinear co-ordinate system.

Consider an elastic body symmetric in properties with respect to one plane, say the x_1–x_2 plane. This symmetry can be expressed by the statement that the Q_{ij}'s are invariant under the transformation $x_1 = x_1'$, $x_2 = x_2'$, and $x_3 = -x_3'$, as shown in Fig. 5.1, together with a table giving the direction cosines, denoted by t, for this transformation. The stresses and strains of the prime co-ordinate systems are related to those of the original co-ordinate system by

$$\epsilon'_{\alpha\beta} = t_{\alpha i} t_{\beta j} \epsilon_{ij} \quad \text{and} \quad \sigma'_{\alpha\beta} = t_{\alpha i} t_{\beta j} \sigma_{ij}$$

Therefore for $i = 1, 2, 3, 6$, $\sigma'_i = \sigma_i$ and $\epsilon'_i = \epsilon_i$. However, $\epsilon'_{23} = t_{22} t_{33} \epsilon_{23} = -\epsilon_{23}$. Therefore, $\epsilon'_4 = -\epsilon_4$. Similarly $\sigma'_4 = -\sigma_4$. Likewise $\epsilon'_{31} = t_{33} t_{11} \epsilon_{31} = -\epsilon_{31}$. Therefore, $\epsilon'_5 = -\epsilon_5$ and also $\sigma'_5 = -\sigma_5$.

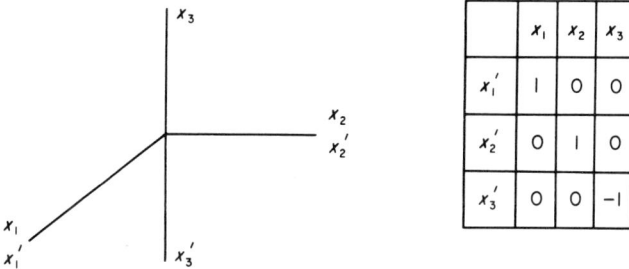

	x_1	x_2	x_3
x_1'	1	0	0
x_2'	0	1	0
x_3'	0	0	-1

Fig. 5.1

Turning now to the first of the equations (5.3), it can be written both in terms of the original co-ordinate system and the transformed (primed) co-ordinate systems, as follows

$$\sigma_1 = c_{11}\epsilon_1 + c_{12}\epsilon_2 + c_{13}\epsilon_3 + c_{14}\epsilon_4 + c_{15}\epsilon_5 + c_{16}\epsilon_6$$

$$\sigma_1' = c_{11}\epsilon_1' + c_{12}\epsilon_2' + c_{13}\epsilon_3' + c_{14}\epsilon_4' + c_{15}\epsilon_5' + c_{16}\epsilon_6'$$

Since $\epsilon'_4 = -\epsilon_4$ and $\epsilon'_5 = -\epsilon_5$, then for $\sigma'_1 = \sigma_1$, requires that $c_{14} = c_{15} = 0$. By similarly considering the other equations of (5.3) such as $\sigma'_2 = \sigma_2$, $\sigma'_3 = \sigma_3$, $\sigma'_4 = -\sigma_4$, $\sigma'_5 = -\sigma_5$ and $\sigma'_6 = \sigma_6$ it is found that

$$c_{24} = c_{25} = c_{34} = c_{35} = c_{64} = c_{65} = 0$$

Remember that $c_{ij} = c_{ji}$. Hence, for materials having *one* plane of symmetry the elasticity tensor has the following array, which involves 13 independent components.

$$c_{ij} = \begin{bmatrix} c_{11} & c_{12} & c_{13} & 0 & 0 & c_{16} \\ c_{21} & c_{22} & c_{23} & 0 & 0 & c_{26} \\ c_{31} & c_{32} & c_{33} & 0 & 0 & c_{36} \\ 0 & 0 & 0 & c_{44} & c_{45} & 0 \\ 0 & 0 & 0 & c_{54} & c_{55} & 0 \\ c_{61} & c_{62} & c_{63} & 0 & 0 & c_{66} \end{bmatrix} \tag{5.5}$$

Materials which have three mutually orthogonal planes of elastic symmetry are called *orthotropic* (a shortened term for orthogonally anisotropic). To consider this class of materials the material must be symmetric also with respect to the x_1–x_3 and the x_2–x_3 plane. Proceeding as before, the diagrams shown in Fig. 5.2 can be made for symmetry with respect to the x_2–x_3 plane

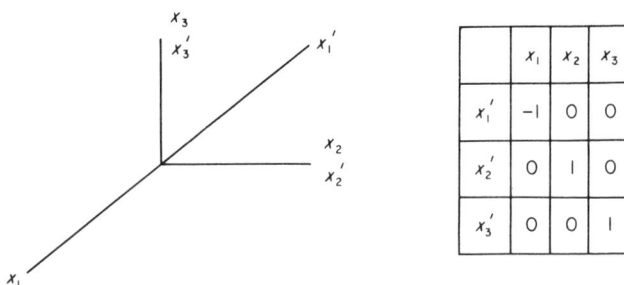

	x_1	x_2	x_3
x_1'	-1	0	0
x_2'	0	1	0
x_3'	0	0	1

Fig. 5.2

Carrying through this transformation, it is found that $c_{16} = c_{26} = c_{36} = c_{45} = 0$ from the x_2–x_3 plane symmetry. Employing symmetry with respect to the x_1–x_3 plane results in these constants being zero also. Therefore for *orthotropic materials* the elasticity tensor is written as follows

$$c_{ij} = \begin{bmatrix} c_{11} & c_{12} & c_{13} & 0 & 0 & 0 \\ c_{21} & c_{22} & c_{23} & 0 & 0 & 0 \\ c_{31} & c_{32} & c_{33} & 0 & 0 & 0 \\ 0 & 0 & 0 & c_{44} & 0 & 0 \\ 0 & 0 & 0 & 0 & c_{55} & 0 \\ 0 & 0 & 0 & 0 & 0 & c_{66} \end{bmatrix} \tag{5.6}$$

where it is seen that for an orthotropic, three-dimensional elastic body there are nine independent components. Hence nine independent elastic properties must be determined.

A *transversely isotropic* material is one which also has three mutually orthogonal planes of elastic symmetry just as an orthotropic material, but in addition is isotropic with respect to one of the three planes. For example if a material were isotropic in the x_1–x_2 plane, $c_{11} = c_{22}$, $c_{13} = c_{23}$ and $c_{44} = c_{55}$, and the transversely isotropic elasticity tensor is written in matrix form as

$$c_{ij} = \begin{bmatrix} c_{11} & c_{12} & c_{13} & 0 & 0 & 0 \\ c_{21} & c_{11} & c_{13} & 0 & 0 & 0 \\ c_{31} & c_{31} & c_{33} & 0 & 0 & 0 \\ 0 & 0 & 0 & c_{44} & 0 & 0 \\ 0 & 0 & 0 & 0 & c_{44} & 0 \\ 0 & 0 & 0 & 0 & 0 & c_{66} \end{bmatrix} \tag{5.7}$$

where the plane of isotropy is the x_1–x_2 plane. Here, there are six independent components, hence six independent elastic properties are required for a three-dimensional transversely isotropic material. Later it will be shown there are really only five.

Returning to the orthotropic elasticity tensor of eqn. (5.6), stress–strain relations for a three-dimensional orthotropic body can be written as

$$\sigma_1 = c_{11}\epsilon_1 + c_{12}\epsilon_2 + c_{13}\epsilon_3$$

$$\sigma_2 = c_{21}\epsilon_1 + c_{22}\epsilon_2 + c_{23}\epsilon_3$$

$$\sigma_3 = c_{31}\epsilon_1 + c_{32}\epsilon_2 + c_{33}\epsilon_3$$

$$\sigma_4 = \sigma_{23} = c_{44}\epsilon_4 = 2c_{44}\epsilon_{23} \tag{5.8}$$

$$\sigma_5 = \sigma_{31} = c_{55}\epsilon_5 = 2c_{55}\epsilon_{31}$$

$$\sigma_6 = \sigma_{12} = c_{66}\epsilon_6 = 2c_{66}\epsilon_{12}$$

The orthotropic elasticity tensor matrix, eqn. (5.6), can be inverted in order to obtain the strains in terms of the stresses as follows

$$\epsilon_1 = a_{11}\sigma_1 + a_{12}\sigma_2 + a_{13}\sigma_3$$

$$\epsilon_2 = a_{21}\sigma_1 + a_{22}\sigma_2 + a_{23}\sigma_3$$

$$\epsilon_3 = a_{31}\sigma_1 + a_{32}\sigma_2 + a_{33}\sigma_3$$

$$\epsilon_4 = 2\epsilon_{23} = a_{44}\sigma_{23} = a_{44}\sigma_4 \tag{5.9}$$

$$\epsilon_5 = 2\epsilon_{31} = a_{55}\sigma_{31} = a_{55}\sigma_5$$

$$\epsilon_6 = 2\epsilon_{12} = a_{66}\sigma_{12} = a_{66}\sigma_6$$

where the matrix of the a_{ij}'s, sometimes called the compliance matrix, is the inverse of the matrix of the c_{ij}'s and is found as follows

$$a_{ij} = \frac{(\text{Co } c_{ij})^T}{|c_{ij}|}$$

In other words, the a_{ij} matrix is the transpose of the cofactor matrix of c_{ij}'s divided by the determinant of the a_{ij} matrix. It is easily shown that $a_{ij} = a_{ji}$. The a_{ij} matrix is of course given by

$$a_{ij} = \begin{bmatrix} a_{11} & a_{12} & a_{13} & 0 & 0 & 0 \\ a_{21} & a_{22} & a_{23} & 0 & 0 & 0 \\ a_{31} & a_{32} & a_{33} & 0 & 0 & 0 \\ 0 & 0 & 0 & a_{44} & 0 & 0 \\ 0 & 0 & 0 & 0 & a_{55} & 0 \\ 0 & 0 & 0 & 0 & 0 & a_{66} \end{bmatrix} \tag{5.10}$$

where

$$\epsilon_i = a_{ij}\sigma_j \qquad [i, j = 1,2,\ldots,6] \tag{5.11}$$

5.2   THE PHYSICAL MEANING OF THE ELEMENTS OF THE ORTHOTROPIC ELASTICITY TENSOR

In the following, several 'mathematical' experiments are performed, that is to say simple tensile tests and pure shear tests, to relate the elements, a_{ij}, to the elastic constants which are familiar to those who are familiar with elementary strength of materials.

Consider a simple tensile test wherein a tensile test specimen is stressed in the x_1 direction. The resulting stress and strain tensors associated with this standard test are

$$\sigma_{ij} = \begin{Vmatrix} \sigma_1 & 0 & 0 \\ 0 & 0 & 0 \\ 0 & 0 & 0 \end{Vmatrix}, \qquad \epsilon_{ij} = \begin{Vmatrix} \epsilon_1 & 0 & 0 \\ 0 & -\nu_{12}\epsilon_1 & 0 \\ 0 & 0 & -\nu_{13}\epsilon_1 \end{Vmatrix}$$

where ν_{ij} is defined as the negative of the ratio of the strain in the x_j direction to the strain in the x_i direction due to a stress in the x_i direction. In other words from the above, $\epsilon_{22} = -\nu_{12}\epsilon_{11}$, or $\nu_{12} = -\epsilon_{22}/\epsilon_{11}$. Since in most materials ϵ_{22} would be negative (i.e. a contraction in the x_2 direction, when a tensile stress is applied in the x_1 direction), ν_{12} would usually be a positive number. The ν_{ij}'s are termed Poisson's ratios. From eqn. (5.9) and the above characteristics of the hypothetical tensile test, and the knowledge of elementary strength of materials, such that the constant of proportionality between stress and strain for a tensile test in the x_1 direction is E_1, called the modulus of elasticity in the x_1 direction, it is seen that

$$\epsilon_1 = a_{11}\sigma_1 = \frac{\sigma_1}{E_1}$$

$$\epsilon_2 = a_{21}\sigma_1 = -\nu_{12}\epsilon_1 = -\frac{\nu_{12}}{E_1}\sigma_1$$

$$\epsilon_3 = a_{31}\sigma_1 = -\nu_{13}\epsilon_1 = -\frac{\nu_{13}}{E_1}\sigma_1$$

Therefore

$$a_{11} = \frac{1}{E_1} \tag{5.12}$$

$$a_{21} = -\frac{\nu_{12}}{E_1} \tag{5.13}$$

$$a_{31} = -\frac{\nu_{13}}{E_1} \tag{5.14}$$

A simple tensile test in the x_2 direction yields the following

$$\sigma_{ij} = \begin{Vmatrix} 0 & 0 & 0 \\ 0 & \sigma_2 & 0 \\ 0 & 0 & 0 \end{Vmatrix}, \qquad \epsilon_{ij} = \begin{Vmatrix} -\nu_{21}\epsilon_2 & 0 & 0 \\ 0 & \epsilon_2 & 0 \\ 0 & 0 & -\nu_{23}\epsilon_2 \end{Vmatrix}$$

$$\epsilon_1 = a_{12}\sigma_2 = -\nu_{21}\epsilon_2 = -\frac{\nu_{21}}{E_2}\sigma_2$$

$$\epsilon_2 = a_{22}\sigma_2 = \frac{\sigma_2}{E_2}$$

$$\epsilon_3 = a_{32}\sigma_2 = -\nu_{23}\epsilon_2 = \frac{\nu_{23}}{E_2}\sigma_2$$

where E_2 is the modulus of elasticity in the x_2 direction.
From the above,

$$a_{12} = -\frac{\nu_{21}}{E_2} \tag{5.15}$$

$$a_{22} = \frac{1}{E_2} \tag{5.16}$$

$$a_{32} = -\frac{\nu_{23}}{E_2} \tag{5.17}$$

Finally, from a tensile test in the x_3 direction, the results are

$$a_{13} = -\frac{\nu_{31}}{E_3} \tag{5.18}$$

$$a_{23} = -\frac{\nu_{32}}{E_3} \tag{5.19}$$

$$a_{33} = \frac{1}{E_3} \tag{5.20}$$

Note that from the above because $a_{ij} = a_{ji}$

$$\frac{\nu_{ij}}{E_i} = \frac{\nu_{ji}}{E_j} \qquad (i, j = 1, 2, 3) \tag{5.21}$$

which is a most important relationship, because having physically measured any three of the four quantities in eqn. (5.21) the fourth is easily calculated.

We now hypothetically consider a simple shear test as depicted below

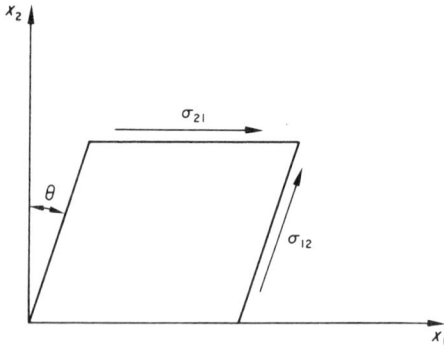

For this the stress, strain and displacement tensors are

$$\sigma_{ij} = \begin{Vmatrix} 0 & \sigma_{12} & 0 \\ \sigma_{21} & 0 & 0 \\ 0 & 0 & 0 \end{Vmatrix}, \qquad \epsilon_{ij} = \begin{Vmatrix} 0 & \epsilon_{12} & 0 \\ \epsilon_{21} & 0 & 0 \\ 0 & 0 & 0 \end{Vmatrix}, \qquad u_{i,j} = \theta = \begin{Vmatrix} 0 & 0 & 0 \\ \dfrac{\sigma_{21}}{G_{21}} & 0 & 0 \\ 0 & 0 & 0 \end{Vmatrix}$$

In the above, u is the displacement $U_{i,j} = \partial U_i / \partial x_j$, and from elementary strength of materials the proportionality constant between shear stress, σ_{21}, and the angle θ is G_{21}, the modulus of rigidity or shear modulus in the x_1–x_2-plane.

Now, from the theory of elasticity

$$\epsilon_{12} = \frac{1}{2}(u_{1,2} + u_{2,1}) = \frac{1}{2}\frac{\sigma_{21}}{G_{21}} = \frac{\tan \theta}{2}$$

Hence, using eqn. (5.9)

$$\epsilon_{12} = \frac{\sigma_{21}}{2G_{21}} = \frac{a_{66}}{2}\sigma_{21}$$

or

$$a_{66} = \frac{1}{G_{21}} = \frac{1}{G_{12}} \qquad (5.22)$$

Similarly, for shear tests in the x_2–x_3-plane and the x_3–x_1-plane, the results are

$$a_{44} = \frac{1}{G_{23}}$$

$$a_{55} = \frac{1}{G_{13}} \qquad (5.23)$$

All a_{ij}'s have now been related to physical material constants. The physical quantities appearing in eqns. (5.12) to (5.20) and (5.22) to (5.24) are E_1, E_2, E_3, G_{12}, G_{23}, G_{31}, ν_{12}, ν_{13}, ν_{21}, ν_{23}, ν_{31} and ν_{32}. However, because of eqn. (5.21) the number of physical quantities to be measured for a three-dimensional orthotropic body are reduced from 12 to 9.

Incidentally, if the material were transversely isotropic such that the material were isotropic in the x_1–x_2-plane, the physical quantities involved are $E_1 = E_2$, E_3, G_{12}, $G_{23} = G_{13}$, $\nu_{12} = \nu_{21}$, $\nu_{13} = \nu_{23}$ and $\nu_{32} = \nu_{31}$. However, because of eqn. (5.21), one of these quantities can be calculated, and because of the isotropic relationship, here written as

$$G_{12} = \frac{E}{2(1 + \nu)}$$

where $E = E_1 = E_2$ and $\nu = \nu_{12} = \nu_{21}$, the total unknown physical quantities are five in number.

So far the elasticity tensor quantities, c_{ij} and a_{ij}, are identical to the notation used by Sokolnikoff.[1] Commencing the following section the notation will be altered somewhat to follow that of the composite materials literature, for example Ref. 2. The stiffness matrix components are denoted by Q_{ij}, and the compliance matrix components by S_{ij}. Specifically these quantities are related to the foregoing by the following

$$Q_{ij} = c_{ij} \qquad (i, j = 1, 2, 6)$$

and

$$S_{ij} = a_{ij} \qquad (i, j = 1, 2, 6)$$

$$Q_{ii} = \frac{c_{ii}}{2} \qquad (i = 4, 5) \qquad (5.24)$$

$$Q_{66} = \frac{c_{66}}{2} - \frac{c_{36}^2}{2c_{33}}$$

5.3 LAMINAE OF ORTHOTROPIC MATERIALS: CLASSICAL THEORY[a]

For the analysis of thin-walled elastic bodies, including beams, membranes, plates and shells, one formulates the equilibrium and elasticity equations for a lamina. Consider an element of a lamina of constant thickness, h, lying symmetrically in the $x-y$-plane as shown in Fig. 5.3.

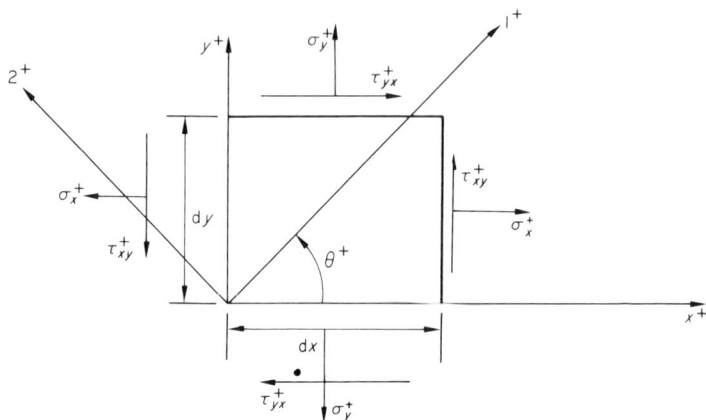

Fig. 5.3

Also shown are the positive directions of all stress components. The $x-y-z$-axes form a right-handed Cartesian co-ordinate system and the $x-y-z$ co-ordinate system is aligned in directions associated with geometry (i.e. in a rectangular plate, they are aligned parallel and perpendicular to the edges, and in a shell of revolution in the directions of principal curvatures). Also shown are the 1-2-3 set of co-ordinate axes, where the 1-2-plane is co-planar with the $x-y$-plane for the present consideration, and also forms a right-handed Cartesian co-ordinate system. In Fig. 5.3, the angle θ^+ is defined as shown, as the angle from the x-axis to the 1-axis rotated counterclockwise when looking from the positive z-axis and the positive 3-axis toward the origin. As it will be seen it is important to remember precisely the definition of θ^+.

[a] Portions of this section are excerpted from Ashton, J. E., Halpin, J. C. and Petit, P. H., *Primer on Composite Materials: Analysis*, Technomic Publishing Co. Reprinted by permission.

For an orthotropic lamina, the 1–2–3-axes form the three orthogonal planes of symmetry, where in the 1–2-plane it is usually customary to make the 1-axis, the one with the larger modulus of elasticity.

Consider the small element taken from the lamina element above, and the equilibrium of forces in the plane (Fig. 5.4).

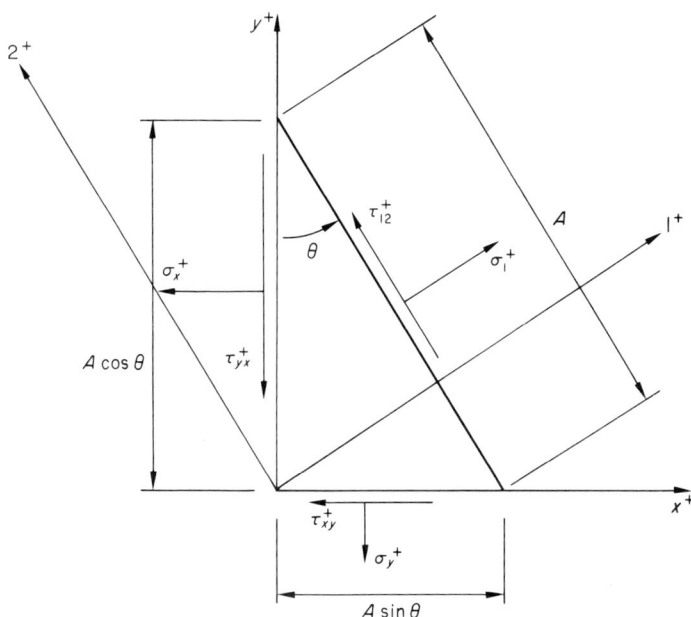

Fig. 5.4

A summation of forces in the 1 direction results in the following equation of equilibrium

$$\sigma_1 = \sigma_x \cos^2 \theta + \sigma_y \sin^2 \theta + \tau_{xy}(2 \sin \theta \cos \theta) \qquad (5.25)$$

Likewise summation of forces in the 2 direction provides

$$\tau_{12} = - \sigma_x (\sin \theta \cos \theta) + \sigma_y \sin \theta \cos \theta +$$
$$+ \tau_{xy}(\cos^2 \theta - \sin^2 \theta) \qquad (5.26)$$

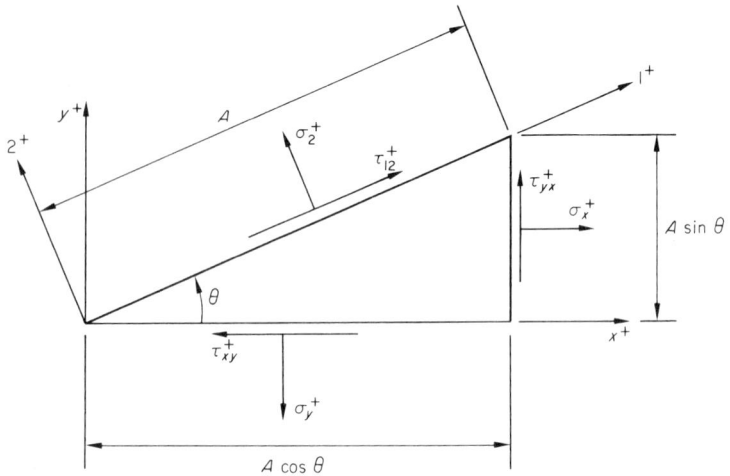

Fig. 5.5

Consider another 'free body' from the original element (Fig. 5.5). Force equilibrium in the 2 direction here provides

$$\sigma_2 = \sigma_x \sin^2 \theta + \sigma_y \cos^2 \theta - \tau_{xy}(2 \sin \theta \cos \theta) \qquad (5.27)$$

Force equilibrium in the 1 direction would again provide eqn. (5.26). These equations can be written succinctly as

$$\begin{bmatrix} \sigma_1 \\ \sigma_2 \\ \tau_{12} \end{bmatrix} = [T] \begin{bmatrix} \sigma_x \\ \sigma_y \\ \tau_{xy} \end{bmatrix} \qquad (5.28)$$

where

$$[T] = \begin{bmatrix} \cos^2 \theta & \sin^2 \theta & 2 \sin \theta \cos \theta \\ \sin^2 \theta & \cos^2 \theta & -2 \sin \theta \cos \theta \\ -\sin \theta \cos \theta & \sin \theta \cos \theta & (\cos^2 \theta - \sin^2 \theta) \end{bmatrix} \qquad (5.29)$$

It is seen that $[T]$ is the transformation relation for a second-rank tensor in two-dimensional space from the x–y co-ordinate system to the 1–2 co-ordinate system. It is nothing more than the Mohr's circle relationships expressed in matrix form. It is completely independent of the material system.

Since the strain is also a second rank tensor, then for two-dimensional space

$$\begin{bmatrix} \epsilon_1 \\ \epsilon_2 \\ \frac{1}{2}\gamma_{12} \end{bmatrix} = [T] \begin{bmatrix} \epsilon_x \\ \epsilon_y \\ \frac{1}{2}\gamma_{xy} \end{bmatrix} \qquad (5.30)$$

where $\epsilon_{12} = \frac{1}{2}\gamma_{12}$ and $\epsilon_{xy} = \frac{1}{2}\gamma_{xy}$.

From the assumptions of *classical* plate and shell theory $\sigma_3 = \sigma_z = \epsilon_3 = \epsilon_z = \tau_{xz} = \tau_{yz} = \tau_{13} = \tau_{23} = 0$, as shown clearly in Section 4.2. Therefore the matrix representation of the equations of elasticity for a lamina in the 1–2 or material oriented co-ordinate system are

$$\begin{bmatrix} \epsilon_1 \\ \epsilon_2 \\ \epsilon_{12} \end{bmatrix} = \begin{bmatrix} S_{11} & S_{12} & 0 \\ S_{12} & S_{22} & 0 \\ 0 & 0 & S_{66}/2 \end{bmatrix} \begin{bmatrix} \sigma_1 \\ \sigma_2 \\ \tau_{12} \end{bmatrix} \qquad (5.31)$$

where the S_{ij}'s are given in eqns. (5.12), (5.13), (5.16), (5.22) and (5.24). Thus eqn. (5.31) is the constitutive relation for a specially orthotropic material in a state of plane stress. Specially orthotropic refers to a lamina of orthotropic material in which the material and geometry axes are aligned.

For filamentary materials, either with unidirectional fibres or woven fibres, the above constitutive relations are used in the analysis of structures of these materials. In other words, macroscopically the composites are assumed to be homogeneously idealised materials, and thus do not account for the details of fibre–resin geometry and interaction. Such considerations comprise an important field called 'micromechanics'.

Equation (5.31) can be inverted, with the result that

$$\begin{bmatrix} \sigma_1 \\ \sigma_2 \\ \tau_{12} \end{bmatrix} = \begin{bmatrix} Q_{11} & Q_{12} & 0 \\ Q_{12} & Q_{22} & 0 \\ 0 & 0 & 2Q_{66} \end{bmatrix} \begin{bmatrix} \epsilon_1 \\ \epsilon_2 \\ \epsilon_{12} \end{bmatrix} = [Q] \begin{bmatrix} \epsilon_1 \\ \epsilon_2 \\ \epsilon_{12} \end{bmatrix} \qquad (5.32)$$

where

$$Q_{11} = \frac{E_{11}}{(1 - \nu_{12}\nu_{21})} \qquad (5.33)$$

$$Q_{22} = \frac{E_{22}}{(1 - \nu_{12}\nu_{21})} \qquad (5.34)$$

$$Q_{12} = Q_{21} = \frac{\nu_{21}E_{11}}{(1 - \nu_{12}\nu_{21})} = \frac{\nu_{12}E_{22}}{(1 - \nu_{12}\nu_{21})} \qquad (5.35)$$

$$Q_{66} = G_{12} \qquad (5.36)$$

Now, in general the principal material axes (1, 2, 3) are not aligned with the geometric axes (x, y, z) as shown in Fig. 5.6 for a unidirectional continuous fibre composite

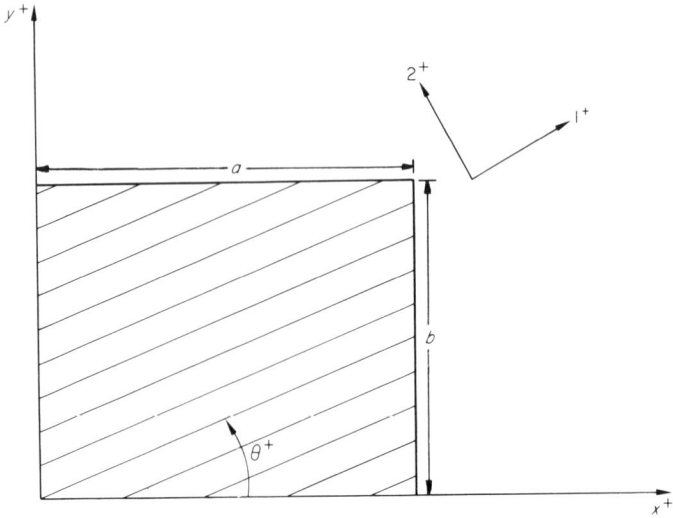

Fig. 5.6.   A generally orthotropic plate.

It is necessary to be able to relate the stresses and strains in both co-ordinate systems. From eqn. (5.28), multiplying both sides by $[T]^{-1}$, and remembering that $[T][T]^{-1} = [I]$, the identity matrix, then

$$\begin{bmatrix} \sigma_x \\ \sigma_y \\ \tau_{xy} \end{bmatrix} = [T]^{-1} \begin{bmatrix} \sigma_1 \\ \sigma_2 \\ \tau_{12} \end{bmatrix} \tag{5.37}$$

Substituting eqn. (5.32) into eqn. (5.37) results in

$$\begin{bmatrix} \sigma_x \\ \sigma_y \\ \tau_{xy} \end{bmatrix} = [T]^{-1}[Q] \begin{bmatrix} \epsilon_1 \\ \epsilon_2 \\ \epsilon_{12} \end{bmatrix} \tag{5.38}$$

Finally, substituting (5.30) into (5.38) results in the constitutive equation for a generally orthotropic lamina, that is to say a lamina composed of an orthotropic material in which the geometric and material axes are not aligned ($\theta \neq 0$, 90°, 180°, 270°).

$$\begin{bmatrix} \sigma_x \\ \sigma_y \\ \tau_{xy} \end{bmatrix} = [T]^{-1}[Q][T] \begin{bmatrix} \epsilon_x \\ \epsilon_y \\ \epsilon_{xy} \end{bmatrix} = [\bar{Q}] \begin{bmatrix} \epsilon_x \\ \epsilon_y \\ \epsilon_{xy} \end{bmatrix} \qquad (5.39)$$

where $[\bar{Q}]$ is the generally orthotropic lamina stiffness matrix, whose components are given by

$$[\bar{Q}] \equiv \begin{bmatrix} \bar{Q}_{11} & \bar{Q}_{12} & 2\bar{Q}_{16} \\ \bar{Q}_{12} & \bar{Q}_{22} & 2\bar{Q}_{26} \\ \bar{Q}_{16} & \bar{Q}_{26} & 2\bar{Q}_{66} \end{bmatrix} \qquad (5.40)$$

where for $m = \cos\theta$ and $n = \sin\theta$

$$\bar{Q}_{11} = Q_{11}m^4 + 2(Q_{12} + 2Q_{66})m^2n^2 + Q_{22}n^4$$
$$\bar{Q}_{22} = Q_{11}n^4 + 2(Q_{12} + 2Q_{66})m^2n^2 + Q_{22}m^4$$
$$\bar{Q}_{12} = (Q_{11} + Q_{22} - 4Q_{66})n^2m^2 + Q_{12}(m^4 + n^4)$$
$$\bar{Q}_{66} = (Q_{11} + Q_{22} - 2Q_{12} - 2Q_{66})m^2n^2 + Q_{66}(m^4 + n^4)$$
$$\bar{Q}_{16} = (Q_{11} - Q_{12} - 2Q_{66})nm^3 + (Q_{12} - Q_{22} + 2Q_{66})n^3m$$
$$\bar{Q}_{26} = (Q_{11} - Q_{12} - 2Q_{66})n^3m + (Q_{12} - Q_{22} + 2Q_{66})nm^3$$

The inclusion of the coefficient 2 in some components of eqn. (5.40) is merely to define the $\bar{Q}_{ij}$'s exactly as done by Ashton et al. in their widely used book.[2] Hence, no confusion will occur when the above is used later in analysing thin walled structures.

Note that the $\bar{Q}$ matrix is now fully populated, as it is in a totally anisotropic material system which would require six material constants to be determined. However, here, $\bar{Q}_{16}$ and $\bar{Q}_{26}$ are linear combinations of the other four relationships. Also because these two components have odd powers of the trigonometric functions one must be consistent in the use of θ as defined before.

By defining certain functions which are independent of θ Tsai and Pagano[3] have rewritten the components of the $\bar{Q}$ matrix as

$$\bar{Q}_{11} = U_1 + U_2\cos(2\theta) + U_3\cos(4\theta)$$
$$\bar{Q}_{22} = U_1 - U_2\cos(2\theta) + U_3\cos(4\theta)$$
$$\bar{Q}_{12} = U_4 - U_3\cos(4\theta)$$
$$\bar{Q}_{66} = U_5 - U_3\cos(4\theta) \qquad (5.41)$$
$$\bar{Q}_{16} = -\tfrac{1}{2}U_2\sin(2\theta) - U_3\sin(4\theta)$$
$$\bar{Q}_{26} = -\tfrac{1}{2}U_2\sin(2\theta) + U_3\sin(4\theta)$$

where $\quad U_1 = \frac{1}{8}(3Q_{11} + 3Q_{22} + 2Q_{12} + 4Q_{66})$

$\qquad U_2 = \frac{1}{2}(Q_{11} - Q_{22})$

$\qquad U_3 = \frac{1}{8}(Q_{11} + Q_{22} - 2Q_{12} - 4Q_{66})$

$\qquad U_4 = \frac{1}{8}(Q_{11} + Q_{22} + 6Q_{12} - 4Q_{66})$

$\qquad U_5 = \frac{1}{8}(Q_{11} + Q_{22} - 2Q_{12} + 4Q_{66})$

Because the U quantities are invariant to the axis rotation they are therefore inherent lamina properties, and can be useful in comparing various material systems in the design of a structure.

5.4 LAMINATES OF COMPOSITE MATERIALS[a]

Having formulated the constitutive relations for a lamina composed of a generally orthotropic material, eqns. (5.39), (5.40) and (5.41), the next step is to derive the constitutive relations accruing from bonding several laminae together, each with any general orientation of material axes and composed of any material system. The notation and approach of Reference 2 is used herein.

Equation (5.39) can be rewritten as follows for the kth lamina of a laminate

$$[\sigma]_k = [\bar{Q}]_k [\epsilon] \qquad (5.42)$$

From previous discussions on assumed forms of displacements and the strain displacements of thin plates and shells, it should be clear that the following are the strain–displacement relations for the laminated plate, analogous to those of Chapter 4 for an isotropic plate.

$$\begin{bmatrix} \epsilon_x \\ \epsilon_y \\ \epsilon_{xy} \end{bmatrix} = \begin{bmatrix} \epsilon_x^{\circ} \\ \epsilon_y^{\circ} \\ \epsilon_{xy}^{\circ} \end{bmatrix} + z \begin{bmatrix} \kappa_x \\ \kappa_y \\ \frac{1}{2}\kappa_{xy} \end{bmatrix} \qquad (5.43)$$

where

$$\epsilon_x^{\circ} = \frac{\partial u_0}{\partial x}, \qquad \epsilon_y^{\circ} = \frac{\partial v_0}{\partial y}, \qquad \epsilon_{xy}^{\circ} = \frac{1}{2}\left(\frac{\partial u_0}{\partial y} + \frac{\partial v_0}{\partial x}\right)$$

and

$$\kappa_x = -\frac{\partial^2 w}{\partial x^2}, \qquad \kappa_y = -\frac{\partial^2 w}{\partial y^2}, \quad \text{and} \quad \kappa_{xy} = -2\frac{\partial^2 w}{\partial x \partial y}$$

and where u_0, v_0 and w are the in-plane middle surface displacement in

[a] Portions of this section are excerpted from Ashston, J. E., Halpin, J. C. and Petit, P. H., *Primer on Composite Materials: Analysis*, Technomic Publishing Co. Reprinted by permission.

the x direction, the in-plane middle surface displacement in the y direction, and the lateral displacement of the laminated plate respectively.

Equation (5.43) can also be written as

$$[\epsilon] = [\epsilon°] + z[\kappa] \qquad (5.44)$$

Substituting (5.44) into (5.42) gives

$$[\sigma]_k = [\bar{Q}]_k [\epsilon°] + z[\bar{Q}]_k [\kappa] \qquad (5.45)$$

Thus, if the middle surface displacements and the lateral displacement are obtained eqn. (5.45) is used to compute the stresses σ_x, σ_y and τ_{xy} at any location in any of the k laminae.

Consider a laminated plate of thickness h as shown in Fig. 5.7.

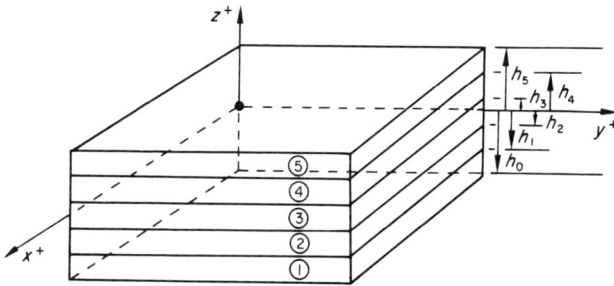

Fig. 5.7. Nomenclature for stacking sequence.

Here h_k is the vectorial distance from the mid-plane, $z = 0$, to the upper surface of the kth lamina.

Remember that for a plate or shell of thickness h, the stress resultants and stress couples are defined as

$$
\begin{bmatrix} N_x \\ N_y \\ N_{xy} \\ Q_x \\ Q_y \end{bmatrix} = \int_{-h/2}^{+h/2} \begin{bmatrix} \sigma_x \\ \sigma_y \\ \sigma_{xy} \\ \sigma_{xz} \\ \sigma_{yz} \end{bmatrix} dz, \qquad
\begin{bmatrix} M_x \\ M_y \\ M_{xy} \end{bmatrix} = \int_{-h/2}^{+h/2} \begin{bmatrix} \sigma_x \\ \sigma_y \\ \sigma_{xy} \end{bmatrix} z \, dz \qquad (5.46)
$$

However, for a laminated plate the stresses that are integrated across the thickness of the plate or shell are the sum of the stresses across each lamina, or, the in-plane stress resultants can be written as below using eqn. (5.45)

$$
\begin{bmatrix} N_x \\ N_y \\ N_{xy} \end{bmatrix} = \sum_{k=1}^{n} \int_{h_{k-1}}^{h_k} \begin{bmatrix} \sigma_x \\ \sigma_y \\ \sigma_{xy} \end{bmatrix}_k \, \mathrm{d}z
$$

$$
= \sum_{k=1}^{n} \left\{ \int_{h_{k-1}}^{h_k} [\bar{Q}]_k \begin{bmatrix} \epsilon_x^0 \\ \epsilon_y^0 \\ \epsilon_{xy}^0 \end{bmatrix} \mathrm{d}z + \int_{h_{k-1}}^{h_k} [\bar{Q}]_k \begin{bmatrix} \kappa_x \\ \kappa_y \\ \tfrac{1}{2}\kappa_{xy} \end{bmatrix} z \, \mathrm{d}z \right\}
$$

However, since $[\epsilon^0]$ and $[\kappa]$ are not functions of z, and within each lamina $[\bar{Q}]_k$ is not a function of z, the above can be written as

$$
\begin{bmatrix} N_x \\ N_y \\ N_{xy} \end{bmatrix} = \sum_{k=1}^{n} \left\{ [\bar{Q}]_k [\epsilon^0] \int_{h_{k-1}}^{h_k} \mathrm{d}z + [\bar{Q}]_k [\kappa] \int_{h_{k-1}}^{h_k} z \, \mathrm{d}z \right\}
$$

or finally

$$
[N] = [A][\epsilon^0] + [B][\kappa] \tag{5.47}
$$

where

$$
A_{ij} = \sum_{k=1}^{n} (\bar{Q}_{ij})_k (h_k - h_{k-1}) \tag{5.48}
$$

$$
B_{ij} = \frac{1}{2} \sum_{k=1}^{n} (\bar{Q}_{ij})_k (h_k^2 - h_{k-1}^2) \tag{5.49}
$$

From eqn. (5.47) it is seen that the in-plane stress resultants for a laminated plate are not only a function of the mid-surface strains, as in a homogeneous plate, but are also in general a function of the curvatures and twisting as well. Stated in another way in-plane forces can cause curvatures of the plate, and there is also an interplay between twisting effects and normal effects as can be seen.

Similarly

$$
\begin{bmatrix} M_x \\ M_y \\ M_{xy} \end{bmatrix} = \int_{-h/2}^{h/2} \begin{bmatrix} \sigma_x \\ \sigma_y \\ \sigma_{xy} \end{bmatrix}_k z \, \mathrm{d}z = \sum_{k=1}^{n} \int_{h_{k-1}}^{h_k} \begin{bmatrix} \sigma_x \\ \sigma_y \\ \sigma_{xy} \end{bmatrix}_k z \, \mathrm{d}z
$$

$$
= \sum_{k=1}^{n} \left\{ \int_{h_{k-1}}^{h_k} [\bar{Q}]_k [\epsilon^0] z \, \mathrm{d}z + \int_{h_{k-1}}^{h_k} [\bar{Q}]_k [\kappa] z^2 \, \mathrm{d}z \right\}
$$

$$
= \left\{ \sum_{k=1}^{n} [\bar{Q}]_k \int_{h_{k-1}}^{h_k} z \, \mathrm{d}z \right\} [\epsilon^0] + \left\{ \sum_{k=1}^{n} [\bar{Q}]_k \int_{h_{k-1}}^{h_k} z^2 \, \mathrm{d}z \right\} [\kappa]
$$

$$
M = [B][\epsilon^0] + [D][\kappa] \tag{5.50}
$$

where B is defined by eqn. (5.49) and

$$D_{ij} = \frac{1}{3} \sum_{k=1}^{n} [\bar{Q}_{ij}]_k (h_k^3 - h_{k-1}^3) \tag{5.51}$$

From eqn. (5.50) it is seen that for a laminated plate, not only are the stress couples a function of curvature, but also are a function of in-plane strains and displacements. Also M_x and M_y are influenced by twisting, and M_{xy} is affected by mid-plane normal strains and curvatures.

Equations (5.47) and (5.50) can be written as

$$\left[\frac{N}{M}\right] = \left[\begin{array}{c|c} A & B \\ \hline B & D \end{array}\right]\left[\frac{\epsilon^\circ}{\kappa}\right] \tag{5.52}$$

$$\begin{bmatrix} N_x \\ N_y \\ N_{xy} \\ \hline M_x \\ M_y \\ M_{xy} \end{bmatrix} = \left[\begin{array}{ccc|ccc} A_{11} & A_{12} & 2A_{16} & B_{11} & B_{12} & 2B_{16} \\ A_{12} & A_{22} & 2A_{26} & B_{12} & B_{22} & 2B_{26} \\ A_{16} & A_{26} & 2A_{66} & B_{16} & B_{26} & 2B_{66} \\ \hline B_{11} & B_{12} & 2B_{16} & D_{11} & D_{12} & 2D_{16} \\ B_{12} & B_{22} & 2B_{26} & D_{12} & D_{22} & 2D_{26} \\ B_{16} & B_{26} & 2B_{66} & D_{16} & D_{26} & 2D_{66} \end{array}\right] \begin{bmatrix} \epsilon_x^\circ \\ \epsilon_y^\circ \\ \epsilon_{xy}^\circ \\ \hline \kappa_x \\ \kappa_y \\ \frac{1}{2}\kappa_{xy} \end{bmatrix} \tag{5.53}$$

Incidentally, in the literature one often finds the strain matrix written with γ_{xy}° instead of ϵ_{xy}° where $\gamma_{xy}^\circ = 2\epsilon_{xy}^\circ$. Care should be taken to avoid confusion on this point.

It is seen that the [A] matrix is the extensional stiffness matrix, [D] is the flexural stiffness matrix, and the [B] matrix is the bending–stretching coupling matrix. It is also seen that a laminate may have bending–stretching coupling even if the various materials comprising the laminae are isotropic. In fact, only when the plate is exactly symmetric about its middle surface is $B = 0$, and this requires symmetry in laminae properties, laminae orientation and distance from the mid-plane.

The stretching–shearing coupling comes about through the A_{16} and A_{26} terms. The twisting–stretching coupling comes about through the B_{16} and B_{26} terms, while the bending–twisting coupling occurs due to the D_{16} and D_{26} terms. Also, the bending–stretching coupling comes about when any B_{ij} component other than B_{16} and B_{26} are not zero.

Consider three practical laminates: cross-ply, angle-ply and uni-ply.

We shall restrict ourselves to the following patterns: $0°/90°/0°/90°$, etc., for the cross-ply and $+\theta/-\theta/+\theta/-\theta$ etc., for the angle-ply.

It can be shown that for the cross-ply laminates, whether there are an even number or an odd number of laminae (i.e. asymmetric or symmetric with respect to the midsurface) $A_{16} = A_{26} = D_{16} = D_{26} = 0$. Hence, there is no twisting–stretching or bending–twisting coupling. However, for an even number of laminae, $B_{ij} \neq 0$, so bending–stretching coupling exists, but the larger the number of laminae the smaller the coupling. Of course if there are an odd number of laminae of equal thickness, resulting in midplane symmetry, $B_{ij} = 0$, and the cross-ply plate behaves as a specially orthotropic single layered plate wherein the proper elastic constants are determined by eqns. (5.48) and (5.51).

For a plate of angle-ply construction defined above, if there are an even number of layers, $A_{16} = A_{26} = 0$ (no stretching–shearing coupling), but D_{16} and D_{26} are zero, resulting in no bending–twisting coupling, and B_{16} and B_{26} are not zero, causing stretching–twisting coupling.

For an angle-ply plate with an odd number of laminae all of the same thickness, midplane symmetry exists, hence, $B_{ij} = 0$. However, A_{16}, A_{26}, D_{16}, and D_{26} are not zero.

From eqn. (5.47)

$$[\epsilon°] = [A]^{-1}[N] - [A]^{-1}[B][\kappa] \tag{5.54}$$

Substituting (5.54) into (5.50) gives another useful expression

$$[M] = [B][A]^{-1}[N] + ([D] - [B][A]^{-1}[B])[\kappa] \tag{5.55}$$

From eqns. (5.54) and (5.55), one can obtain

$$\begin{bmatrix} \epsilon° \\ M \end{bmatrix} = \begin{bmatrix} A* & B* \\ C* & D* \end{bmatrix} \begin{bmatrix} N \\ \kappa \end{bmatrix} \tag{5.56}$$

where

$$
\begin{aligned}
[A*] &= [A]^{-1} \\
[B*] &= -[A]^{-1}[B] \\
[C*] &= [B][A]^{-1} \\
[D*] &= [D] - [B][A]^{-1}[B]
\end{aligned}
\tag{5.57}
$$

The $D*$ matrix is seen to relate the curvature matrix to the stress couple matrix and is known as the reduced flexural modulus matrix, reduced due to bending–stretching coupling (the B matrix). If $B = 0$ then $D^* = D$ and there is no reduction in flexural stiffness. Thus in the analysis of laminated beams, plates and shells, one must use D^* as the flexural stiffness matrix whenever bending–stretching coupling exists to obtain accurate values of stresses, displacements, natural frequencies and buckling loads.

To briefly recapitulate, given the elastic constants E_{11}, E_{22}, G_{12} and either ν_{12} or ν_{21} for each lamina material, the values Q_{ij} can be obtained for each lamina by using eqns. (5.33) to (5.36). Then knowing the angle θ, as defined, the general $\bar{Q}$ matrix can be determined for each lamina using eqns. (5.40) and (5.41). Substituting these into eqns. (5.48), (5.49) and (5.51), the A_{ij}, B_{ij} and D_{ij} components are computed. Hence, the constitutive relations for any generally orthotropic laminate are obtained from eqns. (5.47) and (5.50) or (5.53). Obviously, a simple computer routine can be developed to eliminate the lengthy calculations of complex composite laminates.

It should be noted that up to this point the laminates discussed employ classical theory, i.e. no transverse shear deformation or transverse normal stress. Since most of the literature involving structures of composite materials employs classical theory, the reader is thus prepared for the bulk of the literature at this point.

5.5 LAMINAE AND LAMINATES OF COMPOSITE MATERIALS ACCOUNTING FOR TRANSVERSE SHEAR DEFORMATION

As will be amply explained in Chapters 6 and 7, it is necessary to include transverse shear deformation in the analysis of most plate and shell structures composed of composite materials, even though the plate or shell thickness is such that were the structure composed of an isotropic material these effects would be negligible. The reason is simple: In an isotropic material

$$2 \leqslant \frac{E}{G} = 2(1 + \nu) \leqslant 3$$

However, in a composite lamina (or laminate) the in-plane modulus of elasticity is many times larger than that of the matrix material (in fact, that is one of the primary reasons for adding the fibres!), while the transverse shear modulus is largely that of the matrix material. Hence, for most practical filamentary composite materials

$$20 \leqslant \frac{E_{11}}{G_{13}} \leqslant 50$$

Hence, for a given in-plane modulus, the plate or shell is very weak in transverse shear resistance, and the effects of transverse shear deformation are significant, and cannot be neglected.
Therefore in this case, eqn. (5.32) is modified to be

$$
\begin{bmatrix} \sigma_1 \\ \sigma_2 \\ \sigma_{23} \\ \sigma_{31} \\ \sigma_{12} \end{bmatrix}
=
\begin{bmatrix}
Q_{11} & Q_{12} & 0 & 0 & 0 & 0 \\
Q_{21} & Q_{22} & 0 & 0 & 0 & 0 \\
0 & 0 & 0 & 2Q_{44} & 0 & 0 \\
0 & 0 & 0 & 0 & 2Q_{55} & 0 \\
0 & 0 & 0 & 0 & 0 & 2Q_{66}
\end{bmatrix}
\begin{bmatrix} \epsilon_1 \\ \epsilon_2 \\ \epsilon_{34} \\ \epsilon_{31} \\ \epsilon_{12} \end{bmatrix}
\qquad (5.58)
$$

Here, Q_{11}, Q_{22}, Q_{12} and Q_{66} are given by eqn (5.33) through (5.36) respectively. From eqns. (5.23) and (5.24) it is seen that

$$Q_{44} = G_{23} \qquad (5.59)$$

$$Q_{55} = G_{31} \qquad (5.60)$$

Proceeding as in the earlier section, the stress–strain relations for a generally orthotropic lamina including transverse shear deformation are, referring to Fig. 5.6 and eqns. (5.39) and (5.40)

$$
\begin{bmatrix} \sigma_x \\ \sigma_y \\ \sigma_{yz} \\ \sigma_{zx} \\ \sigma_{xy} \end{bmatrix}_k
=
\begin{bmatrix}
\bar{Q}_{11} & \bar{Q}_{12} & 0 & 0 & 0 & 2\bar{Q}_{16} \\
\bar{Q}_{12} & \bar{Q}_{22} & 0 & 0 & 0 & 2\bar{Q}_{26} \\
0 & 0 & 0 & 2\bar{Q}_{44} & 2\bar{Q}_{45} & 0 \\
0 & 0 & 0 & 2\bar{Q}_{45} & 2\bar{Q}_{55} & 0 \\
\bar{Q}_{16} & \bar{Q}_{26} & 0 & 0 & 0 & 2\bar{Q}_{66}
\end{bmatrix}_k
\begin{bmatrix} \epsilon_x \\ \epsilon_y \\ \epsilon_{yz} \\ \epsilon_{zx} \\ \epsilon_{xy} \end{bmatrix}
\qquad (5.61)
$$

The majority of the $\bar{Q}$ quantities were defined on page 215, the others being,

$$
\left.
\begin{aligned}
\bar{Q}_{11} &= Q_{44}m^2 + Q_{55}n^2 \\
\bar{Q}_{55} &= Q_{44}n^2 + Q_{55}m^2 \\
\bar{Q}_{45} &= (Q_{55} - Q_{44})mn
\end{aligned}
\right\}
\qquad (5.62)
$$

where $m = \cos\theta$ and $n = \sin\theta$, and θ is positive as shown in Fig. 5.6.
With these changes, all stress resultants and stress couples can be expressed as in eqn. (5.53) for a laminate composed of k laminae.

However, in finding the transverse shear forces Q_x and Q_y, it is assumed that the transverse shear stresses are distributed parabolically across the laminate thickness. In spite of the discontinuities at the interfaces between laminae, a continuous function is assumed as a weighting function $f(z)$, consistent with that used by Reissner[4]

$$f(z) = \frac{5}{4}\left[1 - \left(\frac{z}{h/2}\right)^2\right]$$ (5.63)

With the absence of any applied surface shear loads, the transverse shear resultants are obtained by substituting the following equations from (5.61) for the kth lamina

$$\sigma_{xz}^{(k)} = 2\bar{Q}_{55}^{(k)}\epsilon_{xz} + 2\bar{Q}_{45}^{(k)}\epsilon_{yz}$$
$$\sigma_{yz}^{(k)} = 2\bar{Q}_{45}^{(k)}\epsilon_{xz} + 2\bar{Q}_{44}^{(k)}\epsilon_{yz}$$

Multiplying each of the above by eqn. (5.63) and substituting the result into the following definitions

$$Q_x = \int_{-h/2}^{h/2} \sigma_{xz}\, dz \quad \text{and} \quad Q_y = \int_{-h/2}^{h/2} \sigma_{yz}\, dz$$

with the following result

$$Q_x = 2(A_{55}\epsilon_{xz} + A_{45}\epsilon_{yz})$$ (5.64)
$$Q_y = 2(A_{45}\epsilon_{xz} + A_{44}\epsilon_{yz})$$ (5.65)

where

$$A_{ij} = \frac{5}{4}\sum_{k=1}^{n} \bar{Q}_{ij}^{(k)}\left[h_k - h_{k-1} - \frac{4}{3}\frac{(h_k^3 - h_{k-1}^3)}{h^2}\right]$$ (where $i, j = 4, 5$) (5.66)

Thus eqns. (5.53), (5.64) and (5.65) comprise the full system of constitutive equations necessary to study generally orthotropic laminates comprising a plate or shell, including the effects of transverse deformation.

5.6 DETERMINATION OF LAMINA PROPERTIES FROM PROPERTIES OF CONSTITUENTS

One final consideration with regard to fibre reinforced composite materials, is determining the properties of composite lamina if the properties of the fibre and the matrix are known, as well as the amount of fibre in the composite. Although there is not unanimous agreement on accuracy, the most generally agreed upon methods for obtaining lamina material properties are formulated by Halpin and Tsai.[5]

Remembering that the 1 direction is in the direction parallel to the reinforcing fibres,

$$E_{11} \simeq E_f v_f + E_m v_m \qquad (5.67)$$

where

E_{11} = Modulus of elasticity of the fibre reinforced composite lamina in the 1 indirection.

E_f = Modulus of elasticity of the fibre material.

E_m = Modulus of elasticity of the matrix material.

v_f = Volume fraction of the fibre material in the lamina.

v_m = Volume fraction of the matrix material in the lamina. ($= 1 - v_f$).

The relationship given in this equation is referred to as the 'rule of mixtures'. Likewise one of the Poisson's ratios, namely ν_{12}, also follows the 'rule of mixtures'

$$\nu_{12} \simeq \nu_f v_f + \nu_m v_m \qquad (5.68)$$

where ν_f and ν_m are the Poisson's ratios of the fibre and the matrix material respectively.

The determination of all other laminae properties is not so straightforward, and is not as accurate as the above. The Halpin–Tsai relations[a] for determining E_{22}, G_{12}, and G_{23} are given as follows:

$$\frac{\bar{p}}{p_m} = \frac{(1 + \zeta \eta v_f)}{1 - \eta v_f} \qquad (5.69)$$

where

$\bar{p} = E_{22}$, G_{12} or G_{13}.

p_m = Corresponding modulus for the matrix material.

p_f = Corresponding modulus of the fibre material.

ζ = A measure of reinforcement which depends on the boundary conditions of the structure.

$\eta = (p_f/p_m - 1)/(p_f/p_m + \zeta)$.

v_f = Volume fraction of the fibre material in the lamina.

Obviously ζ is a factor used to assist the above expression to better fit the test data, and depends on the geometry of any inclusions, packing geometry, loading, and boundary conditions. It is found either by reference to appropriate elastic micromechanics solutions or fitting to test data when available (see Reference 2, pp. 77–81). Whitney[6] states that for G_{12} or G_{13}, $\zeta = 1$, for E_{22}, ζ is 1 or 2, for hexagonal or square arrays respectively.

[a] Excerpted from Ashton, J. E., Halpin, J. C. and Petit, P. H., *Primer on Composite Materials: Analysis*, Technomic Publishing Co. Reprinted by permission.

The above equations together with the basic orthotropy relation shown previously in Section 5.2

$$\frac{\nu_{ij}}{E_i} = \frac{\nu_{ji}}{E_j} \qquad (5.70)$$

enable the analyst and designer to obtain all lamina elastic properties from knowledge of the constituent properties and proportions.

5.7 PROBLEMS

5.1. Consider a lamina composed of an epoxy matrix reinforced by boron fibres such that the laminae properties are given by

$$E_{11} = 35 \times 10^6 \text{ psi} = 2 \cdot 47 \times 10^6 \text{ kg/cm}^2$$
$$E_{22} = 3 \cdot 5 \times 10^6 \text{ psi} = 0 \cdot 247 \times 10^6 \text{ kg/cm}^2$$
$$G_{12} = G_{21} = 1 \cdot 5 \times 10^6 \text{ psi} = 1 \cdot 056 \times 10^5 \text{ kg/cm}^2$$
$$\nu_{12} = 0 \cdot 3$$

Using the relations contained herein determine for an arbitrary value of θ

    a. $[Q]$
    b. $U_i \ (i = 1 \ldots 5)$
    c. $[\bar{Q}]$

5.2. Consider a laminate composed of graphite fibre reinforced polyimide, with a fibre volume fraction of 50 per cent. If the laminate thickness is 1 inch, for the various number of layers and fibre orientations calculate the A, B and D matrices. The fibre and matrix properties are

	Graphite	Polyimide
	$E = 40 \times 10^6 \text{ psi}$	$E = 0 \cdot 4 \times 10^6 \text{ psi}$
	$\nu = 0 \cdot 2$	$\nu = 0 \cdot 33$

No. of Layers	Fibre Orientation	Layer Thickness (in.)
2	0, 90	$\frac{1}{2}$
3	0, 120, 240	$\frac{1}{3}$
4	0, 90, 180, 270	$\frac{1}{4}$
5	0, 72, 144, 216, 288	$\frac{1}{5}$
6	0, 60, 120, 180, 240, 300	$\frac{1}{6}$
8	0, 45, 90, 135, 180, 225, 270, 315	$\frac{1}{8}$
1	0	1
2	+45, −45	$\frac{1}{2}$
3	0, 90, 180	$\frac{1}{3}$

5.3. Consider a lamina composed of a fibre-reinforced material as shown below, in which the fibres are oriented as shown. If the lamina is subjected to a uniaxial tension σ_x, due to the angle θ between the lamina axes x–y and the principal material axis, 1, there will be a shear deformation ϵ_{xy} introduced due to the tension stress σ_x. This relationship can be expressed as

$$\epsilon_{xy} = \psi \sigma_x$$

Find ψ. (*Note:* for an isotropic material $\psi = 0$).

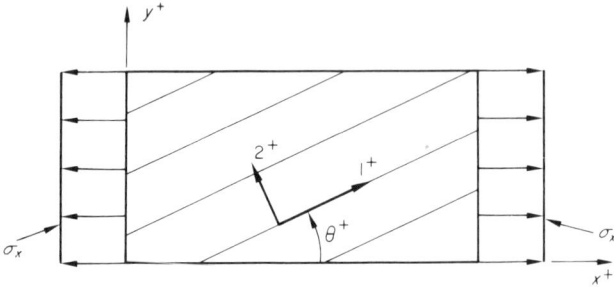

5.4. Consider the lamina in Problem 5.2 to be subjected to an in-plane shear stress $\tau_{xy} = \tau_{yx}$. Due to the angle θ there will be a dilatational strain, ϵ_x, induced into the plate due to this shear stress. This relationship can be expressed as

$$\epsilon_x = \phi \tau_{xy}$$

Find ϕ. (*Note:* for an isotropic material $\phi = 0$).

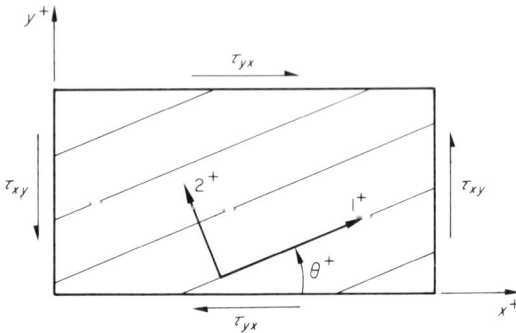

REFERENCES

1. Sokolnikoff, I. S., *Mathematical Theory of Elasticity*, McGraw-Hill Book Company, New York (1956).
2. Ashton, J. E., Halpin, J. C., and Petit, P. H., *Primer on Composite Materials: Analysis*, Technomic Pub. Co. Inc., Stamford, Conn. (1969).
3. Tsai, S. W. and Pagano, N. J., 'Invariant Properties of Composite Materials', *Composite Materials Workshop*, Technomic Publ. Co. (January 1968), pp. 233–53.
4. Reissner, E., 'On a Variational Theorem in Elasticity', *J. Math. and Physics*, **29**, 90 (1950).
5. Halpin, J. C. and Tsai, S. W., 'Environmental Factors in Composite Materials Design', AFML TR 67-423.
6. Whitney, J. M., 'Analytical and Experimental Methods in Composite Mechanics', *J. Structural Division*, ASCE (Januray 1973), pp. 113–29.

CHAPTER 6

ANALYSIS OF PLATES COMPOSED OF COMPOSITE MATERIALS

6.1 HISTORY OF ANISOTROPIC PLATES UTILISING CLASSICAL PLATE THEORY

The behaviour of anisotropic plates and shells generally differs markedly from that of isotropic plates and shells. Many of the early analyses were devoted to the study of crystals and the analysis of wooden structures. The impetus for the study of anisotropic plates and shells today is in response to the rapid development of composite materials competitive to conventional isotropic materials during the last two decades. Activity and effort in this field seem to be increasing, especially for laminated constructions, which are the most often used form in which composite materials are employed. Laminated structures exhibit properties considerably different from those of the single-layer structures. In addition, the great difference in elastic properties between fibre filament and matrix materials leads to a high ratio of in-plane Young's modulus to transverse shear modulus for most of the composite materials developed to date. This results in the fact that the classical theory of plates and shells, based on neglecting transverse shear deformation, is invalid for most composite structures, even those which are geometrically thin. Unfortunately, most of the literature on the subject available today is based on the classical theory. Thus, in order to provide reliable criteria for designers, a more accurate theory including transverse shear deformation is necessary for the analysis of structures composed of composite materials.

Early attempts in the analysis of anisotropic plates and shells were restricted to the specially orthotropic cases. The earliest work in this field was carried out by Gehring[1] in 1860 when he presented his theory for a specially orthotropic plate under static load conditions. As to vibrations of orthotropic plates, perhaps Hearmon[2] is the first investigator (1946) who dealt with the fundamental frequency of vibration of

228

rectangular, wood and plywood plates. Using algebraic polynomials as the assumed mode shape, he applied the Rayleigh–Ritz method to specially orthotropic plates with various clamped and simply supported boundary conditions. The accuracy of the method was also discussed and the frequencies compared with experimental results. A detailed review of research in orthotropic plates prior to 1961 is found in a dissertation by Vinson.[3] The highlight of publications dealing with vibrations is a survey paper on the free vibrations of elastic plates in general by Leissa[4] in 1969. He included twenty-seven pertinent papers up to 1964 under the subtitle of anisotropic plates, although they dealt only with specially orthotropic plates.

In 1957, Lekhnitskii[5] published his comprehensive book on the subject of anisotropic plates, the English translation of which did not appear until 1968. In this book the theory of single-layer anisotropic plates based upon classical plate theory is presented systematically treating the bending, stability and vibration problems of specially orthotropic plates. One problem of general orthotropic (or anisotropic) plates was solved. This was for a plate under uniform lateral loading. The solution was found using the Kantorovich first iteration method (in contrast to the general Kantorovich iteration method due to Kerr[6]). It is of interest that he included transverse shear deformation in the derivation of the governing equations for the bending of a specially cylindrically orthotropic plate, but no solutions were attempted.

In 1959, Stavsky[7] formulated a theory of anisotropic laminated plates. Reissner and Stavsky[8] 1961, paid particular attention to the effect of stretching–bending coupling for a laminated anisotropic plate when they reinvestigated the earlier works of Smith[9] and Lekhnitskii.[5] On the basis of this research, a series of papers by Stavsky[10–13] appeared. The example problems of the above publications dealt entirely with specially orthotropic laminates, cylindrical bending type, or $(\theta, -\theta)$ angle-ply laminates. Stavsky also collected his previous pertinent work in a chapter co-authored with Hoff[14] in the book entitled *Composite Engineering Laminates*, edited by Dietz[15] 1969.

Waddoups[16] in 1965, dealt with the vibration response of layered, specially orthotropic plates, both experimentally and analytically. Since 1967, the *Journal of Composite Materials* has devoted much of its attention in the problem of laminated anisotropic materials. In 1968, a book entitled *Composite Material Workshop* edited by Tsai *et al.*[17] provided chapters written by various authors associated with fundamental problems of composite materials and structures.

In 1968, Mayberry[18] presented his Master's thesis on the vibrations of laminated anisotropic panels. Both experimental and analytical investigations were made. Also, Ashton and Anderson[19] studied the natural modes of vibration of boron–epoxy plates. Using a Rayleigh–Ritz formulation, the analytical results for natural frequencies and mode shapes compared favourably with experimental investigations for laminated anisotropic plates that are symmetric with respect to the middle surface, with fully clamped boundary conditions.

The work associated with anisotropic plates with arbitrary numbers of layers reached a flood in 1969. Whitney and Leissa[20] formulated the governing equations of generally laminated anisotropic plates analogous to the von Karman plate equation including stretching–bending coupling and in-plane rotatory inertia coupling. Solutions to the problems of natural frequencies, critical buckling loads, and displacements due to lateral static loading were presented using a one-term trigonometric function after linearising the general von Karman equations. The plates considered were $(0°, 90°, 0°, 90°, \ldots)$ cross-ply or $(\theta, -\theta, \theta, -\theta, \ldots)$ angle-ply laminates which are in fact specially orthotropic or quasi-isotropic, and were simply supported on the boundaries. Two later publications by Whitney[21] and by Whitney and Leissa[22] also dealt with specially orthotropic or quasi-isotropic plates qualitatively similar to their previous work[20] except that a full Fourier trigonometric series was employed and discussed in the latter.

Also in 1969, Ashton and Waddoups[23] presented an energy formulation for the problem of a single-layer anisotropic plate including a linear stability analysis, the calculation of frequencies and mode shapes, and the analysis of displacements due to lateral loading. It was noted that the single-plate equation can only characterise the behaviour of a laminated plate which possesses symmetrical properties with respect to the middle surface. Solutions were also presented for symmetric layered plates. The convergence of solutions using characteristic beam functions was also presented numerically. Ashton[24] extended this work to the problem in which the stiffness rigidities can be expressed as a polynomial series in x and y. This analysis of frequencies and mode shapes was compared to the results of an experimental investigation.[25] Ashton[26] also determined that the use of the reduced bending stiffness in the single-layer plate equation, instead of considering stretching–bending coupling in the unsymmetrically laminated plate equations, gives excellent results in comparison with those obtained by Whitney.[21] However, Whitney and Leissa[22] pointed out that the use of reduced bending stiffness is generally valid except for the case of $\pm 15°$ angle-ply laminates under transverse load where the difference between the two is greater than 20 per cent. An idea of applying isotropic skew plates analysis to rectangular anisotropic plates was also proposed by Ashton,[27] but the use of the analogy is restricted to a few coincident cases.

Kicher and Mandell[28] also in 1969, experimentally investigated the critical buckling load for layered anisotropic plates. In order to provide a comparison for experimental results, they analysed the stability using the classical orthotropic plate equation with the reduced bending rigidity. Numerical results obtained on the basis of Galerkin's method agreed most favourably with the experimental results except for those plates using extremely highly anisotropic materials such as 'Thornel-50'. They showed that the analytical results are usually too high for those unfavourable values. Chamis[29] in a 1969 paper, also presented critical buckling loads of uniaxial compression, in-plane shear, or combined loading for a laminated anisotropic plate using Galerkin's method in connection with the use of reduced bending rigidity. Instead of plotting the well-known cusp-shaped buckling curves, tables of so-called critical buckling loads are listed. The analytical results are satisfactory compared with the experimental results by Mandell.[30] He also concluded: For uniaxial in-plane compressive loading, twisting–stretching coupling can result in a 50 per cent reduction in buckling load; composite plates can have the same critical load as plates of aluminium of the same thickness with 33 per cent weight reduction; neglecting the unbalanced edge moments gives a conservative buckling prediction; the use of a symmetric coefficient matrix, which is the average of the coefficient matrix and its transpose, gives satisfactory results. In 1969, Bert and Mayberry[31] also published a paper on the free vibration of fully clamped anisotropic plates with an arbitrary number of layers, using the Rayleigh–Ritz method. Their analytical results are favourable in comparison with experimentally determined resonant frequencies. Whitney,[32] also in 1969, presented a solution obtained using a cylindrical bending approach for a laminated generally aniso-tropic plate, which was found to be favourably comparable with the previous results obtained by him and Leissa from a full bending theory.

In 1970, Ashton[33] discussed the boundary conditions imposed on an anisotropic plate. He concluded that in employing the Rayleigh–Ritz method, the use of a series of characteristic beam functions is excellent in all cases for a specially orthotropic plate; it is also satisfactory for determining the buckling load, natural frequencies, mode shapes, and displacements but not for stress resultants, moments, and corner reactions for strongly anisotropic plates. The number of terms used is up to 9 in the graphs presented. Also in 1970, Whitney[34] employed Reissner and Stavsky's[8] governing equation treating the unsymmetrically laminated anisotropic plates under transverse loading with clamped and simply supported edges. Trigonometric Fourier series are used as exact solutions, however, the numerical results must be found solving simultaneous linear algebraic equations for the coefficients of an assumed sufficient number of terms of Fourier series. He concluded that the boundary conditions do not change the amount of the stretching–bending coupling effects. These are usually dependent on the number of layers and the ratio of E_{11}/E_{22}; also that for sufficiently high values of E_{11}/E_{22}, the strip approach gives reasonable answers; for some orientations of anti-symmetric angle-ply ($\pm 15°$, $\pm 25°$, $\pm 35°$), membrane boundary conditions can significantly influence plate response; and the reduced bending stiffness method does not provide good results for clamped boundary conditions for the angle-ply orientations which are sensitive to membrane boundary conditions. Recently, Reuter[35] extended his previous work to examine the elastic long-wavelength wave propagation for a homogeneous semi-infinite generally anisotropic plate. Also Mohan[36] in 1970, obtained natural frequencies and mode shapes using Galerkin's method for a homogeneous anisotropic plate.

In 1969, Chung and Testa[37] studied the elastic stability of unidirectional, equally spaced fibres in the plane of a composite plate wherein they treated the fibres as beams, and assumed the matrix to be in a state of plane stress. Critical values of buckling loads and the corresponding buckling modes were evaluated numerically.

In April 1969, Ashton and Love[38] provided the results of an experimental study of the stability of rectangular plates of boron–epoxy composite. The plates were clamped on to the loaded edges and either damped or simply-supported on the unloaded edges. Buckling loads were determined by Southwell plots, and these were compared to the analytical results of Reference 23, which showed good agreement.

The publications mentioned above are based upon the classical theory for anisotropic thin plates. The exact solution for a generally orthotropic plate is considerably difficult to find even in the case of simply-supported boundary conditions. The Rayleigh–Ritz formulation or Galerkin's method is reliable for engineering purposes for the calculation of critical buckling loads, frequencies, mode shapes and displacements, and has been used for most attempted problems in which classical plate theory assumptions are employed.

The most extensive work on plates of composite materials in one volume is the book by Ashton and Whitney[39] published in 1970.

6.2   HISTORY OF ANISOTROPIC PLATES INCLUDING TRANSVERSE SHEAR DEFORMATION

Perhaps the consideration of transverse shear deformation in the vibration analysis of anisotropic plates started in 1960, when Kaczkowski[40] investigated the effect of transverse shear deformation and rotatory inertia on the natural frequencies of specially orthotropic plates. Ambartsumyan's book,[41] published in 1961, presented the governing equations for the analysis of vibration for homogeneous generally anisotropic plates and layered specially orthotropic shallow shells, including the effect of transverse shear deformation.

In 1967, Ambartsumyan[42] collected his previous works associated with the analysis of anisotropic plates, including the effect of transverse shear deformation, and published a book entitled, *Theory of Anisotropic Plates*, which was translated by Ashton and Cheron[43] in 1969. In this book, he presented systematically the governing equations for the analysis of laminated specially orthotropic anisotropic plates involving static distributed loadings, vibration and stability problems under either small deflection or geometrically large deflection. The example problems, however, are restricted to specially orthotropic or transversely isotropic plates.

It should be noted that Ambartsumyan showed that the agreement between the results obtained using the theory including transverse shear deformation and that from a point of view of three-dimensional theory is excellent even for a very thick plate rather than a moderately thick plate as outlined by Love. As is well known, the effect of transverse normal stress is negligible for the problems of plates in vibration, stability, wave propagation, and uniform static loadings. Consequently, in dealing with these problems it is logical that there is no limit to the thickness-to-edge-length ratio for the applicability of this refined theory, providing the plate remains a plate.

Since 1966 a considerable amount of work in investigating the effect of transverse shear deformation and rotatory inertia has been conducted by Vinson[44-59] and his co-workers at Delaware. Their work was originally oriented toward structures of pyrolytic graphite type materials, but has been extended more recently to composite material constructions. Both single lamina and laminated plates and shells have been considered. They concluded that due to the high ratio of in-plane Young's modulus to transverse shear modulus for pyrolytic graphite-type materials and composite materials, the use of classical plate and shell theory for analysing the structures of such materials can lead to significant errors.

In 1967, Pagano[60] included transverse shear deformation in the analysis of a bidirectional composite beam. The agreement between experimental data and theoretical results was found to be excellent. He also drew attention to the fact that the effect of transverse shear is significant even for beams with a span-to-depth ratio as high as 30.

In 1969, Wu and Vinson[47-50,54] studied the vibration analysis for plates composed of pyrolytic graphite materials as well as those composed of composite materials. They discovered that classical theory can predict frequencies that are erroneous by a factor of nearly two compared with the more accurate solution evinced when considering transverse shear deformation. As expected, the rotatory inertia effect was shown to be negligible.

Whitney,[61] also in 1969, extended Ambartsumyan's[43] work on the analysis of laminated generally anisotropic plates. Whitney assumed transverse shear stress functions for the laminated unsymmetric general anisotropic plate analogous to those used by Ambartsumyan for homogeneous specially orthotropic plates. His solution for infinite two-layer plates is comparable to the corresponding exact beam solution by Pagano.[60] All numerical curves were calculated using a parabolic function which Ambartsumyan used for homogeneous plates. This function leads to a shear correction factor of $\frac{2}{3}$ (the same as that used by Timoshenko[62]) unlike $\frac{5}{6}$ due to Reissner's variation method[63] and $\pi^2/12$ from Mindlin,[64] where the Reissner and Mindlin constants are almost identical.

Pagano published two papers in 1970,[65,66] one dealing with exact solutions for composite laminated beams in cylindrical bending and the other dealing with rectangular bidirectional composite layered plates. In both papers, he indicated that the curve of the magnitude of transverse shear at any point is of course a function of the thickness co-ordinate, and is discontinuous in its first derivative at the interface between layers. This behaviour is a result of different shear moduli in two adjacent layers. To find an assumed function for transverse shear stress satisfying these discontinuities is somewhat difficult. An alternative approach to the above was suggested by Vinson[67] in 1958, that an accurate solution can be found by treating each single layer individually employing shear stress, normal stress and displacement boundary conditions between each lamina. The advantages of this treatment are twofold: first, one is able to investigate the shear stress as well as the transverse normal stress at each interface between layers; second, the solution function for the shear stress throughout the thickness automatically satisfies the boundary conditions at each interface. This approach has been used by Summers,[44] Mehta,[55] Zukas,[56] Renton[59] and Waltz[77] in their research.

The use of a parabolic function from the homogeneous plate theory for the shear stress distribution across the plate can predict the behaviour of plates sufficiently accurately, as outlined by Whitney[61] and Ambartsumyan.[43] Based on this, one is able to develop reliable analysis methods for laminated anisotropic plates and shells.

It is worthwhile noting that although the problem of anisotropic plates and shells has been investigated for almost a century, the analyses of unsymmetric, laminated, generally orthotropic cases are still in their infancy. One work, by Whitney,[61] deals with unsymmetric, laminated, generally orthotropic plates, including transverse shear deformation. As mentioned many times above, the effects of transverse shear deformation are significant and it is necessary that they be considered when dealing with the analysis of composite structures. Since only the surface of composite structural analysis has been scratched, the development of more accurate methods of analysis is urgent. What follows in the remainder of this chapter will provide the reader with some of the solutions that are available, and the techniques used in obtaining them. Hopefully it will stimulate additional solutions.

6.3 REISSNER'S VARIATIONAL THEOREM AND ITS APPLICATION

Because of its broad application in the analysis of isotropic or anisotropic plates, accounting for transverse shear deformation and transverse normal stress, Reissner's variational theorem is dealt with here, and its application to a beam structure is presented as a simple example. In addition, its use with Hamilton's principle deals with the vibration problem. The following comprises the bulk of

Reference 68. A general discussion of the variational principle is presented, followed by a treatment of the theory of moderately thick beams, which represents a striking example of the power of this technique. The first application is the development of the governing equations for the static deformations of moderately thick rectangular beams, including the effects of transverse shear deformation and transverse normal stress. The second application involves the use of the theorem, together with Hamilton's Principle, in developing a theory of beam vibrations, including rotatory inertia, in addition to the other effects listed above.

The Calculus of Variations has long been recognised as a powerful mathematical tool in many branches of mathematical physics and engineering. Variational principles are found to constitute the central core of many of the most useful techniques in such fields as dynamics, optics and continuum mechanics. The utility of such principles is twofold: first, they provide a very convenient method for the derivation of the governing equations of complex problems and, second, they provide the mathematical foundation required to produce consistent approximate theories. It is in this second role that variational methods have been most useful in the theory of elasticity. There are two variational principles in the classical theory of elasticity, namely the Principle of Minimum Potential Energy and the Principle of Minimum Complementary Energy. It will be useful to discuss these two principles very briefly here, because it was certain of their features which led Reissner, in 1950, to investigate the existence of a third, more general, variational theorem.

The Principle of Minimum Potential Energy was discussed in Section 4.11. It was noted that, in carrying out the variations to minimise V, the class of admissible variations are displacements satisfying the boundary conditions, and the appropriate stress–strain relations have to be obtained by other means. The resulting Euler–Lagrange equations of the variational problem are then equilibrium equations. When the principle is used to formulate approximate theories (i.e. beam, plate or shell theory), it can therefore only yield appropriate equilibrium equations and the stress–strain or stress–displacement relations must be obtained independently.

Consider the Principle of Minimum Complementary Energy, which may be stated as follows: *of all the stress systems satisfying equilibrium and the stress boundary conditions, that which satisfies the compatibility conditions corresponds to a minimum of the complementary energy V^**

defined as

$$V^* = \int_R W \, dR - \int_{S_u} T_i u_i \, dS \qquad (6.1)$$

where S_u denotes that part of the boundary S on which displacements are prescribed.

It is emphasised that, in the Principle of Minimum Complementary Energy, the class of admissible variables are stresses which must satisfy equilibrium everywhere, as well as the stress boundary conditions. The Euler–Lagrange equations of the variational problem are now compatibility equations or stress–displacement relations which insure the satisfaction of the compatibility requirements. Thus, when this principle is used in developing approximate theories, only the stress–displacement relations are obtained and the equilibrium relations must be derived independently.

It should be pointed out that, in the language of structural analysis, the Principle of Minimum Potential Energy corresponds to the Principle of Virtual Displacements, while the Principle of Minimum Complementary Energy corresponds to the Theorem of Castigliano.

It should be clear from the previous discussion that no approximate theory can be obtained in its entirety by the use of either of these principles. In fact, one must either satisfy the stress–strain relations exactly and formulate approximate equilibrium conditions or vice versa. As a result, any approximate theory obtained by such means runs the risk of inconsistency. These considerations led Reissner in 1950 to search for a third variational Theorem of Elasticity which would yield as its Euler–Lagrange equations *both* the equilibrium equations *and* the stress–displacement relations. Clearly, if such a principle could be discovered, its use would yield approximate theories which would satisfy both requirements to the same degree and would, therefore, have the advantage of consistency. The result of this investigation was the Reissner Variational Theorem, which may be stated as follows.

Of all the stress and displacement states satisfying the boundary conditions, those which also satisfy the equilibrium equations and the stress–displacement relations correspond to a minimum of the functional ψ defined as,

$$\psi = \int_R H \, dR - \int_R F_i u_i \, dR - \int_{S_t} T_i u_i \, dS \qquad (6.2)$$

where

S_t = *Portion of* S *on which stresses are prescribed.*

$H = \sigma_{ij}\epsilon_{ij} - W(\sigma_{ij})$.

$W(\sigma_{ij})$ = *Strain energy function in terms of stresses only.*

In a rectangular co-ordinate system, $W(\sigma_{ij})$ in general is written as follows. For an isotropic material:

$$W = \frac{1}{2E}\left[\sigma_x^2 + \sigma_y^2 + \sigma_z^2 - 2\nu(\sigma_x\sigma_y + \sigma_y\sigma_z + \sigma_z\sigma_x) + \right.$$
$$\left. + 2(1 + \nu)(\sigma_{xy}^2 + \sigma_{yz}^2 + \sigma_{zx}^2)\right] \tag{6.3}$$

We now present the proof of the Theorem before considering typical applications. The tensor notation is standard, and can be found in many references, hence it will not be explained here. Also, the operations used herein in taking variations are identical to those of partial differentiation, such as $\delta(\sigma_{ij}^2) = 2\sigma_{ij}\delta\sigma_{ij}$, $\delta(\sigma_{ij}\epsilon_{ij}) = \sigma_{ij}\delta\epsilon_{ij} + \epsilon_{ij}\delta(\sigma_{ij})$.

Taking the variation of ψ and equating it to zero, we obtain

$$\delta\psi = \int_R \left[\sigma_{ij}\delta\epsilon_{ij} + \epsilon_{ij}\delta\sigma_{ij} - \frac{\partial W}{\partial\sigma_{ij}}\delta\sigma_{ij}\right] dR -$$
$$- \int_R F_i\delta u_i\, dR - \int_{S_t} T_i\delta u_i\, dS = 0 \tag{6.4}$$

where

$$\epsilon_{ij} = \frac{1}{2}\left(\frac{\partial u_i}{\partial x_j} + \frac{\partial u_j}{\partial x_i}\right)$$

It should be noted that all stress-and-strain components have been varied, while F_i and T_i, which are prescribed functions, are not. Rearranging the above expression, we obtain

$$\delta\psi = \int_R \left\{\left[\epsilon_{ij} - \frac{\partial W}{\partial\sigma_{ij}}\right]\delta\sigma_{ij} + \frac{1}{2}\sigma_{ij}\left[\frac{\partial}{\partial x_j}(\delta u_i) + \frac{\partial}{\partial x_i}(\delta u_j)\right]\right\} dR -$$
$$- \int_R F_i\delta u_i\, dR - \int_{S_t} T_i u_i\, dS = 0 \tag{6.5}$$

The terms

$$\sigma_{ij}\frac{\partial}{\partial x_j}(\delta u_i) \quad \text{and} \quad \sigma_{ij}\frac{\partial}{\partial x_i}(\delta u_j)$$

are symmetrical with respect to i and j, and we may, therefore, interchange these indices and obtain

$$\delta\psi = \int_R \left\{\left[\epsilon_{ij} - \frac{\partial W}{\partial\sigma_{ij}}\right]\delta\sigma_{ij} + \sigma_{ij}\frac{\partial}{\partial x_j}(\delta u_i)\right\} dR -$$

$$-\int_R F_i \delta u_i \, dR - \int_{S_t} T_i \delta u_i \, dS = 0 \qquad (6.6)$$

We note that

$$\frac{\partial}{\partial x_j} (\sigma_{ij} \delta u_i) = \sigma_{ij} \frac{\partial}{\partial x_j} (\delta u_i) + \frac{\partial \sigma_{ij}}{\partial x_j} \delta u_i$$

so that eqn. (6.6) may be written

$$\delta \psi = \int_R \left\{ \left[\epsilon_{ij} - \frac{\partial W}{\partial \sigma_{ij}} \right] \delta \sigma_{ij} + \frac{\partial}{\partial x_j} (\sigma_{ij} \delta u_i) - \frac{\partial \sigma_{ij}}{\partial x_j} \delta u_i \right\} dR -$$

$$- \int_R F_i \delta u_i \, dR - \int_{S_t} T_i \delta u_i \, dS = 0$$

Applying the divergence theorem and remembering that $\delta u_i = 0$, on all surfaces where displacements are prescribed, we obtain

$$\int_R \frac{\partial}{\partial x_j} (\sigma_{ij} \delta u_i) \, dR = \int_S \sigma_{ij} \nu_j \delta u_i \, dS = \int_{S_t} T_i \delta u_i \, dS \qquad (6.7)$$

where here ν_i is the direction cosine.

Finally, substituting eqn. (6.7) into (6.6) yields the equation

$$\delta \psi = \int_R \left\{ \left[\epsilon_{ij} - \frac{\partial W}{\partial \sigma_{ij}} \right] \delta \sigma_{ij} - \left[\frac{\partial \sigma_{ij}}{\partial x_j} + F_i \right] \delta u_i \right\} dR = 0 \qquad (6.8)$$

Since $\delta \sigma_{ij}$ and δu_i are arbitrary variations, eqn. (6.8) is satisfied only if the stresses σ_{ij} and strains ϵ_{ij} satisfy the equations

$$\frac{\partial \sigma_{ij}}{\partial x_j} + F_i = 0 \qquad (6.9)$$

$$\epsilon_{ij} = \frac{\partial W}{\partial \sigma_{ij}} \qquad (6.10)$$

Equations (6.9) and (6.10) are the equilibrium and stress–strain relations of elasticity. Thus, the variational theorem is found to be equivalent to the three-dimensional equations of elasticity and is, therefore, established. We now consider typical applications of the theorem to the derivation approximate theories for the static and dynamic deformations of beams.

6.3.1 Static Deformation of Moderately Thick Beams

As a first illustration, we consider the development of a theory for the static deformation of moderately thick beams in which the effects of transverse shear deformation and transverse normal stress must be accounted for. We consider a beam of rectangular cross-section of

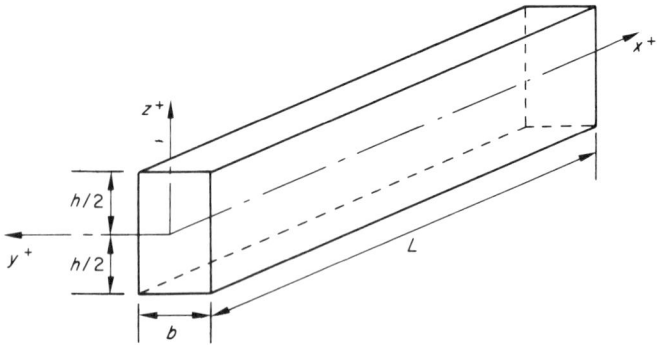

Fig. 6.1

width b, height h and length L, as shown in Fig. 6.1, and subjected to a distributed load $q(x)$ acting on the surface $z = +h/2$.

In order to apply the variational theorem, we first assume for the stresses the following expressions

$$\sigma_x = \frac{Mz}{I}, \quad \text{where} \quad I = \frac{bh^3}{12},$$

$$\sigma_{xz} = \frac{3Q}{2A}\left[1 - \left(\frac{z}{h/2}\right)^2\right], \quad \text{where} \quad A = bh,$$

$$\sigma_z = \frac{3q}{4b}\left[\frac{z}{h/2} + \frac{2}{3} - \frac{1}{3}\left(\frac{z}{h/2}\right)^3\right],$$

$$\sigma_y = \sigma_{yx} = \sigma_{yz} = 0$$

(6.11)

It should be noted that the form of the stress components σ_x and σ_{xz} is identical to that of classical theory; the form of the transverse normal stress σ_z may easily be derived from the stress equations of equilibrium as a consequence of the assumptions made for σ_x and σ_{xz}. The stress-couple M and shear resultant Q are defined in the usual manner by the equations

$$M(x) = \int_{-h/2}^{h/2} b\sigma_x z \, dz$$

$$Q(x) = \int_{-h/2}^{h/2} b\sigma_{xz} \, dz$$

(6.12)

It should be noted that the eqns. (6.11) satisfy all the stress boundary conditions.

As in classical beam-bending theory, we assume that beam cross-sections undergo translation and rotation but no deformation in their own planes, so that the displacements are of the following form for bending only (no stretching)

$$u = z\alpha(x)$$
$$w = w(x)$$

(6.13)

It should be noted that the cross-sections will not be assumed to remain normal to the deformed middle surface; this assumption, made in classical beam theory, is equivalent to the neglect of transverse shear deformation, and will not be made here.

The appropriate strain displacement relations may be written

$$\epsilon_x = z\alpha'(x)$$

$$\epsilon_{xz} = \frac{1}{2}(\alpha + w')$$

(6.14)

$$\epsilon_z = \frac{\partial w}{\partial z} = 0$$

where the primes denote differentiation with respect to x.

For the present case, the functional ψ (eqn. (6.2)) takes the form,

$$\psi = \int_0^L \int_{-h/2}^{h/2} b \left\{ \sigma_x z \alpha' + \sigma_{xz}(\alpha + w') - \right.$$

$$\left. - \frac{1}{2E} [\sigma_x^2 + \sigma_z^2 - 2\nu\sigma_x\sigma_z + 2(1 + \nu)\sigma_{xz}^2] \right\} dz\ dx - \int_0^L wq\ dx \quad (6.15)$$

Substituting eqn. (6.11) in eqn. (6.15) and carrying out the integrations with respect to z, we obtain,

$$\psi = \int_0^L \left\{ M\alpha' + Q(\alpha + w') - \frac{M^2}{2EI} + \frac{6\nu qM}{5EA} - \frac{3Q^2}{5GA} - qw \right\} dx +$$

$$+ \int_0^L \int_{-h/2}^{h/2} \frac{\sigma_z^2}{2E} b\ dz\ dx \quad (6.16)$$

It should be noted that the integration of the term in σ_z^2 has not been carried out because this term depends only on q and not on the basic unknown stresses and displacements, α, w, M and Q. Thus, when we take variations to minimise ψ, the term in σ_z^2 will not contribute to the result. We may now obtain the governing equations we seek by minimising the functional ψ of eqn. (6.16). Taking the variation of this equation gives,

$$\delta\psi = \int_0^L \left\{ M\delta(\alpha') + \alpha'\delta M + Q[\delta\alpha + \delta(w')] + (\alpha + w')\delta Q - \frac{M}{EI}\delta M + \right.$$

$$\left. + \frac{6\nu q}{5EA}\delta M - \frac{6Q}{5GA}\delta Q - q\delta w \right\} dx = 0 \tag{6.17}$$

Integrating by parts and rearranging, eqn. (6.17) may be written in the form,

$$\delta\psi = [M\delta\alpha + Q\delta w]_0^L + \int_0^L \left\{ [Q - M']\delta\alpha - [Q' + q]\delta w + \right.$$

$$\left. + \left[\alpha' - \frac{M}{EI} + \frac{6\nu q}{5EA} \right]\delta M + \left[\alpha + w' - \frac{6Q}{5GA} \right]\delta Q \right\} dx = 0 \tag{6.18}$$

Setting the first term equal to zero yields the natural boundary conditions for the beam. It is seen that, either $M = 0$ or α must be prescribed at $x = 0$, and L, and either $Q = 0$ or w must be prescribed at $x = 0$, and L.

Finally, since the variations $\delta\alpha$, δw, δM and δQ are all independent arbitrary functions of x, the only way in which the definite integral of eqn. (6.18) can be made to vanish is by requiring the unknowns M, Q, α and w to satisfy the equations,

$$-\frac{dM}{dx} + Q = 0 \tag{6.19}$$

$$\frac{dQ}{dx} + q = 0 \tag{6.20}$$

$$\frac{d\alpha}{dx} - \frac{M}{EI} + \frac{6\nu q}{5EA} = 0 \tag{6.21}$$

$$\alpha + \frac{dw}{dx} - \frac{6Q}{5GA} = 0 \tag{6.22}$$

We note that eqns. (6.19) and (6.20) are identical with the equilibrium equations of classical beam theory. This is as expected since no new stress resultants or stress couples were introduced. Considering eqn. (6.22), it is seen that the quantity $\alpha + w'$ is precisely the change in the angle between the beam cross-section and the middle surface occurring during the deformation; eqn. (6.22) shows that this angular change, which is a measure of the shear deformation, must be proportional to Q/A which is the average shear stress. In addition, we note that as $G \to \infty$, the shear deformation tends to vanish as assumed in classical beam theory. Finally, we observe that the third term of eqn. (6.21) depends on the lateral load q and the Poisson's ratio ν; this term would vanish if we assumed $\nu = 0$ as in classical beam theory, and it is identified as the effect of the transverse normal stress σ_z which is proportional to q, according to our initial assumptions (see eqn. (6.11)). Solutions of eqns. (6.19) to (6.22) may easily be obtained for typical loading and boundary conditions; these solutions reveal that the effects of transverse shear deformation and transverse normal stress are negligible for sufficiently large values of L/h and become important as L/h decreases and becomes of order unity.

6.3.2 Vibrations of Moderately Thick Beams

As a second example, a theory of free vibrations for moderately thick beams of rectangular cross-section is treated; this will include the effects of transverse shear deformation and rotatory inertia.

In order to derive the equations of motion, we now apply Hamilton's principle in conjunction with the Reissner Variational Theorem. It will be remembered that Hamilton's principle is nothing but a variational statement of Newton's Laws of Motion. Thus, we may state that the motion of the beam of Fig. 6.1 will be such as to minimise the integral

$$\Phi = \int_{t_1}^{t_2} (T - \psi) \, dt \tag{6.23}$$

where

T = Kinetic energy of the system.
ψ = Reissner functional.
t = Time.

The equations of motion will now be obtained from the condition

$$\delta \Phi = 0 \tag{6.24}$$

and it must be remembered that all stresses, strains and displacements are now functions of time, as well as the space co-ordinates x and z.

The kinetic energy for the beam of Fig. 6.1 may be written in the form

$$T = \int_0^L \int_{-h/2}^{h/2} \frac{\rho b}{2} \left[\left(\frac{\partial u}{\partial t} \right)^2 + \left(\frac{\partial w}{\partial t} \right)^2 \right] dz \, dx \tag{6.25}$$

where ρ is the mass density of the beam material. Substituting eqn. (6.13) in eqn. (6.25) and integrating with respect to z gives

$$T = \int_0^L \frac{\rho}{2} \left[I \left(\frac{\partial \alpha}{\partial t} \right)^2 + A \left(\frac{\partial w}{\partial t} \right)^2 \right] dx \tag{6.26}$$

The substitution of eqns. (6.16) and (6.26) in eqn. (6.23) then yields

$$\Phi = \int_{t_1}^{t_2} \int_0^L \left\{ \frac{\rho}{2} \left[I \left(\frac{\partial \alpha}{\partial t} \right)^2 + A \left(\frac{\partial w}{\partial t} \right)^2 \right] - M \frac{\partial \alpha}{\partial x} - Q \left(\alpha + \frac{\partial w}{\partial x} \right) + \right.$$
$$\left. + \frac{M^2}{2EI} - \frac{6\nu q M}{5EA} + \frac{3Q^2}{5GA} + qw \right\} dx \, dt \tag{6.27}$$

where the term in σ_z^2 has been dropped since it will not contribute to the variations (as explained previously).

The governing equations are then obtained by taking the variation of eqn. (6.27) and setting the result equal to zero. It is found that the natural boundary conditions are the same as for the static case, while the initial deflection and velocity must also be specified. The equations of motion are obtained in the form,

$$Q - \frac{\partial M}{\partial x} + \rho I c \frac{\partial^2 \alpha}{\partial t^2} = 0 \tag{6.28}$$

$$\frac{\partial Q}{\partial x} - \rho A \frac{\partial^2 w}{\partial t^2} + q(x, t) = 0 \tag{6.29}$$

$$\frac{\partial \alpha}{\partial x} - \frac{M}{EI} + \frac{6\nu q}{5EA} = 0 \tag{6.30}$$

$$\alpha + \frac{\partial w}{\partial x} - \frac{6Qk}{5GA} = 0 \tag{6.31}$$

In the above equations, we have introduced two tracing constants c and k for the purpose of identifying terms. We note that eqn. (6.28) is identical to the corresponding moment equilibrium condition of classical beam theory, except for the term $\rho I c (\partial^2 \alpha / \partial t^2)$ which represents the contribution of rotatory inertia. Thus, when we set $c = 1$ in the resulting solutions, we include rotatory inertia, and when we set $c = 0$, we obtain a theory which neglects the effect of rotatory inertia. Equation (6.29) is identical to the classical beam theory equation for transverse force equilibrium. Equation (6.30) exhibits the term $6\nu q / 5EA$, which is the contribution of transverse normal stress; since this is the only term in which ν appears explicitly, setting $\nu = 0$ is equivalent to neglecting the transverse normal stress. Equation (6.31) is identical to the corresponding stress–strain relation of classical theory and the term $6Qk/5GA$ represents the effect of shear deformation which is included when we set $k = 1$ and neglected when $k = 0$. We now consider a simple application of this theory.

6.3.3 Natural Frequencies of a Simply-Supported Beam

For free vibrations, eqns. (6.28) to (6.31) reduce to,

$$
\left.
\begin{aligned}
Q - \frac{\partial M}{\partial x} + \rho Ic \frac{\partial^2 \alpha}{\partial t^2} &= 0 \\
\frac{\partial Q}{\partial x} - \rho A \frac{\partial^2 w}{\partial t^2} &= 0 \\
\frac{\partial \alpha}{\partial x} - \frac{M}{EI} &= 0 \\
\alpha + \frac{\partial w}{\partial x} - \frac{6Qk}{5GA} &= 0
\end{aligned}
\right\}
\tag{6.32}
$$

It is convenient to reduce these equations to a system of two equations in the unknown displacements w and α.

From the first and the third of eqn. (6.32), we obtain,

$$
M = EI \frac{\partial \alpha}{\partial x}
$$

$$
Q = EI \frac{\partial^2 \alpha}{\partial x^2} - \rho Ic \frac{\partial^2 \alpha}{\partial t^2}
$$

and the substitution of these expressions in the second and fourth of eqn. (6.32) yields

$$
\left.
\begin{aligned}
\frac{\partial^3 \alpha}{\partial x^3} - \frac{\rho c}{E} \frac{\partial^3 \alpha}{\partial x \partial t^2} - \frac{\rho A}{EI} \frac{\partial^2 w}{\partial t^2} &= 0 \\
\alpha + \frac{\partial w}{\partial x} - \frac{kh^2}{10} \left[\frac{E}{G} \frac{\partial^2 \alpha}{\partial x^2} - \frac{\rho c}{G} \frac{\partial^2 \alpha}{\partial t^2} \right] &= 0
\end{aligned}
\right\}
\tag{6.33}
$$

For a simply supported beam of length L, the boundary conditions are

$$
w = M = 0 \quad \text{for} \quad x = 0, L
$$

and when the beam is oscillating in a normal mode, the motion is harmonic so that the solutions for α and w may be taken in the form

$$
\left.
\begin{aligned}
w &= W_n \sin \frac{n\pi x}{L} \cos \omega_n t \\
\alpha &= \Gamma_n \cos \frac{n\pi x}{L} \cos \omega_n t
\end{aligned}
\right\}
\tag{6.34}
$$

where W_n and Γ_n are the amplitudes of the translation and rotation respectively, and ω_n is the natural frequency of the nth mode of vibration. It is easily verified that these expressions satisfy the boundary conditions. The substitution of eqn. (6.34) for eqn. (6.33) yields two simultaneous homogeneous algebraic equations for the amplitude W_0 and Γ; these are

$$\left[\left(\frac{n\pi}{L}\right)^3 - \frac{\rho c}{E}\left(\frac{n\pi}{L}\right)\omega_n^2\right]\Gamma_n + \frac{\rho A}{EI}\omega_n^2 W_n = 0$$

$$\left[1 + \frac{kh^2}{10}\left(\frac{E}{G}\frac{n^2\pi^2}{L^2} - \frac{\rho c}{G}\omega_n^2\right)\right]\Gamma_n + \left(\frac{n\pi}{L}\right)W_n = 0$$

(6.35)

Since eqn. (6.35) forms a homogeneous system, the condition for a non-trivial solution is that the determinant of the system vanishes; this yields the frequency equation, while the amplitude ratios are obtained from satisfying either of the two equations. Thus, the amplitude ratio is given by

$$\Gamma_n = -\frac{(\rho A/EI)\omega_n^2}{(n\pi/L)^3 - (\rho c/E)(n\pi/L)\omega_n^2}W_n$$

(6.36)

while the frequency equation may be written in the form

$$\omega_n^4 - \left[\frac{10G}{\rho h^2 kc} + \left(\frac{n\pi}{L}\right)^2\left(\frac{E}{\rho c} + \frac{5}{6}\frac{G}{\rho k}\right)\right]\omega_n^2 + \frac{10}{h^2 kc}\left(\frac{G}{\rho}\right)\left(\frac{EI}{\rho A}\right)\left(\frac{n\pi}{L}\right)^4 = 0$$

(6.37)

The natural frequencies for the case where both shear and rotatory inertia are included may be obtained by solving eqn. (6.37) with $k = c = 1$. The frequency equation is then of the form

$$\omega_n^4 - \left[\frac{10G}{\rho h^2} + \left(\frac{n\pi}{L}\right)^2\left(\frac{E}{\rho} + \frac{5}{6}\frac{G}{\rho}\right)\right]\omega_n^2 + \frac{10}{h^2}\left(\frac{G}{\rho}\right)\left(\frac{EI}{\rho A}\right)\left(\frac{n\pi}{L}\right)^4 = 0 \quad (6.38)$$

To obtain a simplified theory neglecting the effect of rotatory inertia, but retaining shear deformation, set $k = 1$ and $c = 0$ after multiplying the frequency equation (6.38) by c; the resulting simplified frequency equation may be written

$$\left[\frac{10G}{\rho h^2} + \left(\frac{n\pi}{L}\right)^2\frac{E}{\rho}\right]\omega_n^2 - \frac{10}{h^2}\left(\frac{G}{\rho}\right)\left(\frac{EI}{\rho A}\right)\left(\frac{n\pi}{L}\right)^4 = 0$$

(6.39)

and the natural frequencies are given by

$$\omega_n^2 = \left\{\left(\frac{EI}{\rho A}\right)\left(\frac{n\pi}{L}\right)^4 \bigg/ \left[1 + \frac{n^2\pi^2}{10}\left(\frac{E}{G}\right)\left(\frac{h}{L}\right)^2\right]\right\}$$

(6.40)

Finally, to obtain a frequency equation in which both shear deformation and rotatory inertia are neglected, we multiply eqn. (6.37) by kc and set $k = c = 0$; the frequencies are then given by

$$\omega_n^2 = \left(\frac{EI}{\rho A}\right)\left(\frac{n\pi}{L}\right)^4 \qquad (6.41)$$

Equation (6.41) is easily recognised to be the well-known solution of classical beam theory for a simply-supported beam. A few calculations using eqns. (6.38) and (6.40) will show that most of the error (approximately 90 per cent) of the classical theory is due to the neglect of transverse shear deformation, so that accurate results may be obtained by using eqn. (6.40) which still neglects rotatory inertia, but has the advantage of simplicity. Comparison of eqns. (6.40) and (6.41) reveals that the effect of shear deformation is to reduce the frequencies by a factor equal to

$$1 + \frac{n^2 \pi^2}{10}\left(\frac{E}{G}\right)\left(\frac{h}{L}\right)^2$$

It is seen that this factor increases with increasing h/L, so that the error in classical theory tends to become large for 'stubby' beams. The factor also increases with n, indicating that classical theory is only adequate for lower modes and becomes increasingly inaccurate for higher modes.

We will now employ the Reissner Variational Theorem in solving the problem of the next section.

6.4   NATURAL VIBRATIONS OF PLATES COMPOSED OF COMPOSITE MATERIALS

Consider the free flexural vibration of a flat, rectangular plate of thickness h, referred to an O-xyz system of Cartesian co-ordinates. $z = + h/2$ and $- h/2$ are taken as the upper and lower faces and, O is at a corner of the plate. The plate material possesses three mutually perpendicular axes of elastic symmetry, two of which lie in the plane of the plate parallel to the geometric axes. The third axis of symmetry is normal to the plane of the plate. Under the assumption that the deflection of the plate is small, the deformation of lineal elements originally perpendicular to the xy-plane may be considered to be a rotation. Thus, the relations among strains, displacements and stresses can be written, analogous to eqn. (6.14),

$$\epsilon_x = z\frac{\partial \alpha}{\partial x} = \frac{1}{E_x}\sigma_x - \frac{\nu_{yx}}{E_y}\sigma_y - \frac{\nu_{zx}}{E_z}\sigma_z$$

$$\epsilon_y = z\frac{\partial \beta}{\partial y} = -\frac{\nu_{xy}}{E_x}\sigma_x + \frac{1}{E_y}\sigma_y - \frac{\nu_{yz}}{E_y}\sigma_z$$

$$\epsilon_{xy} = \frac{1}{2}\left(\frac{\partial \alpha}{\partial y} + \frac{\partial \beta}{\partial x}\right) = \frac{1}{2G_{xy}}\,\sigma_{xy}$$

$$\epsilon_{yz} = \frac{1}{2}\left(\frac{\partial w}{\partial y} + \beta\right) = \frac{1}{2G_{yz}}\,\sigma_{yz}$$

$$\epsilon_{xz} = \frac{1}{2}\left(\frac{\partial w}{\partial x} + \alpha\right) = \frac{1}{2G_{xz}}\,\sigma_{xz} \tag{6.42}$$

$$\epsilon_z = 0$$

Note that w, α, and β are assumed as functions of x, y and t. Additionally, the thickness is assumed to be unchanged during the deformation procedure, and the elements normal to the middle surface before deformation are clearly not perpendicular to the deformed middle plane.

As a matter of fact, for the free vibrations case, we may neglect the transverse normal stress since there is no applied lateral load (see eqn. (6.21)). Applying Reissner's Variational Theorem, the following expressions, analogous to eqn. (6.11) are used

$$(\sigma_x, \sigma_y, \sigma_{xy}) = \frac{12z}{h^3}(M_x, M_y, M_{xy})$$

$$(\sigma_{xz}, \sigma_{yz}) = \frac{3}{2h}\left[1 - \left(\frac{z}{h/2}\right)^2\right](Q_x, Q_y) \tag{6.43}$$

$$\sigma_z = 0$$

The expression H for the strain energy as used by Reissner is given by

$$H = \sigma_{ij}\epsilon_{ij} - W(\sigma_{ij}) \tag{6.44}$$

where for the case of a rectangular plate composed of a specially orthotropic material

$$\sigma_{ij}\epsilon_{ij} = \sigma_x z \frac{\partial \alpha}{\partial x} + \sigma_y z \frac{\partial \beta}{\partial y} + \sigma_z \frac{\partial w}{\partial z} + \sigma_{xy}\left(\frac{\partial \alpha}{\partial y} + \frac{\partial \beta}{\partial x}\right)z +$$

$$+ \sigma_{yz}\left(\frac{\partial w}{\partial y} + \beta\right) + \sigma_{xz}\left(\frac{\partial w}{\partial x} + \alpha\right)$$

$$W(\sigma_{ij}) = \frac{1}{2}\left[\left(\frac{\sigma_x^2}{E_x} + \frac{\sigma_y^2}{E_y} + \frac{\sigma_z^2}{E_z}\right) - \right.$$

$$- 2\left(\frac{\nu_{xy}}{E_x}\,\sigma_x\sigma_y + \frac{\nu_{yz}}{E_y}\,\sigma_y\sigma_z + \frac{\nu_{xz}}{E_x}\,\sigma_x\sigma_z\right)$$

$$\left. + \frac{\sigma_{yz}^2}{G_{yz}} + \frac{\sigma_{xz}^2}{G_{xz}} + \frac{\sigma_{xy}^2}{G_{xy}}\right]$$

Referring to eqn. (6.2), neglecting body forces, and since no surface forces are concerned, the Reissner functional, ψ, is expressed as

$$\psi = \int_0^a \int_0^b \int_{-h/2}^{+h/2} H \, dx \, dy \, dz \tag{6.45}$$

The kinetic energy of the plate is analogous to eqn. (6.25)

$$T = \int_0^a \int_0^b \int_{-h/2}^{h/2} \frac{\rho}{2} \left[z^2 \left(\frac{\partial \alpha}{\partial t} \right)^2 + z^2 \left(\frac{\partial \beta}{\partial t} \right)^2 + \left(\frac{\partial w}{\partial t} \right)^2 \right] dx \, dy \, dz \tag{6.46}$$

Equation (6.23) for this case is written as

$$\Phi = \int_{t_1}^{t_2} \int_0^a \int_0^b \left\{ \frac{\rho h^3}{24} \left[\left(\frac{\partial \alpha}{\partial t} \right)^2 + \left(\frac{\partial \beta}{\partial t} \right)^2 \right] + \frac{\rho h}{2} \left(\frac{\partial w}{\partial t} \right)^2 - \right.$$

$$- M_x \frac{\partial \alpha}{\partial x} - M_y \frac{\partial \beta}{\partial y} - M_{xy} \left(\frac{\partial \alpha}{\partial y} + \frac{\partial \beta}{\partial x} \right) - Q_x \left(\alpha + \frac{\partial w}{\partial x} \right) -$$

$$- Q_y \left(\beta + \frac{\partial w}{\partial y} \right) + \frac{6M_x^2}{E_x h^3} + \frac{6M_y^2}{E_y h^3} + \frac{6M_{xy}^2}{G_{xy} h^3} -$$

$$\left. - \frac{12\nu_{xy} M_x M_y}{E_x h^3} + \frac{3}{5} \left(\frac{Q_x^2}{G_{xz} h} + \frac{Q_y^2}{G_{yz} h} \right) \right\} dt \, dx \, dy \tag{6.47}$$

According to Hamilton's Principle, as in eqn. (6.24)

$$\delta \Phi = \delta \int_{t_1}^{t_2} (T - \psi) \, dt = 0 \tag{6.48}$$

or

$$\int_0^a \int_0^b \left[\frac{\rho h^3}{12} \left(\frac{\partial \alpha}{\partial t} \delta\alpha + \frac{\partial \beta}{\partial t} \delta\beta \right) + \rho h \frac{\partial w}{\partial t} \delta w \right]_{t_1}^{t_2} dx \, dy -$$

$$- \int_{t_1}^{t_2} \int_0^a [M_{xy} \delta\alpha + M_y \delta\beta + Q_y \delta w]_0^b \, dt \, dx -$$

$$- \int_{t_1}^{t_2} \int_0^b [M_x \delta\alpha + M_{xy} \delta\beta + Q_x \delta w]_0^a \, dt \, dy -$$

$$- \int_{t_1}^{t_2} \int_0^a \int_0^b \left\{ \left[-\frac{\partial \alpha}{\partial x} + \frac{12}{E_x h^3} M_x - \frac{12\nu_{xy}}{E_x h^3} M_y \right] \delta M_x + \right.$$

$$+ \left[-\frac{\partial \beta}{\partial y} + \frac{12}{E_y h^3} M_y - \frac{12\nu_{yx}}{E_y h^3} M_x \right] \delta M_y +$$

$$+ \left[-\frac{\partial \beta}{\partial x} - \frac{\partial \alpha}{\partial y} + \frac{12}{G_{xy} h^3} M_{xy} \right] \delta M_{xy} +$$

$$+ \left[-\frac{\partial w}{\partial x} - \alpha + \frac{6Q_x}{5G_{xz} h} \right] \delta Q_x +$$

$$+ \left[-\frac{\partial w}{\partial y} - \beta + \frac{6Q_y}{5G_{yz}h} \right] \delta Q_y +$$

$$+ \left[\frac{\partial M_x}{\partial x} + \frac{\partial M_{xy}}{\partial y} - Q_x - \frac{\rho h^3}{12} \frac{\partial^2 \alpha}{\partial t^2} \right] \delta \alpha +$$

$$+ \left[\frac{\partial M_{xy}}{\partial x} + \frac{\partial M_y}{\partial y} - Q_y - \frac{\rho h^3}{12} \frac{\partial^2 \beta}{\partial t^2} \right] \delta \beta +$$

$$+ \left[\frac{\partial Q_x}{\partial x} + \frac{\partial Q_y}{\partial y} - \rho h \frac{\partial^2 w}{\partial t^2} \right] \delta w \Big\} \, dt \, dx \, dy = 0 \qquad (6.49)$$

Setting the first three integrals in the above equation equal to zero, the natural boundary conditions are found and the initial conditions are specified. From eqn. (6.49), the Euler–Lagrangian differential equations of the variational problem are found to be as the following

$$\frac{\partial \alpha}{\partial x} - \frac{12}{E_x h^3} M_x + \frac{12 \nu_{xy}}{E_x h^3} M_y = 0 \qquad (6.50)$$

$$\frac{\partial \beta}{\partial y} - \frac{12}{E_y h^3} M_y + \frac{12 \nu_{yx}}{E_y h^3} M_x = 0 \qquad (6.51)$$

$$\frac{\partial \alpha}{\partial y} + \frac{\partial \beta}{\partial x} - \frac{12}{G_{xy} h^3} M_{xy} = 0 \qquad (6.52)$$

$$\frac{\partial w}{\partial x} + \alpha - \frac{6k}{5G_{xz}h} Q_x = 0 \qquad (6.53)$$

$$\frac{\partial w}{\partial y} + \beta - \frac{6k}{5G_{yz}h} Q_y = 0 \qquad (6.54)$$

$$\frac{\partial M_x}{\partial x} + \frac{\partial M_{xy}}{\partial y} - Q_x - \frac{\rho h^3 f}{12} \frac{\partial^2 \alpha}{\partial t^2} = 0 \qquad (6.55)$$

$$\frac{\partial M_{xy}}{\partial x} + \frac{\partial M_y}{\partial y} - Q_y - \frac{\rho h^3 f}{12} \frac{\partial^2 \beta}{\partial t^2} = 0 \qquad (6.56)$$

$$\frac{\partial Q_x}{\partial x} + \frac{\partial Q_y}{\partial y} - \rho h \frac{\partial^2 w}{\partial t^2} = 0 \qquad (6.57)$$

In the above equations, there are two tracing constants f and k introduced for the purpose of identifying the effects of rotatory inertia and transverse shear deformation respectively. It is useful to investigate the plate oscillation, if we reduce eqns. (6.50) to (6.57) to a system of three equations in the unknown displacements w, α and β. The results are

$$\alpha + \frac{\partial w}{\partial x} - \frac{6k}{5G_{xz}h}\left[D_x\frac{\partial^2\alpha}{\partial x^2} + D_{xy}\frac{\partial^2\alpha}{\partial y^2} - \frac{\rho h^3 f}{12}\frac{\partial^2\alpha}{\partial t^2} + \right.$$
$$\left. + (\bar{H} - D_{xy})\frac{\partial^2\beta}{\partial x\partial y}\right] = 0$$

$$\beta + \frac{\partial w}{\partial y} - \frac{6k}{5G_{yz}h}\left[D_y\frac{\partial^2\beta}{\partial y^2} + D_{xy}\frac{\partial^2\beta}{\partial x^2} - \frac{\rho h^3 f}{12}\frac{\partial^2\beta}{\partial t^2} + \right. \qquad (6.58)$$
$$\left. + (\bar{H} - D_{xy})\frac{\partial^2\alpha}{\partial x\partial y}\right] = 0$$

$$G_{xz}\frac{\partial\alpha}{\partial x} + G_{yz}\frac{\partial\beta}{\partial y} + G_{xz}\frac{\partial^2 w}{\partial x^2} + G_{yz}\frac{\partial^2 w}{\partial y^2} + \frac{6\rho k}{5}\frac{\partial^2 w}{\partial t^2} = 0$$

where

$$D_x = \frac{E_x h^3}{12(1 - \nu_{xy}\nu_{yx})}, \qquad D_y = \frac{E_y h^3}{12(1 - \nu_{xy}\nu_{yx})}, \qquad D_{xy} = \frac{G_{xy}h^3}{12}$$
$$\bar{H} = D_x\nu_{yx} + 2D_{xy}$$

We now look at the example for a plate with all edges simply supported. The boundary conditions are

$$w = M_x = \beta = 0 \qquad \text{on } x = 0, a$$
$$w = M_y = \alpha = 0 \qquad \text{on } y = 0, b \qquad (6.59)$$

No difficulty is encountered in effecting a separation of space and time-dependent variables for the coupled homogeneous differential form of eqn. (6.58). When the plate is oscillating in a normal mode, the solutions for w, α and β may be taken in a form of a harmonic function of time with circular natural frequency ω_{mn}, such as

$$w = W_0 \sin\frac{n\pi x}{a}\sin\frac{m\pi y}{b}\cos\omega_{mn}t$$

$$\alpha = \Gamma\cos\frac{n\pi x}{a}\sin\frac{m\pi y}{b}\cos\omega_{mn}t \qquad (6.60)$$

$$\beta = \Lambda\sin\frac{n\pi x}{a}\cos\frac{m\pi y}{b}\cos\omega_{mn}t$$

One can easily prove that the expressions in (6.60) satisfy the boundary conditions of (6.59). The initial conditions are specified. The substitution of eqn. (6.60) in eqn. (6.58) yields

$$\lambda_n W_0 + \left[1 + \frac{6k}{5G_{xz}h}\left(D_x\lambda_n^2 + D_{xy}\delta_m^2 - \frac{\rho h^3 f}{12}\omega_{mn}^2\right)\right]\Gamma +$$

$$+ \frac{6k}{5G_{xz}h}(\bar{H} - D_{xy})\lambda_n\delta_m\Lambda = 0$$

$$\delta_m W_0 + \left[1 + \frac{6k}{5G_{yz}h}\left(D_y\delta_m^2 + D_{xy}\lambda_n^2 - \frac{\rho h^3 f}{12}\omega_{mn}^2\right)\right]\Lambda +$$

$$+ \frac{6k}{5G_{yz}h}(\bar{H} - D_{xy})\lambda_n\delta_m\Gamma = 0$$

$$\left(G_{xz}\lambda_n^2 + G_{yz}\delta_m^2 - \frac{6\rho k}{5}\omega_{mn}^2\right)W_0 + G_{xz}\lambda_n\Gamma + G_{yz}\delta_m\Lambda = 0$$

(6.61)

where

$$\lambda_n = \frac{n\pi}{a}, \qquad \delta_m = \frac{m\pi}{b}$$

For a non-trivial solution of eqn. (6.61), the determinant of the coefficients of W, Γ and Λ vanishes; this yields the frequency equation as the following

$$\omega_{mn}^6 - \left\{\frac{10}{\rho h^2 f}(G_{xz} + G_{yz}) + \left(\frac{5}{6\rho k}\right)(G_{xz}\lambda_n^2 + G_{yz}\delta_m^2) + \right.$$

$$+ \frac{12}{\rho h^3 f}[(D_x + D_{xy})\lambda_n^2 + (D_y + D_{xy})\delta_m^2]\Big\}\omega_{mn}^4 +$$

$$+ \left\{\frac{100G_{xz}G_{yz}}{\rho^2 h^4 f^2 k^2} + \frac{25}{3}\frac{G_{xz}G_{yz}}{\rho^2 h^2 fk^2}(\lambda_n^2 + \delta_m^2) + \right.$$

$$+ \frac{120}{\rho^2 h^3 f^2 k}[(G_{yz}D_x + G_{xz}D_{xy})\lambda_n^2 + (G_{xz}D_y + G_{yz}D_{xy})\delta_m^2] +$$

$$+ \frac{10}{\rho^2 h^3 fk}[G_{xz}(D_x + D_{xy})\lambda_n^4 + G_{yz}(D_y + D_{xy})\delta_m^4] +$$

$$+ \frac{144D_{xy}}{\rho^2 h^6 f^2}(D_x\lambda_n^4 + D_y\delta_m^4) +$$

$$+ \left[\frac{10G_{yz}}{\rho^2 h^3 fk}(D_x + D_{xy}) + \frac{10G_{xz}}{\rho^2 h^3 fk}(D_y + D_{xy}) + \right.$$

$$+ \frac{144}{\rho^2 h^6 f^2}(D_xD_y - \bar{H}^2 + 2\bar{H}D_{xy})\Big]\lambda_n^2\delta_m^2\Big\}\omega_{mn}^2 -$$

$$- \left\{\frac{100G_{xz}F_{yz}}{\rho^3 h^5 k^2 f^2}(D_x\lambda_n^4 + 2\bar{H}\lambda_n^2\delta_m^2 + D_y\delta_m^4) + \right.$$

$$+ \frac{120}{\rho^3 h^6 k f^2} [D_x D_{xy} G_{xz} \lambda_n^6 + D_y D_{xy} G_{yz} \delta_m^6 +$$

$$+ (D_x D_{xy} G_{yz} + D_x D_y G_{xz} - \bar{H}^2 G_{xz} + 2\bar{H} D_{xy} G_{xz}) \lambda_n^4 \delta_m^2 +$$

$$+ (D_y D_{xy} G_{xz} + D_x D_y G_{yz} - \bar{H}^2 G_{yz} + 2\bar{H} D_{xy} G_{yz}) \lambda_n^2 \delta_m^4] \} = 0 \quad (6.62)$$

Solving eqn. (6.62) with $k = f = 1$, three principal natural frequencies of the case where both transverse shear deformation and rotatory inertia are included can be obtained. Two of them are very much higher than the other one which corresponds to the primarily lateral motion. The higher frequencies are named as the motions of thickness-shear and thickness-twist respectively by Mindlin;[64] one is associated with the idea that the thickness-shear deformation predominates over the lateral deformation, the other contains no lateral deflection, and the two components of rotation are so related in phase as to form a twist about the normal to the plate. If we multiply eqn. (6.62) by k^2 and set $k = 0$, $f = 1$, the frequency equation which omits the transverse shear deformation is obtained.

$$\left[1 + \frac{h^2}{12} (\lambda_n^2 + \delta_m^2) \right] \omega_{mn}^2 - \frac{1}{\rho h} (D_x \lambda_n^4 + 2\bar{H} \lambda_n^2 \delta_m^2 + D_y \delta_m^4) = 0 \quad (6.63)$$

Equation (6.63) is identical to eqn. (39) in Reference 64 except for the term characterising the orthotropy of the plate material. A simplified frequency equation which neglects the effects of the rotatory inertia, but retains transverse shear deformation can be obtained by setting $k = 1$ and $f = 0$ after multiplying eqn. (6.62) by f^2. It is seen that

$$\left\{ 100 \frac{G_{xz} G_{yz}}{\rho^2 h^4} + \frac{120}{\rho^2 h^5} [(G_{yz} D_x + G_{xz} D_{xy}) \lambda_n^2 + (G_{xz} D_y + G_{yz} D_{xy}) \delta_m^2] + \right.$$

$$+ \frac{144}{\rho^2 h^6} (D_x \lambda_n^4 + D_y \delta_m^4) + \frac{144}{\rho^2 h^6} (D_x D_y - \bar{H}^2 + 2\bar{H} D_{xy}) \lambda_n^2 \delta_m^2 \Big\} \omega_{mn}^2 -$$

$$- \left\{ \frac{100 G_{xz} G_{yz}}{\rho^3 h^5} (D_x \lambda_n^4 + 2\bar{H} \lambda_n^2 \delta_m^2 + D_y \delta_m^4) + \right.$$

$$+ \frac{120}{\rho^3 h^6} [D_x D_{xy} G_{xz} \lambda_n^6 + D_y D_{xy} G_{yz} \delta_m^6 +$$

$$+ (D_x D_{xy} G_{yz} + D_x D_y G_{xz} + 2\bar{H} D_{xy} G_{xz} - \bar{H}^2 G_{xz}) \lambda_n^4 \delta_m^2 +$$

$$+ (D_y D_{xy} G_{xz} + D_x D_y G_{yz} + 2\bar{H} D_{xy} G_{yz} - \bar{H}^2 G_{yz}) \lambda_n^2 \delta_m^4] \} = 0 \quad (6.64)$$

Finally, to obtain a frequency equation in which both transverse shear deformation and rotatory inertia are neglected, we multiply eqn.

(6.62) by k^2f^2 and set $k = f = 0$, the result is then given by

$$\bar{\omega}^2_{mn} = \frac{1}{\rho h}(D_x\lambda^4_n + 2\bar{H}\lambda^2_n\delta^2_m + D_y\delta^2_m) \tag{6.65}$$

the well-known solution of classical plate theory of a simply-supported orthotropic plate.

That eqn. (6.64) may be used as an approximately correct solution for the free vibration of a simply-supported orthotropic plate with primarily lateral motion will be discussed later. In order to show the errors that result in the use of classical orthotropic plate theory to describe the vibrational behaviour of plates made of composite materials for which the ratios of in-plane modulus of elasticity to transverse shear modulus have very high values, the ratios of the present plate frequency equations to the corresponding classical theory equations are used. Geometrically, the dimensional ratios of plates have been taken which are considered as 'thin' for plate constructions.

Dividing eqn. (6.64) by (6.65), the dimensionless frequency ratio is obtained as

$$R = \frac{\omega^2_{mn}}{\bar{\omega}^2_{mn}} = \left[1 \Big/ 1 + \left(\frac{6}{5h}\right)\left(\frac{D_x}{G_{xz}}\lambda^2_n + \frac{D_y}{G_{yz}}\delta^2_m + B\right)\right] \tag{6.66}$$

where

$$B = \left[1 + \left(\frac{6}{5h}\right)\left(\frac{D_{xy}}{G_{yz}}\lambda^2_n + \frac{D_{xy}}{G_{xz}}\delta^2_m + A\right)\right]^{-1} \times$$

$$\times \left\{\left(\frac{6}{5h}\right)\left(\frac{D_x}{G_{xz}}\frac{D_y}{G_{yz}} - \frac{\bar{H}^2}{G_{xz}G_{yz}} + \frac{2\bar{H}D_{xy}}{G_{xz}G_{yz}} - \frac{D_xD_{xy}}{G^2_{xz}} - \frac{D_yD_{xy}}{G^2_{yz}}\right)\lambda^2_n\delta^2_m - \right.$$

$$\left. - A\left[1 + \left(\frac{6}{5h}\right)\left(\frac{D_x}{G_{xz}}\lambda^2_n + \frac{D_y}{G_{yz}}\delta^2_m\right)\right]\right\} \tag{6.67}$$

and

$$A = (D_x\lambda^4_n + 2\bar{H}\lambda^2_n\delta^2_m + D_y\delta^4_m)^{-1}(D_xD_y - \bar{H}^2) \times$$

$$\times \left(\frac{1}{G_{xz}}\lambda^2_n + \frac{1}{G_{yz}}\delta^2_m\right)\lambda^2_n\delta^2_m \tag{6.68}$$

The foregoing results for plates of rectangular orthotropic material necessarily apply to the special case of the transversely isotropic as well as isotropic material. The mutually correspondent elastic constants are as listed.

$$E_x, E_y \;\rightarrow\; E; \qquad \nu_{xy}, \nu_{yx} \;\rightarrow\; \nu; \qquad G_{xy} \;\rightarrow\; G = \frac{E}{2(1 + \nu)}$$

$$D_x, D_y, \bar{H} \;\rightarrow\; D = \frac{Eh^3}{12(1 - \nu^2)}; \qquad D_{xy} \;\rightarrow\; \frac{1 - \nu}{2}D$$

$$G_{xz}, G_{yz} \;\rightarrow\; G_c \text{ (transversely isotropic)} \;\rightarrow\; G \text{ (isotropic)}$$

In the transversely isotropic case (such as plate made of pyrolytic graphite type materials), eqns. (6.67) and (6.68) vanish. Also, all results reduce to the solutions of the isotropic case which may be compared with those in References 64 and 69. Note that in Reference 69, Martincek's results are incorrect, because several terms which occur in eqn. (6.64) were omitted.

A numerical example utilises the elastic constants for a composite lamina of boron filaments (56·1% by volume) embedded in an epoxy matrix. The constitutive material properties, as well as the macroscopic properties for this composite material determined through methods of Section 5.6 are as follows:

$$Filament$$
$$E_f = 57 \cdot 6 \times 10^6 \text{ psi} = 4 \cdot 06 \times 10^6 \text{ kg/cm}^2$$
$$\nu_f = 0 \cdot 2$$
$$G_f = 23 \cdot 95 \times 10^6 \text{ psi} = 1 \cdot 655 \times 10^6 \text{ kg/cm}^2$$
$$\rho_f = 0 \cdot 221 \times 10^{-2} \text{ kg/cm}^3$$

$$Matrix$$
$$E_m = 0 \cdot 5 \times 10^6 \text{ psi} = 3 \cdot 53 \times 10^4 \text{ kg/cm}^2$$
$$\nu_m = 0 \cdot 35$$
$$G_m = 0 \cdot 185 \times 10^6 \text{ psi} = 1 \cdot 305 \times 10^4 \text{ kg/cm}^2$$
$$\rho_m = 0 \cdot 270 \times 10^{-2} \text{ kg/cm}^3$$

$$Composite$$
$$E_x = 32 \cdot 5 \times 10^6 \text{ psi} = 2 \cdot 29 \times 10^6 \text{ kg/cm}^2$$
$$E_y = 1 \cdot 84 \times 10^6 \text{ psi} = 0 \cdot 13 \times 10^6 \text{ kg/cm}^2$$
$$\nu_{xy} = 0 \cdot 256$$
$$\nu_{yx} = 0 \cdot 0146$$
$$G_{xy} = G_{xz} = 0 \cdot 642 \times 10^6 \text{ psi} = 4 \cdot 52 \times 10^4 \text{ kg/cm}^2$$
$$G_{yz} = 0 \cdot 361 \times 10^6 \text{ psi} = 2 \cdot 545 \times 10^4 \text{ kg/cm}^2$$
$$\rho = 0 \cdot 244 \times 10^{-2} \text{ kg/cm}^3$$

The curves given in Fig. 6.2 of the dimensionless frequency ratio, R, defined by eqn. (6.66) as a function of geometry show that the effect of transverse shear deformation is very significant even for the fundamental and other lower wave number modes of oscillations for a plate composed of this typical boron–epoxy composite material. For example a plate in which $(h/a) = (h/b) = \frac{1}{10}$, transverse shear deformation causes a reduction of 9 per cent in the fundamental natural frequency. Even for a very thin plate $(h/a) = (h/b) = \frac{1}{40}$, the reduction in a low mode $(m = n = 3)$ of vibration is 6 per cent. As expected, the higher the mode number, the greater the effect. It is seen that for geometri-

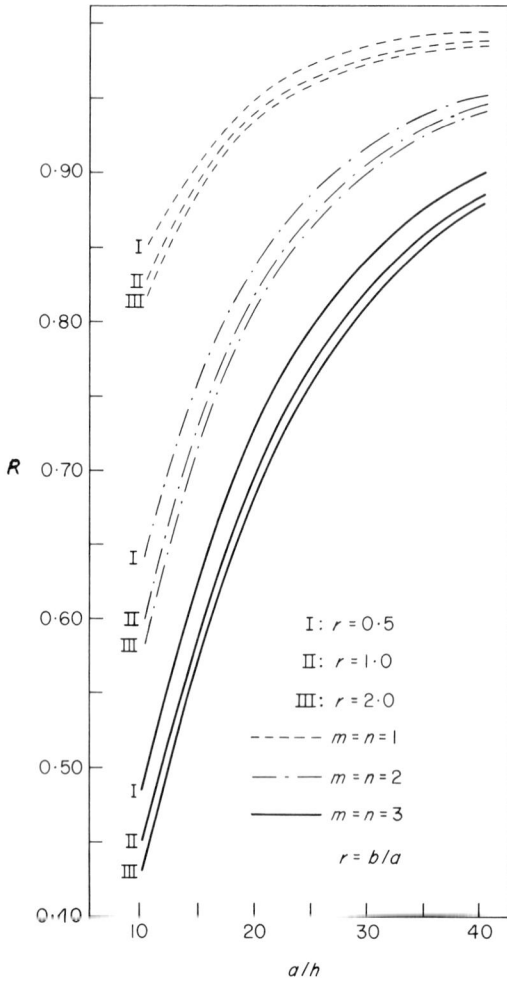

Fig. 6.2.   *Frequency ratio as a function of geometry.*

cally thin plates, the use of classical plate theory to predict natural frequencies for this composite material can lead to serious errors. Note for $m = n = 3$, and $(h/a) = \frac{1}{10}$, $r = 2\cdot0$, the use of classical theory would overestimate the frequency by greater than a factor of $1\cdot5$.

It is concluded that when many anisotropic composite materials are used in any geometrically thin plate or shell structure, the effects of transverse shear deformation (as well as transverse normal stress and rotatory inertia, where applicable) cannot be ignored when calculating stresses, deformations, natural frequencies, and buckling loads for the majority of structural applications.

The work described herein is derived from Reference 49.

6.5   NON-LINEAR OSCILLATIONS OF PLATES COMPOSED OF COMPOSITE MATERIALS[a]

This section provides a logical follow-on to the previous section in introducing the complexities of the non-linear problem, but employing the same technique of solution as used in Section 6.4.

Consider once again the plate of Section 6.4. However, now to investigate the large amplitude, non-linear vibrations, one needs to include additional terms involving w in the strain–displacement relations as was done in eqn. (4.11).

$$\epsilon_{jk} = \frac{1}{2}[u_{j,k} + u_{k,j} + u_{i,j}u_{i,k}] \qquad (6.69)$$

where $i, j, k = x, y, z$.

Now, even though one is considering lateral vibrations, whenever large amplitude oscillations are involved, there is a coupling between the in-plane displacements and the lateral displacements, that is to say bending and stretching are inextricably coupled. Hence, it is now assumed that the displacement components are of the form

$$u = u_0(x, y, t) + z\alpha(x, y, t) \qquad (6.70)$$

$$v = v_0(x, y, t) + z\beta(x, y, t) \qquad (6.71)$$

$$w = w(x, y, t) \qquad (6.72)$$

where the subscript 0 is used to associate with the middle surface. It is noted that the in-plane displacements take the usual form of being the sum of a translation and the rotation of a linear element through the thickness.

[a] Excerpted from Reference 50 by permission of the Technomic Publishing Co.

The explicit forms of the strain–displacement relations are obtained by substituting (6.70) to (6.72) into (6.69); hence

$$\epsilon_x = \frac{\partial u_0}{\partial x} + z\frac{\partial \alpha}{\partial x} + \frac{1}{2}\left(\frac{\partial w}{\partial x}\right)^2$$

$$\epsilon_y = \frac{\partial v_0}{\partial y} + z\frac{\partial \beta}{\partial y} + \frac{1}{2}\left(\frac{\partial w}{\partial y}\right)^2$$

$$\epsilon_{xy} = \frac{1}{2}\left(\frac{\partial u_0}{\partial y} + \frac{\partial v_0}{\partial x} + z\frac{\partial \alpha}{\partial y} + z\frac{\partial \beta}{\partial x}\right) + \frac{1}{2}\left(\frac{\partial w}{\partial x}\right)\left(\frac{\partial w}{\partial y}\right) \tag{6.73}$$

$$\epsilon_{yz} = \frac{1}{2}\left(\frac{\partial w}{\partial y} + \beta\right); \qquad \epsilon_{xz} = \frac{1}{2}\left(\frac{\partial w}{\partial x} + \alpha\right); \qquad \epsilon_z = 0$$

Neither the stress–strain relations nor the equilibrium equations change, however, for the case of non-linear large amplitude vibrations. Thus, the stress–strain relations remain as in (6.42),

$$\epsilon_x = \frac{\sigma_x}{E_x} - \frac{\nu_{yx}}{E_y}\sigma_y - \frac{\nu_{zx}}{E_z}\sigma_z$$

$$\epsilon_y = -\frac{\nu_{xy}}{E_x}\sigma_x + \frac{\sigma_y}{E_y} - \frac{\nu_{zy}}{E_z}\sigma_z$$

$$\epsilon_z = -\frac{\nu_{xz}}{E_x}\sigma_x - \frac{\nu_{yz}}{E_y}\sigma_y + \frac{\sigma_z}{E_z} \tag{6.74}$$

$$\epsilon_{xy} = \frac{\sigma_{xy}}{2G_{xy}}; \qquad \epsilon_{xz} = \frac{\sigma_{xz}}{2G_{xz}}; \qquad \epsilon_{yz} = \frac{\sigma_{yz}}{2G_{yz}}$$

For the free vibrations of a plate, the effect of transverse normal stress are neglected since there is no applied lateral load. One may therefore eliminate the terms including σ_z from eqn. (6.74). Hence, eqn. (6.74) reduces to

$$\epsilon_x = \frac{\sigma_x}{E_x} - \frac{\nu_{yz}}{E_y}\sigma_y$$

$$\epsilon_y = -\frac{\nu_{xy}}{E_x}\sigma_x + \frac{\sigma_y}{E_y}$$

$$\epsilon_{xy} = \frac{\sigma_{xy}}{2G_{xy}}; \qquad \epsilon_{xz} = \frac{\sigma_{xz}}{2G_{xz}} \tag{6.75}$$

$$\epsilon_{yz} = \frac{\sigma_{yz}}{2G_{yz}}; \qquad \epsilon_z = 0$$

Again, because bending and stretching are coupled, the stresses are written as below rather than the simpler form of eqn. (6.43).

$$(\sigma_x, \sigma_y, \sigma_z) = \frac{1}{h} (N_x, N_y, N_{xy}) + \frac{12z}{h^3} (M_x, M_y, M_{xy})$$

$$(\sigma_{xz}, \sigma_{yz}) = \frac{3}{2h} \left[1 - \left(\frac{z}{h/2} \right)^2 \right] (Q_x, Q_y); \qquad \sigma_z = 0 \tag{6.76}$$

The membrane stress resultants in terms of strains are obtained by substituting (6.73) and (6.75) into (6.76) and integrating across the thickness of the plate. The results are

$$N_x = \frac{E_x h}{(1 - \nu_{xy}\nu_{yx})} (\epsilon_{x_0} + \nu_{yx}\epsilon_{y_0})$$

$$N_y = \frac{E_y h}{(1 - \nu_{xy}\nu_{yx})} (\epsilon_{y_0} + \nu_{xy}\epsilon_{x_0}) \tag{6.77}$$

$$N_{xy} = 2G_{xy}h\epsilon_{x_0 y_0}$$

where ϵ_{x_0}, ϵ_{y_0} are normal strains of the middle surface in the x and y direction, respectively, and $\epsilon_{x_0 y_0}$ is the middle surface shear strain. From (6.73) it is seen that the middle surface strains $(z = 0)$ are

$$\epsilon_{x_0} = \frac{\partial u_0}{\partial x} + \frac{1}{2} \left(\frac{\partial w}{\partial x} \right)^2$$

$$\epsilon_{y_0} = \frac{\partial v_0}{\partial y} + \frac{1}{2} \left(\frac{\partial w}{\partial y} \right)^2$$

$$\epsilon_{x_0 y_0} = \frac{1}{2} \left(\frac{\partial u_0}{\partial y} + \frac{\partial v_0}{\partial x} \right) + \frac{1}{2} \left(\frac{\partial w}{\partial x} \right) \left(\frac{\partial w}{\partial y} \right)$$

Again, one can now apply Reissner's Variational Theorem. The function H is defined again as in eqn. (6.44), and the Reissner Functional is written again as in eqn. (6.45).

The strain energy is obtained by substituting (6.73), (6.76) and (6.77) in (6.45), and integrating across the thickness of plate. Proceeding as in the last Section, it is seen that

$$\psi = \int_0^a \int_0^b \left\{ \frac{E_x h}{2(1 - \nu_{xy}\nu_{yx})} [\bar{I}_e^2 - 2(\gamma - \nu_{yx})\overline{II}_e^2] + M_x \frac{\partial \alpha}{\partial x} + \right.$$

$$+ M_y \frac{\partial \beta}{\partial y} + M_{xy} \left(\frac{\partial \alpha}{\partial y} + \frac{\partial \beta}{\partial x} \right) + Q_x \left(\alpha + \frac{\partial w}{\partial x} \right) + Q_y \left(\beta + \frac{\partial w}{\partial y} \right) -$$

$$- \frac{6M_x^2}{E_x h^3} - \frac{6M_y^2}{E_y h^3} - \frac{6M_{xy}^2}{G_{xy}h^3} + \frac{12\nu_{xy}M_x M_y}{E_x h^3} -$$

$$\left. - \frac{3Q_x^2}{5G_{xz}h} - \frac{3Q_y^2}{5G_{yz}h} \right\} dy \, dx \tag{6.78}$$

where $\gamma^2 = D_y/D_x$, and

$$\bar{I}_e = \epsilon_{x_0} + \gamma\epsilon_{y_0}$$

and   (6.79)

$$\overline{II}_e = \epsilon_{x_0}\epsilon_{y_0} - \frac{2G_{xy}(1 - \nu_{xy}\nu_{yx})}{E_x(\gamma - \nu_{yx})}\epsilon_{x_0y_0}^2$$

$\bar{I}_e$ and $\overline{II}_e$ play the role of the first and second invariants of the middle surface strains of the plate composed of an orthotropic material.[70]

The kinetic energy of the plate, after integrating across the thickness becomes, analogous to (6.46),

$$T = \int_0^a \int_0^b \left\{ \frac{\rho h}{2}\left[\left(\frac{\partial u_0}{\partial t}\right)^2 + \left(\frac{\partial v_0}{\partial t}\right)^2 + \left(\frac{\partial w}{\partial t}\right)^2\right] + \right.$$
$$\left. + \frac{\rho h^3}{24}\left[\left(\frac{\partial \alpha}{\partial t}\right)^2 + \left(\frac{\partial \beta}{\partial t}\right)^2\right]\right\} \mathrm{d}y\,\mathrm{d}x \quad (6.80)$$

Again, Hamilton's Principle (eqn. (6.48)) applies. By virtue of the hypothesis of Berger,[71] which suggests that the second invariant of the middle surface strains in the plate can be neglected, the substitution of eqns. (6.78) and (6.80) into (6.48) with $\overline{II}_e = 0$ yields, upon taking variations,

$$\delta\Phi = \int_0^a \int_0^b \left[\frac{\rho h^3}{12}\left(\frac{\partial \alpha}{\partial t}\delta\alpha + \frac{\partial \beta}{\partial t}\delta\beta\right)\right.$$
$$\left. + \rho h\left(\frac{\partial u_0}{\partial t}\delta u_0 + \frac{\delta v_0}{\partial t}\delta v_0 + \frac{\partial w}{\partial t}\delta w\right)\right]_{t_1}^{t_2} \mathrm{d}y\,\mathrm{d}x -$$
$$- \int_{t_1}^{t_2}\int_0^b \left[M_x\delta\alpha + M_{xy}\delta\beta + \left(Q_x + \frac{E_x h}{(1 - \nu_{xy}\nu_{yx})}\bar{I}_e\frac{\partial w}{\partial x}\right)\delta w + \right.$$
$$\left. + \frac{E_x h}{(1 - \nu_{xy}\nu_{yx})}\bar{I}_e\delta u_0\right]_0^a \mathrm{d}y\,\mathrm{d}t$$
$$- \int_{t_1}^{t_2}\int_0^a \left[M_{xy}\delta\alpha + M_y\delta\beta + \left(Q_y + \frac{E_x h}{(1 - \nu_{xy}\nu_{yx})}\gamma\bar{I}_e\frac{\partial w}{\partial y}\right)\delta w + \right.$$
$$\left. + \frac{E_x h}{(1 - \nu_{xy}\nu_{yx})}\gamma\bar{I}_e\delta v_0\right]_0^b \mathrm{d}x\,\mathrm{d}t$$
$$- \int_{t_1}^{t_2}\int_0^a\int_0^b \left\{\left[\frac{\partial \alpha}{\partial x} - \frac{12}{E_x h^3}M_x + \frac{12\nu_{xy}}{E_x h^3}M_y\right]\delta M_x + \right.$$
$$\left. + \left[\frac{\partial \beta}{\partial y} - \frac{12M_y}{E_y h^3} + \frac{12\nu_{yx}M_x}{E_y h^3}\right]\delta M_y + \left[\frac{\partial \alpha}{\partial y} + \frac{\partial \beta}{\partial x} - \frac{12M_{xy}}{G_{xy}h^3}\right]\delta M_{xy} + \right.$$
$$\left. + \left[\frac{\partial M_x}{\partial x} + \frac{\partial M_{xy}}{\partial y} - Q_x - \frac{\rho h^3}{12}\frac{\partial^2\alpha}{\partial t^2}\right]\delta\alpha + \right.$$

$$+ \left[\frac{\partial M_{xy}}{\partial x} + \frac{\partial M_y}{\partial y} - Q_y - \frac{\rho h^3}{12} \frac{\partial^2 \beta}{\partial t^2} \right] \delta \beta +$$

$$+ \left[\frac{\partial Q_x}{\partial x} + \frac{\partial Q_y}{\partial y} + \frac{E_x h}{(1 - \nu_{xy}\nu_{yx})} \frac{\partial}{\partial x} \left(\bar{I}_e \frac{\partial w}{\partial x} \right) + \right.$$

$$+ \frac{E_x h}{(1 - \nu_{xy}\nu_{yx})} \frac{\partial}{\partial y} \left(\bar{I}_e \gamma \frac{\partial w}{\partial y} \right) - \rho h \frac{\partial^2 w}{\partial t^2} \bigg] \delta w +$$

$$+ \left[\alpha + \frac{\partial w}{\partial x} - \frac{6 Q_x}{5 G_{xz} h} \right] \delta Q_x + \left[\beta + \frac{\partial w}{\partial y} - \frac{6 Q_y}{5 G_{yz} h} \right] \delta Q_y +$$

$$+ \left[\frac{E_x h}{(1 - \nu_{xy}\nu_{yx})} \frac{\partial \bar{I}_e}{\partial x} - \rho h \frac{\partial^2 u_0}{\partial t^2} \right] \delta u_0 +$$

$$+ \left[\frac{E_x h \gamma}{(1 - \nu_{xy}\nu_{yx})} \frac{\partial \bar{I}_e}{\partial y} - \rho h \frac{\partial^2 v_0}{\partial t^2} \right] \delta v_0 \bigg\} \, \mathrm{d}y \, \mathrm{d}x \, \mathrm{d}t = 0 \tag{6.81}$$

As stated previously, it is due to the merit of the variational principle that one is able to find explicitly the natural boundary conditions which are involved in the second and third integrals in eqn. (6.81) above, while the first integral specifies the initial conditions. Further, because the contents of every bracket in eqn. (6.81) must be set equal to zero, the Euler–Lagrangian equations of motion are obtained as in the following

$$\frac{\partial M_x}{\partial x} + \frac{\partial M_{xy}}{\partial y} - Q_x - \frac{\rho h^3}{12} f \frac{\partial^2 \alpha}{\partial t^2} = 0 \tag{6.82}$$

$$\frac{\partial M_{xy}}{\partial x} + \frac{\partial M_y}{\partial y} - Q_y - \frac{\rho h^3}{12} f \frac{\partial^2 \beta}{\partial t^2} = 0 \tag{6.83}$$

$$\frac{\partial Q_x}{\partial x} + \frac{\partial Q_y}{\partial y} + \frac{E_x h}{(1 - \nu_{xy}\nu_{yx})} \left[\frac{\partial}{\partial x} \left(\bar{I}_e \frac{\partial w}{\partial x} \right) + \gamma \frac{\partial}{\partial y} \left(\bar{I}_e \frac{\partial w}{\partial y} \right) \right] - \rho h \frac{\partial^2 w}{\partial t^2} = 0 \tag{6.84}$$

$$\frac{\partial \alpha}{\partial x} - \frac{12 M_x}{E_x h^3} + \frac{12 \nu_{xy}}{E_x h^3} M_y = 0 \tag{6.85}$$

$$\frac{\partial \beta}{\partial y} - \frac{12 M_y}{E_y h^3} + \frac{12 \nu_{yx}}{E_y h^3} M_x = 0 \tag{6.86}$$

$$\frac{\partial \alpha}{\partial y} + \frac{\partial \beta}{\partial x} - \frac{12}{G_{xy} h^3} M_{xy} = 0 \tag{6.87}$$

$$\alpha + \frac{\partial w}{\partial x} - \frac{6k}{5 G_{xz} h} Q_x = 0 \tag{6.88}$$

$$\beta + \frac{\partial w}{\partial y} - \frac{6k}{5 G_{yz} h} Q_y = 0 \tag{6.89}$$

$$\frac{E_x h}{(1 - \nu_{xy}\nu_{yx})} \frac{\partial \bar{I}_e}{\partial x} = \rho h \frac{\partial^2 u_0}{\partial t^2} \tag{6.90}$$

$$\frac{\gamma E_x h}{(1 - \nu_{xy}\nu_{yx})} \frac{\partial \bar{I}_e}{\partial y} = \rho h \frac{\partial^2 v_0}{\partial t^2} \qquad (6.91)$$

Again, two tracing constants k and f are introduced for the purpose of identifying terms which characterise the effects of transverse shear deformation and rotatory inertia, respectively. It is seen that eqns. (6.82) to (6.89), except (6.84) which involves the membrane effects as well as non-linear behaviour of strains, are the same form as those of the linear theory, eqns. (6.50) to (6.57). Equations (6.90) and (6.91) are the simplified equations for in-plane force equilibrium through the use of Berger's hypothesis.

It is hypothesised that the inertia effect of the in-plane motion can be neglected in the study of primarily lateral vibration. This assumption is tantamount to following the procedure of previous investigators (References 72 to 78), in which they eliminated u_0 and v_0 from the equations at the outset. Hence, the right-hand side of (6.90) and (6.91) are set equal to zero, from which it is seen that

$$\frac{\partial \bar{I}_e}{\partial x} = \frac{\partial \bar{I}_e}{\partial y} = 0 \qquad (6.92)$$

These equations imply that $\bar{I}_e$ is independent of the space coordinates x and y. Hence we may assume, using (6.79)

$$\bar{I}_e = \frac{\partial u_0}{\partial x} + \gamma \frac{\partial v_0}{\partial y} + \frac{1}{2}\left(\frac{\partial w}{\partial x}\right)^2 + \frac{\gamma}{2}\left(\frac{\partial w}{\partial y}\right)^2 = \frac{\bar{\alpha}^2 h^2}{12} \qquad (6.93)$$

where $\bar{\alpha}$ is a coupling parameter of the system. From (6.82) to (6.89) and (6.93), eliminating M_x, M_y, M_{xy}, Q_x and Q_y, one can obtain the following three governing equations

$$\alpha + \frac{\partial w}{\partial x} - \left(\frac{6k}{5G_{xz}h}\right)\left[D_x \frac{\partial^2 \alpha}{\partial x^2} + D_{xy} \frac{\partial^2 \alpha}{\partial y^2} - \right.$$
$$\left. \frac{\rho h^3 f}{12} \frac{\partial^2 \alpha}{\partial t^2} + (\bar{H} - D_{xy}) \frac{\partial^2 \beta}{\partial x \partial y}\right] = 0 \qquad (6.94)$$

$$\beta + \frac{\partial w}{\partial y} - \left(\frac{6k}{5G_{yz}h}\right)\left[D_y \frac{\partial^2 \beta}{\partial y^2} + D_{xy} \frac{\partial^2 \beta}{\partial x^2} - \right.$$
$$\left. \frac{\rho h^3 f}{12} \frac{\partial^2 \beta}{\partial t^2} + (\bar{H} - D_{xy}) \frac{\partial^2 \alpha}{\partial x \partial y}\right] = 0 \qquad (6.95)$$

$$G_{xz}\left(\frac{\partial^2 w}{\partial x^2} + \frac{\partial \alpha}{\partial x}\right) + G_{yz}\left(\frac{\partial^2 w}{\partial y^2} + \frac{\partial \beta}{\partial y}\right) +$$
$$+ \left(\frac{6k}{5h}\right)\frac{E_x h \bar{I}_e}{(1 - \nu_{xy}\nu_{yx})}\left(\frac{\partial^2 w}{\partial x^2} + \gamma \frac{\partial^2 w}{\partial y^2}\right) - \left(\frac{6\rho k}{5}\right)\frac{\partial^2 w}{\partial t^2} = 0 \qquad (6.96)$$

Equations (6.94) to (6.96) are the approximate equations that govern the motion of plates which consists of three independent families of modes of vibration: flexural, thickness-shear and thickness-twist. Here, we are primarily interested in the flexural motion. As pointed out previously in References 64 and 79, the effect of rotatory inertia can be neglected when the plate vibrates primarily laterally in the small deflection linear vibration case. We hypothesise that the effect of rotatory inertia is also small when considering the plate undergoing large amplitude non-linear lateral vibrations. Hence, one can set $f = 0$ among (6.94) and (6.95) above.

Consider a simply-supported plate for an example, in which in-plane displacements are not allowed at the boundaries. [*Note*: Recently, Nowinski,[81] has shown that if in-plane displacements are not constrained at the boundaries, the Berger's Hypothesis can lead to inaccurate and sometimes ridiculous results.]

$$\beta = u_0 = w = M_x = 0 \qquad \text{along } x = 0, a$$
$$\alpha = v_0 = w = M_y = 0 \qquad \text{along } y = 0, b \tag{6.97}$$

Using these conditions for u_0 and v_0, the integration of (6.93) over the plate gives

$$\bar{\alpha}^2 = \frac{6}{h^2 ab} \int_0^a \int_0^b \left[\left(\frac{\partial w}{\partial x} \right)^2 + \gamma \left(\frac{\partial w}{\partial y} \right)^2 \right] \mathrm{d}x \, \mathrm{d}y \tag{6.98}$$

The problem is now to solve eqns. (6.94) to (6.96) subject to the remaining boundary conditions in (6.97). To satisfy the boundary conditions, solutions are assumed of the following form, analogous to (6.60)

$$w = \sum_m \sum_n W_{mn} \sin \frac{n\pi x}{a} \sin \frac{m\pi y}{b} \tau(t)$$

$$\alpha = \sum_m \sum_n \Gamma_{mn} \cos \frac{n\pi x}{a} \sin \frac{m\pi y}{b} \tau(t) \tag{6.99}$$

$$\beta = \sum_m \sum_n \Lambda_{mn} \sin \frac{n\pi x}{a} \cos \frac{m\pi y}{b} \tau(t)$$

The normal mode solution is exploited since the system of governing differential equations is linear in the space functions and non-linear only in the time-dependent function which is not defined in the solution (6.99). If we consider the fundamental mode of oscillation, one may use only the first term of each expression given by eqn. (6.99) as an approximation to the true solution. With this in mind, the substitution of the expression for w into (6.98) yields

$$\bar{\alpha}^2 = \frac{3W_{11}^2}{2h^2}(\lambda_1^2 + \gamma\delta_1^2)\tau^2 \qquad (6.100)$$

in which $\lambda_1 = \pi/a$, $\delta_1 = \pi/b$. If one substitutes eqn. (6.99) with $m = n = 1$, into (6.94) to (6.96) having $f = 0$ and $k = 1$, one obtains

$$\lambda_1 W_{11} + \left[1 + \left(\frac{6}{5G_{xz}h}\right)(D_x\lambda_1^2 + D_{xy}\delta_1^2)\right]\Gamma_{11} +$$

$$+ \left(\frac{6}{5G_{xz}h}\right)(\bar{H} - D_{xy})\tau_1\delta_1\Lambda_{11} = 0$$

$$\delta_1 W_{11} + \left[1 + \left(\frac{6}{5G_{yz}h}\right)(D_y\delta_1^2 + D_{xy}\lambda_1^2)\right]\Lambda_{11} + \qquad (6.101)$$

$$+ \left(\frac{6}{5G_{yz}h}\right)(\bar{H} - D_{xy})\lambda_1\delta_1\Gamma_{11} = 0$$

$$\left\{\left[G_{xz} + \bar{\alpha}^2 D_x\left(\frac{6}{5h}\right)\right]\lambda_1^2\tau + \left[G_{yz} + \bar{\alpha}^2\gamma D_x\left(\frac{6}{5h}\right)\right]\delta_1^2\tau + \right.$$

$$\left. + \left(\frac{6\rho}{5}\right)\frac{\partial^2\tau}{\partial t^2}\right\} W_{11} + G_{xz}\lambda_1\tau\Gamma_{11} + G_{yz}\delta_1\tau\Lambda_{11} = 0$$

For the non-trivial solution, the determinant of coefficients of W_{11}, Γ_{11} and Λ_{11} in (6.101) must vanish. With (6.100) the result is the familiar Duffing Equation, which occurs so often in describing non-linear behaviour in physical systems

$$\alpha^*\ddot{\tau} + \beta^*\tau + \delta^*\tau^3 = 0 \qquad (6.102)$$

where, in this case

$$\alpha^* = \frac{6\rho}{5}\left\{1 + \left(\frac{6}{5G_{xz}h}\right)(D_x\lambda_1^2 + D_{xy}\delta_1^2) + \left(\frac{6}{5G_{yz}h}\right)(D_{xy}\lambda_1^2 + D_y\delta_1^2) + \right.$$

$$+ \left(\frac{6}{5G_{xz}h}\right)\left(\frac{6}{5G_{yz}h}\right)[D_xD_{xy}\lambda_1^4 + D_yD_{xy}\delta_1^4 +$$

$$\left. + (D_xD_y - \bar{H}^2 + 2\bar{H}D_{xy})\lambda_1^2\delta_1^2]\right\} \qquad (6.103)$$

$$\beta^* = \left(\frac{6}{5h}\right)(D_x\lambda_1^4 + 2\bar{H}\lambda_1^2\delta_1^2 + D_y\delta_1^4) +$$

$$+ \left(\frac{6}{5G_{xz}h}\right)\left(\frac{6}{5G_{yz}h}\right)(D_xD_{xy}G_{xz}\lambda_1^6 + D_yD_{xy}G_{yz}\delta_1^6) +$$

$$+ \left(\frac{6}{5G_{xz}h}\right)\left(\frac{6}{5G_{yz}h}\right)(D_xD_{xy}G_{yz} + D_xD_yG_{xz} - \bar{H}^2G_{xz} +$$

$$+ 2\bar{H}D_{xy}G_{xz})\lambda_1^4\delta_1^2 +$$

$$+ \left(\frac{6}{5G_{xz}h}\right)\left(\frac{6}{5G_{yz}h}\right)(D_y D_{xy} G_{xz} + D_x D_y G_{yz} - \bar{H}^2 G_{yz}$$
$$+ 2\bar{H} D_{xy} G_{yz})\lambda_1^2 \delta_1^4 \quad (6.104)$$

$$\delta^* = \frac{3}{2\rho h}\frac{W_{11}^2}{h^2} D_x(\lambda_1^4 + 2\gamma\lambda_1^2\delta_1^2 + \gamma^2\delta_1^4)\alpha^* \quad (6.105)$$

If the plate is subjected to the initial conditions of

$$\tau = 1, \qquad \dot{\tau} = 0 \quad \text{at } t = 0 \quad (6.106)$$

the well-established solution of eqn. (6.102) is

$$\tau = cn(\omega^* t, p) \quad (6.107)$$

wherein cn is an elliptic cosine with

$$\omega^{*2} = \frac{1 + \frac{3}{2}\bar{\beta}^2\lambda_1\left[1 + \left(\frac{6}{5h}\right)\left(\frac{D_x}{G_{xz}}\lambda_1^2 + \frac{D_y}{G_{yz}}\delta_1^2 + B\right)\right]}{1 + \left(\frac{6}{5h}\right)\left(\frac{D_x}{G_{xz}}\lambda_1^2 + \frac{D_y}{G_{yz}}\delta_1^2 + B\right)}$$

$$\quad (6.108)$$

$$p^2 = \frac{\frac{3}{4}\bar{\beta}^2\gamma_1\left[1 + \left(\frac{6}{5h}\right)\left(\frac{D_x}{G_{xz}}\lambda_1^2 + \frac{D_y}{G_{yz}}\delta_1^2 + B\right)\right]}{1 + \frac{3}{2}\bar{\beta}^2\gamma_1\left[1 + \left(\frac{6}{5h}\right)\left(\frac{D_x}{G_{xz}}\lambda_1^2 + \frac{D_y}{G_{yz}}\delta_1^2 + B\right)\right]}$$

in which

$$\bar{\beta} = \frac{W_{11}}{h}, \quad \gamma_1 = \frac{D_x\lambda_1^4 + 2D_x\gamma\lambda_1^2\delta_1^2 + D_y\delta_1^4}{D_x\lambda_1^4 + 2\bar{H}\lambda_1^2\delta_1^2 + D_y\delta_1^4} \quad (6.109)$$

$$B = \left[1 + \left(\frac{6}{5h}\right)\left(\frac{D_{xy}}{G_{yz}}\lambda_1^2 + \frac{D_{xy}}{G_{xz}}\delta_1^2 + A\right]^{-1}\left\{\left(\frac{6}{5h}\right)\times\right.$$
$$\times\left(\frac{D_x D_y}{G_{xz}G_{yz}} - \frac{\bar{H}^2}{G_{xz}G_{yz}} + \frac{2\bar{H}D_{xy}}{G_{xz}G_{yz}} - \frac{D_x D_{xy}}{G_{xz}^2} - \frac{D_y D_{xy}}{G_{yz}^2}\right)\lambda_1^2\delta_1^2 -$$
$$\left. - A\left[1 + \left(\frac{6}{5h}\right)\left(\frac{D_x}{G_{xz}}\lambda_1^2 + \frac{D_y}{G_{yz}}\delta_1^2\right)\right]\right\} \quad (6.110)$$

$$A = (D_x\lambda_1^4 + 2\bar{H}\lambda_1^2\delta_1^2 + D_y\delta_1^4)^{-1}(D_x D_y - \bar{H}^2)\left(\frac{\lambda_1^2}{G_{xz}} + \frac{\delta_1^2}{G_{yz}}\right)\lambda_1^2\delta_1^2 \quad (6.111)$$

The period T^* of $cn(\omega^* t, p)$ is

$$T^* = \frac{4K(p)}{\omega^*} \quad (6.112)$$

where $K(p)$ is the complete elliptic integral of the first kind with parameter p. The corresponding linear period for a classical orthotropic plate not including transverse shear deformation is

$$\bar{T} = \frac{2\pi}{\bar{\omega}}$$

wherein

$$\bar{\omega}^2 = \frac{1}{\rho h}[D_x\lambda_1^4 + 2H\lambda_1^2\delta_1^2 + D_y\delta_1^4] \qquad (6.113)$$

The frequency ratio is defined here by the ratio of the frequency determined by (6.112) to that obtained for a classical plate by (6.113), and is

$$R^* = \frac{f^*}{\bar{f}} = \frac{\pi}{2K(p)}\left\{\frac{1 + \frac{3}{2}\bar{\beta}^2\gamma_1\left[1 + \left(\frac{6}{5h}\right)\left(\frac{D_x}{G_{xz}}\lambda_1^2 + \frac{D_y}{G_{yz}}\delta_1^2 + B\right)\right]}{1 + \left(\frac{6}{5h}\right)\left(\frac{D_x}{G_{xz}}\lambda_1^2 + \frac{D_y}{G_{yz}}\delta_1^2 + B\right)}\right\}^{1/2} \qquad (6.114)$$

where $f^* = 1/T^*$, $\bar{f} = 1/\bar{T}$ are the frequencies of vibrations.

In this context it should be remembered that only the lowest, or fundamental, frequency is being investigated. One would use the same procedure to investigate higher modes, however, simply by taking other integer values of m and n.

As an illustration, numerical calculations have been made for the large amplitude lateral oscillations of simply-supported plates composed of the boron–epoxy composite material of Section 6.4.

The results are plotted in Fig. 6.3, where the ordinate R^*, defined by (6.114), is the ratio of the fundamental frequency using the methods developed herein to the fundamental frequency for the same orthotropic plate using classical linear theory with no provision for transverse shear deformation. The abscissa is a geometric ratio of length to thickness, varying from a moderately thick plate to a relatively thin plate. The ratio b/a is a width to length ratio, and $\bar{\beta}$ is the ratio of amplitude of vibration to plate thickness.

Three separate and distinct effects are obvious. First, the deviation of all curves from the horizontal lucidly reflects the effect of transverse shear deformation upon the natural frequencies. The largest *percentage* effect occurs at lower values of $\bar{\beta}$ indicating that the larger the amplitude of the plate vibration (i.e. the further the plate vibrates into the non-linear, large amplitude regime) the effects of transverse shear deformation are increasingly suppressed. This is because the further the plate goes into the large amplitude regime the more it behaves as a membrane rather than a plate, and membrane action involves only in-plane or extensional stiffness and strains, and no flexural or shearing

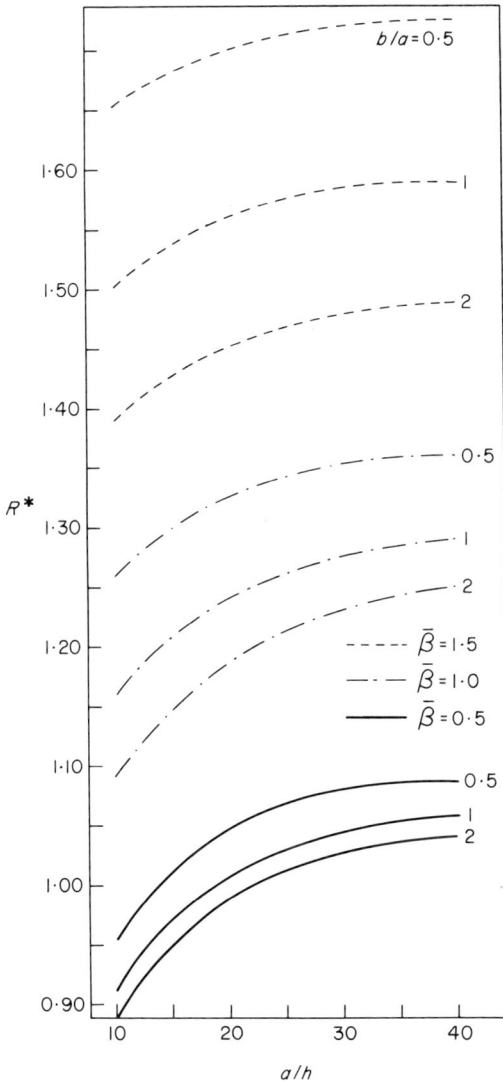

Fig. 6.3.   Frequency ratio as a function of geometry and amplitude.

action. Also referring to Fig. 6.2 the dashed lines ($m = n = 1$) could be superimposed on Fig. 6.3. For the small amplitude linear vibrations of Fig. 6.2, they correspond in the limit to a $\bar{\beta} = 0\cdot0$. However, it is clear that for many fibre-reinforced composite materials transverse shear deformation effects must be included.

Second, it is also clear that the non-linearity that occurs for the orthotropic plate is of the hardening spring type, and that the non-linear effects are very important. It is seen that for a thin plate ($a/h = 40$), classical methods (linear, small deformation orthotropic plate theory, neglecting transverse shear deformation) can predict a fundamental frequency differing from the correct solution by a factor of $1\cdot7$ for amplitudes of vibration only $1\cdot5$ times the plate thickness, and $b/a = 0\cdot5$.

Third, it is seen that when the plate is longer in the stiffer direction (the direction of the filaments), the natural frequencies are significantly higher than when the reverse is true. This, of course, is expected.

6.6 NATURAL VIBRATIONS OF LAMINATED PLATES OF COMPOSITE MATERIALS UNDER VARIOUS BOUNDARY CONDITIONS

This Section is included to illustrate the effects that laminates introduce, as well as to show the use of the widely used Galerkin procedure.

For laminated plates, the methods of analysis used in the previous sections provide an accurate solution for natural frequencies if the reduced flexural modulus matrix, $[D^*]$, is used instead of the flexural moduli defined in Sections 6.4 and 6.5 (see Reference 26). These are defined in and discussed subsequent to eqn. (5.57), and repeated here.

$$[D^*] = [D] - [B][A]^{-1}[B] \qquad (6.115)$$

The quantities $[A]$, $[B]$, and $[D]$ are defined by (5.48), (5.49) and (5.51). Thus, this approximation automatically accounts for the effect of stretching–bending coupling for an arbitrarily laminated composite material plate. Hence, the in-plane displacements can be neglected from the displacement equations for small amplitude lateral vibrations

$$u = z\alpha(x, y, t)$$
$$v = z\beta(x, y, t) \qquad (6.116)$$
$$w = w(x, y, t)$$

where z is measured from the geometric midsurface of the laminated plate. Then, analogous to eqns. (6.50) to (6.57) it is straightforward to evolve respectively the governing equations given below

$$M_x = D_{11}^* \frac{\partial \alpha}{\partial x} + D_{12}^* \frac{\partial \beta}{\partial y} + D_{16}^* \left(\frac{\partial \alpha}{\partial y} + \frac{\partial \beta}{\partial x} \right) \tag{6.117}$$

$$M_y = D_{12}^* \frac{\partial \alpha}{\partial x} + D_{22}^* \frac{\partial \beta}{\partial y} + D_{26}^* \left(\frac{\partial \alpha}{\partial y} + \frac{\partial \beta}{\partial x} \right) \tag{6.118}$$

$$M_{xy} = D_{16}^* \frac{\partial \alpha}{\partial x} + D_{26}^* \frac{\partial \beta}{\partial y} + D_{66}^* \left(\frac{\partial \alpha}{\partial y} + \frac{\partial \beta}{\partial x} \right) \tag{6.119}$$

$$Q_x = A_{55} \left(\alpha + \frac{\partial w}{\partial x} \right) + A_{45} \left(\beta + \frac{\partial w}{\partial y} \right) \tag{6.120}$$

$$Q_y = A_{45} \left(\beta + \frac{\partial w}{\partial x} \right) + A_{44} \left(\beta + \frac{\partial w}{\partial y} \right) \tag{6.121}$$

$$\frac{\partial M_x}{\partial x} + \frac{\partial M_{xy}}{\partial y} - Q_x = I \frac{\partial^2 \alpha}{\partial t^2} \tag{6.122}$$

$$\frac{\partial M_{xy}}{\partial x} + \frac{\partial M_y}{\partial y} - Q_y = I \frac{\partial^2 \beta}{\partial t^2} \tag{6.123}$$

$$\frac{\partial Q_x}{\partial x} + \frac{\partial Q_y}{\partial y} = -p(x, y) + \bar{\rho} \frac{\partial^2 w}{\partial t^2} - N_x^i \frac{\partial^2 w}{\partial x^2} - 2N_{xy}^i \frac{\partial^2 w}{\partial x \partial y} - N_y^i \frac{\partial^2 w}{\partial y^2} \tag{6.124}$$

Some explanation will clarify the terms above. In the moment curvature relations (6.117) to (6.119) of the laminate construction, if the laminae are such that they are exact mirror images above and below the midsurface where $[B] = 0$, then $D_{ij}^* = D_{ij}$. If, further, the construction were cross-ply such that the stacking is $0°$ and $\pm 90°$, then $D_{16} = D_{26} = 0$, that is, no twisting–bending coupling, and these expressions would reduce to (6.50) to (6.52).

Likewise, under certain constructions in which $A_{45} = 0$, (6.120) and (6.121) reduce to (6.53) and (6.54).

In (6.122) and (6.123), it is rather easy to see that

$$I = \sum_{k=1}^{n} \int_{h_{k-1}}^{h_k} \rho_0^{(k)} z^2 \, dz = \frac{1}{3} \sum_{k=1}^{n} \rho_0^{(k)} (h_k^3 - h_{k-1}^3) \tag{6.125}$$

where I is the moment of inertia of the plate construction with respect to the plate midsurface per unit planform area, and $\rho_0^{(k)}$ is the mass density of the kth lamina, using the notation of Fig. 5.2.

Equation (6.124) has been written in the most general form, including the effect of a laterally distributed load, $p(x, y)$, dynamic inertia loading ($\bar{\rho} \partial^2 w / \partial t^2$), and the effects of in-plane loading, as discussed in Chapter 4 concerning buckling due to in-plane loads.

$\bar{\rho}$ is the mass per unit planform area, in this case

$$\bar{\rho} = \sum_{k=1}^{n} \int_{h_{k-1}}^{h_k} \rho_0^{(k)}\, dz = \sum_{k=1}^{n} \rho_0^{(k)}(h_k - h_{k-1}) \qquad (6.126)$$

Also, N_x^i, N_{xy}^i and N_y^i are the initial in-plane forces per unit edge distances that are acting on the plate. (Note: consistent with Chapter 4 positive values of these resultants produce tensile stresses.)

Substituting (6.117) to (6.121) into (6.122) to (6.124) results in the following governing differential equations

$$[L_1(D_{ij}^*) - A_{55}]\alpha + [L_2(D_{ij}^*) - A_{45}]\beta - L_7 w = I\frac{\partial^2 \alpha}{\partial t^2} \qquad (6.127)$$

$$[L_2(D_{ij}^*) - A_{45}]\alpha + [L_4(D_{ij}^*) - A_{44}]\beta - L_8 w = I\frac{\partial^2 \beta}{\partial t^2} \qquad (6.128)$$

$$L_7\alpha + L_8\beta + L_{55}w = -p(x, y) + \bar{\rho}\frac{\partial^2 w}{\partial t^2} - N_x^i\frac{\partial^2 w}{\partial x^2} -$$
$$- 2N_{xy}^i\frac{\partial^2 w}{\partial x\partial y} - N_y^i\frac{\partial^2 w}{\partial y^2} \qquad (6.129)$$

In these the operators are defined as

$$L_1(D_{ij}^*) = D_{11}^*\frac{\partial^2}{\partial x^2} + 2D_{16}^*\frac{\partial^2}{\partial x\partial y} + D_{66}^*\frac{\partial^2}{\partial y^2} \qquad (6.130)$$

$$L_2(D_{ij}^*) = D_{16}^*\frac{\partial^2}{\partial x^2} + (D_{12}^* + D_{66}^*)\frac{\partial^2}{\partial x\partial y} + D_{26}^*\frac{\partial^2}{\partial y^2} \qquad (6.131)$$

$$L_4(D_{ij}^*) = D_{66}^*\frac{\partial^2}{\partial x^2} + 2D_{26}^*\frac{\partial^2}{\partial x\partial y} + D_{22}^*\frac{\partial^2}{\partial y^2} \qquad (6.132)$$

$$L_7 = A_{55}\frac{\partial}{\partial x} + A_{45}\frac{\partial}{\partial y} \qquad (6.133)$$

$$L_8 = A_{45}\frac{\partial}{\partial x} + A_{44}\frac{\partial}{\partial y} \qquad (6.134)$$

$$L_{55} = A_{55}\frac{\partial^2}{\partial x^2} + 2A_{45}\frac{\partial^2}{\partial x\partial y} + A_{44}\frac{\partial^2}{\partial y^2} \qquad (6.135)$$

Just as in (6.49) the natural boundary conditions are

either $M_n = 0$ or β_n are prescribed

and either $Q_n = 0$ or w are prescribed

and either $M_{nt} = 0$ or β_t are prescribed

on any boundary where n refers to the direction normal to the edge and t refers to the direction tangential to the edge.

Exact solutions to (6.127) to (6.129) are difficult, and very seldom are found in closed form. However, for crossply construction ($0°$, $90°$, $0°$, $90°$...) or angle-ply construction ($+\theta°$, $-\theta°$, $+\theta°$, $-\theta°$...) in which $A_{45} = D_{16}^* = D_{26}^* = 0$, resulting in a specially orthotropic construction these equations are reduced to the following

$$D_{11}^* \frac{\partial^2 \alpha}{\partial x^2} + D_{66}^* \frac{\partial^2 \alpha}{\partial y^2} + (D_{12}^* + D_{66}^*)\frac{\partial^2 \beta}{\partial x \partial y} - A_{55}\left(\alpha + \frac{\partial w}{\partial x}\right) = I\frac{\partial^2 \alpha}{\partial t^2} \quad (6.136)$$

$$(D_{12}^* + D_{66}^*)\frac{\partial^2 \alpha}{\partial x \partial y} + D_{66}^* \frac{\partial^2 \beta}{\partial x^2} + D_{22}^* \frac{\partial^2 \beta}{\partial y^2} - A_{44}\left(\beta + \frac{\partial w}{\partial y}\right) = I\frac{\partial^2 \beta}{\partial t^2} \quad (6.137)$$

$$A_{55}\left(\frac{\partial \alpha}{\partial x} + \frac{\partial^2 w}{\partial x^2}\right) + A_{44}\left(\frac{\partial \beta}{\partial y} + \frac{\partial^2 w}{\partial y^2}\right) = -p(x,y) - N_x^i \frac{\partial^2 w}{\partial x^2} -$$
$$- 2N_{xy}^i \frac{\partial^2 w}{\partial x \partial y} - N_y^i \frac{\partial^2 w}{\partial y^2} + \bar{\rho}\frac{\partial^2 w}{\partial t^2} \quad (6.138)$$

The exact solutions for the problem including critical buckling loads, natural frequencies and deflections due to lateral loads for *simply-supported boundary* conditions are found easily analogous to the procedures of Sections 4.8, 4.9, 4.10, 6.4 and 6.5.

However, for generality, the procedures of Section 6.5 can be employed to derive the governing equations analogous to (6.136) to (6.138) for the large amplitude, non-linear vibrations of a specially orthotropic laminated plate of composite material, including transverse shear deformations and rotatory inertia. The results are

$$L_1(w, \alpha, \beta) = A_{55}\left(\alpha + \frac{\partial w}{\partial x}\right) - D_{11}^* \frac{\partial^2 \alpha}{\partial x^2} - D_{12}^* \frac{\partial^2 \alpha}{\partial y^2} +$$
$$+ I\frac{\partial^2 \alpha}{\partial t^2} - (D_{12}^* + D_{66}^*)\frac{\partial^2 \beta}{\partial x \partial y} = 0 \quad (6.139)$$

$$L_2(w, \alpha, \beta) = A_{44}\left(\beta + \frac{\partial w}{\partial y}\right) - D_{12}^* \frac{\partial^2 \beta}{\partial x^2} - D_{22}^* \frac{\partial^2 \beta}{\partial y^2} +$$
$$+ I\frac{\partial^2 \beta}{\partial t^2} - (D_{12}^* + D_{66}^*)\frac{\partial^2 \alpha}{\partial x \partial y} = 0 \quad (6.140)$$

$$L_3(w, \alpha, \beta) = A_{55}\left(\frac{\partial \alpha}{\partial x} + \frac{\partial^2 w}{\partial x^2}\right) + A_{44}\left(\frac{\partial^2 w}{\partial y^2} + \frac{\partial \beta}{\partial y}\right) +$$
$$+ A_{11}\frac{h^2 \bar{\alpha}^2}{12}\left(\frac{\partial^2 w}{\partial x^2} + \gamma\frac{\partial^2 w}{\partial y^2}\right) - \rho\frac{\partial^2 w}{\partial t^2} = 0 \quad (6.141)$$

$$\frac{\partial u_0}{\partial x} + \gamma \frac{\partial v_0}{\partial y} + \frac{1}{2}\left(\frac{\partial w}{\partial x}\right)^2 + \frac{\gamma}{2}\left(\frac{\partial w}{\partial y}\right)^2 = \frac{\bar{\alpha}^2 h^2}{12} \qquad (6.142)$$

where again $\bar{\alpha}$ is a time-dependent coupling parameter, and $\gamma^2 = A_{22}/A_{11}$. For the single lamina plate these equations reduce to those of the previous Section, 6.5.

In solving these equations, we now utilise Galerkin's method which is treated in general in Reference 82, and will not be explained here in detail.

Again as in the previous Section, it is assumed that the effect of coupling among the individual vibration modes is not significant, one assumes the following form for the vibration modes

$$w = W\phi_{wm}(x)\phi_{wn}(y)\tau(t)$$
$$\alpha = \Gamma\phi_{\alpha m}(x)\phi_{\alpha n}(y)\tau(t) \qquad (6.143)$$
$$\beta = \Lambda\phi_{\beta m}(x)\phi_{\beta n}(y)\tau(t)$$

Here the ϕ's are characteristic beam functions, analogous to those used by Warburton,[83] and catalogued very conveniently by Young and Felgar.[84,85] The beam functions are the vibration modes of beams under various boundary conditions and as such form a complete and orthogonal set of functions—hence usable in Galerkin's method.

Employing Galerkin's method results in the following three algebraic equations.

$$\int_0^a \int_0^b L_1(w, \alpha, \beta)\phi_{\alpha m}(x)\phi_{\alpha n}(y)\,\mathrm{d}y\,\mathrm{d}x = 0 \qquad (6.144)$$

$$\int_0^a \int_0^b L_2(w, \alpha, \beta)\phi_{\beta m}(x)\phi_{\beta n}(y)\,\mathrm{d}y\,\mathrm{d}x = 0 \qquad (6.145)$$

$$\int_0^a \int_0^b L_3(w, \alpha, \beta)\phi_{wm}(x)\phi_{wn}(y)\,\mathrm{d}y\,\mathrm{d}x = 0 \qquad (6.146)$$

and

$$\bar{\alpha}^2 = \frac{6\bar{\beta}^2\tau^2(t)}{ab}\int_0^a \int_0^b [\phi_{wm}'^2(x)\psi_{wn}^2(y) + \gamma\phi_{wm}^2(x)\phi_{wn}'^2(y)]\,\mathrm{d}y\,\mathrm{d}x \qquad (6.147)$$

where $\bar{\beta}$ is again the ratio of the vibration amplitude to the plate thickness, and primes denote differentiation to the appropriate variable. In (6.147) the fact that $u_0 = v_0 = 0$ on all edges has been employed. (If this were not so, Berger's Hypothesis could not have been employed as discussed in Reference 81.)

where in (6.148) μ_m are the roots of the following:

$$\tan\left(\frac{\mu_m}{2}\right) + \tanh\left(\frac{\mu_m}{2}\right) = 0 \quad \text{and} \quad \eta_m = \frac{\sin\left(\mu_m/2\right)}{\sinh\left(\mu_m/2\right)}$$

As an explicit example, consider a plate clamped along all four edges. Then

$$u_0 = v_0 = w = \alpha = \beta = 0 \quad \text{on} \quad x = 0, a \quad \text{and} \quad y = 0, b$$

For this case the clamped–clamped beam functions are written as follows in the x-direction. Changing x to y, m to n, a to b and the interchange of α and β provides the proper functions in the y-direction: For $m = 1, 3, 5, \ldots$ all odd numbers,

$$\phi_{wm}(x) = \cos \mu_m\left(\frac{x}{a} - \frac{1}{2}\right) + \eta_m \cosh \mu_m\left(\frac{x}{a} - \frac{1}{2}\right)$$

$$\phi_{\alpha m}(x) = \sin \mu_m\left(\frac{x}{a} - \frac{1}{2}\right) - \eta_m \sinh \mu_m\left(\frac{x}{a} - \frac{1}{2}\right) \qquad (6.148)$$

$$\phi_{\beta m}(x) = \cos \mu_m\left(\frac{x}{a} - \frac{1}{2}\right) + \eta_m \cosh \mu_m\left(\frac{x}{a} - \frac{1}{2}\right)$$

For $m = 2, 4, 6, \ldots$ all even numbers

$$\phi_{wm}(x) = \sin \mu_m\left(\frac{x}{a} - \frac{1}{2}\right) + \eta_m \sinh \mu_m\left(\frac{x}{a} - \frac{1}{2}\right)$$

$$\phi_{\alpha m}(x) = \cos \mu_m\left(\frac{x}{a} - \frac{1}{2}\right) + \eta_m \cosh \mu_m\left(\frac{x}{a} - \frac{1}{2}\right) \qquad (6.149)$$

$$\phi_{\beta m}(x) = \sin \mu_m\left(\frac{x}{a} - \frac{1}{2}\right) + \eta_m \sinh \mu_m\left(\frac{x}{a} - \frac{1}{2}\right)$$

Substituting (6.148) or (6.149) and the corresponding functions of y into (6.139) to (6.146) one obtains, for the case of a laminate symmetric with respect to the mid-plane (i.e. $D_{ij}^* = D_{ij}$)

$$[-\lambda_m d_{21} A_{55}\tau]W + [D_{12} + D_{66}]\lambda_m \delta_n d_{22}\tau \Lambda +$$
$$+ [A_{55}d_{21} + D_{11}\lambda_m^2 d_{11}\tau + D_{66}\delta_m^2 d_{22}\tau + Id_{21}\ddot{\tau}]\Gamma = 0 \qquad (6.150)$$

$$[-\delta_n d_{12} A_{44}\tau]W + [D_{12} + D_{66}]\lambda_m \delta_n d_{22}\tau \Gamma +$$
$$+ [A_{44}d_{12}\tau + D_{22}\delta_n^2 d_{21}\tau + D_{66}\lambda_m^2 d_{22}\tau - Id_{12}\ddot{\tau}]\Lambda = 0 \qquad (6.151)$$

where in (6.149) μ_m are the roots of the following:

$$\tan\left(\frac{\mu_m}{2}\right) - \tanh\left(\frac{\mu_m}{2}\right) = 0 \quad \text{and} \quad \eta_m = -\frac{\sin\left(\mu_m/2\right)}{\sinh\left(\mu_m/2\right)}$$

$$-\left[\left(A_{55}+\frac{A_{11}h^2\bar{\alpha}^2}{12}\right)\lambda_m^2 d_{21}\tau+\left(A_{44}+\frac{A_{11}h^2\bar{\alpha}^2\gamma}{12}\right)\delta_n^2 d_{12}\tau+\right.$$

$$\left.+\rho d_{11}\ddot{\tau}\right]W_0+A_{55}\lambda_m d_{21}\tau\Gamma+A_{44}\delta_n d_{12}\tau\Lambda=0 \quad (6.152)$$

and

$$\bar{\alpha}^2=1\cdot5\bar{\beta}^2(d_{21}\lambda_m^2+d_{12}\delta_n^2)\tau^2(t) \quad (6.153)$$

where

$$\lambda_m=\frac{\mu_m}{a}, \qquad \delta_n=\frac{\mu_n}{b}, \qquad d_{ij}=\theta_{im}\theta_{jn} \qquad (i,j=1,2)$$

$$\theta_{1k}=1-(-1)^k\eta_k^2$$

and

$$\theta_{2k}=1+(-1)^k\left(\eta_k^2-\frac{2}{\mu_k}\sin\mu_k\right) \qquad (k=1,2,3\ldots)$$

with

$$\eta_k=(-1)^{k+1}\frac{\sin(\mu_k/2)}{\sinh(\mu_k/2)}$$

and

$$\mu_1=1\cdot506\pi; \qquad \mu_k\simeq\left(k+\frac{1}{2}\right)\pi \quad \text{as} \quad k\geqslant2$$

With the neglect of rotatory inertia ($I=0$), the characteristic equation of motion can be obtained by setting the determinant of the coefficients of eqns. (6.150) through (6.152) to zero. Again one obtains the Duffing equation (see eqn. (6.102)):

$$\alpha^*\ddot{\tau}(t)+\beta^*\tau(t)+\delta^*\tau^3(t)=0, \quad (6.154)$$

where in this case,

$$\alpha^*=1+\left(\frac{D_{11}}{A_{55}}\frac{d_{11}}{d_{21}}+\frac{D_{66}}{A_{44}}\frac{d_{22}}{d_{12}}\right)\lambda_m^2+\left(\frac{D_{66}}{A_{55}}\frac{d_{21}}{d_{11}}+\frac{D_{22}}{A_{44}}\frac{d_{11}}{d_{12}}\right)\delta_n^2+$$

$$+\frac{D_{11}}{A_{55}}\frac{D_{66}}{A_{44}}\lambda_m^4+\frac{D_{22}}{A_{44}}\frac{D_{66}}{A_{55}}\delta_n^4+$$

$$+\left(\frac{D_{11}}{A_{55}}\frac{D_{22}}{A_{44}}\frac{d_{11}}{d_{22}}-\frac{D_{12}^2+2D_{12}D_{66}}{A_{55}A_{44}}\frac{d_{22}}{d_{11}}\right)\lambda_m^2\delta_n^2 \quad (6.155)$$

$$\beta^*=\frac{1}{\bar{\rho}}\left[D_{11}\lambda_m^4+2(D_{12}+2D_{66})\frac{d_{22}}{d_{11}}\lambda_m^2\delta_n^2+D_{22}\delta_n^4+\right.$$

$$+\frac{D_{11}D_{66}}{A_{44}}\frac{d_{22}}{d_{12}}\lambda_m^6+\frac{D_{22}D_{66}}{A_{55}}\cdot\frac{d_{22}}{d_{21}}\delta_n^6+$$

$$+\left(\frac{D_{11}D_{22}}{A_{44}}\frac{d_{11}}{d_{12}}+\frac{D_{11}D_{66}}{A_{55}}\frac{d_{22}}{d_{21}}-\frac{D_{12}^2+2D_{12}D_{66}}{A_{44}}\frac{d_{22}^2}{d_{11}d_{12}}\right)\lambda_m^4\delta_n^2+$$

$$+ \left(\frac{D_{11}D_{22}}{A_{55}} \frac{d_{11}}{d_{21}} + \frac{D_{22}D_{66}}{A_{44}} \frac{d_{22}}{d_{12}} - \frac{D_{12}^2 + 2D_{12}D_{66}}{A_{55}} \frac{d_{22}^2}{d_{11}d_{21}} \right) \lambda_m^2 \delta_n^4 \right]$$

(6.156)

$$\delta^* = \frac{\bar{\beta}^2 A_{11} h^2}{8\bar{\rho}} (d_{21}\lambda_m^2 + \gamma d_{12}\delta_n^2) \alpha^*$$

(6.157)

If the plate is subjected to the initial conditions: $\tau(0) = 1$ and $\dot{\tau}(0) = 0$, the solution to eqn. (6.154) is an elliptic cosine function, as was obtained in (6.107),

$$\tau(t) = cn(\omega^* t, p)$$

(6.158)

with

$$\omega^* = \bar{\omega} \left[\frac{3}{2} \bar{\beta}^2 \gamma_1 + \left(1 + \frac{D_{11}}{A_{55}} \frac{d_{11}}{d_{21}} \lambda_m^2 + \frac{d_{11}}{d_{12}} \delta_n^2 + B \right) \right]^{-1}$$

(6.159)

$$p = \left[\frac{3}{4} \bar{\beta}^2 \gamma_1 \left(\frac{\bar{\omega}}{\omega^*} \right) \right]^{1/2}$$

(6.160)

where

$$\bar{\omega} = \left\{ \frac{1}{\bar{\rho}} \left[D_{11}\lambda_m^4 + 2(D_{12} + 2D_{66}) \frac{d_{22}}{d_{11}} \lambda_m^2 \delta_n^2 + D_{22}\delta_n^4 \right] \right\}^{1/2}$$

(6.161)

$$\gamma_1 = \frac{A_{11} h^2}{12\bar{\rho}\bar{\omega}^2} (d_{21}\lambda_m^2 + \gamma d_{12}\delta_n^2)^2$$

(6.162)

$$B = \left(1 + \frac{D_{66}}{A_{44}} \lambda_m^2 + \frac{D_{66}}{A_{55}} \delta_n^2 + A \right)^{-1} \left[\left(\frac{D_{11}D_{22}}{A_{44}A_{55}} \frac{d_{11}}{d_{22}} - \right. \right.$$

$$- \frac{D_{12}^2 + 2D_{12}D_{66}}{A_{44}A_{55}} \frac{d_{22}}{d_{11}} - \frac{D_{22}D_{66}}{A_{44}^2} \frac{d_{21}}{d_{12}} - \frac{D_{11}D_{66}}{A_{55}^2} \frac{d_{12}}{d_{21}} \right) \lambda_m^2 \delta_n^2 -$$

$$\left. - A \left(1 + \frac{D_{11}}{A_{55}} \frac{d_{11}}{d_{21}} \lambda_m^2 + \frac{D_{22}}{A_{44}} \frac{d_{11}}{d_{12}} \delta_n^2 \right) \right]$$

(6.163)

$$A = \left[\lambda_m^4 + \frac{D_{22}}{D_{11}} \delta_n^4 + \frac{2(D_{12} + 2D_{66})}{D_{11}} \frac{d_{22}}{d_{11}} \lambda_m^2 \delta_n^2 \right]^{-1} \times$$

$$\times \left[\left(\frac{D_{22}}{A_{44}} \frac{d_{11}}{d_{12}} - \frac{(D_{12} + 2D_{66})^2}{D_{11}A_{44}} \frac{d_{22}^2}{d_{11}d_{12}} \right) \lambda_m^2 + \right.$$

$$\left. + \left(\frac{D_{22}}{A_{55}} \frac{d_{11}}{d_{21}} - \frac{(D_{12} + 2D_{66})^2}{D_{11}A_{55}} \frac{d_{22}^2}{d_{11}d_{21}} \right) \delta_n^2 \right] \lambda_m^2 \delta_n^2$$

(6.164)

We note that $\bar{\omega}$ is the natural frequency obtained from the classical theory which is equivalent to setting $\bar{\beta} = 0$ which implies small amplitude linear vibration and requiring that all terms including A_{44} and A_{55} vanish (which implies neglect of transverse shear deformation). Equation (6.161) for the value of $\bar{\omega}$ is the same as that due to Hearmon[86] for orthotropic plates.

Finally the frequency ratio defined as the inverse ratio of the period due to the present non-linear theory to that from the classical theory for orthotropic plates is

$$R^* = \frac{\pi}{2K(p)} \left[\frac{3}{2} \bar{\beta}^2 \gamma_1 + \left(1 + \frac{D_{11}}{A_{55}} \frac{d_{11}}{d_{21}} \lambda_m^2 + \frac{D_{22}}{A_{44}} \frac{d_{11}}{d_{12}} \delta_n^2 + B \right)^{-1} \right]^{1/2} \quad (6.165)$$

where $K(p)$ is the complete elliptic integral of first kind with a parameter p. For the case of small amplitude $\bar{\beta} = 0$ and $K(p) = \pi/2$, (6.165) reduces to

$$R = \left(1 + \frac{D_{11}}{A_{55}} \frac{d_{11}}{d_{21}} \lambda_m^2 + \frac{D_{22}}{A_{44}} \frac{d_{11}}{d_{12}} \delta_n^2 + B \right)^{-1/2} \quad (6.166)$$

which is the ratio of the small amplitude linear vibration frequency taking into consideration the effect of transverse shear deformation to that of the classical theory for a fully clamped orthotropic plate.

It should be noted that the above solution of natural frequency was derived for a completely clamped plate only. However, for other boundary conditions, namely, combinations of clamped and simply-supported edges, if one follows the same procedures, the solution can be obtained by selecting the appropriate assumed functions analogous to those due to Warburton.[83] It is found that we will arrive at solutions of the same form as eqns. (6.165) and (6.166) but with different values of boundary constants, which are presented in Table 6.1. The frequency ratio is therefore defined as the ratio of the frequency obtained from the present theory to that from the classical theory for the same plate with the same boundary conditions.

Notice that if we replace D_{ij} by D_{ij}^*, which is the reduced bending rigidity as defined in (6.115), eqns. (6.165) and (6.166) can be used also for an arbitrarily laminated cross-ply and angle-ply plate. Of course, one can systematically reduce the equations to provide for one lamina, small deflection, and even isotropy through straightforward simplifications. Using Table 6.1, one can also determine these frequencies for all combinations of simple support and clamped boundary conditions.

Of course, these beam modes can also be used with (6.127) to (6.129) to find deflections and stresses due to lateral loads, as well as buckling loads.

TABLE 6.1

CONSTANTS BASED ON BOUNDARY CONDITIONS

Boundary conditions	d_{ij}	λ_m	δ_n
1	$\theta_{im}\theta_{jn}$	$\dfrac{\mu_m}{a}$	$\dfrac{\mu_n}{b}$
2	$\theta_{im}\theta_{j(2n)}$	$\dfrac{\mu_m}{a}$	$\dfrac{\mu_{2n}}{2b}$
3	θ_{im}	$\dfrac{\mu_m}{a}$	$\dfrac{n\pi}{b}$
4	$\theta_{i(2m)}\theta_{j(2n)}$	$\dfrac{\mu_{2m}}{2a}$	$\dfrac{\mu_{2n}}{2b}$
5	$\theta_{i(2m)}$	$\dfrac{\mu_{2m}}{2a}$	$\dfrac{n\pi}{b}$
6	1	$\dfrac{m\pi}{a}$	$\dfrac{n\pi}{b}$

- - - - - - Simply-supported edge.
///////// Clamped edge.

Once more we turn to the boron–epoxy composite material of Section 6.4 to provide a numerical example in order to obtain some physical insight as to the behaviour of plates of composite materials.

The following curves are designed such that the ordinates are the ratio of the fundamental frequency using the method presented herein to the fundamental frequency for the same plate with the same boundary conditions using classical linear theory neglecting transverse shear deformation. The abscissae are the geometric ratio of length to thickness or the ratio of amplitude to thickness where the deflection is of the same order of magnitude as the plate thickness.

Figure 6.4 shows the effect of the boundary conditions on the frequency ratio for a four-layer symmetric cross-ply (0°, 90°, 90°, 0°) square plate. Figure 6.5 shows the effect of the number of layers on the frequency ratio for a cross-ply (0°, 90°, 0°, 90°, ...) plate with fully clamped edges. Finally, the effect of transverse shear deformation for the large amplitude vibration of a fully clamped, two-layer (0°, 90°) square plate is shown in Fig. 6.6 for various ratios of length to thickness.

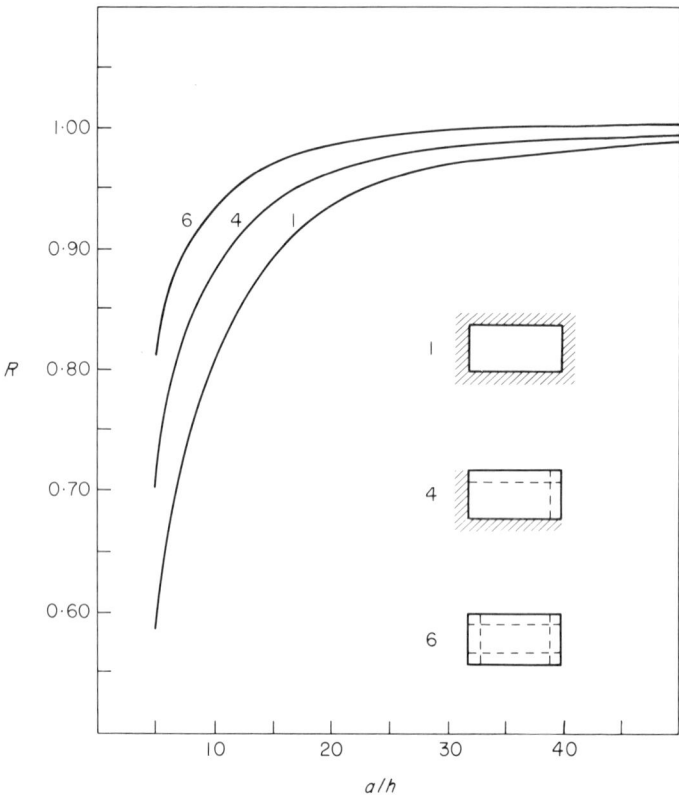

Fig. 6.4. *Fundamental frequency ratio for a four-layer* (0°, 90°, 90°, 0°), *square plate with various boundary conditions under small amplitude vibrations.*

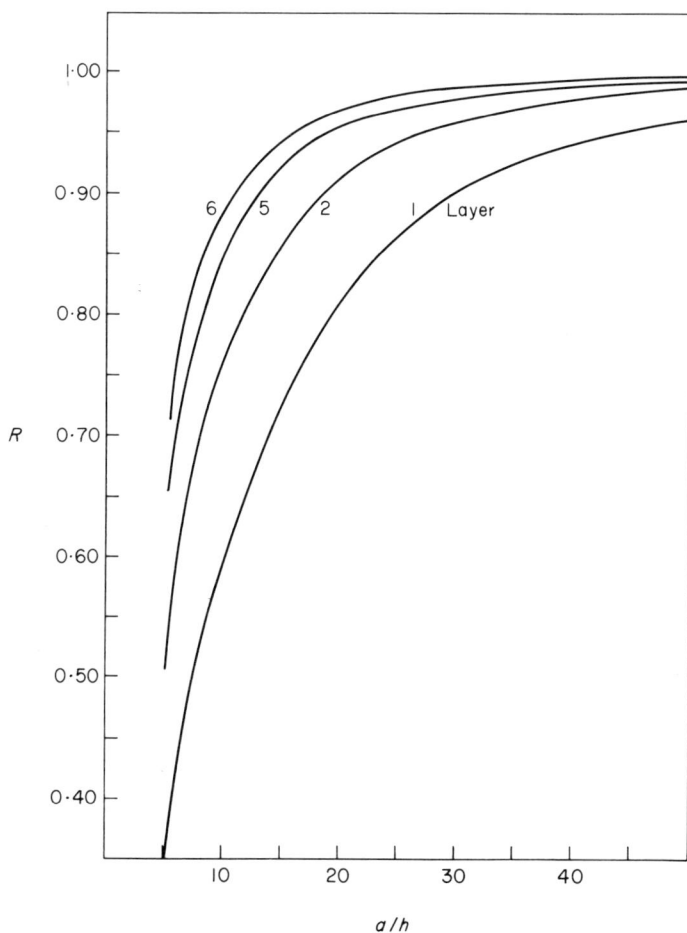

Fig. 6.5. Fundamental frequency ratio for a fully clamped, square plate with various numbers of layers (0°, 90°, 0°, 90°,...) under small amplitude vibrations.

The following conclusions are drawn from the numerical curves:

1. The effect of transverse shear deformation, in terms of the percentage reduction of the corresponding frequency due to classical theory, for a fully clamped plate is greater than that for a simply-supported plate.

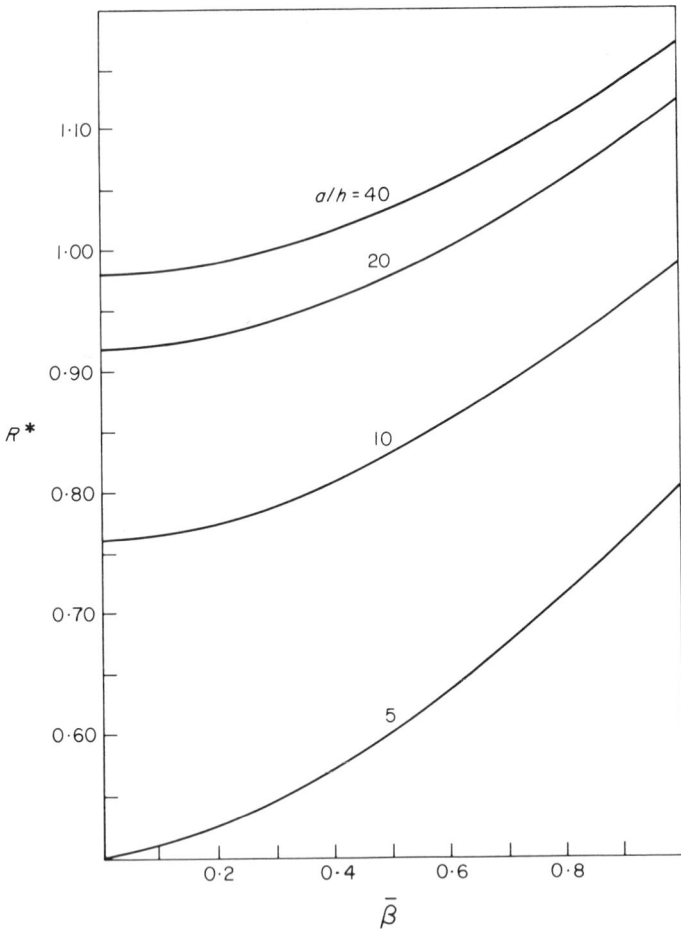

Fig. 6.6.   Fundamental frequency ratio for a two-layer (0°, 90°), fully clamped, square plate under large amplitude vibrations.

2. The effect of transverse shear deformation decreases when the number of layers for the cross-ply (0°, 90°, 0°, 90°, ...) plates is increased.

3. For a plate composed of highly anisotropic composite layers with either clamped or simply-supported edges, the effect of transverse shear deformation is significant for both small and large amplitude vibration.

This Section is taken in part from Reference 54.

6.7 BEHAVIOUR OF SIMPLY-SUPPORTED, CROSS-PLY
LAMINATED PLATES UNDER STATIC LATERAL
AND IN-PLANE LOADS[a]

In the previous Sections, the natural frequencies have been determined
for a variety of cases in order to illustrate as simply as possible the
effects of anisotropy, laminations, and large deformation. Of course,
this only involved the homogeneous governing differential equations.
Whitney[61] has studied in detail the effects of transverse shear
deformation on the bending and buckling of simply-supported, cross-
ply laminated plates of composite materials.

Starting with an equivalent formulation of equations: (6.136) to
(6.138), Whitney treated the case of the plate being subjected to a
lateral load, $q(x, y)$, in the following form

$$q(x, y) = q_0 \sin \frac{m\pi x}{a} \sin \frac{n\pi y}{b} \qquad (6.167)$$

where m and n are integers, and $N_x^i = N_{xy}^i = N_y^i = \partial^2( )/\partial t^2 = 0$.

With the plate simply supported, the Navier type solutions can be
utilised in which

$$w(x, y) = A \sin \frac{m\pi x}{a} \sin \frac{n\pi y}{b}$$

$$\alpha(x, y) = B \cos \frac{m\pi x}{a} \sin \frac{n\pi y}{b} \qquad (6.168)$$

$$\beta(x, y) = C \sin \frac{m\pi x}{a} \cos \frac{n\pi y}{b}$$

Whitney graphically shows the results of including transverse shear
deformation, by plotting a non-dimensionalised maximum deflection,
$w_{max} = w(a/2, a/2)$ for a square plate of length and width a, as a
function of a/h, where h is the laminated plate thickness. Figures 6.7
and 6.8 show Whitney's results for cross-ply plates symmetrical about
the mid-surface, in which the former is for a graphite–epoxy composite
in which $E_{11}/E_{22} = 40$, and the latter is for a glass–epoxy composite
wherein $E_{11}/E_{22} = 3$. In each case the construction is cross-ply of (0°,
90°, 90°, 0) construction. It is seen from Fig. (6.7) that transverse
deformation increases the plate maximum deflection by more than 15
per cent even when the plate is of geometry $a/h = 20$, which is certainly
in the regime of classically thin plate theory. The effects are understand-
ably less in Fig. 6.8.

[a] Excerpted from Reference 61 by permission of the Technomic Publishing Co.

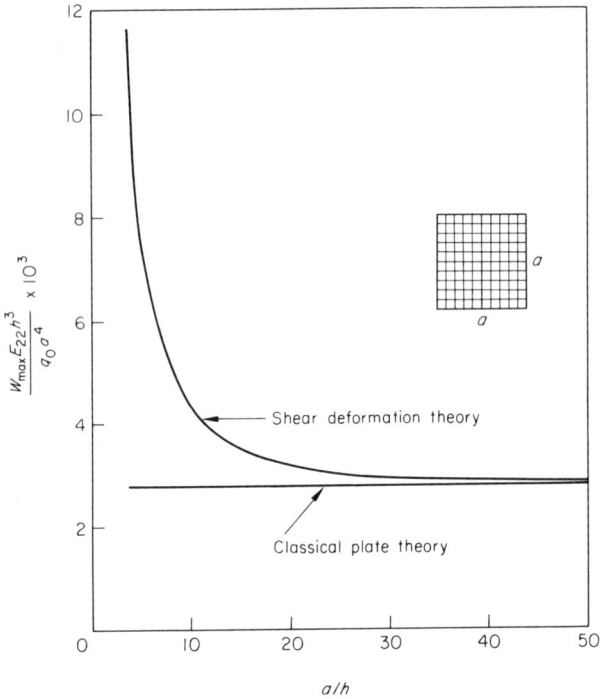

Fig. 6.7. Bending of a four-layer, symmetric cross-ply, square plate under the load $q_0 \sin(\pi x/a) \sin(\pi y/b)$ with $E_{11}/E_{22} = 40$, $G_{12}/E_{22} = 0.6$, $G_{23}/E_{22} = 0.5$, $\nu_{12} = 0.25$.

Similarly, Whitney studied the elastic stability of simply-supported, symmetric laminates, involving in-plane loads N_x^i and N_y^i. Again the Navier approach was used and results analogous to those in Section 4.10 can be easily obtained. Again using eqns. (6.136) to (6.138), with $p(x, y) = N_{xy}^i = \partial^2(\ )/\partial t^2 = 0$ three simultaneous equations are obtained, and the determinant of the coefficients of A, B, and C of (6.167) give the combination of N_x and N_y causing an instability. Specifically, Whitney provided graphically the effect of transverse shear deformation on the buckling load $N_x = -N_0$, $N_y = 0$ for a square plate composed of an infinite number of layers of $\pm 45°$ angle-ply construction (where an infinite number of layers eliminates unwanted coupling effects). The results are shown in Fig. 6.9 for a typical graphite–epoxy materials system. Again the effect of transverse shear deformations is obvious for this case when elastic buckling occurs at $m = n = 1$ due to the geometry. In the next Section, one sees that transverse shear

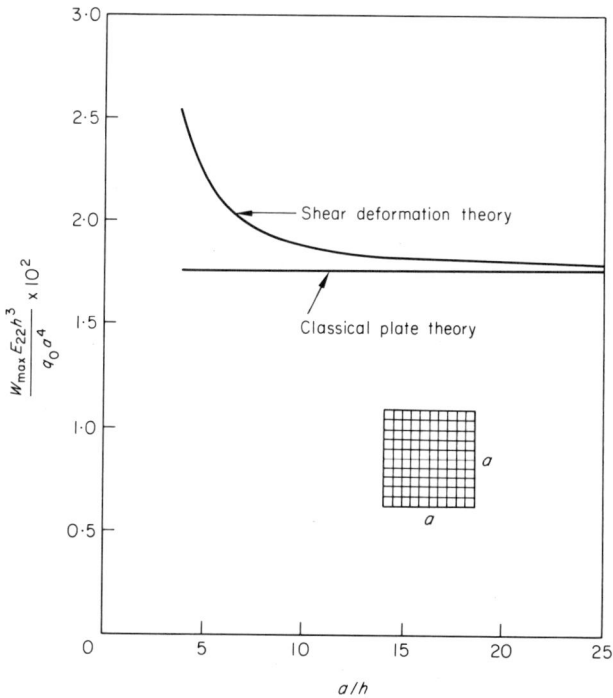

Fig. 6.8. *Bending of a four-layer, symmetric cross-ply square plate under the load* $q_0 \sin(\pi x/a) \sin(\pi y/b)$ *with* $E_{11}/E_{22} = 3$, $G_{12}/E_{22} = 0\cdot6$, $G_{23}/E_{22} = 0\cdot5$, $\nu_{12} = 0\cdot25$.

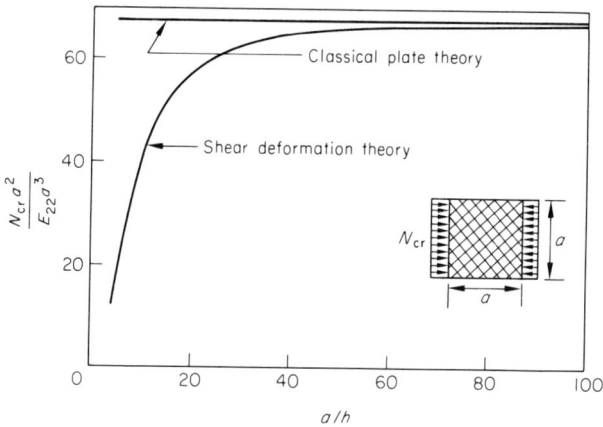

Fig. 6.9. *Buckling of a square* $\pm45°$ *plate with an infinite number of layers under uniform compression*; $E_{11}/E_{22} = 40$, $G_{12}/E_{22} = 0\cdot6$, $G_{23}/E_{22} = 0\cdot5$, $\nu_{12} = 0\cdot25$.

deformation affects not only the value of the buckling load but can also affect the wave number (the number of buckles) for various plate geometries.

6.8   THE ELASTIC STABILITY OF SPECIALLY ORTHOTROPIC PLATES DUE TO IN-PLANE LOADS FOR VARIOUS BOUNDARY CONDITIONS

This problem can be approached by utilising the Reissner functional, ψ, defined by eqns. (6.45) and (6.44), as the strain energy expression of the plate in the Theorem of Minimum Potential Energy, eqn. (4.119). The term involving the surface tractions (the second term in (4.119)), in this case the work done by the in-plane stress resultants N_x and N_y, is precisely that given in (4.133) involving these terms, namely

$$\int_{S_t} T_i u_i \, ds = -\frac{1}{2} \int_0^a \int_0^b \left[N_x \left(\frac{\partial w}{\partial x} \right)^2 + N_y \left(\frac{\partial w}{\partial y} \right)^2 \right] dy \, dx$$

Therefore the Potential Energy expression is

$$V = \int_0^a \int_0^b \left[-\frac{\bar{N}_x}{2} \left(\frac{\partial w}{\partial x} \right)^2 - \frac{\bar{N}_y}{2} \left(\frac{\partial w}{\partial y} \right)^2 + M_x \frac{\partial \alpha}{\partial x} + M_y \frac{\partial \beta}{\partial y} + \right.$$
$$+ M_{xy} \left(\frac{\partial \alpha}{\partial y} + \frac{\partial \beta}{\partial x} \right) + Q_x \left(\alpha + \frac{\partial w}{\partial x} \right) + Q_y \left(\beta + \frac{\partial w}{\partial y} \right) -$$
$$- \frac{6M_x^2}{E_x h^3} - \frac{6M_y^2}{E_y h^3} - \frac{6M_{xy}^2}{G_{xy} h^3} + \frac{12\nu_{xy}}{E_x h^3} M_x M_y -$$
$$\left. - \frac{3}{5} \left(\frac{Q_x^2}{G_{xz} h} + \frac{Q_y^2}{G_{yz} h} \right) \right] dy \, dx \tag{6.169}$$

where $\bar{N}_x \equiv -N_x$, and $\bar{N}_y \equiv -N_y$, that is to say compressive resultants. Taking variations with respect to all variables, (6.169) becomes:

$$0 = \int_0^a [M_y \delta\beta + M_{xy} \delta\alpha + Q_y \delta w]_0^b \, dx +$$
$$+ \int_0^b \left[-\bar{N}_x \frac{\partial w}{\partial x} \delta w + M_x \delta\alpha + M_{xy} \delta\beta + Q_x \delta w - \bar{N}_y \frac{\partial w}{\partial y} \delta w \right]_0^\eta \, dy +$$
$$+ \int_0^a \int_0^b \left\{ \left[\bar{N}_x \frac{\partial^2 w}{\partial x^2} + \bar{N}_y \frac{\partial^2 w}{\partial y^2} - \frac{\partial Q_x}{\partial x} - \frac{\partial Q_y}{\partial y} \right] \delta w + \right.$$
$$+ \left[-\frac{\partial M_{xy}}{\partial y} + Q_x - \frac{\partial M_x}{\partial x} \right] \delta\alpha + \left[-\frac{\partial M_y}{\partial y} - \frac{\partial M_{xy}}{\partial x} + Q_y \right] \delta\beta +$$
$$+ \left[\frac{\partial \alpha}{\partial x} - \frac{12M_x}{E_x h^3} + \frac{12\nu_{xy} M_y}{E_x h^3} \right] \delta M_x + \left[\frac{\partial \beta}{\partial y} - \frac{12M_y}{E_y h^3} + \frac{12\nu_{xy} M_x}{E_x h^3} \right] \delta M_y +$$

$$+ \left[\frac{\partial \alpha}{\partial y} + \frac{\partial \beta}{\partial x} - \frac{12 M_{xy}}{G_{xy} h^3}\right] \delta M_{xy} + \left[\alpha + \frac{\partial w}{\partial x} - \frac{6}{5} \frac{Q_x}{G_{xz} h}\right] \delta Q_x +$$

$$+ \left[\beta + \frac{\partial w}{\partial y} - \frac{6}{5} \frac{Q_y}{G_{yz} h}\right] \delta Q_y \Big\} \, dy \, dx \tag{6.170}$$

Note the similarities between terms in (6.170) and (6.49). From the Euler–Lagrange equations, which are the bracketed coefficients of the variations in the double integral quantity set equal to zero, after manipulation one can express the stress couples and shear stress resultants as

$$M_x = D_x \left(\frac{\partial \alpha}{\partial x} + \nu_{yx} \frac{\partial \beta}{\partial y}\right)$$

$$M_y = D_y \left(\nu_{xy} \frac{\partial \alpha}{\partial x} + \frac{\partial \beta}{\partial y}\right)$$

$$M_{xy} = \frac{G_{xy} h^3}{12} \left(\frac{\partial \alpha}{\partial y} + \frac{\partial \beta}{\partial x}\right) \tag{6.171}$$

$$Q_x = \frac{5 G_{xz} h}{6} \left(\alpha + \frac{\partial w}{\partial x}\right)$$

$$Q_y = \frac{5 G_{yz} h}{6} \left(\beta + \frac{\partial w}{\partial y}\right)$$

Substituting (6.171) into (6.169) results in the potential energy being expressed in terms of w, α and β only. The result is

$$V = \int_0^a \int_0^b \left\{ \left(-\frac{\bar{N}_x}{2} + \frac{5}{12} G_{xz} h\right)\left(\frac{\partial w}{\partial x}\right)^2 + \frac{5}{6} \alpha \frac{\partial w}{\partial x} G_{xz} h + \right.$$

$$+ \left(\frac{5}{12} G_{yz} h - \frac{\bar{N}_y}{2}\right)\left(\frac{\partial w}{\partial y}\right)^2 + \frac{5}{6} \beta \frac{\partial w}{\partial y} G_{yz} h +$$

$$+ \frac{D_x}{2} \left(\frac{\partial \alpha}{\partial x}\right)^2 + D_x \nu_{yx} \left(\frac{\partial \alpha}{\partial x}\right)\left(\frac{\partial \beta}{\partial y}\right) + \frac{D_y}{2} \left(\frac{\partial \beta}{\partial y}\right)^2 +$$

$$+ \frac{G_{xy} h^3}{24} \left[\left(\frac{\partial \alpha}{\partial y}\right)^2 + 2 \frac{\partial \alpha}{\partial y} \frac{\partial \beta}{\partial x} + \left(\frac{\partial \beta}{\partial x}\right)^2\right] +$$

$$\left. + \frac{5}{12} G_{xz} h \alpha^2 + \frac{5}{12} G_{yz} h \beta^2 \right\} \, dy \, dx \tag{6.172}$$

The Rayleigh method will be employed here to obtain solutions for buckling loads. One way to proceed is to assume the form of the lateral deflection, w, as well as the rotations, α and β, that satisfy the boundary conditions, or at least the geometric boundary conditions. In that case all three variables would contain an undetermined amplitude, that is to say for a simply-supported plate we could assume (6.168) wherein A, B and C are the undetermined amplitudes with which to take variations.

However, an alternative approach is used herein to obtain expressions of α and β, the rotations in terms of w, the lateral deflection, such that the potential energy is in terms of the variable w only. This is accomplished as follows: Consider a beam in the x-direction, such that $\beta = \partial(\ )/\partial y = 0$. Hence the Euler–Lagrange equations of (6.170) are, for the beam of unit width,

$$
\left.
\begin{aligned}
\bar{N}_x \frac{d^2 w}{dx^2} - \frac{dQ_x}{dx} &= 0 \\[4pt]
-\frac{dM_x}{dx} + Q_x &= 0 \\[4pt]
\frac{d\alpha}{dx} - \frac{12 M_x}{E_x h^3} &= 0 \\[4pt]
\alpha + \frac{dw}{dx} - \frac{6}{5} \frac{Q_x}{G_{xz} h} &= 0
\end{aligned}
\right\}
\tag{6.173}
$$

Note the similarities between (6.173) and eqns. (6.19) to (6.22), which are for a beam of width b, $A = bh$, $I = bh^3/12$, and of course an isotropic material. From (6.173) it is found that

$$
\alpha(x) = \frac{E_x h^2}{10 G_{xz}} \frac{d^3 w}{dx^3} \left(\frac{6 \bar{N}_x}{5 G_{xz} h} - 1 \right) - \frac{dw}{dx}
\tag{6.174}
$$

For the plate, this and the corresponding rotation β are taken as

$$
\alpha(x, y) = \frac{E_x h^2}{10 G_{xz}} \frac{\partial^3 w}{\partial x^3} \left(\frac{6 \bar{N}_x}{5 G_{xz} h} - 1 \right) - \frac{\partial w}{\partial x}
\tag{6.175}
$$

$$
\beta(x, y) = \frac{E_y h^2}{10 G_{yz}} \frac{\partial^3 w}{\partial y^3} \left(\frac{6 \bar{N}_y}{5 G_{yz} h} - 1 \right) - \frac{\partial w}{\partial y}
\tag{6.176}
$$

Note that if transverse shear deformation is neglected, i.e. if G_{xz} and G_{yz} are infinity, then the rotations assume the classical values of $\alpha = -\partial w/\partial x$ and $\beta = -\partial w/\partial y$.

Substituting (6.175) and (6.176) into (6.172) results in the potential energy expression involving only the variable $w(x, y)$, as follows

$$
V = \int_0^a \int_0^b \left\{ -\frac{\bar{N}_x}{2} \left(\frac{\partial w}{\partial x} \right)^2 - \frac{\bar{N}_y}{2} \left(\frac{\partial w}{\partial y} \right)^2 + 2 D_{xy} \left(\frac{\partial^2 w}{\partial x \partial y} \right)^2 \right.
$$

$$
- 2 D_{xy} \frac{E_x h^2}{10 G_{xz}} \left(\frac{6 \bar{N}_x}{5 G_{xz} h} - 1 \right) \frac{\partial^4 w}{\partial x^3 \partial y} \frac{\partial^2 w}{\partial x \partial y} -
$$

$$
-2D_{xy}\frac{E_yh^2}{10G_{yz}}\left(\frac{6\bar{N}_y}{5G_{yz}h}-1\right)\frac{\partial^4 w}{\partial x\partial y^3}\frac{\partial^2 w}{\partial x\partial y}+
$$

$$
+\frac{D_{xy}}{2}\left[\frac{E_xh^2}{10G_{xz}}\frac{\partial^4 w}{\partial x^3\partial y}\left(\frac{6\bar{N}_x}{5G_{xz}h}-1\right)+\frac{E_yh^2}{10G_{yz}}\frac{\partial^4 w}{\partial x\partial y^3}\left(\frac{6\bar{N}_y}{5G_{yz}h}-1\right)\right]^2+
$$

$$
+\frac{D_x}{2}\left[\frac{E_xh^2}{10G_{xz}}\frac{\partial^4 w}{\partial x^4}\left(\frac{6\bar{N}_x}{5G_{xz}h}-1\right)-\frac{\partial^2 w}{\partial x^2}\right]^2+
$$

$$
+D_x\nu_{yx}\left[\frac{E_xh^2}{10G_{xz}}\left(\frac{6\bar{N}_x}{5G_{xz}h}-1\right)\frac{\partial^4 w}{\partial x^4}-\frac{\partial^2 w}{\partial x^2}\right]\times
$$

$$
\times\left[\frac{E_yh^2}{10G_{yz}}\left(\frac{6\bar{N}_y}{5G_{yz}h}-1\right)\frac{\partial^4 w}{\partial y^4}-\frac{\partial^2 w}{\partial y^2}\right]+
$$

$$
+\frac{D_y}{2}\left[\frac{E_yh^2}{10G_{yz}}\left(\frac{6\bar{N}_y}{5G_{yz}h}-1\right)\frac{\partial^4 w}{\partial y^4}-\frac{\partial^2 w}{\partial y^2}\right]^2+
$$

$$
+\frac{5}{1200}\frac{E_x^2h^5}{G_{xz}}\left(\frac{6\bar{N}_x}{5G_{xz}h}-1\right)^2\left(\frac{\partial^3 w}{\partial x^3}\right)^2+
$$

$$
+\frac{5}{1200}\frac{E_y^2h^5}{G_{yz}}\left(\frac{6\bar{N}_y}{5G_{yz}h}-1\right)\left(\frac{\partial^3 w}{\partial y^3}\right)^2\Bigg\}\,dy\,dx \qquad (6.177)
$$

(6.177) is the general form of the potential energy function to be minimised for a specially orthotropic plate including transverse shear deformation under in-plane compressive loads $\bar{N}_x$ and $\bar{N}_y$. If transverse shear deformation were not included, (6.177) would be reduced to

$$
V=\int_0^a\int_0^b\left\{-\frac{\bar{N}_x}{2}\left(\frac{\partial w}{\partial x}\right)^2-\frac{\bar{N}_y}{2}\left(\frac{\partial w}{\partial y}\right)^2+2D_{xy}\left(\frac{\partial^2 w}{\partial x\partial y}\right)^2+\right.
$$

$$
\left.+\frac{D_x}{2}\left[\left(\frac{\partial^2 w}{\partial x^2}\right)^2+2\nu_{yx}\frac{\partial^2 w}{\partial x^2}\frac{\partial^2 w}{\partial y^2}+\frac{\nu_{yx}}{\nu_{xy}}\left(\frac{\partial^2 w}{\partial y^2}\right)^2\right]\right\}\,dy\,dx \quad (6.178)
$$

This simplification is made by letting $G_{xz}=G_{yz}=\infty$ in (6.177).

One can now proceed to use the Rayleigh method to obtain buckling loads for any set of boundary conditions by assuming a form of the lateral deflection $w(x,y)$, which if it

(1) Satisfies all boundary conditions and satisfies the governing differential equations, is the exact buckling load.
(2) Satisfies all the boundary conditions, provides a very accurate value for the buckling load.
(3) Satisfies only the geometric boundary conditions (i.e. the lateral deflection and its first derivative), provides a satisfactory value for the buckling load.

In cases (2) and (3) the approximate buckling loads are higher than the exact buckling load for the problem.

We will now proceed to solve several problems involving different boundary conditions and different planform geometries.

6.8.1 The Elastic Stability of Specially Orthotropic Plates with x-edges and One y-edge Simply-Supported and the other y-edge Free

In this case a suitable approximation for the lateral displacement is

$$w(x, y) = Ay \sin \frac{m\pi x}{a} \tag{6.179}$$

where A is the undetermined amplitude. This expression satisfies all geometric boundary conditions (as well as most of the stress boundary conditions), and hence will provide accurate buckling loads.

Substituting (6.179) into (6.177) and remembering that

$$\int_0^a \int_0^b \sin^2 \frac{m\pi x}{a} \, dy \, dx = \int_0^a \int_0^b \cos^2 \frac{m\pi x}{a} \, dy \, dx = \frac{ab}{2}$$

$$\int_0^a \int_0^b y^2 \cos^2 \frac{m\pi x}{a} \, dx \, dy = \int_0^a \int_0^b y^2 \sin^2 \frac{m\pi x}{a} \, dx \, dy = \frac{ab^3}{6}$$

taking the variation of V with respect to A results in

$$
\begin{aligned}
0 = \hat{N}_x^2 &\left[\frac{3}{50} \left(\frac{G_{xy}}{G_{xz}}\right) \left(\frac{E_x}{G_{xz}}\right)^2 \left(\frac{h}{a}\right)^4 m^6 + \frac{1}{5} \left(\frac{E_x}{G_{xz}}\right)^2 \left(\frac{h}{a}\right)^2 \left(\frac{b}{a}\right)^2 m^6 + \right. \\
&\left. + \frac{3\pi^2}{50\bar{\nu}} \left(\frac{E_x}{G_{xz}}\right)^3 \left(\frac{h}{a}\right)^4 \left(\frac{b}{a}\right)^2 m^8 \right] + \\
+ \hat{N}_x &\left[2 \left(\frac{G_{xy}}{G_{xz}}\right) \left(\frac{E_x}{G_{xz}}\right) \left(\frac{h}{a}\right)^2 m^4 + \frac{\pi^2}{3\bar{\nu}} \left(\frac{E_x}{G_{xz}}\right)^2 \left(\frac{h}{a}\right)^2 \left(\frac{b}{a}\right)^2 m^6 - \right. \\
&- \frac{3\pi^2}{10} \left(\frac{G_{xy}}{G_{xz}}\right) \left(\frac{E_x}{G_{xz}}\right)^2 \left(\frac{h}{a}\right)^4 m^6 - \frac{\pi^4}{30\bar{\nu}} \left(\frac{E_x}{G_{xz}}\right)^3 \left(\frac{h}{a}\right)^4 \left(\frac{b}{a}\right)^2 m^8 - \\
&\left. - \frac{\pi^2}{3} \left(\frac{E_x}{G_{xz}}\right)^2 \left(\frac{h}{a}\right)^2 \left(\frac{b}{a}\right)^2 m^6 - \frac{50}{3\pi^2} \left(\frac{a}{h}\right)^2 \left(\frac{b}{a}\right)^2 m^6 - \frac{50}{\pi^4} \left(\frac{a}{h}\right)^2 \gamma \right] + \\
+ &\left[\frac{50}{3} \left(\frac{G_{xy}}{G_{xz}}\right) m^2 + \frac{25\,\pi^2}{18\,\bar{\nu}} \left(\frac{E_x}{G_{xz}}\right) \left(\frac{b}{a}\right)^2 m^4 - \right. \\
&- \frac{5\pi^2}{3} \left(\frac{G_{xy}}{G_{xz}}\right) \left(\frac{E_x}{G_{xz}}\right) \left(\frac{h}{a}\right)^2 m^4 - \frac{5\pi^4}{18\bar{\nu}} \left(\frac{E_x}{G_{xz}}\right)^2 \left(\frac{h}{a}\right)^2 \left(\frac{b}{a}\right)^2 m^6 + \\
&+ \frac{\pi^4}{24} \left(\frac{G_{xy}}{G_{xz}}\right) \left(\frac{E_x}{G_{xz}}\right)^2 \left(\frac{h}{a}\right)^4 m^6 + \frac{\pi^6}{72\bar{\nu}} \left(\frac{E_x}{G_{xz}}\right)^3 \left(\frac{h}{a}\right)^4 \left(\frac{b}{a}\right)^2 m^8 + \\
&\left. + \frac{5\pi^4}{36} \left(\frac{E_x}{G_{xz}}\right)^2 \left(\frac{h}{a}\right)^2 \left(\frac{b}{a}\right)^2 m^6 \right]
\end{aligned}
\tag{6.180}
$$

wherein

$$\bar{N}_y = \alpha \bar{N}_x, \qquad \hat{N}_x = \frac{\bar{N}_x \pi^2}{G_{xz} h}$$

and

$$\bar{\nu} = 1 - \nu_{xy} \nu_{yx}$$

This quadratic equation provides the critical buckling load for a specially orthotropic ($\theta = 0°$ or $90°$) plate including transverse shear deformations for the subject boundary conditions.

If $\bar{N}_y = 0$, and (6.179) is substituted into (6.178), the result is

$$(\hat{N}_x)_{\text{Classical}} = \frac{E_x}{G_{xz}} \frac{\pi^4}{12\bar{\nu}} \left(\frac{h}{a}\right)^2 m^2 + \left(\frac{G_{xy}}{G_{xz}}\right)\left(\frac{h}{a}\right)^2 \pi^2 \left(\frac{a}{b}\right)^2 \qquad (6.181)$$

It is seen that in (6.181) $(\bar{N}_x)_{\min}$ occurs at $m = 1$ for any ratio of (a/b), and one concludes that for these boundary conditions the critical

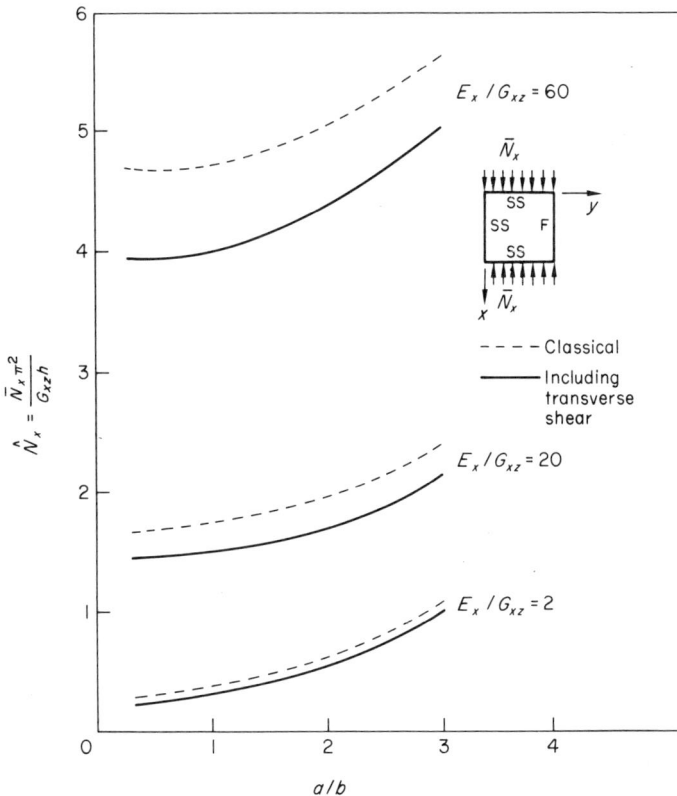

Fig. 6.10.   *Buckling load versus a/b for h/a = 1/10.*

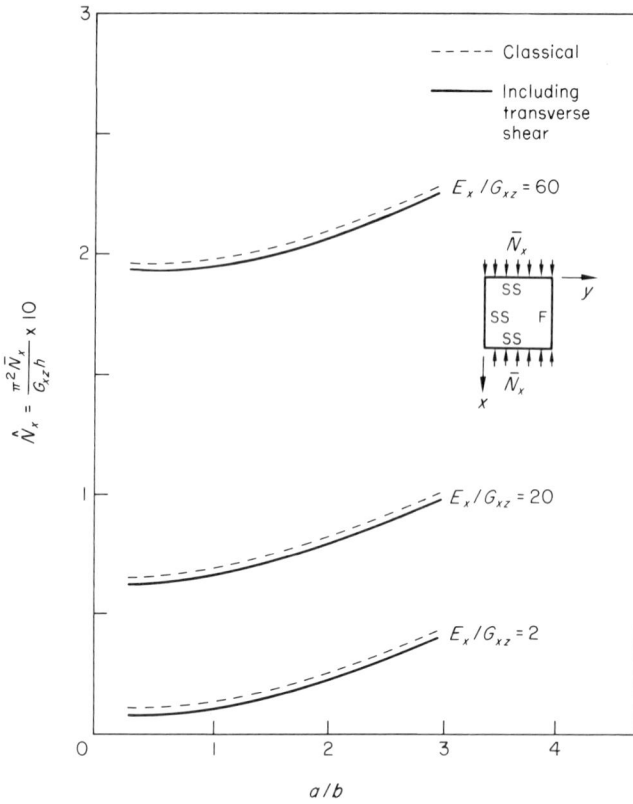

Fig. 6.11.   Buckling load versus a/b for $h/a = 1/50$.

buckling load will always occur for a mode shape of one-half sine wave in the x-direction. One can also conclude that that is the case for the plate including transverse shear deformation.

Now consider a unidirectional fibre composite plate wherein the fibres are in the x direction such that $E_x > E_y$ and $G_{xy} = G_{xz}$, and further consider the case of $\bar{N}_y = 0$. To cover the range of possible values for practical plates and practical material systems

$$2 \leqslant E_x/G_{xz} \leqslant 60$$

$$\tfrac{1}{3} \leqslant a/b \leqslant 3$$

$$1/100 \leqslant h/a \leqslant 1/10$$

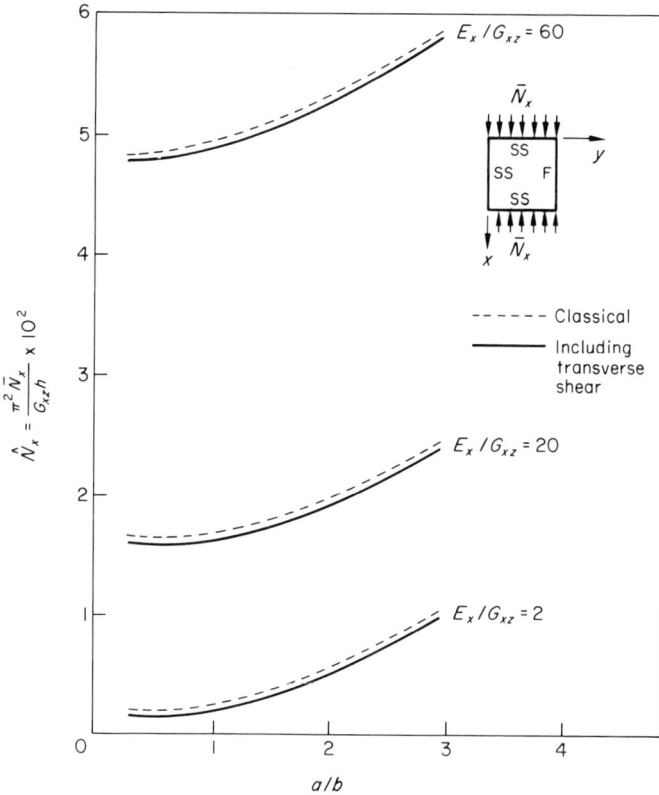

Fig. 6.12. Buckling load versus a/b for h/a = 1/100.

Results of solving eqns. (6.180) and (6.181) are shown in Figs. 6.10 to 6.12 for the boron–epoxy composite material of Section 6.4.

6.8.2 The Elastic Stability of Specially Orthotropic Plates with the y-edges Clamped and the x-edges Simply-Supported

In this case let the lateral deflection function be

$$w(x, y) = A_{mn} \sin \frac{m\pi x}{a} \left[1 - \cos \left(\frac{2n\pi y}{b} \right) \right] \qquad (6.182)$$

which satisfies all boundary conditions. Substituting (6.182) into (6.177) gives upon taking the variation with respect to A_{mn}

$$0 = \hat{N}_x^2 \left[\frac{3\pi^2}{500} \left(\frac{G_{xy}}{G_{xz}} \right) \left(\frac{E_x}{G_{xz}} \right)^2 \left(\frac{a}{b} \right)^2 n^2 + \frac{9\pi^2}{2000\bar{\nu}} \left(\frac{E_x}{G_{xz}} \right)^3 m^2 + \frac{9}{200} \left(\frac{E_x}{G_{xz}} \right)^2 \left(\frac{a}{h} \right)^2 \right]$$

$$+ \hat{N}_x \left[-\frac{15}{4\pi^2} \left(\frac{a}{h} \right)^6 \frac{1}{m^4} + \frac{\pi^2}{5} \left(\frac{G_{xy}}{G_{xz}} \right) \left(\frac{E_x}{G_{xz}} \right) \left(\frac{a}{b} \right)^2 \frac{n^2}{m^2} - \right.$$

$$- \frac{\pi^4}{100} \left(\frac{G_{xy}}{G_{xz}} \right) \left(\frac{E_x}{G_{xz}} \right)^2 \left(\frac{a}{b} \right)^2 n^2 - \frac{\pi^4}{25} \left(\frac{G_{xy}}{G_{xz}} \right) \left(\frac{E_x}{G_{xz}} \right) \left(\frac{E_y}{G_{yz}} \right) \left(\frac{a}{b} \right)^4 \frac{n^4}{m^2} -$$

$$- \frac{3\pi^4}{400\bar{\nu}} \left(\frac{E_x}{G_{xz}} \right)^3 m^2 + \frac{3\pi^2}{40\bar{\nu}} \left(\frac{E_x}{G_{xz}} \right) \left(\frac{a}{h} \right)^2 (1 - \bar{\nu}) -$$

$$- \frac{\pi^4}{25\bar{\nu}} \nu_{yx} \left(\frac{E_x}{G_{xz}} \right)^2 \left(\frac{E_y}{G_{yz}} \right) \left(\frac{a}{b} \right)^4 \frac{n^4}{m^2} + \frac{\pi^2}{10\bar{\nu}} \nu_{yx} \left(\frac{E_x}{G_{xz}} \right)^2 \left(\frac{a}{b} \right)^2 \left(\frac{a}{h} \right)^2 \frac{n^2}{m^2} \right] +$$

$$+ \left[\frac{5\pi^2}{3} \left(\frac{G_{xy}}{G_{xz}} \right) \left(\frac{a}{b} \right)^2 \left(\frac{a}{h} \right)^4 \frac{n^2}{m^4} - \frac{\pi^4}{6} \left(\frac{G_{xy}}{G_{xz}} \right) \left(\frac{E_x}{G_{xz}} \right) \left(\frac{a}{b} \right)^2 \left(\frac{a}{h} \right)^2 \frac{n^2}{m^2} - \right.$$

$$- \frac{2\pi^4}{3} \left(\frac{G_{xy}}{G_{xz}} \right) \left(\frac{E_y}{G_{yz}} \right) \left(\frac{a}{b} \right)^4 \left(\frac{a}{h} \right)^2 \frac{n^4}{m^4} + \frac{\pi^6}{240} \left(\frac{G_{xy}}{G_{xz}} \right) \left(\frac{E_x}{G_{xz}} \right)^2 \left(\frac{a}{b} \right)^2 n^2 +$$

$$+ \frac{\pi^6}{30} \left(\frac{G_{xy}}{G_{xz}} \right) \left(\frac{E_x}{G_{xz}} \right) \left(\frac{E_y}{G_{yz}} \right) \left(\frac{a}{b} \right)^4 \frac{n^4}{m^2} + \frac{\pi^6}{15} \left(\frac{G_{xy}}{G_{xz}} \right) \left(\frac{E_y}{G_{yz}} \right)^2 \left(\frac{a}{b} \right)^6 \frac{n^6}{m^4} +$$

$$+ \frac{\pi^6}{320\bar{\nu}} \left(\frac{E_x}{G_{xz}} \right)^3 m^2 - \frac{\pi^4}{16\bar{\nu}} \left(\frac{E_x}{G_{xz}} \right)^2 \left(\frac{a}{h} \right)^2 + \frac{5\pi^2}{16\bar{\nu}} \left(\frac{E_x}{G_{xz}} \right) \left(\frac{a}{h} \right)^4 \frac{1}{m^2} +$$

$$+ \frac{\pi^6}{30\bar{\nu}} \nu_{yx} \left(\frac{E_x}{G_{xz}} \right)^2 \left(\frac{E_y}{G_{yz}} \right) \left(\frac{a}{b} \right)^4 \frac{n^4}{m^2} - \frac{\pi^4}{12\bar{\nu}} \nu_{yx} \left(\frac{E_x}{G_{xz}} \right)^2 \left(\frac{a}{b} \right)^2 \left(\frac{a}{h} \right)^2 \frac{n^2}{m^2} -$$

$$- \frac{\pi^4}{3\bar{\nu}} \nu_{yx} \left(\frac{E_x}{G_{xz}} \right) \left(\frac{E_y}{G_{yz}} \right) \left(\frac{a}{b} \right)^4 \left(\frac{a}{h} \right)^2 \frac{n^4}{m^4} + \frac{5\pi^2}{6\bar{\nu}} \nu_{yx} \left(\frac{E_x}{G_{xz}} \right) \left(\frac{a}{b} \right)^2 \left(\frac{a}{h} \right)^4 \frac{n^2}{m^4} +$$

$$+ \frac{\pi^4}{32} \left(\frac{E_x}{G_{xz}} \right)^2 \left(\frac{a}{h} \right)^2 + \frac{4\pi^6}{15\bar{\nu}} \frac{\nu_{yx}}{\nu_{xy}} \left(\frac{E_x}{G_{xz}} \right) \left(\frac{E_y}{G_{yz}} \right)^2 \left(\frac{a}{b} \right)^8 \frac{n^8}{m^6} -$$

$$- \frac{4\pi^4}{3\bar{\nu}} \frac{\nu_{yx}}{\nu_{xy}} \left(\frac{E_x}{G_{xz}} \right) \left(\frac{E_y}{G_{yz}} \right) \left(\frac{a}{b} \right)^6 \left(\frac{a}{h} \right)^2 \frac{1}{m^6} + \frac{5\pi^2}{3\bar{\nu}} \frac{\nu_{yx}}{\nu_{xy}} \left(\frac{E_x}{G_{xz}} \right) \left(\frac{a}{b} \right)^4 \left(\frac{a}{h} \right)^4 \frac{n^4}{m^6} +$$

$$+ \frac{2\pi^4}{3} \frac{\nu_{yx}}{\nu_{xy}} \left(\frac{E_x}{G_{xz}} \right) \left(\frac{E_y}{G_{yz}} \right) \left(\frac{a}{b} \right)^6 \left(\frac{a}{h} \right)^2 \frac{n^6}{m^6} \right] \tag{6.183}$$

Similarly, substituting (6.182) into (6.178) results in

$$(\hat{N}_x)_{\text{Classical}} = \frac{\pi^4}{12\bar{\nu}} \left(\frac{E_x}{G_{xz}} \right) \left(\frac{h}{a} \right)^2 \left[m^2 + \frac{8}{3} \nu_{yx} \left(\frac{a}{b} \right)^2 n^2 + \frac{16}{3} \frac{\nu_{yx}}{\nu_{xy}} \left(\frac{a}{b} \right)^4 \frac{n^4}{m^2} \right] +$$

$$+ \frac{4\pi^4}{9} \left(\frac{G_{xy}}{G_{xz}} \right) \left(\frac{h}{a} \right)^2 \left(\frac{a}{b} \right)^2 n^2 \tag{6.184}$$

Looking at (6.184) it is obvious that the minimum buckling load occurs when $n = 1$; this will also be true for (6.183). However, the value of m, the wave number in the load direction will vary with the aspect ratio, a/b.

Consider a composite material considered in Section 6.4, a boron–epoxy composite. Graphical results are given in Figs. 6.13 to 6.16. It is seen that not only does the inclusion of transverse shear deformation have a significant effect upon the value of the buckling load, but there is a significant effect upon the wave number—the number of buckles in the load direction, as the h/a value increases.

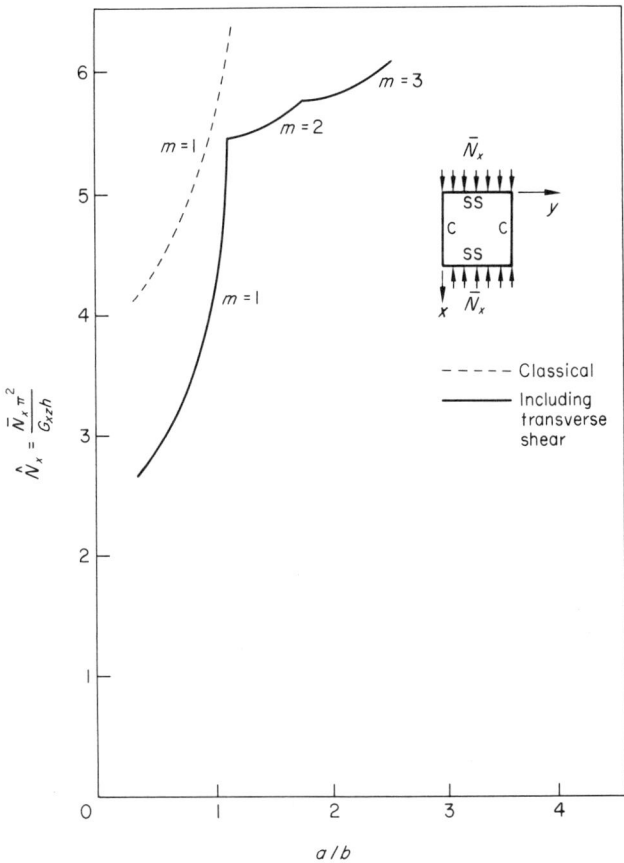

Fig. 6.13.   *Buckling load versus a/b for $h/a = 1/10$. Plate: 56·1% boron fibres in epoxy matrix.*

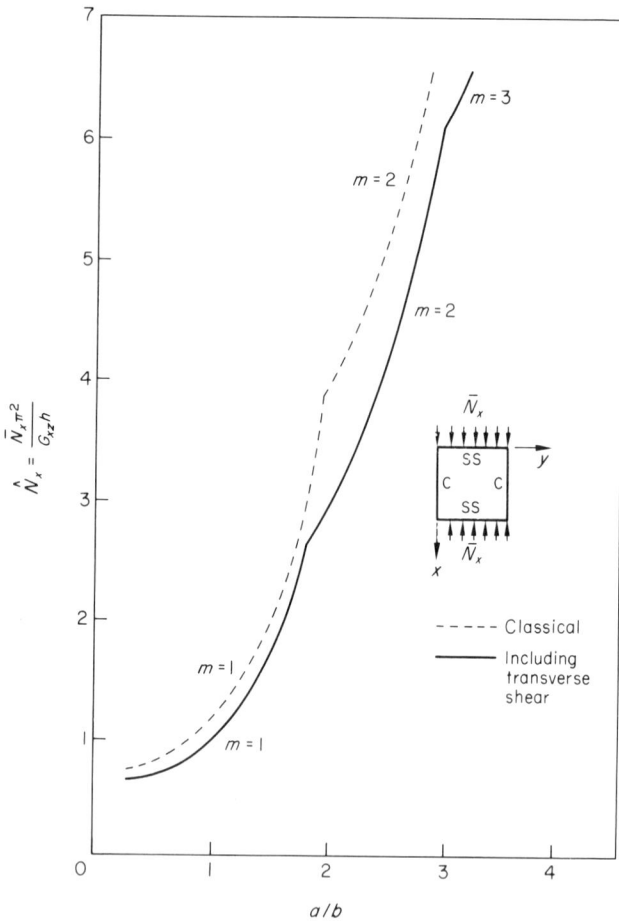

Fig. 6.14. Buckling load versus a/b for h/a = 1/25. Plate: 56·1% boron fibres in epoxy matrix.

6.8.3 The Elastic Stability of Specially Orthotropic Plates with All Four Edges Simply-Supported

This is the same problem treated by Whitney, discussed in Section 6.7. It is included here purely for completeness and to present the results for aspect ratios other than as square plate. Obviously, the form for the lateral deflection can be the Navier form of Section 4.8.

$$w(x, y) = A_{mn} \sin\left(\frac{m\pi x}{a}\right) \sin\left(\frac{n\pi y}{b}\right) \tag{6.185}$$

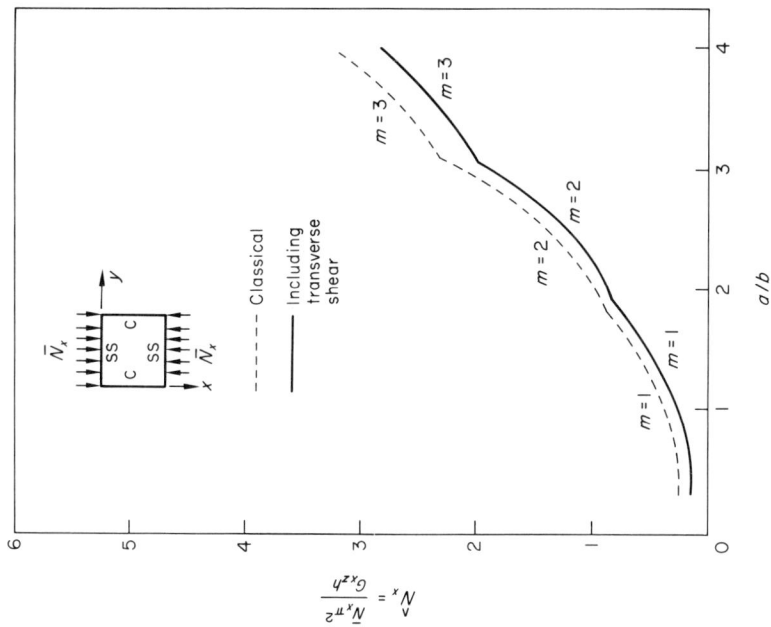

Fig. 6.16. Buckling load versus a/b for $h/a = 1/100$. Plate:
56·1% boron fibres in epoxy matrix.

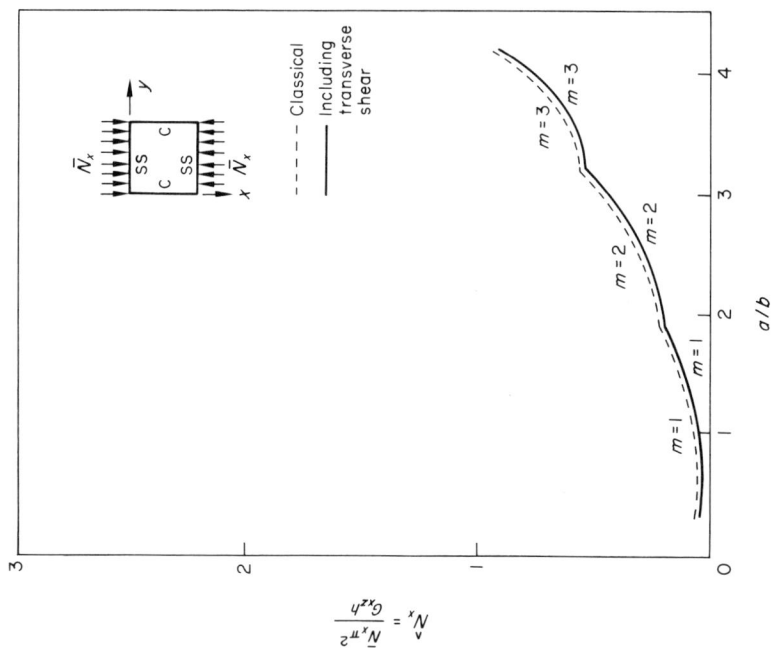

Fig. 6.15. Buckling load versus a/b for $h/a = 1/50$. Plate:
56·1% boron fibres in epoxy matrix.

In the case of the $\bar{N}_x$ load acting only, the non-dimensional buckling load $\hat{N}_x$ is given through substituting (6.185) into (6.177) and taking variations with respect to A_{mn}. The result is

$$0 = \hat{N}_x^2 \left[\frac{3\pi^2}{500} \left(\frac{G_{xy}}{G_{xz}}\right)\left(\frac{E_x}{G_{xz}}\right)^2 \left(\frac{a}{b}\right)^2 n^2 + \frac{9\pi^2}{1500\bar{\nu}} \left(\frac{E_x}{G_{xz}}\right)^3 m^2 + \frac{3}{50} \left(\frac{E_x}{G_{xz}}\right)^2 \left(\frac{a}{h}\right)^2 \right] +$$

$$+ \hat{N}_x \left[-\frac{5}{\pi^2} \left(\frac{a}{h}\right)^6 \frac{1}{m^4} + \frac{\pi^2}{5} \left(\frac{G_{xy}}{G_{xz}}\right)\left(\frac{E_x}{G_{xz}}\right)\left(\frac{a}{b}\right)^2 \left(\frac{a}{h}\right)^2 \frac{n^2}{m^2} - \right.$$

$$- \frac{\pi^4}{100} \left(\frac{G_{xy}}{G_{xz}}\right)\left(\frac{E_x}{G_{xz}}\right)^2 \left(\frac{a}{b}\right)^2 n^2 - \frac{\pi^4}{100} \left(\frac{G_{xy}}{G_{xz}}\right)\left(\frac{E_x}{G_{xz}}\right)\left(\frac{E_y}{G_{yz}}\right)\left(\frac{a}{b}\right)^4 \frac{n^4}{m^2} -$$

$$- \frac{\pi^4}{100\bar{\nu}} \left(\frac{E_x}{G_{xz}}\right)^3 m^2 + \frac{\pi^2}{10\bar{\nu}} \left(\frac{E_x}{G_{xz}}\right)^2 \left(\frac{a}{h}\right)^2 -$$

$$- \frac{\pi^4}{100\bar{\nu}} \nu_{yx} \left(\frac{E_y}{G_{yz}}\right)\left(\frac{E_x}{G_{xz}}\right)\left(\frac{a}{b}\right)^4 \frac{n^4}{m^2} +$$

$$+ \frac{\pi^2}{10\bar{\nu}} \nu_{yx} \left(\frac{E_x}{G_{xz}}\right)^2 \left(\frac{a}{b}\right)^2 \left(\frac{a}{h}\right)^2 \frac{n^2}{m^2} - \frac{\pi^2}{10} \left(\frac{E_x}{G_{xz}}\right)^2 \left(\frac{a}{h}\right)^2 \right] +$$

$$+ \left[\frac{5\pi^2}{3} \left(\frac{G_{xy}}{G_{xz}}\right)\left(\frac{a}{b}\right)^2 \left(\frac{a}{h}\right)^4 \frac{n^2}{m^4} - \frac{\pi^4}{6} \left(\frac{G_{xy}}{G_{xz}}\right)\left(\frac{E_x}{G_{xz}}\right)\left(\frac{a}{b}\right)^2 \left(\frac{a}{h}\right)^2 \frac{n^2}{m^2} - \right.$$

$$- \frac{\pi^4}{6} \left(\frac{G_{xy}}{G_{xz}}\right)\left(\frac{E_y}{G_{yz}}\right)\left(\frac{a}{b}\right)^4 \left(\frac{a}{h}\right)^2 \frac{n^4}{m^4} + \frac{\pi^6}{240} \left(\frac{G_{xy}}{G_{xz}}\right)\left(\frac{E_x}{G_{xz}}\right)^2 \left(\frac{a}{b}\right)^2 n^2 +$$

$$+ \frac{\pi^6}{120} \left(\frac{G_{xy}}{G_{xz}}\right)\left(\frac{E_x}{G_{xz}}\right)\left(\frac{E_y}{G_{yz}}\right)\left(\frac{a}{b}\right)^4 \frac{n^4}{m^2} + \frac{\pi^6}{240} \left(\frac{G_{xy}}{G_{xz}}\right)\left(\frac{E_y}{G_{yz}}\right)^2 \left(\frac{a}{b}\right)^6 \frac{n^6}{m^4} +$$

$$+ \frac{\pi^6}{240\bar{\nu}} \left(\frac{E_x}{G_{xz}}\right)^3 m^2 - \frac{\pi^4}{12\bar{\nu}} \left(\frac{E_x}{G_{xz}}\right)^2 \left(\frac{a}{h}\right)^2 + \frac{5\pi^2}{12\bar{\nu}} \left(\frac{E_x}{G_{xz}}\right)\left(\frac{a}{h}\right)^4 \frac{1}{m^2} +$$

$$+ \frac{\pi^6}{120\bar{\nu}} \nu_{yx} \left(\frac{E_x}{G_{xz}}\right)^2 \left(\frac{E_y}{G_{yz}}\right)\left(\frac{a}{b}\right)^4 \frac{n^4}{m^2} - \frac{\pi^4}{12\bar{\nu}} \nu_{yx} \left(\frac{E_x}{G_{xz}}\right)^2 \left(\frac{a}{b}\right)^2 \left(\frac{a}{h}\right)^2 \frac{n^2}{m^2} -$$

$$- \frac{\pi^4}{12\bar{\nu}} \nu_{yx} \left(\frac{E_x}{G_{xz}}\right)\left(\frac{E_y}{G_{yz}}\right)\left(\frac{a}{b}\right)^4 \left(\frac{a}{h}\right)^2 \frac{n^4}{m^4} + \frac{5\pi^2}{6\bar{\nu}} \nu_{yx} \left(\frac{E_x}{G_{xz}}\right)\left(\frac{a}{b}\right)^2 \left(\frac{a}{h}\right)^4 \frac{n^2}{m^4} +$$

$$+ \frac{\pi^6}{240\bar{\nu}} \frac{\nu_{yx}}{\nu_{xy}} \left(\frac{E_x}{G_{xz}}\right)\left(\frac{E_y}{G_{yz}}\right)^2 \left(\frac{a}{b}\right)^8 \frac{n^8}{m^6} -$$

$$- \frac{\pi^4}{12\bar{\nu}} \frac{\nu_{yx}}{\nu_{xy}} \left(\frac{E_x}{G_{xz}}\right)\left(\frac{E_y}{G_{yz}}\right)\left(\frac{a}{b}\right)^6 \left(\frac{a}{h}\right)^2 \frac{n^6}{m^6} + \frac{5\pi^2}{12\bar{\nu}} \frac{\nu_{yx}}{\nu_{xy}} \left(\frac{E_x}{G_{xz}}\right)\left(\frac{a}{b}\right)^4 \left(\frac{a}{h}\right)^4 \frac{n^4}{m^6} +$$

$$+ \frac{\pi^4}{24} \left(\frac{E_x}{G_{xz}}\right)^2 \left(\frac{a}{h}\right)^2 + \frac{\pi^4}{24} \frac{\nu_{yx}}{\nu_{xy}} \left(\frac{E_x}{G_{xz}}\right)\left(\frac{E_y}{G_{yz}}\right)\left(\frac{a}{b}\right)^6 \left(\frac{a}{h}\right)^2 \frac{n^6}{m^6} \right] \qquad (6.186)$$

Without transverse shear deformation the substitution of (6.185) into (6.178) results in the much more simple expression

$$(\hat{N}_x)_{\text{Classical}} = \frac{\pi^4}{12\,\bar{\nu}} \left(\frac{E_x}{G_{xz}}\right)\left(\frac{h}{a}\right)^2 \left[m^2 + 2\nu_{yx} \left(\frac{a}{b}\right)^2 n^2 + \frac{\nu_{yx}}{\nu_{xy}} \left(\frac{a}{b}\right)^4 \frac{n^4}{m^2} \right] +$$

$$+ \frac{\pi^4}{3} \left(\frac{G_{xy}}{G_{xz}}\right)\left(\frac{h}{a}\right)^2 \left(\frac{a}{b}\right)^2 \quad (6.187)$$

From (6.187) it is obvious that the lowest critical buckling load occurs with $n = 1$, and this is also the case with (6.186). Calculating with (6.186) and (6.187) for the boron–epoxy material used throughout this Chapter, Figs. 6.17 to 6.20 result. Again, notice that the effects of shear deformation become more and more pronounced as h/a increases.

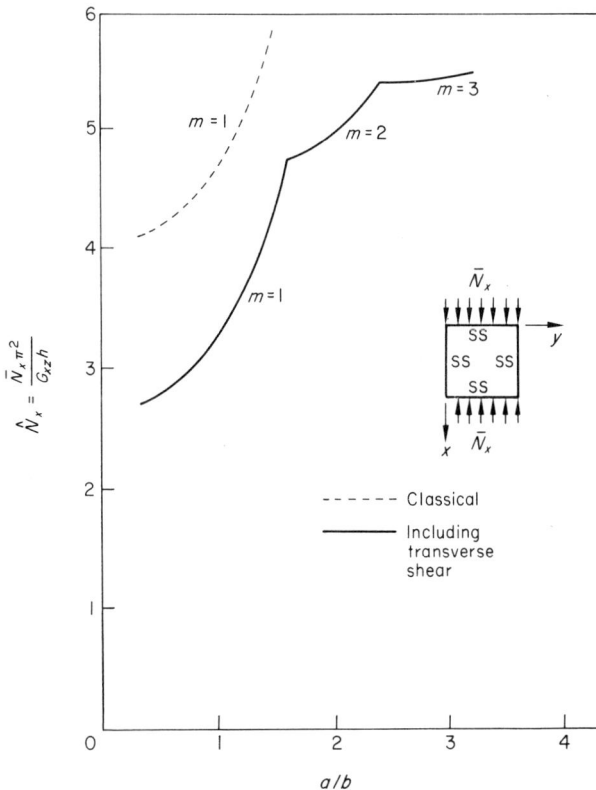

Fig. 6.17.   *Buckling load versus a/b for $h/a = 1/10$. Plate: 56·1% boron fibres in epoxy matrix.*

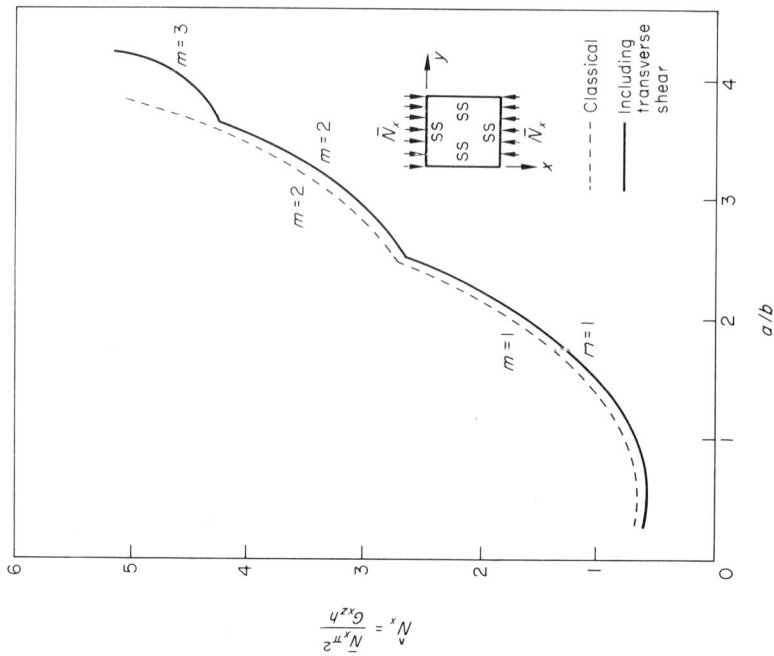

Fig. 6.19. Buckling load versus a/b for $h/a = 1/50$. Plate: 56·1% boron fibres in epoxy matrix.

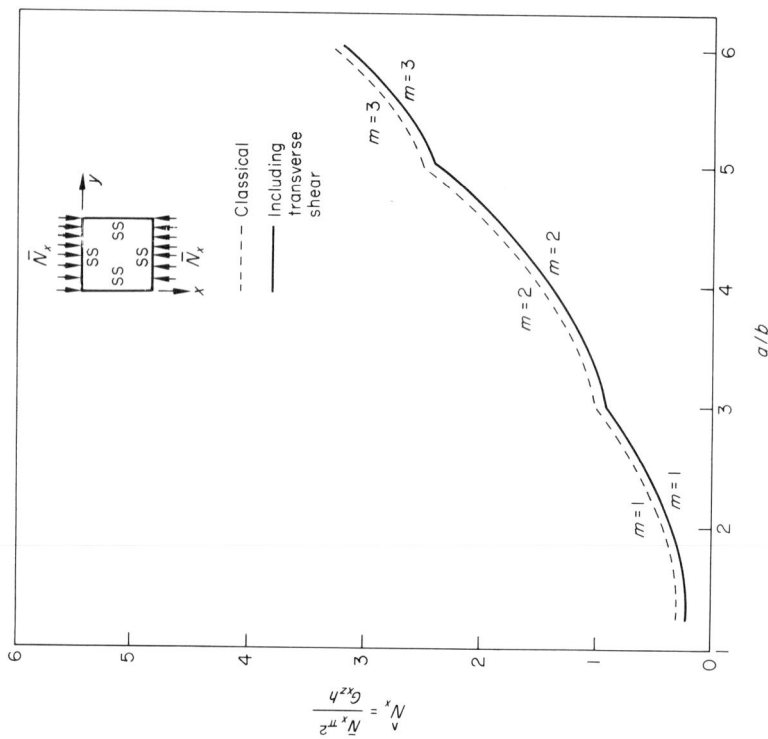

Fig. 6.18. Buckling load versus a/b for $h/a = 1/25$. Plate: 56·1% boron fibres in epoxy matrix.

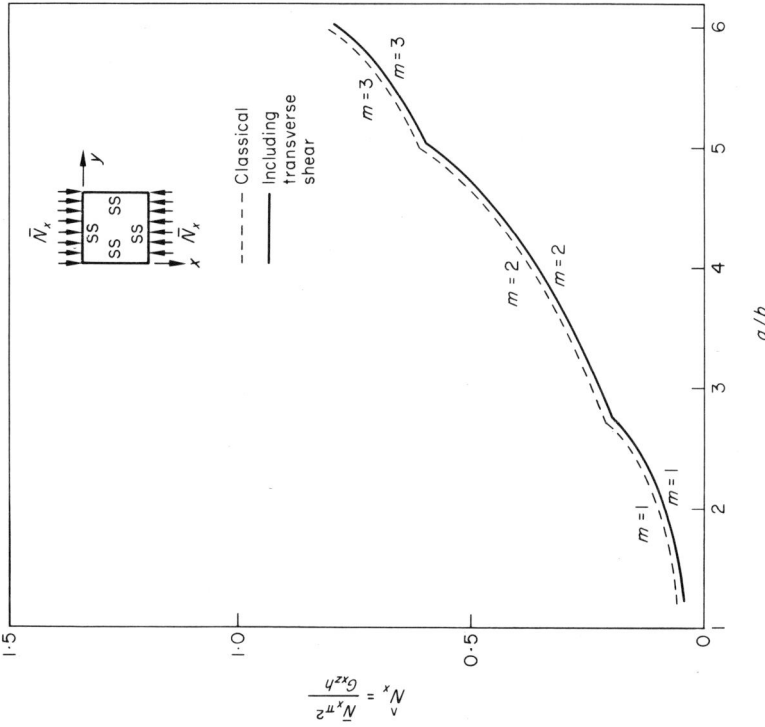

Fig. 6.20.   *Buckling load versus* a/b *for* $h/a = 1/100$. *Plate: 56·1% boron fibres in epoxy matrix.*

6.8.4   The Elastic Stability of Specially Orthotropic Plates with the x-edges and One y-edge Simply-Supported and the Other y-edge Clamped

In this case a deflection function is assumed as follows which satisfies all boundary conditions:

$$w(x, y) = A_m \sin\left(\frac{m\pi x}{a}\right)(b^3 y - 3by^3 + 2y^4) \qquad (6.188)$$

Substituting this form into eqn. (6.177) and taking variations with respect to A_m provides values of the buckling load $\hat{N}_x$ (where $\bar{N}_y = 0$), as shown below

$$
\begin{aligned}
0 = \hat{N}_x^2 &\left[0 \cdot 001 \left(\frac{G_{xy}}{G_{xz}}\right)\left(\frac{E_x}{G_{xz}}\right)^2\left(\frac{h}{a}\right)^2\left(\frac{a}{b}\right)^2 + \frac{0 \cdot 00087}{\bar{\nu}}\left(\frac{E_x}{G_{xz}}\right)^3\left(\frac{h}{a}\right)^2 m^2 + \right. \\
&\left. + 0 \cdot 0085\left(\frac{E_x}{G_{xz}}\right)^2\left(\frac{h}{a}\right)^4 \right] + \hat{N}_x\left[-0 \cdot 0075\left(\frac{a}{h}\right)^4\frac{1}{m^4} + \right. \\
&+ 0 \cdot 033\left(\frac{G_{xy}}{G_{xz}}\right)\left(\frac{E_x}{G_{xz}}\right)\left(\frac{a}{b}\right)^2\frac{1}{m^2} - 0 \cdot 0165\left(\frac{G_{xy}}{G_{xz}}\right)\left(\frac{E_x}{G_{xz}}\right)^2\left(\frac{h}{a}\right)^2\left(\frac{a}{b}\right)^2 - \\
&- 0 \cdot 0345\left(\frac{G_{xy}}{G_{xz}}\right)\left(\frac{E_x}{G_{xz}}\right)\left(\frac{E_y}{G_{yz}}\right)\left(\frac{h}{a}\right)^2\left(\frac{a}{b}\right)^4\frac{1}{m^2} - \frac{0 \cdot 0143}{\bar{\nu}}\left(\frac{E_x}{G_{xz}}\right)^3\left(\frac{h}{a}\right)^2 m^2 + \\
&+ \frac{0 \cdot 0144}{\bar{\nu}}\left(\frac{E_x}{G_{xz}}\right)^2 - \frac{0 \cdot 0348}{\bar{\nu}}\left(\frac{E_x}{G_{xz}}\right)^2\left(\frac{E_y}{G_{yz}}\right)\nu_{yx}\left(\frac{h}{a}\right)^2\left(\frac{a}{b}\right)^4\frac{1}{m^2} + \\
&\left. + \frac{0 \cdot 0168}{\bar{\nu}}\left(\frac{E_x}{G_{xz}}\right)^2\nu_{yx}\left(\frac{a}{b}\right)^2\frac{1}{m^2} - 0 \cdot 0139\left(\frac{E_x}{G_{xz}}\right)^2 \right] + \\
&+ \left[0 \cdot 276\left(\frac{G_{xy}}{G_{xz}}\right)\left(\frac{a}{b}\right)^2\left(\frac{a}{h}\right)^2\frac{1}{m^4} - 0 \cdot 272\left(\frac{G_{xy}}{G_{xz}}\right)\left(\frac{E_x}{G_{xz}}\right)\left(\frac{a}{b}\right)^2\frac{1}{m^2} - \right. \\
&- 0 \cdot 59\left(\frac{G_{xy}}{G_{xz}}\right)\left(\frac{E_y}{G_{yz}}\right)\left(\frac{a}{b}\right)^4\frac{1}{m^4} + 0 \cdot 068\left(\frac{G_{xy}}{G_{xz}}\right)\left(\frac{E_x}{G_{xz}}\right)^2\left(\frac{h}{a}\right)^2\left(\frac{a}{b}\right)^2 + \\
&+ 0 \cdot 29\left(\frac{G_{xy}}{G_{xz}}\right)\left(\frac{E_x}{G_{xz}}\right)\left(\frac{E_y}{G_{yz}}\right)\left(\frac{h}{a}\right)^2\left(\frac{a}{b}\right)^4\frac{1}{m^2} + \\
&+ 0 \cdot 468\left(\frac{G_{xy}}{G_{xz}}\right)\left(\frac{E_y}{G_{yz}}\right)^2\left(\frac{h}{a}\right)^2\left(\frac{a}{b}\right)^6\frac{1}{m^4} - \\
&- \frac{0 \cdot 0585}{\bar{\nu}}\left(\frac{E_x}{G_{xz}}\right)^3\left(\frac{h}{a}\right)^2 m^2 - \frac{0 \cdot 118}{\bar{\nu}}\left(\frac{E_x}{G_{xz}}\right)^2 + \\
&+ \frac{0 \cdot 06}{\bar{\nu}}\left(\frac{E_x}{G_{xz}}\right)\left(\frac{a}{h}\right)^2\frac{1}{m^2} + \frac{0 \cdot 285}{\bar{\nu}}\left(\frac{E_x}{G_{xz}}\right)^2\left(\frac{E_y}{G_{yz}}\right)\nu_{yx}\left(\frac{h}{a}\right)^2\left(\frac{a}{b}\right)^4\frac{1}{m^2} - \\
&- \frac{0 \cdot 138}{\bar{\nu}}\nu_{yx}\left(\frac{E_x}{G_{xz}}\right)^2\left(\frac{a}{b}\right)^2\frac{1}{m^2} - \frac{0 \cdot 3}{\bar{\nu}}\nu_{yx}\left(\frac{E_x}{G_{xz}}\right)\left(\frac{E_y}{G_{yz}}\right)\left(\frac{a}{b}\right)^4\frac{1}{m^4} + \\
&+ \frac{0 \cdot 14}{\bar{\nu}}\nu_{yx}\left(\frac{E_x}{G_{xz}}\right)\left(\frac{a}{b}\right)^2\left(\frac{a}{h}\right)^2\frac{1}{m^4} + 0 \cdot 057\left(\frac{E_x}{G_{xz}}\right)^2 +
\end{aligned}
$$

$$+ \frac{0 \cdot 48}{\bar{\nu}} \frac{\nu_{yx}}{\nu_{xy}} \left(\frac{E_x}{G_{xz}}\right) \left(\frac{E_y}{G_{yz}}\right)^2 \left(\frac{h}{a}\right)^2 \left(\frac{a}{b}\right)^8 \frac{1}{m^6} -$$

$$- \frac{0 \cdot 2}{\bar{\nu}} \frac{\nu_{yx}}{\nu_{xy}} \left(\frac{E_x}{G_{xz}}\right) \left(\frac{E_y}{G_{yz}}\right) \left(\frac{a}{b}\right)^6 \frac{1}{m^6} + \frac{0 \cdot 15}{\bar{\nu}} \frac{\nu_{yx}}{\nu_{xy}} \left(\frac{E_x}{G_{xz}}\right) \left(\frac{a}{h}\right)^2 \left(\frac{a}{b}\right)^4 \frac{1}{m^6} +$$

$$+ 0 \cdot 475 \frac{\nu_{yx}}{\nu_{xy}} \left(\frac{E_x}{G_{xz}}\right) \left(\frac{E_y}{G_{yz}}\right) \left(\frac{a}{b}\right)^6 \frac{1}{m^6} \Bigg] \tag{6.189}$$

From the case of no transverse shear deformation, using (6.178), one obtains

$$(\hat{N}_x)_{\text{Classical}} = \frac{8 \cdot 1}{\bar{\nu}} \left(\frac{E_x}{G_{xz}}\right) \left(\frac{h}{a}\right)^2 \left[m^2 + 2 \cdot 34 \nu_{yx} \left(\frac{a}{b}\right)^2 + 2 \cdot 5 \frac{\nu_{yx}}{\nu_{xy}} \left(\frac{a}{b}\right)^4 \frac{1}{m^2}\right] +$$

$$+ 37 \cdot 4 \left(\frac{G_{xy}}{G_{xz}}\right) \left(\frac{h}{a}\right)^2 \left(\frac{a}{b}\right)^2 \tag{6.190}$$

The results for the boron–epoxy material are shown in Figs. 6.21 to 6.24. The values and the effects are clear.

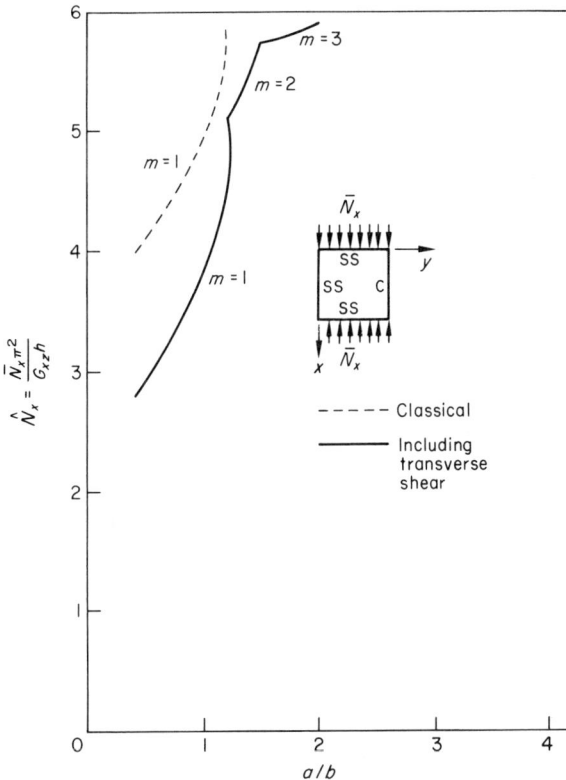

Fig. 6.21.   *Buckling load versus a/b for $h/a = 1/10$. Plate: 56·1% boron fibres in epoxy matrix.*

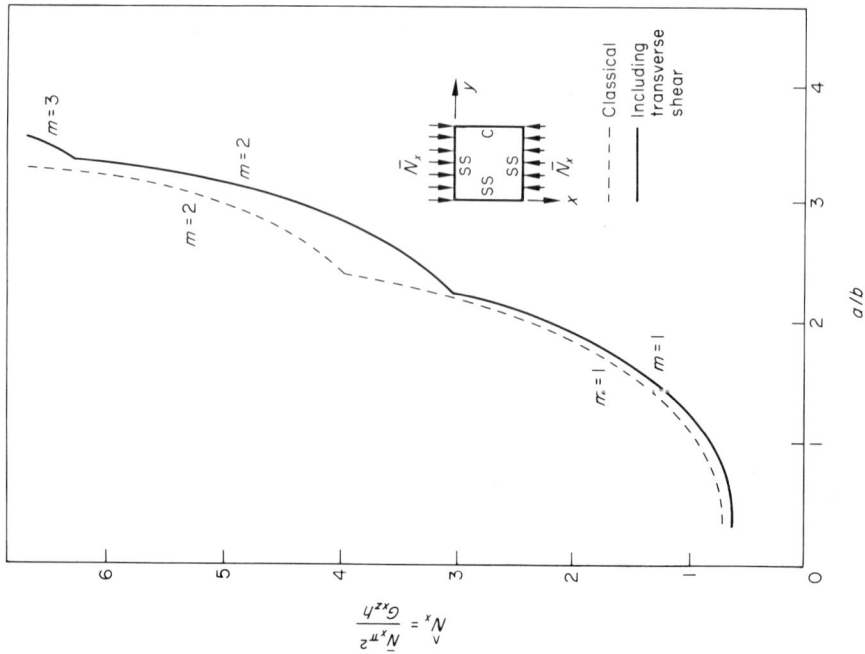

Fig. 6.22.   Buckling load versus a/b for h/a = 1/25. Plate:
SS 1/0, b = aa, fibres in ....:

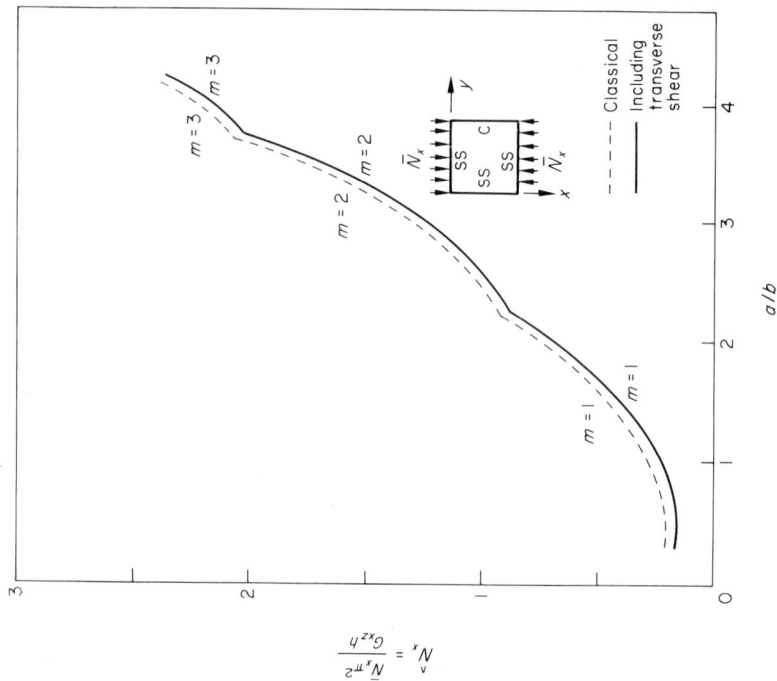

Fig. 6.23.   Buckling load versus a/b for h/a = 1/50. Plate:
SS 1/0, b = aa, fibres in ....:

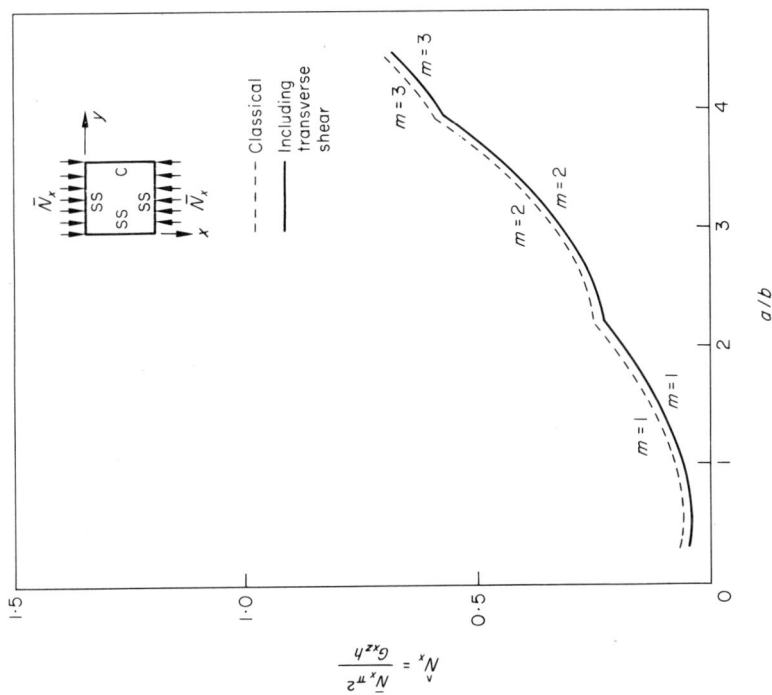

Fig. 6.24.   Buckling load versus a/b for h/a = 1/100. Plate: 56·1% boron fibres in epoxy matrix.

6.8.5   Remarks

In this Section, several buckling problems have been treated in detail. Solutions have been presented in order to obtain some physical insight for several important effects. The general approach of using the minimum Potential Energy Theorem and the selection of appropriate functions should now be well understood. The same approach even with the same functions can be used for solving problems under lateral loads, or employing Hamilton's Principle to obtain natural frequencies as an alternative to that used in a previous Section. Of course the beam functions of Section 6.6 could have been employed in this Section and in some cases were.

It should be pointed out that Brunelle[87] recently investigated the problem of a Mindlin plate (which includes transverse shear deformation) for a transversely isotropic materials system. The equations developed were used to investigate static deflections, free and forced vibrations, and also elastic stability. Thus, Brunelle's results are obtained when in this Section the following restrictions are made

$$E_x = E_y = E, \qquad \nu_{yx} = \nu_{yx} = \nu$$

$$G_{xy} = \frac{E}{2(1 - \nu)}, \qquad G_{xz} = G_{yz}$$

The material in this Section was obtained by Smith.[58] More recently, Linsenmann[80] has obtained similar solutions for generally orthotropic plates of composite materials.

6.9   PROBLEMS

6.1. Consider a beam composed of the boron–epoxy material described on page 255, such that the fibres are parallel to the x-axis. In that case, looking at eqn. (6.40), $E_x = E$, $G_{xz} = G$, and $\rho = 0\cdot244 \times 10^{-2}$ kg/cm^3, plot the ratio of the natural frequencies for a beam including transverse shear deformations (eqn. 6.40) to that of a beam without considering transverse shear deformation (eqn. 6.41) as a function of L/h.

6.2. Add to Fig. 6.2 by solving for R, for $a/h = 10, 20, 30, 40$; $r = 0\cdot5, 1\cdot0, 2\cdot0$, for (1) $m = 1$ and $n = 2$; (2) $m = 2$ and $n = 1$.

6.3. What is the fundamental frequency ($m = n = 1$) in Hertz (cps) for a boron–epoxy plate of dimensions $a = b = 20$ in. $= 50\cdot8$ cm, $h = 1$ in. $= 2\cdot54$ cm, vibrating at small amplitude in the linear range? For $\bar{\beta} = 0\cdot5, 1\cdot0$, and $1\cdot5$? Use the properties on page 255.

6.4. For a plate of dimensions $a/b = 2$, and $h/a = 1/50$, composed of the boron–epoxy material given on page 255, compare the non-dimensionalised buckling load per inch of edge distance, $\hat{N}_x$, for each of the boundary conditions treated in Sections 6.8.1 to 6.8.4. Which is the highest, which is the lowest?

REFERENCES

1. Gehring, F., 'De Aequationibus Differentialibus, Quibus Aequilibrium et Motus Laminae Crystallinae Deffiniuntur', unpublished dissertation, Berlin, 1860.
2. Hearmon, R., 'The Elastic Constants of Anisotropic Materials', *Revs. Mod. Phys.*, **18**, 409 (1946).
3. Vinson, J., 'New Techniques of Solutions for Problems in the Theory of Orthotropic Plates', Doctoral Dissertation, University of Pennsylvania, Philadelphia, 1961.
4. Leissa, A. W., 'Vibration of Plates', NASA SP-160 (1969).
5. Lekhntiskii, S., *Anisotropic Plates*, Translated from Russian by American Iron and Steel Institute, 1956. The second edition was translated by S. W. Tsai and T. Cheron, Gordon and Breach, New York (1968).
6. Kerr, A., 'An Extended Kantorovich Method for the Solution of Eigenvalue Problems', *Inter. J. Sol. Stru.*, **5**, No. 6 (June 1969), p. 559.
7. Stavsky, Y., 'On the Theory of Heterogeneous Anisotropic Plates', Doctoral Dissertation, MIT Press, Cambridge, Mass. (1959).
8. Reissner, E. and Stavsky, Y., 'Bending and Stretching of Certain Types of Heterogeneous Aeolotropic Elastic Plates, *J. Appl. Mech.*, **28**, 402 (1961).
9. Smith, C., 'Some New Types of Orthotropic Plates Laminated of Orthotropic Materials', *Trans. ASME*, **75**, p. 286.
10. Stavsky, Y., 'Bending and Stretching of Laminated Aeolotropic Plates', *Proc. ASCE J. Eng. Mech. Div.*, Vol. 87, EM6 (1963), p. 31, also *Trans. ASCE*, Vol. 127.
11. Stavsky, Y. and Roy, J., 'Some Equilibrium and Stability Solution for Laminated Aelotropic Plates', MIT Report ARDC Cont, No. AF 33 (616)-6280, 15 May 1960; also in 'Investigation of Mechanics of Reinforced Plastics', WADD-TR-60-746, Part II.
12. Stavsky, Y., 'On the General Theory of Heterogeneous Aeolotropic Plates', *AIAA Journal*, **1**, 221 (1963).
13. Stavsky, Y., 'Thermoelastic Vibrations of Heterogeneous Membranes and Inextensional Plates', *AIAA Journal*, **1**, 722 (1963).
14. Stavsky, Y. and Hoff, N., 'Mechanics of Composite Structures', *Composite Engineering Laminates*, A. G. H. Dietz, ed., MIT Press, Cambridge, Mass. (1969), p. 1.
15. Dietz, A., *Composite Engineering Laminates*, MIT Press, Cambridge, Mass. (1969).
16. Waddoups, M., 'The Vibration Response of Laminated Orthotropic Plates', M. S. Thesis, Department of Mechanical Engineering, Brigham Young University, 1965.
17. Tsai, S., Halpin J., and Pagano, J., *Composite Material Workshop*, Technomic Publ. Co. Inc., Stamford, Conn. (1968).
18. Mayberry, B., 'Vibration of Layered Anisotropic Panels', unpublished Master's Thesis, School of Engineering, University of Oklahoma, Norman, June 1968.
19. Ashton, J. and Anderson, J., 'The Natural Modes of Vibrations of Boron–Epoxy Plates', *Bulletin of 39th Shock and Vibration Symposium*, Pacific Grave, California (1969).

20. Whitney, J. and Leissa, A., 'Analysis of Heterogeneous Anisotropic Plates', *J. Appl. Mech.*, **36**, 261 (1969).
21. Whitney, J., 'Bending-Extension Coupling in Laminated Plates, Under Transverse Loading', *J. Comp. Mater.*, **3**, 20 (1969).
22. Whitney, J. and Leissa, A., 'Analysis of a Simply-Supported Laminated Anisotropic Plate', AIAA/ASME 10th Structures, Structural Dynamics, and Material Conference, New Orleans, Louisiana, April 1969; a revised version published in *AIAA Journal*, **8**, No. 1 (1970), p. 28.
23. Ashton, J. and Waddoups, M., 'Analysis of Anisotropic Plates', *J. Comp. Mater.*, **3**, 148 (1969).
24. Ashton, J., 'Analysis of Anisotropic Plates II', *J. Composite Materials* (July 1969), p. 470.
25. Ashton, J. and Waddoups, M., 'Dynamic Response of Anisotropic Plates', General Dynamics Corp., Fort Worth Div., Contract No. AF33 (615)-5257, Report FZM-5088, March 1969.
26. Ashton, J., 'Approximation Solution for Unsymmetric Laminated Plates', *J. Comp. Mater.* (January 1969), p. 189.
27. Ashton, J., 'An Analogy for Certain Anisotropic Plates', *J. Comp. Mater.* (April 1969), p. 359.
28. Kicher, T. and Mandell, J., 'An Experimental Study of the Buckling of Anisotropic Plates', AIAA/ASME 10th Structure, Structural Dynamics and Materials Conference, New Orleans, Louisiana (April 1969).
29. Chamis, C., 'Buckling of Anisotropic Composite Plates', *J. Stuc. Div.*, *Proceeding of ASCE*, Vol. 95, St. 10 (October 1969), p. 2119.
30. Mandell, J., 'An Experimental Study of the Buckling of Anisotropic Plates', Case Western Reserve Univ., SMSMD Report No. 23 (June 1968).
31. Bert, C. and Mayberry, B., 'Free Vibration of Unsymmetrically Laminated Anisotropic Plates with Clamped Edges', *J. Comp. Mater.* (April 1969), p. 282.
32. Whitney, Jr, 'Cylindrical Bending of Unsymmetrically Laminated Plates', *J. Comp. Mater.*, **3**, 715 (1969).
33. Ashton, J., 'Anisotropic Plate Analysis-Boundary Conditions', *J. Comp. Mater.*, **4**, 162 (1970).
34. Whitney J., 'The Effect of Boundary Conditions on the Response of Laminated Composite', *J. Comp. Mater.*, **4**, 192 (1970).
35. Reuter, R., 'On the Plate Velocity of a Generally Orthotropic Plate', *J. Comp. Mater.* **4**, 129 (1970).
36. Mohan, D., 'Free Vibrations of Anisotropic Plates', Master's Thesis, Mechanical and Aerospace Engineering Department, University of Delaware, June 1970.
37. Chung, Wen-Yi and Testa, R. B., 'Elastic Stability of Fibers in a Composite Plate', *J. Comp Mater,*, **3**, No. 1 (1969).
38. Ashton, J. E. and Love, T. S., 'Experimental Study of the Stability of Composite Plates', *J. Comp. Mater.*, **3**, No. 2 (1969).
39. Ashton, J. E. and Whitney, J. M., *Theory of Laminated Plates*, Technomic Publ. Co. Inc., Stamford, Conn. (1970).
40. Kaczkowski, Z., 'The Influence of Shear and Rotatory Inertia on the Frequencies of an Anisotropic Vibrating Plate', *Bull, Acad. Polonaise Sci.*, **8**, No. 7 (1960), p. 343.
41. Ambartsumyan, S., 'Theory of Anisotropic Shells', NASA TTF-118 (1964).
42. Ambartsumyan, S., *Anisotropic Plates*, Nauka, Moscow (1967).

43. Ambartsumyan, S., *Theory of Anisotropic Plates*, translated from Russian edition by T. Cheron and edited by J. E. Ashton, Technomic Publ. Co., Stamford, Conn. (1969).
44. Summers, G. D. and Vinson, J. R., 'Theoretical Analysis of Rectangular Orthotropic Laminated Plates Subjected to Transverse Loading', Transactions, 5th U.S. National Congress of Applied Mechanics, 1966.
45. Raju, P., 'Shallow Shells of Pyrolytic Graphite Type Materials', Ph.D. Dissertation, University of Delaware, 1968.
46. Kliger, H. and Vinson, J., 'Truncated Cones of Pyrolytic Graphite Type Materials', AIAA Paper 68-295.
47. Wu, C. I. and Vinson, J., 'Free Vibrations of Plates and Beams of Pyrolytic Graphite Type Materials', *AIAA Journal*, **8**, No. 2 (1970), pp. 246–51.
48. Wu, C. I. and Vinson, J., 'The Natural Vibrations of Plates Composed of Composite Materials', *Fibre Sci. and Tech.*, **2**, 97–102 (1969).
49. Wu, C. I. and Vinson, J., 'Influences of Large Amplitudes, Transverse Shear Deformation, and Rotatory Inertia on Lateral Vibrations of Transversely Isotropic Plates', *J. Appl. Mech.*, **36**, 254–260 (1969); MMAE thesis, University of Delaware, 1968.
50. Wu, C. I. and Vinson, J., 'On the Nonlinear Oscillations of Plates Composed of Composite Materials,' *J. Comp. Mater.* **3**, 548–561 (1969).
51. Daugherty, R. and Vinson, J., 'Asymptotic Solution for Noncircular Cylinders of Pyrolytic Graphite', *Proceed. of the 11th Midwest Mechanics Conference*, **5**, 563–576 (1969).
52. Daugherty, R. L., Kliger, H. S., and Vinson, J. R., 'Governing Equations for Shells of Pyrolytic Graphite Type Materials', published in the *AIAA Journal*, **9**, No. 3 (1971) pp. 508–10.
53. Kliger, H. S. and Vinson, J. R., 'Response of Spherical Shells of Composite Materials to Localized Loads', presented at the Third International Conference of Space Technology, Rome 3–8 May 1971, to be published in the *ASME J. of Pressure Vessel Technology* (1974).
54. Wu, C. I. and Vinson, J. R., 'Nonlinear Oscillations of Laminated Specially Orthotropic Plates with Clamped and Simply Supported Edges', *J. of Acoustical Soc. of Am.*, **49**, No. 5, Pt. 2 (1971), pp. 1561–7; Ph.D. dissertation, University of Delaware, 1971.
55. Mehta, V. and Vinson, J. R., 'Thermoelastic Analysis of Rectangular Plates of Pyrolytic Graphite Type Material', *Iranian J. of Sci. and Tech.*, **1**, No. 1 (1971), pp. 61–81.
56. Zukas, J. A. and Vinson, J. R., 'Laminated Transversely Isotropic Cylindrical Shells', *J. Appl. Mech.* **38**, 400–407 (1971).
57. Kliger, H. S., Forristall, C. Z. and Vinson, J. R., 'Stress in Circular Cylindrical Shells of Composite Materials Subjected to Localized Loads', AFOSR TR 73-0494, January (1973).
58. Smith, Alton P., Jr, 'The Effect of Transverse Shear Deformation on the Elastic Stability of Orthotropic Plates due to Inplane Loads', MMAE Thesis, University of Delaware, May 1973.
59. Renton, W. James and Vinson, J. R., 'The Analysis and Design of Composite Material Bonded Joints under Static and Fatigue Loadings', AFOSR TR 73-1627, August (1973).

60. Pagano, N. J., 'Analysis of the Flexure Test of Bidirectional Composites', *J. Comp. Mater.*, 1, No. 4 (1967), pp. 336–343.
61. Whitney, J. M., 'The Effect of Transverse Shear Deformation on Bending of Laminated Plates', *J. Comp. Mater.*, 3 (1969).
62. Timoshenko, S., 'On the Correction for Shear of the Differential Equations for Transverse Vibrations of Prismatic Bars', *Phil. Mag.*, 41, Ser. 5 (1921), p. 742.
63. Reissner, E., 'On a Variational Theorem in Elasticity', *J. Math. Phys.*, 29 (1950), p. 90.
64. Mindlin, R., 'Influence of Rotatory Inertia and Shear on Flexural Motions of Isotropic, Elastic Plates', *J. Appl. Mech.*, *Trans. ASME*, 73, No. 1 (1951), pp. 31–38.
65. Pagano, N., 'Exact Solutions for Rectangular Bi-directional Composites and Sandwich Plates', *J. Composite Materials*, 4, 20 (1970).
66. Pagano, N. J., 'Influence of Shear Coupling in Cylindrical Bending of Anisotropic Laminates', *J. Comp. Mater.*, 4, 330–43 (1970).
67. Vinson, J. R., 'Thermal Stresses in Laminated, Circular Plates', *Transactions, 3rd U.S. National Congress of Applied Mechanics* (1958), p. 467.
68. Brull, M. A. and Vinson, J. R., 'On Reissner's Variational Principle and Its Application to the Theory of Moderately Thick Beams', Dyna/Structures Inc. Report No. 62-A-017, January 1962.
69. Martincek, G., 'Influence of Shear and Rotatory Inertia on the Vibration of Plates', (in Czech.), Strojnicky Casopis (15.4.1964), pp. 337–57.
70. Nowinski, J. L., 'Some Static and Dynamic Problems Concerning Nonlinear Behavior of Plates and Shallow Shells with Discontinuous Boundary Conditions', Invited paper presented to the Congress of Scholars and Scientists of Polish Background, New York, November 1965.
71. Berger, H. M., 'A New Approach to the Analysis of Large Deflections of Plates', *J. Appl. Mech.* 22, 465–72 (1955).
72. Herrmann, 'Influence of Large Amplitudes on Flexural Motions of Elastic Plates', NACA TN 3578, 1956.
73. Chu, H. and Herrmann, G., 'Influence of Large Amplitudes on Free Flexural Vibrations of Rectangular Elastic Plates', *J. Appl. Mech.*, 23, 532–40 (1945).
74. Yamaki, N., 'Influence of Large Amplitudes on Flexural Vibrations of Elastic Plates', *ZAMM, Berlin*, 41, 501–10 (1961).
75. Nash, W. A. and Modeer, J., 'Certain Approximate Analysis of the Nonlinear Behavior of Plates and Shallow Shells', *Proc. Symp. Thin Elastic Shells*, Interscience, New York (1959).
76. Wah, T., 'Large Amplitude Flexural Vibration of Rectangular Plates', *Int. J. Mech. Sci.*, 5, No. 6 (1963), pp. 425–38.
77. Waltz, T, 'Interlaminar Stresses in Laminated Cylindrical Shells', MMAE thesis, University of Delaware, 1975.
78. Gajendar, N., 'Large Amplitude Vibrations of Plates on Elastic Foundations', *Int. J. Nonlinear Mech.* 2, 163–72 (1967).
79. Wu, C. I., 'On the Lateral Vibration of Rectangular Plates of Pyrolytic Graphite Type Materials', Master Thesis, Department of Mechanical and Aerospace Engineering, University of Delaware, 1968.
80. Linsenmann, D. R., 'Stability of Plates of Composite Materials', MMAE thesis, University of Delaware, 1974.
81. Nowinski, J. L., 'On Certain Inconsistencies in Berger Equations for Plates with Large Deflections', *Int. J. Mech. Sci.*, 14, 165 (1972).

82. Sokolnikoff, I. S., *Mathematical Theory of Elasticity*, McGraw-Hill Book Company, Inc., New York 1961.
83. Warburton, G., 'The Vibration of Rectangular Plates', *Proceedings of the Institute of Mechanical Engineers*, **168**, 371 (1954).
84. Young, D. and Felgar, R. Jr, 'Tables of Characteristic Functions Representing Normal Modes of Vibration of a Beam', The University of Texas Publication No. 4913, July 1944.
85. Felgar, R. Jr, *Formulas for Integrals Containing Characteristic Functions of a Vibrating Beam*, Bureau of Engineering Research, The University of Texas, 1950.
86. Hearmon, R., 'The Frequency of Flexural Vibrations of Rectangular Orthotropic Plates with Clamped or Supported Edges', *J. Appl. Mech.*, **26**, 537 (1959).
87. Brunelle, E. J., 'Buckling of Transversely Isotropic Mindlin Plates', *AIAA Journal*, **9**, No. 6 (June 1971).

CHAPTER 7

ANISOTROPIC SHELLS

In this Chapter, methods of analysis are developed for thin-walled shells composed of anisotropic materials. Because this is a basic teaching text, only the simplest shell geometry—the circular cylindrical shell—will be dealt with. To cover shells of general shape would require lengthy treatments of differential geometry, and for generally anisotropic materials the mere length of the expressions involved form a deterrent to inclusion here. General treatment of more general shell geometries has been published by Wu,[1] Daugherty,[2] Zukas[64] and Raju.[42]

Also, the most general material system will be treated that is practical: a laminated construction of generally orthotropic laminae. To be realistic, the effects of transverse shear deformation, stretching–bending coupling, stretching–shearing coupling, bending–twisting coupling and the coupling of extensional-rotational motions are included, because of their significance in treating either static or dynamic behaviour of laminated anisotropic thin shells.

Subsequent to the derivation for the most general material system, the equations will be developed for specially orthotropic fibre-reinforced composite materials, and transversely isotropic materials, such as pyrolytic graphite.

An excellent discussion of research performed in anisotropic shells to date is found in References 1 and 2, from which the following is partially excerpted.

7.1 HISTORY OF ANISOTROPIC SHELLS UTILISING CLASSICAL SHELL THEORY

In 1924 Shtaerman[3] published the first paper treating anisotropic shells, in which he treated specially orthotropic shells of revolution under axially symmetric static loads. However, it was not until 1955 that the first paper treated vibrations of orthotropic shells, when Chakrovorty treated the case of the spherical shell.[4] Since then the literature has increased as is seen from the survey papers by Bert and Egle[5] and Ambartsumyan.[6]

In 1953 Ambartsumyan discovered the important effect of bending–stretching coupling that can exist in shells of anisotropic laminae.[7] In 1961 he published his comprehensive book, which was translated into English in 1964.[8] The book is almost completely restricted to specially orthotropic shells both in derivation as well as solutions, including single-layer and symmetrical laminates. Only in the last chapter does he include transverse shear deformation.

In 1962, Dong et al.[9] analysed layered, anisotropic shells using classical shell theory. Using Donnell's assumptions they treated a two-layered, angle-ply, quasi-isotropic cylindrical shell under axially symmetric loads, which showed the effects of bending–stretching coupling.

Papers by Becker and Gerard,[10] Cheng and Kuenzi,[11] Hess,[12] Hedgepeth and Hall,[13] and Thielemann et al.[14] deal with the stability of orthotropic cylinders. The buckling strength of filament-wound cylinders, and the effects of heterogeneity on the stability of composite cylindrical shells were studied by Tasi et al.,[15] and Tasi.[16] Cheng and Ho treated the stability of laminated anisotropic cylindrical shells under combined loading of lateral pressure, axial compression, and torsional loading[17,18] in 1963, using Flügge's shell theory.

Also in 1963, Dong and Dong[19] used the perturbation technique of Vinson and Brull[20] to reduce the anisotropic shell equations to successive systems of specially orthotropic shells, but no numerical results were published.

In 1965, Kingsbury[21] treated the homogeneous anisotropic cylindrical shell and shells of revolution under axially symmetric loading. In the cylindrical shell, edge load solutions were obtained and plots of stresses and deformations were made as functions of elastic constants and fibre directions. Solutions for shells of revolution were found using a perturbation scheme in conjunction with an asymptotic integration technique.

Bert[22] developed a theory in 1967 for the analysis of a general anisotropic shell with an arbitrary number of layers using Vlasov's shell theory. Also in 1967, Kunukkasseril[23] carried out a vibration analysis of a multi-layered generally anisotropic cylindrical shell. For the shell of infinite length he employed a helical mode solution, but for shells of finite length, he considered only axially symmetric and inextensional motion.

Other research includes the work of Franklin,[24] who obtained membrane solutions of fibre-reinforced conjugated shells of revolution. These solutions, although analytical, not only do not include transverse shear deformation, but do not include any bending effects; hence they cannot provide accuracy for cases of practical importance.

Homogeneous solutions for orthotropic toroidal shells have been obtained by Bessarabov and Kraus,[25,26] without accounting for transverse shear deformation.

In 1969, Bert et al.[27] developed equations of motion for a thin, arbitrarily layered, generally anisotropic cylindrical shell using classical bending theory, without using the Donnell's approximation. Natural frequencies for a two-layer specially orthotropic cylindrical shell were shown, and compared with Dong's earlier results.[28] The significance of bending–stretching coupling is shown again. Exact solutions were found for axially symmetric motions, and they showed how to analyse asymmetric vibrations approximately using helical solutions (which will be used later in this chapter).

Again in 1969, Reuter[29] investigated elastic, long wave length wave-propagation for axially symmetric motions of a thin, homogeneous, semi-infinite, generally orthotropic, circular cylindrical shell, resulting from a uniform axial stress pulse on the free end. In the same year, Matin[30] studied the free vibrations of homogeneous, anisotropic conical shells. A computer solution was obtained by using Galerkin's method.

Also in 1969, two books were published. One, by Calcote[31] derives equations for laminated plates and shells, but provides no example problems. The book by Librescu[32] provides detailed static and dynamic analyses for layered anisotropic plates and shells. In 1970, Reuter[33] extended his previous work to treat wave propagation in a thin, laminated, generally orthotropic, circular cylindrical shell under axially symmetric membrane motion, including the radial motion.

As to anisotropic shells, most research has dealt with the specially orthotropic case. Perhaps, only three papers to date deal with the vibrations of the general anisotropic material system using bending shell theory, and all of these use classical thin-shell theory. For structures composed of composite materials, where the ratio of in-plane Young's modulus to transverse shear modulus is large, transverse shear deformation must be included even for shells of classical thinness.

7.2 HISTORY OF ANISOTROPIC SHELLS INCLUDING TRANSVERSE SHEAR DEFORMATION

For several decades, transverse shear deformation has been included in the analysis of isotropic plates and shells of moderate thickness, but it was only recently that it was realised that transverse shear deformation must be included for shells and plates of classical thinness for those material systems where the ratio of an in-plane modulus of elasticity to the transverse shear modulus is high. Such material systems include fibre-reinforced composite materials and pyrolitic graphite type material systems.

Very early, Hildebrand et al.[34] derived a linear, specially orthotropic shell theory which included the effects of transverse shear deformation and transverse normal stress. In 1964, Ambartsumyan's book[8] presented the governing equations for the vibration of layered specially orthotropic shallow shells including transverse shear deformation. In 1964 Kalnins[35] developed a computer technique which could be applied to a generally anisotropic shell of revolution under axially symmetric loads including transverse shear deformation. Also in 1964 Mirsky[36] included transverse shear and rotatory inertia in his study of wave propagation in an infinite, specially orthotropic, thick, circular, cylindrical shell. In 1967 Gulati and Essenburg[37] studied the effects of anisotropy and transverse shear deformation in a generally anisotropic, circular, cylindrical shell under an axially symmetric static loading. Also in 1967, Ahmed[38] used the Rayleigh–Ritz approach to determine the axisymmetric motion of a homogeneous, specially orthotropic, truncated, conical shell of tapered thickness, including transverse shear deformation.

In studying the effects of temperature on pyrolytic graphite, Garber[39] and McDonough[40] showed the necessity of including transverse normal strain (thermal thickening) in the analysis of thin-walled shells of this material, because the coefficient of thermal expansion in the thickness direction is many times greater than that in the in-plane directions.

McDonough[40] was also the first to show that it is necessary to include transverse shear deformation in the analysis of thin shells of pyrolytic graphite. This is necessary because in pyrolytic graphite the ratio of in-plane modulus of elasticity to transverse shear modulus varies between 20 and 50, compared with a maximum of 3 for an isotropic material (see Section 2.6.2).

Kliger,[41] Raju,[42] and Daugherty[43] included both thermal thickening and transverse shear deformation in their study of pyrolytic graphite when used for conical shells, shallow, spherical shells and non-circular, cylindrical shells, respectively. A general set of governing equations for shells of general shape composed of pyrolytic graphite was presented in Reference 44.

Zukas[45,64] clearly showed how important both transverse shear deformation and thermal thickening are in the analysis of laminated shells, including those composed of pyrolytic graphite.

Daugherty,[2] however, was the first to show that in composite materials, wherein the coefficient of thermal expansion in the thickness direction is largely a function of the matrix material while in the fibre directions it is largely influenced by the fibres (using the rule of mixtures, Section 5.6), it is necessary to include 'thermal thickening' in analysing thin-walled shells of composite materials for some material systems when they are subjected to thermal loads. Daugherty[2] provides solutions for symmetrical deformations of orthotropic thin shells of revolution including transverse shear deformation, mechanical and thermal loadings and 'thermal thickening'. Governing equations for these shells of arbitrary shape with the above considerations are also included.

In 1967, Pagano[46] in his study of a bidirectional composite beam, included transverse shear deformation. The agreement between theory and experiment was excellent. He pointed out that the effect of transverse shear deformation is significant in composite materials even for beams with span-to-depth ratios as high as 30.

In 1967 also, Schipper[47] derived the governing equations for thick, orthotropic shells, including mechanical and thermal loads and temperature-dependent material properties. In the same year, Kalnins[48] developed the governing equations for shells of specially orthotropic materials.

In 1968, Vasilev,[49] having discussed the fact that composite materials can have elastic moduli in the plane of the structure substantially greater than the transverse shear modulus, also included transverse shear deformation in his study of orthotropic laminar cylindrical shells.

Kliger and Vinson[50-53] have studied in detail spherical shells of composite materials, with bending and transverse shear deformation, subjected to localised loadings—a problem of great practical importance. It is interesting to note that with spherical shells of orthotropic materials there is a singularity in the solution at the apex when the modulus of elasticity in the meridional direction is greater than that in the circumferential direction.

Kliger *et al.* also have provided solutions for cylindrical shells of composite materials subjected to localised loadings.[54]

It should be noted that Whitney *et al.* have, since 1968, aggressively pursued research,[55-60] toward characterising the mechanical properties of fibre-reinforced composites through tests using a thin-walled tube test-piece.

It has been shown conclusively that methods of analysis which do not include transverse shear deformation can lead to very erroneous results in analysing plates and shells of pyrolytic graphite or fibre-reinforced composite materials. This conclusion greatly reduces the solutions available for the accurate analysis of thin-walled shells involving these material systems. The rest of this chapter will deal with some of the accurate methods for cylindrical shells that can be employed.

7.3 ANISOTROPIC LAMINATED CYLINDRICAL SHELLS

In Reference 1, Wu develops rigorous methods of analysis for shells, including the effects of transverse shear deformation, stretching–bending coupling (due to the $[B]$ matrix), extension-shearing coupling (due to A_{16}, A_{26} terms), bending–twisting coupling (due to D_{16}, D_{26} terms), and the coupling of extensional and rotational motions. Governing equations are derived for arbitrarily laminated, generally anisotropic shells. The governing equations for laminated anisotropic shells of revolution and cylindrical shells are treated as special cases. The governing equations are formulated in terms of displacements only (instead of incorporating an Airy stress function). The development for the circular cylindrical shells follows.

Consider the circular cylindrical shell shown in Figs. 4.11 and 4.12. For the general case the positive values of all stress and displacement quantities are shown in Fig. 7.1.

The strain-displacement relations for a circular, cylindrical shell of any material system, including the effects of transverse shear deformation, are given by

$$\epsilon_x^0 = \frac{\partial u_0}{\partial x} \tag{7.1}$$

$$\epsilon_\theta^0 = \frac{\partial v_0}{\partial s} + \frac{w}{R} \tag{7.2}$$

$$\gamma_{x\theta}^0 = 2\epsilon_{x\theta}^0 = \frac{\partial v_0}{\partial x} + \frac{\partial u_0}{\partial s} \tag{7.3}$$

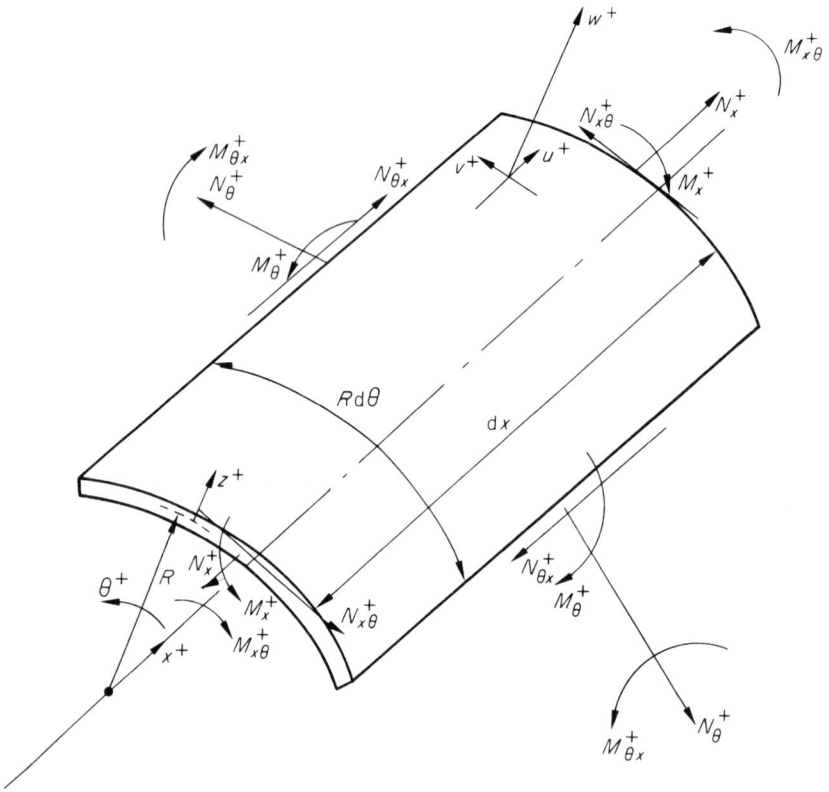

Fig. 7.1.   *Positive direction for all stress and displacement quantities.*

$$\kappa_x = \frac{\partial \beta_x}{\partial x} \tag{7.4}$$

$$\kappa_\theta = \frac{\partial \beta_\theta}{\partial s} \tag{7.5}$$

$$\kappa_{x\theta} = \frac{\partial \beta_\theta}{\partial x} + \frac{\partial \beta_x}{\partial s} \tag{7.6}$$

$$\gamma_{xz} = \beta_x + \frac{\partial w}{\partial x} = 2\epsilon_{xz} \tag{7.7}$$

$$\gamma_{\theta z} = \beta_\theta + \frac{\partial w}{\partial s} - \frac{v_0}{R} = 2\epsilon_{\theta z} \tag{7.8}$$

where $\partial s \equiv R\partial \theta$, the superscript 0 refers to values at the shell midsurface ($z = 0$), and β_x and β_θ are rotations in the x and θ

directions, respectively, of a linear element that was normal to the shell middle surface prior to loading.

The ϵ are strains, u_0 and v_0 are the in-plane midsurface deflections, w is the radical deflection, and the κ are the curvature quantities.

In these equations, if R were allowed to go to infinity these would be the strain-displacement equations for a flat plate, including transverse shear deformation. Note the similarity under those conditions between eqns. (4.10), (4.11) and (4.13) with eqns. (7.1) to (7.3), which are restricted to the middle surface ($z = 0$), (hence the superscript 0). Also note that in (7.7) and (7.8) if transverse shear deformations were ignored, $\epsilon_{\theta z} = \epsilon_{xz} = 0$, that equation becomes identical to eqn. (4.148). These equations can be derived from general three-dimensional equations of elasticity, and their derivation is beyond the scope of those who have not had a graduate course in the theory of shells.

The integrated stress–strain and moment curvature relations for the cylindrical shell, which are also called the constitutive relations, are given by eqns. (5.53), (5.64) and (5.65), which are repeated below for the shell co-ordinate system.

$$
\begin{bmatrix} N_x \\ N_\theta \\ N_{x\theta} \\ \hline M_x \\ M_\theta \\ M_{x\theta} \end{bmatrix} =
\left[\begin{array}{ccc|ccc}
A_{11} & A_{12} & 2A_{16} & B_{11} & B_{12} & 2B_{16} \\
A_{12} & A_{22} & 2A_{26} & B_{12} & B_{22} & 2B_{26} \\
A_{16} & A_{26} & 2A_{66} & B_{16} & B_{26} & 2B_{66} \\
\hline
B_{11} & B_{12} & 2B_{16} & D_{11} & D_{12} & 2D_{16} \\
B_{12} & B_{22} & 2B_{26} & D_{12} & D_{22} & 2D_{26} \\
B_{16} & B_{26} & 2B_{66} & D_{16} & D_{26} & 2D_{66}
\end{array}\right]
\begin{bmatrix} \epsilon_x^0 \\ \epsilon_\theta^0 \\ \epsilon_{x\theta}^0 \\ \kappa_x \\ \kappa_\theta \\ \frac{1}{2}\kappa_{x\theta} \end{bmatrix}
\tag{7.9}
$$

$$
\begin{bmatrix} Q_x \\ Q_\theta \end{bmatrix} = \begin{bmatrix} A_{55} & A_{45} \\ A_{45} & A_{44} \end{bmatrix} \begin{bmatrix} 2\epsilon_{xz} \\ 2\epsilon_{\theta z} \end{bmatrix}
\tag{7.10}
$$

As discussed previously, if the material system is made of one lamina of an isotropic material, (7.9), utilising (7.1) through (7.8) with $R = \infty$ and $\epsilon_{xz} = \epsilon_{\theta z} = 0$, reduces very straightforwardly to (4.47) through (4.49) and (4.52) through (4.54) with $y = R\theta = s$.

Also, if the shell were composed of one isotropic lamina, and if the shell were subjected to axially symmetric loadings, hence $\partial(\ )/\partial\theta = v = 0$, (7.9) and (7.10) reduce straightforwardly to (4.129) through (4.133).

The equilibrium equations for a circular cylindrical shell are presented below without derivation.

$$
\frac{\partial N_x}{\partial x} + \frac{\partial N_{x\theta}}{\partial s} = -q_x + \bar{\rho}\frac{\partial^2 u_0}{\partial t^2} + Q\frac{\partial^2 \beta_x}{\partial t^2}
\tag{7.11}
$$

$$
\frac{\partial N_{x\theta}}{\partial x} + \frac{\partial N_\theta}{\partial s} + \frac{Q_\theta}{R} = -q_\theta + \bar{\rho}\frac{\partial^2 v_0}{\partial t^2} + Q\frac{\partial^2 \beta_\theta}{\partial t^2}
\tag{7.12}
$$

$$\frac{\partial Q_x}{\partial x} + \frac{\partial Q_\theta}{\partial s} - \frac{N_\theta}{R} = -p(x, \theta) + \bar{\rho}\frac{\partial^2 w}{\partial t^2} \qquad (7.13)$$

$$\frac{\partial M_x}{\partial x} + \frac{\partial M_{x\theta}}{\partial s} - Q_x = -m_x + I\frac{\partial^2 \beta_x}{\partial t^2} + Q\frac{\partial^2 u_0}{\partial t^2} \qquad (7.14)$$

$$\frac{\partial M_{x\theta}}{\partial x} + \frac{\partial M_\theta}{\partial s} - Q_\theta = -m_\theta + I\frac{\partial^2 \beta_\theta}{\partial t^2} + Q\frac{\partial^2 v_0}{\partial t^2} \qquad (7.15)$$

In the above, the stress resultants (N_x, N_θ, $N_{x\theta}$), the stress couples (M_x, M_θ, $M_{x\theta}$) and the transverse shear resultants (Q_x and Q_θ) are defined exactly as in the plates of Chapter 4 when Love's First Approximation ($h/R \ll 1$) has been employed when there is only one lamina. For laminated shells, wherein the kth lamina is bounded by the surfaces $z = h_{k-1}$ and $z = h_k$, if the shell has n laminae, stress couples and resultants are found by simply summing the contribution made by each individual lamina. The result is

$$\left\{\begin{array}{c} N_x \\ N_\theta \\ N_{x\theta} \end{array}\right\} = \sum_{k=1}^{n} \int_{h_{k-1}}^{h_k} \left\{\begin{array}{c} \sigma_x \\ \sigma_\theta \\ \sigma_{x\theta} \end{array}\right\} dz \qquad (7.16)$$

$$\left\{\begin{array}{c} M_x \\ M_\theta \\ M_{x\theta} \end{array}\right\} = \sum_{k=1}^{n} \int_{h_{k-1}}^{h_k} \left\{\begin{array}{c} \sigma_x \\ \sigma_\theta \\ \sigma_{x\theta} \end{array}\right\} z \, dz \qquad (7.17)$$

$$\left\{\begin{array}{c} Q_x \\ Q_\theta \end{array}\right\} = \sum_{k=1}^{n} \int_{h_{k-1}}^{h_k} \left\{\begin{array}{c} \sigma_{xz} \\ \sigma_{\theta z} \end{array}\right\} dz \qquad (7.18)$$

where z is measured from the shell midsurface.

On the right-hand side of eqns. (7.13) to (7.15) are the forcing functions that can occur. The first term on the right of each equation is due to the surface traction on the inner and outer shell surfaces

$$q_x \equiv \sigma_{xz}]_{-h/2}^{h/2} = \sigma_{xz}(h/2) - \sigma_{xz}(-h/2)$$

$$q_\theta \equiv \sigma_{\theta z}]_{-h/2}^{h/2} = \sigma_{\theta z}(h/2) - \sigma_{\theta z}(-h/2)$$

$$m_x = z\sigma_{xz}]_{-h/2}^{h/2} = \frac{h}{2}[\sigma_{xz}(h/2) + \sigma_{xz}(-h/2)] \qquad (7.19)$$

$$m_\theta \equiv z\sigma_{\theta z}]_{-h/2}^{h/2} = \frac{h}{2}[\sigma_{\theta z}(h/2) - \sigma_{\theta z}(-h/2)]$$

$$p \equiv \sigma_z]_{-h/2}^{h/2} = \sigma_z(h/2) - \sigma_z(-h/2)$$

The first two of the above are the loads caused by surface shear stresses; the second two are couples caused by these surface shear stresses; the last are the normal stresses caused by laterally distributed loads, such as internal and external pressures.

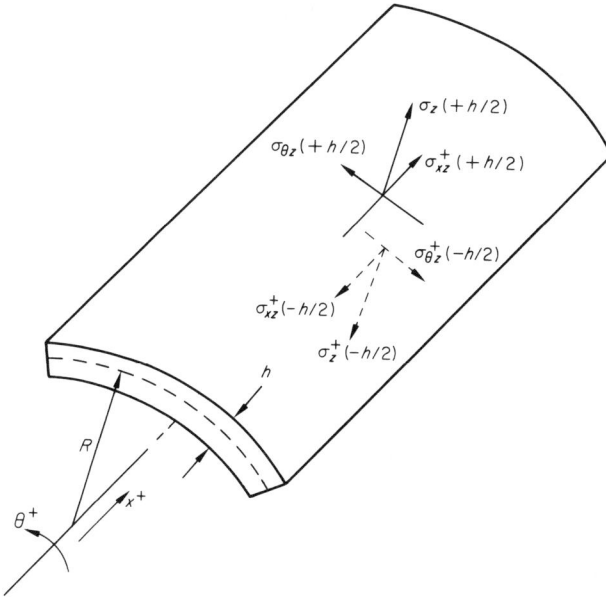

Fig. 7.2. Positive directions of surface tractions.

The positive directions of these quantities are illustrated in Fig. 7.2.
The remaining terms on the right-hand sides of eqns. (7.11) to (7.15)
are inertia terms involving the following quantities, two of which were
defined previously in eqns. (6.125) and (6.126)

$$\bar{\rho} = \sum_{k=1}^{n} \int_{h_{k-1}}^{h_k} \rho_0^{(k)} \, dz = \sum_{k=1}^{n} \rho_0^{(k)}(h_k - h_{k-1}) \tag{7.20}$$

$$Q = \sum_{k=1}^{n} \int_{h_{k-1}}^{h_k} \rho_0^{(k)} z \, dz = \frac{1}{2} \sum_{k=1}^{n} \rho_0^{(k)}(h_k^2 - h_{k-1}^2) \tag{7.21}$$

$$I = \sum_{k=1}^{n} \int_{h_{k-1}}^{h_k} \rho_0^{(k)} z^2 \, dz = \frac{1}{3} \sum_{k=1}^{n} \rho_0^{(k)}(h_k^3 - h_{k-1}^3) \tag{7.22}$$

In the above, $\rho_0^{(k)}$ is the mass density of the kth layer. $\bar{\rho}$ and I are the
mass and the moment of inertia with respect to the middle surface per
unit area, and Q is the coupling between extensional and rotational
motions, a quantity which does not appear in equations for a
homogeneous shell.

Note also that in eqns. (7.11) to (7.15), if $R = \infty$, these equations
become the equilibrium equations for a laminated plate with x and s
co-ordinates, as in Chapter 6.

All of the above equations can be manipulated in a manner similar to those for a single-layer isotropic shell. For studying the free vibrations only, wherein $q_x = q_\theta = m_x = m_\theta = p = 0$, the equations can be written in the following form

$$
\begin{bmatrix}
L_{11} & L_{12} & L_{13} & L_{14} & L_{15} \\
L_{21} & L_{22} & L_{23} & L_{24} & L_{25} \\
L_{31} & L_{32} & L_{33} & L_{34} & L_{35} \\
L_{41} & L_{42} & L_{43} & L_{44} & L_{45} \\
L_{51} & L_{52} & L_{53} & L_{54} & L_{55}
\end{bmatrix}
\begin{bmatrix} u_0 \\ v_0 \\ \beta_x \\ \beta_\theta \\ w \end{bmatrix}
=
\begin{bmatrix}
\bar\rho & 0 & Q & 0 & 0 \\
0 & \bar\rho & 0 & Q & 0 \\
Q & 0 & I & 0 & 0 \\
0 & Q & 0 & I & 0 \\
0 & 0 & 0 & 0 & \bar\rho
\end{bmatrix}
\frac{\partial^2}{\partial t^2}
\begin{bmatrix} u_0 \\ v_0 \\ \beta_x \\ \beta_\theta \\ w \end{bmatrix}
\tag{7.23}
$$

where

$$L_{11} = A_{11}\frac{\partial^2}{\partial x^2} + 2A_{16}\frac{1}{R}\frac{\partial^2}{\partial\theta\partial x} + A_{66}\frac{1}{R^2}\frac{\partial^2}{\partial\theta^2}$$

$$L_{12} = A_{16}\frac{\partial^2}{\partial x^2} + (A_{12}+A_{66})\frac{1}{R}\frac{\partial^2}{\partial x\partial\theta} + A_{26}\frac{1}{R^2}\frac{\partial^2}{\partial\theta^2}$$

$$L_{13} = B_{11}\frac{\partial^2}{\partial x^2} + 2B_{16}\frac{1}{R}\frac{\partial^2}{\partial x\partial\theta} + B_{66}\frac{1}{R^2}\frac{\partial^2}{\partial\theta^2}$$

$$L_{14} = B_{16}\frac{\partial^2}{\partial x^2} + (B_{12}+B_{66})\frac{1}{R}\frac{\partial^2}{\partial x\partial\theta} + B_{26}\frac{1}{R^2}\frac{\partial^2}{\partial\theta^2}$$

$$L_{15} = A_{12}\frac{1}{R}\frac{\partial}{\partial x} + A_{26}\frac{1}{R^2}\frac{\partial}{\partial\theta}$$

$$L_{21} = A_{16}\frac{\partial^2}{\partial x^2} + (A_{12}+A_{66})\frac{1}{R}\frac{\partial^2}{\partial x\partial\theta} + A_{26}\frac{1}{R^2}\frac{\partial^2}{\partial\theta^2}$$

$$L_{22} = A_{66}\frac{\partial^2}{\partial x^2} + 2A_{26}\frac{1}{R}\frac{\partial^2}{\partial x\partial\theta} + A_{22}\frac{1}{R^2}\frac{\partial^2}{\partial\theta^2} - A_{44}\frac{1}{R^2}$$

$$L_{23} = B_{16}\frac{\partial^2}{\partial x^2} + (B_{12}+B_{66})\frac{1}{R}\frac{\partial^2}{\partial x\partial\theta} + B_{26}\frac{1}{R^2}\frac{\partial^2}{\partial\theta^2} + A_{45}\frac{1}{R}$$

$$L_{24} = B_{66}\frac{\partial^2}{\partial x^2} + 2B_{26}\frac{1}{R}\frac{\partial^2}{\partial x\partial\theta} + B_{22}\frac{1}{R^2}\frac{\partial^2}{\partial\theta^2} + A_{44}\frac{1}{R}$$

$$L_{25} = (A_{12}+A_{55})\frac{1}{R}\frac{\partial}{\partial x} + (A_{26}+A_{45})\frac{1}{R^2}\frac{\partial}{\partial\theta}$$

$$L_{31} = B_{11}\frac{\partial^2}{\partial x^2} + 2B_{16}\frac{1}{R}\frac{\partial^2}{\partial x\partial\theta} + B_{66}\frac{1}{R^2}\frac{\partial^2}{\partial\theta^2}$$

$$L_{32} = B_{16}\frac{\partial^2}{\partial x^2} + (B_{12}+B_{66})\frac{1}{R}\frac{\partial^2}{\partial x\partial\theta} + B_{26}\frac{1}{R^2}\frac{\partial^2}{\partial\theta^2}$$

$$L_{33} = D_{11}\frac{\partial^2}{\partial x^2} + 2D_{16}\frac{1}{R}\frac{\partial^2}{\partial x\partial\theta} + D_{66}\frac{1}{R^2}\frac{\partial^2}{\partial\theta^2} - A_{55}$$

$$L_{34} = D_{16} \frac{\partial^2}{\partial x^2} + (D_{12} + D_{66}) \frac{1}{R} \frac{\partial^2}{\partial x \partial \theta} + D_{26} \frac{1}{R^2} \frac{\partial^2}{\partial \theta^2} - A_{45}$$

$$L_{35} = \left(\frac{B_{12}}{R} - A_{55} \right) \frac{\partial}{\partial x} + \left(\frac{B_{26}}{R} - A_{45} \right) \frac{\partial}{\partial \theta}$$

$$L_{41} = B_{16} \frac{\partial^2}{\partial x^2} + (B_{12} + B_{66}) \frac{1}{R} \frac{\partial^2}{\partial x \partial \theta} + B_{26} \frac{1}{R^2} \frac{\partial^2}{\partial \theta^2}$$

$$L_{42} = B_{66} \frac{\partial^2}{\partial x^2} + 2B_{26} \frac{1}{R} \frac{\partial^2}{\partial x \partial \theta} + B_{22} \frac{1}{R^2} \frac{\partial^2}{\partial \theta^2} + A_{44} \frac{1}{R}$$

$$L_{43} = D_{16} \frac{\partial^2}{\partial x^2} + (D_{12} + D_{66}) \frac{1}{R} \frac{\partial^2}{\partial x \partial \theta} + D_{26} \frac{1}{R^2} \frac{\partial^2}{\partial \theta^2} - A_{45}$$

$$L_{44} = D_{66} \frac{\partial^2}{\partial x^2} + 2D_{26} \frac{1}{R} \frac{\partial^2}{\partial x \partial \theta} + D_{22} \frac{1}{R^2} \frac{\partial^2}{\partial \theta^2} - A_{44}$$

$$L_{45} = \left(\frac{B_{26}}{R} - A_{45} \right) \frac{\partial}{\partial x} + \left(\frac{B_{22}}{R} - A_{44} \right) \frac{1}{R} \frac{\partial}{\partial \theta}$$

$$L_{51} = - A_{12} \frac{1}{R} \frac{\partial}{\partial x} - A_{26} \frac{1}{R^2} \frac{\partial}{\partial \theta}$$

$$L_{52} = - (A_{26} + A_{45}) \frac{1}{R} \frac{\partial}{\partial x} - (A_{22} + A_{44}) \frac{1}{R^2} \frac{\partial}{\partial \theta}$$

$$L_{53} = \left(A_{55} - \frac{B_{12}}{R} \right) \frac{\partial}{\partial x} + \left(A_{45} - \frac{B_{26}}{R} \right) \frac{1}{R} \frac{\partial}{\partial \theta}$$

$$L_{54} = \left(A_{45} - \frac{B_{26}}{R} \right) \frac{\partial}{\partial x} + \left(A_{44} - \frac{B_{22}}{R} \right) \frac{1}{R} \frac{\partial}{\partial \theta}$$

$$L_{55} = A_{55} \frac{\partial^2}{\partial x^2} + 2A_{45} \frac{1}{R} \frac{\partial^2}{\partial x \partial \theta} - A_{22} \frac{1}{R^2}$$

It is clear that great simplifications will result in (7.23) if the shell is specially orthotropic, symmetric with respect to the middle surface, single layered, or combinations of these.

If one considers a complete anisotropic cylindrical shell, $0 \leq \theta \leq 2\pi$, under asymmetric free vibrations, even when the elastic axes of symmetry do not coincide with the shell geometric axes, the dependent variables retain a periodicity in the θ direction. However, unfortunately, sine and cosine functions do not represent the mode shapes in the circumferential direction for a closed, generally orthotropic shell.

For the shell with both ends fully clamped the boundary conditions are found to be

$$w = u_0 = v_0 = \beta_x = \beta_\theta = 0 \quad \text{at } x = 0 \text{ and } L \qquad (7.24)$$

That is to say that at each end, clamping does not permit any displacements or rotations.

Bert et al.[27] studied this problem in detail, and following them, the following functions are assumed for the solution

$$\begin{Bmatrix} u_0 \\ v_0 \\ \beta_x \\ \beta_\theta \\ w \end{Bmatrix} = \sum_{n=1}^{N} \begin{Bmatrix} U_n \\ V_n \\ \Lambda_n \\ \Gamma_n \\ W_n \end{Bmatrix} \left\{ \left[1 - \cos \frac{2n\pi x}{L} \right] \sin n\theta + \sin \frac{n\pi x}{L} \cos n\theta \right\} e^{i\omega t} \qquad (7.25)$$

(7.25) satisfies the periodicity in the circumferential direction as well as the end conditions in the axial direction.

Using Galerkin's method, a characteristic equation in matrix form can be obtained for the eigenvalue problem, and natural frequencies and mode shapes can be found. The process is straightforward, but involved, and will not be treated here.

It is useful to have eqn. (7.23) for easy reference because its left-hand side is the most general set of governing equations for an isothermal laminated anisotropic material system. The right-hand side, as it is shown is for free vibration. However, using eqns. (7.11) to (7.15) the suitable forms for forcing functions are easily found.

7.4 STRESSES IN A CIRCULAR CYLINDRICAL SHELL SUBJECTED TO A LOCALISED LOAD

This problem is treated here to provide insight into another analytical technique that can be used in a wide variety of problems, and to provide some physical insight from the results presented.

Kliger et al.[54] consider the stresses that occur in a single-layer, circular cylindrical shell of a specially orthotropic composite material when the shell is subjected to localised rectangular loads normal to the shell surface. The problem is of great importance when considering the supports or other localised conditions in a practical situation. Actually, the single-layer results presented here can be generalised for a specially multilayer shell that is symmetrical with respect to its middle surface.

In this case the general equations of the previous Section can undergo considerable simplification, because of the single layer, the special orthotropy, the normal loading, and the static condition.

The equilibrium equations for this asymmetric problem are found from eqns. (7.11) to (7.15) by letting all forcing function terms be zero, except $p(x, s)$. The result is

$$\frac{\partial N_x}{\partial x} + \frac{\partial N_{x\theta}}{\partial s} = 0 \qquad (7.26)$$

$$\frac{\partial N_{x\theta}}{\partial x} + \frac{\partial N_\theta}{\partial s} + \frac{Q_\theta}{R} = 0 \qquad (7.27)$$

$$\frac{\partial Q_x}{\partial x} + \frac{\partial Q_\theta}{\partial s} - \frac{N_\theta}{R} + p(x, \theta) = 0 \qquad (7.28)$$

$$\frac{\partial M_x}{\partial x} + \frac{\partial M_{x\theta}}{\partial s} - Q_x = 0 \qquad (7.29)$$

$$\frac{\partial M_{x\theta}}{\partial x} + \frac{\partial M_\theta}{\partial s} - Q_\theta = 0 \qquad (7.30)$$

The strain–displacement relations, being purely kinematic, remain as in (7.1) to (7.8). However, it is convenient to merely list the strain displacement relations throughout the shell wall thickness as

$$\epsilon_x = \epsilon_x^0 + z\kappa_x \qquad (7.31)$$

$$\epsilon_\theta = \epsilon_\theta^0 + z\kappa_\theta \qquad (7.32)$$

$$\epsilon_{x\theta} = \epsilon_{x\theta}^0 + z\kappa_{x\theta} \qquad (7.33)$$

where all terms are defined in eqns. (7.1) to (7.8).

Following Naghdi,[65] with transverse shear effects included

$$\gamma_{xz} = 2\epsilon_{xz} = \frac{6}{5G_{xz}h} Q_x \qquad (7.34)$$

$$\gamma_{\theta z} = 2\epsilon_{\theta z} = \frac{6}{5G_{\theta z}h} Q_\theta \qquad (7.35)$$

From (7.34), (7.35), (7.7), and (7.8), the transverse shear resultants are found to be

$$Q_x = \frac{5}{6} G_{xz}h \left(\beta_x + \frac{\partial w}{\partial x} \right) \qquad (7.36)$$

$$Q_\theta = \frac{5}{6} G_{\theta z}h \left(\beta_\theta - \frac{v_0}{R} + \frac{\partial w}{\partial s} \right) \qquad (7.37)$$

Without transverse shear deformation effects included, the terms in parentheses in (7.36) and (7.37) would be set equal to zero, thus providing explicit relations for the rotations β_x and β_θ in terms of the displacements v_0 and w.

These can also be derived from the simplification of (7.10). Similarly from (7.9), the integrated stress–strain and moment curvature relations can be found for the single layer specially orthotropic case

$$N_x = K_x \left[\frac{\partial u_0}{\partial x} + \nu_{\theta x} \frac{\partial v_0}{\partial s} + \nu_{\theta x} \frac{w}{R} \right] \qquad (7.38)$$

$$N_\theta = K_\theta \left[\frac{\partial v_0}{\partial s} + \frac{w}{R} + \nu_{x\theta} \frac{\partial u_0}{\partial x} \right] \qquad (7.39)$$

$$N_{x\theta} = G_{xz}h \left[\frac{\partial u_0}{\partial s} + \frac{\partial v_0}{\partial x} \right] \qquad (7.40)$$

$$M_x = D_x \left[\frac{\partial \beta_x}{\partial x} + \nu_{\theta x} \frac{\partial \beta_\theta}{\partial \theta} \right] \qquad (7.41)$$

$$M_\theta = D_\theta \left[\nu_{x\theta} \frac{\partial \beta_x}{\partial x} + \frac{\partial \beta_\theta}{\partial s} \right] \qquad (7.42)$$

$$M_{x\theta} = M_{\theta x} = \frac{G_{x\theta}h^3}{12} \left[\frac{\partial \beta_x}{\partial s} + \frac{\partial \beta_\theta}{\partial x} \right] \qquad (7.43)$$

where $K_i = E_i h /(1 - \nu_{x\theta}\nu_{\theta x})$ and $D_i = E_i h^3/12(1 - \nu_{x\theta}\nu_{\theta x})$, $i = x, \theta$.

Again by letting $R = \infty$ in the above equations, the result is the stress–strain relations for a specially orthotropic plate with co-ordinates x and s.

Substituting (7.36) to (7.43) into (7.26) to (7.30) provides five equations in terms of displacements and rotations only. The resulting equations are as follows, but it should be noted that by making the proper simplifications to the left-hand side of (7.23) and replacing the right-hand side by $p(x, \theta)$ in the proper equations, the following would also have been obtained

$$K_x \left[\frac{\partial^2 u_0}{\partial x^2} + \frac{\nu_{\theta x}}{R} \frac{\partial^2 v_0}{\partial x \partial \theta} + \frac{\nu_{\theta x}}{R} \frac{\partial w}{\partial x} \right] + \frac{G_{x\theta}h}{R} \left[\frac{1}{R} \frac{\partial^2 u_0}{\partial \theta^2} + \frac{\partial^2 v_0}{\partial x \partial \theta} \right] = 0 \qquad (7.44)$$

$$G_{x\theta}h \left[\frac{1}{R} \frac{\partial^2 u_0}{\partial x \partial \theta} + \frac{\partial^2 v_0}{\partial x^2} \right] + \frac{K_\theta}{R} \left[\frac{1}{R} \frac{\partial^2 v_0}{\partial \theta^2} + \frac{1}{R} \frac{\partial w}{\partial \theta} + \nu_{x\theta} \frac{\partial^2 u_0}{\partial x \partial \theta} \right] +$$

$$+ \frac{1}{R} \left(\frac{5}{6} \right) G_{\theta z}h \left[\beta_\theta - \frac{v_0}{R} + \frac{1}{R} \frac{\partial w}{\partial \theta} \right] = 0 \qquad (7.45)$$

$$\frac{5}{6} G_{xz}h \left[\frac{\partial \beta_x}{\partial x} + \frac{\partial^2 w}{\partial x^2} \right] + \frac{5}{6} \frac{G_{\theta z}h}{R} \left[\frac{\partial \beta_\theta}{\partial \theta} - \frac{1}{R} \frac{\partial v_0}{\partial \theta} + \frac{1}{R} \frac{\partial^2 w}{\partial \theta^2} \right] -$$

$$- \frac{K_\theta}{R} \left[\frac{1}{R} \frac{\partial v_0}{\partial \theta} + \frac{w}{R} + \nu_{x\theta} \frac{\partial u_0}{\partial x} \right] + p(x, \theta) = 0 \qquad (7.46)$$

ANISOTROPIC SHELLS 325

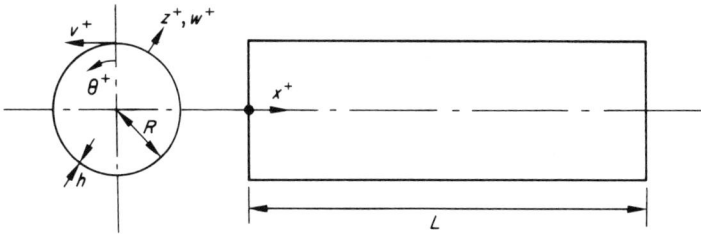

Fig. 7.3.   Shell geometry.

$$D_x \left[\frac{\partial^2 \beta_x}{\partial x^2} + \frac{\nu_{\theta x}}{R} \frac{\partial^2 \beta_\theta}{\partial x \partial \theta} \right] + \frac{G_{x\theta} h^3}{12R} \left[\frac{1}{R} \frac{\partial^2 \beta_x}{\partial \theta^2} + \frac{\partial^2 \beta_\theta}{\partial x \partial \theta} \right] -$$

$$- \frac{5}{6} G_{xz} h \left[\beta_x + \frac{\partial w}{\partial x} \right] = 0 \quad (7.47)$$

$$\frac{G_{x\theta} h^3}{12} \left[\frac{1}{R} \frac{\partial^2 \beta_x}{\partial \theta \partial x} + \frac{\partial^2 \beta_\theta}{\partial x^2} \right] + \frac{D_\theta}{R} \left[\nu_{x\theta} \frac{\partial^2 \beta_x}{\partial x \partial \theta} + \frac{1}{R} \frac{\partial^2 \beta_\theta}{\partial \theta^2} \right] -$$

$$- \frac{5}{6} G_{\theta z} h \left[\beta_\theta - \frac{v_0}{R} + \frac{1}{R} \frac{\partial w}{\partial \theta} \right] = 0 \quad (7.48)$$

The shell to be analysed and its loading is shown in Figs. 7.3 and 7.4.

To analyse the effect of a rectangularly shaped localised loading some distance away from either end of the cylindrical shell, it is quite satisfactory to assume that the ends of the shell are both simply-supported. This is rational because as long as the loading is greater than an axial distance from the edge equal to $4\sqrt{[(E_x/E_\theta)^{1/2} Rh]}$, the length of the bending boundary layer, then all effects of edge shear resultants and stress couples have become negligible. Hence, the solution sought,

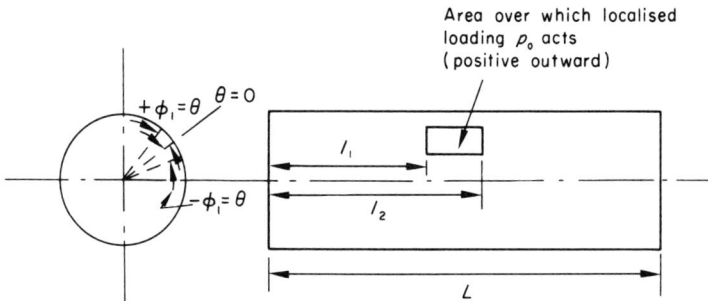

Fig. 7.4.   Details of localised loading.

namely the localised stresses due to the localised loading is in effect independent of the shell boundary conditions. Incidentally, if the shell does not extend the full 360 degrees, as long as the localised loading is further than $4\sqrt{[(E_\theta/E_x)^{1/2}Rh]}$ away from a $\theta =$ constant edge the following solution will also apply.

All displacements and rotations can be placed in terms of a doubly infinite trigonometric series which do satisfy the end boundary conditions of simple support.

$$
\begin{Bmatrix} u_0 \\ v_0 \\ w \\ \beta_\theta \\ \beta_x \\ p \end{Bmatrix} = \sum_{n=0}^{\infty} \sum_{m=1}^{\infty} \begin{Bmatrix} U_{nm} \cos(n\theta)\cos(m\pi x/L) \\ V_{nm} \sin(n\theta)\sin(m\pi x/L) \\ W_{nm} \cos(n\theta)\sin(m\pi x/L) \\ B_{nm} \sin(n\theta)\sin(m\pi x/L) \\ T_{nm} \cos(n\theta)\cos(m\pi x/L) \\ P_{nm} \cos(n\theta)\sin(m\pi x/L) \end{Bmatrix} \tag{7.49}
$$

Substituting (7.49) into eqns. (7.44) to (7.48) results in

$$
\sum_{n=0}^{\infty} \sum_{m=1}^{\infty} \left\{ \left[\!\left[U_{nm}\left[-K_x\left(\frac{m\pi}{L}\right)^2 - \frac{G_{x\theta}hn^2}{R^2} \right] + V_{nm}\left[\left(\frac{n}{R}\right)\!\left(\frac{m\pi}{L}\right)(K_x\nu_{\theta x} + G_{x\theta}h) \right] \right.\right.
$$
$$
\left.\left. + W_{nm}\left[K_x\frac{\nu_{\theta x}}{R}\left(\frac{m\pi}{L}\right) \right] \right]\!\right] \cos(n\theta)\cos(m\pi x/L) \right\} = 0 \quad (7.50)
$$

$$
\sum_{n=0}^{\infty} \sum_{m=1}^{\infty} \left\{ \left[\!\left[U_{nm}\left[\left(\frac{n}{R}\right)\!\left(\frac{m\pi}{L}\right)(G_{x\theta}h + K_\theta\nu_{x\theta}) \right] + \right.\right.\right.
$$
$$
+ V_{nm}\left[-G_{x\theta}h\left(\frac{m\pi}{L}\right)^2 - \frac{K_\theta}{R^2}n^2 - \frac{1}{R^2}\left(\frac{5}{6}\right)G_{\theta z}h \right] +
$$
$$
+ W_{nm}\left[\left(\frac{n}{R^2}\right)\!\left(-K_\theta - \frac{5}{6}1G_{\theta z}h\right) \right] +
$$
$$
\left.\left.+ B_{nm}\left[\frac{1}{R}\left(\frac{5}{6}\right)G_{\theta z}h \right] \right]\!\right] \sin(n\theta)\sin(m\pi x/L) \right\} = 0 \tag{7.51}
$$

$$
\sum_{n=0}^{\infty} \sum_{m=1}^{\infty} \left\{ \left[\!\left[U_{nm}\left[\frac{K_\theta}{R}\nu_{x\theta}\left(\frac{m\pi}{L}\right) \right] + V_{nm}\left[-\frac{5}{6}\frac{G_{\theta z}hn}{R^2} - \frac{K_\theta n}{R^2} \right] + \right.\right.\right.
$$
$$
+ W_{nm}\left[-\frac{5}{6}G_{xz}h\left(\frac{m\pi}{L}\right)^2 - \frac{5}{6}\frac{G_{\theta z}hn^2}{R^2} - \frac{K_\theta}{R^2} \right] + B_{nm}\left[\frac{5}{6}\frac{G_{\theta z}hn}{R} \right] +
$$
$$
\left.\left.+ T_{nm}\left[-\frac{5}{6}G_{xz}h\left(\frac{m\pi}{L}\right) \right] \right]\!\right] \cos(n\theta)\sin(m\pi x/L) \right\}
$$
$$
= -\sum_{n=0}^{\infty} \sum_{m=1}^{\infty} P_{nm}\cos(n\theta)\sin(m\pi x/L) \tag{7.52}
$$

$$\sum_{n=0}^{\infty} \sum_{m=1}^{\infty} \left\{ \left[\!\left[W_{nm} \left[-\frac{5}{6} G_{xz} h \left(\frac{m\pi}{L}\right) \right] + B_{nm} \left[\left(\frac{m\pi}{L}\right)\left(\frac{n}{R}\right)\left(D_x \nu_{\theta x} + \frac{G_{x\theta} h^3}{12}\right) \right] + \right.\right.$$

$$\left.\left. + T_{nm} \left[-D_x \left(\frac{m\pi}{L}\right)^2 - \frac{G_{x\theta} h^3 n^2}{12 R^2} - \frac{5}{6} G_{xz} h \right]\right]\!\right] \cos(n\theta) \cos\left(\frac{m\pi x}{L}\right) \right\} = 0$$

$$(7.53)$$

$$\sum_{n=0}^{\infty} \sum_{m=1}^{\infty} \left\{ \left[\!\left[V_{nm} \left[\frac{5}{6} G_{\theta z} \frac{h}{R} \right] + W_{nm} \left[\frac{5}{6} G_{\theta z} h \left(\frac{n}{R}\right) \right] + \right.\right.$$

$$\left. + B_{nm} \left[-\frac{G_{x\theta} h^3}{12} \left(\frac{m\pi}{L}\right)^2 - \frac{D_\theta n^2}{R^2} - \frac{5}{6} G_{\theta z} h \right] + \right.$$

$$\left.\left. + T_{nm} \left[\left(\frac{m\pi}{L}\right)\left(\frac{n}{R}\right)\left(\frac{G_{x\theta} h^3}{12} + D_\theta \nu_{x\theta}\right) \right]\right]\!\right] \sin(n\theta) \sin(m\pi x / L) \right\} = 0$$

$$(7.54)$$

These are the five governing differential equations to be solved. However, before proceeding it is necessary to explicitly expand the Euler coefficients of the applied localised load. The details are given in Fig. 7.4 where it is seen that the load is in the region $l_1 \leqslant x \leqslant l_2$ and $-\phi_1 \leqslant \theta \leqslant +\phi_1$. If the load intensity $p(x, \theta) = p_0$, a constant, the Euler coefficients for the last of eqn. (7.49) are

$$P_{nm} = 0 \quad \text{for} \quad m = 0$$

$$P_{nm} = \frac{2p_0 \phi_1}{m\pi^2} \left[\cos\left(\frac{m\pi l_1}{L}\right) - \cos\left(\frac{m\pi l_2}{L}\right) \right]$$
$$\text{for } n = 0 \text{ and } m \neq 0 \qquad (7.55)$$

$$P_{nm} = \frac{4p_0}{\pi^2} \left(\frac{1}{nm}\right) \left[\cos\left(\frac{m\pi l_1}{L}\right) - \cos\left(\frac{m\pi l_2}{L}\right) \right] \sin(n\phi_1)$$
$$\text{for } n \neq 0 \text{ and } m \neq 0$$

Finally, eqns. (7.50) to (7.55) can be placed in the following array: (it should be noted that eqn. (7.52) appears first)

$$\begin{bmatrix} A_{11} & A_{12} & A_{13} & A_{14} & A_{15} \\ A_{21} & A_{22} & A_{23} & A_{24} & A_{25} \\ A_{31} & A_{32} & A_{33} & A_{34} & A_{35} \\ A_{41} & A_{42} & A_{43} & A_{44} & A_{45} \\ A_{51} & A_{52} & A_{53} & A_{54} & A_{55} \end{bmatrix} \begin{bmatrix} U_{nm} \\ V_{nm} \\ W_{nm} \\ B_{nm} \\ T_{nm} \end{bmatrix} = \begin{bmatrix} AA_1 \\ AA_2 \\ AA_3 \\ AA_4 \\ AA_5 \end{bmatrix} \qquad (7.56)$$

where

$$A_{11} = \frac{K_\theta}{R} \nu_{x\theta} \left(\frac{m\pi}{L}\right)$$

$$A_{12} = -\frac{n}{R^2} \left(\frac{5}{6} G_{\theta z}h + K_\theta\right)$$

$$A_{13} = -\frac{5}{6} G_{xz}h \left(\frac{m\pi}{L}\right)^2 - \frac{5}{6} G_{\theta z}h \left(\frac{n}{R}\right)^2 - \frac{K_\theta}{R^2}$$

$$A_{14} = \frac{5}{6} G_{\theta z} \frac{hn}{R}$$

$$A_{15} = -\frac{5}{6} G_{xz}h \left(\frac{m\pi}{L}\right)$$

$$A_{21} = -K_x \left(\frac{m\pi}{L}\right)^2 - \frac{G_{x\theta}hn^2}{R^2}$$

$$A_{22} = \frac{n}{R} \left(\frac{m\pi}{L}\right) (K_x\nu_{\theta x} + G_{x\theta}h)$$

$$A_{23} = K_x \frac{\nu_{\theta x}}{R} \left(\frac{m\pi}{L}\right)$$

$$A_{24} = A_{25} = 0$$

$$A_{31} = \left(\frac{n}{R}\right)\left(\frac{m\pi}{L}\right) (G_{x\theta}h + K_\theta\nu_{x\theta})$$

$$A_{32} = -G_{x\theta}h \left(\frac{m\pi}{L}\right)^2 - K_\theta \left(\frac{n}{R}\right)^2 - \frac{5}{6}\frac{1}{R^2}G_{\theta z}h$$

$$A_{33} = \frac{n}{R^2} \left(-K_\theta - \frac{5}{6}1G_{\theta z}h\right)$$

$$A_{34} = \frac{5}{6}\frac{1}{R} G_{\theta z}h$$

$$A_{35} = A_{41} = A_{42} = 0$$

$$A_{43} = -\left(\frac{m\pi}{L}\right)\frac{5}{6} G_{xz}h$$

$$A_{44} = \left(\frac{m\pi}{L}\right)\left(\frac{n}{R}\right)\left(D_x\nu_{\theta x} + \frac{G_{x\theta}h^3}{12}\right)$$

$$A_{45} = -\left(\frac{m\pi}{L}\right)^2 D_x - \frac{G_{x\theta}h^3 n^2}{12R^2} - \frac{5}{6} G_{xz}h$$

$$A_{51} = 0$$

$$A_{52} = \frac{5}{6} G_{\theta z} \frac{h}{R}$$

$$A_{53} = \left(\frac{n}{R}\right)\frac{5}{6}G_{\theta z}h$$

$$A_{54} = G_{\theta x}\frac{h^3}{12}\left(\frac{m\pi}{L}\right)^2 - D_\theta\left(\frac{n}{R}\right)^2 - \frac{5}{6}G_{\theta z}h$$

$$A_{55} = \left(\frac{m\pi}{L}\right)\left(\frac{n}{R}\right)\left(\frac{G_{x\theta}h^3}{12} + D_\theta\nu_{x\theta}\right)$$

$$AA_1 = -\frac{4p_0\phi_1}{m\pi^2}\sin m\pi\left(\frac{l_2+l_1}{2L}\right)\sin m\pi\left(\frac{l_2-l_1}{2L}\right) \quad \text{for } n = 0$$

$$AA_1 = -\frac{8p_0}{\pi^2 mn}\sin(n\phi_1)\sin m\pi\left(\frac{l_2+l_1}{2L}\right)\sin m\pi\left(\frac{l_2-l_1}{2L}\right) \quad \text{for } n \neq 0$$

$$AA_2 = AA_3 = AA_4 = AA_5 = 0$$

To solve eqn. (7.56) it is convenient to develop and utilise a digital computer program. This has been done and is given in detail in Reference 54. Because of the characteristics of the localised load one can predict that many terms of the Fourier series are required to produce a high degree of accuracy. It is interesting to note that 41 terms in the circumferential (θ) portion and 101 terms in the axial (x) portion were used.

Parameters of various orthotropic material properties, various geometries of the localised loading, various sizes of loading, the effects of transverse shear deformation, and the wall thickness-to-radius ratio effects have been investigated. Also, the program was used to duplicate a sufficient number of isotropic cases performed by Bijlaard[61] to confirm that the solution and the computer program are correct.

In the following examples, it is assumed that

When $E_x > E_\theta$, then $G_{zx} = G_{x\theta}$, and $E_\theta/G_{\theta z} = 2.5$

When $E_\theta > E_x$, then $G_{x\theta} = G_{\theta z}$, and $E_x/G_{xz} = 2.5$

The equality of the shear moduli shown above would be true if the material is a unidirectional fibre-reinforced composite. However, the ratio shown for E/G is chosen arbitrarily. In addition, it should be noted that the stress quantities given below are at the centre of the loaded region. Also, in all cases a shell of radius 10 in., and length 100 in. with the loading symmetrically placed midway between the ends was studied. However, one check of shells of larger size confirmed identity of results for shells having the same h/R ratio.

In each of the following Figures, the two numbers associated with each curve are, first, the ratio of E_θ/E_x, while the second is the

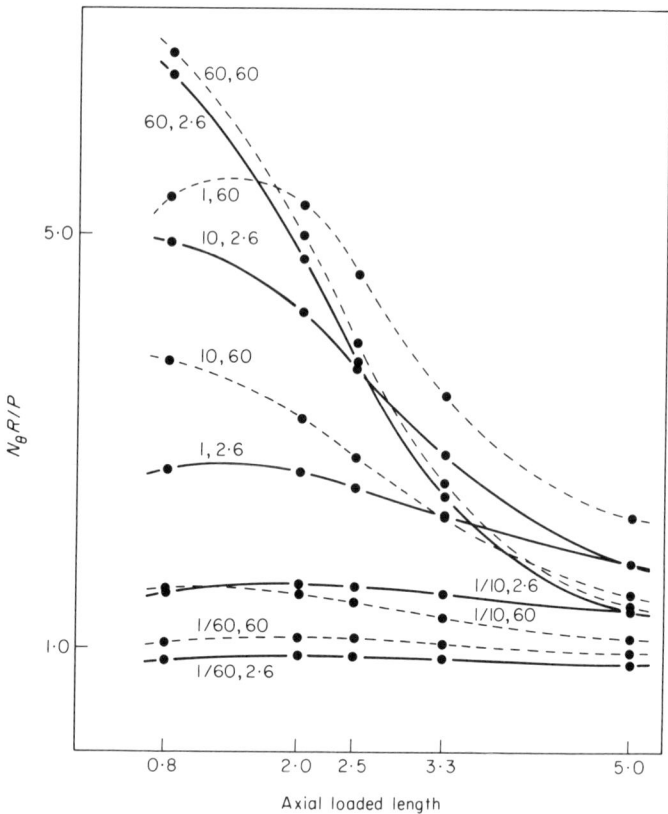

*Fig. 7.5. Dimensionless circumferential stress resultant parameter as a func-
tion of axial loaded length. Loaded area = 4 in.^{2} $h/R = 1/15$. For $E_i > E_j$
$(i, j = x, \theta)$.* ————, $E_i/G_{iz} = 2 \cdot 6$; -----, $E_i/G_{iz} = 60$.

pertinent E/G ratio, namely when $E_\theta > E_x$ it is $E_\theta/G_{\theta z}$, and when
$E_x > E_\theta$ it is F_m/G_{xz}.

  Figures 7.5 to 7.8 investigate the effects of the shape of the loaded
area. In each curve the thicker shell ($h/R = 1/15$) is utilised, and in
each case the area of the loaded surface is 4 in.^{2} The solid curves are
those material systems with high shear resistance ($E/G = 2 \cdot 6$); the
dashed curves are those weak in shear ($E/G = 60$). It is generally seen
that short axial lengths of loading cause high values of N_θ, while
whenever the loaded area is longer in the x direction, the values of N_x

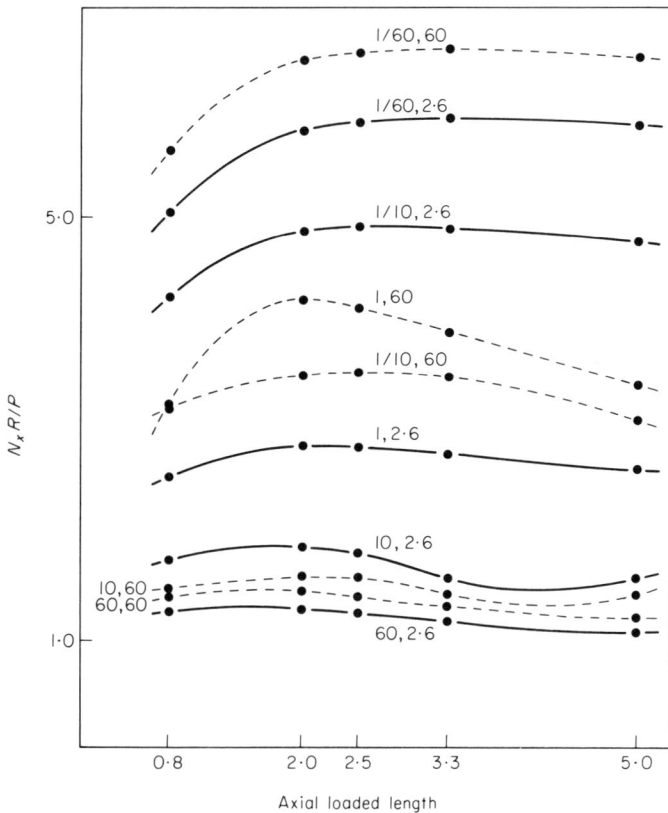

Fig. 7.6. Dimensionless axial stress resultant parameter as a function of axial loaded length. Loaded area = 4 in.² h/R = 1/15. For $E_i > E_j$ (i, j = x, θ). ——, $E_i/G_{iz} = 2·6$; -----, $E_i/G_{iz} = 60$.

are insensitive to the load geometry. It is interesting that the maximum values of the bending moments M_x and M_θ occur for nearly square loadings, regardless of the material system.

In Fig. 7.5 it is seen that for materials with higher E_θ/E_x ratios, N_θ increases. From Fig. 7.6, the reverse is true for N_x, as would be expected. From Figs. 7.7 and 7.8 analogous statements are made about M_θ and M_x.

Upon calculating the stress resultants N_x and N_θ, and the stress couples M_x and M_θ, the stresses in the shell are found by the usual

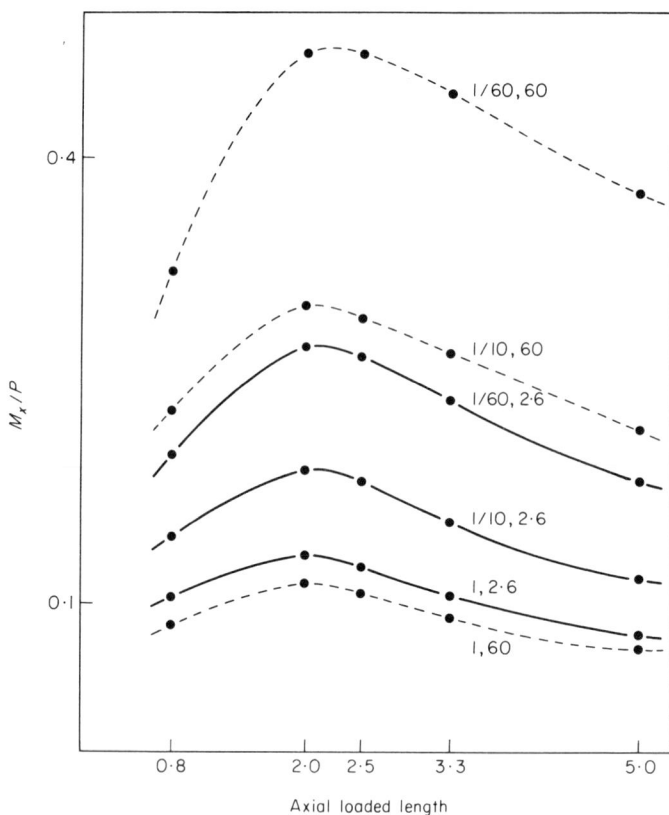

Fig. 7.7   Dimensionless axial stress couple parameter as a function of axial loaded length.   Loaded   area = 4 in.^{2} h/R = 1/15.   For $E_i > E_j$ $(i, j = x, \theta)$.
————, $E_i/G_{iz} = 2.6$; ------, $E_i/G_{iz} = 60$.

relations

$$\sigma_x(x, \theta, z) = \frac{N_x(x, \theta)}{h} + \frac{zM_x(x, \theta)}{h^3/12} \tag{7.57}$$

$$\sigma_\theta(x, \theta, z) = \frac{N_\theta(x, \theta)}{h} + \frac{zM_\theta(x, \theta)}{h^3/12} \tag{7.58}$$

From the total parametric study some of the conclusions are:

1. When $E_x > E_\theta$, $M_x > M_\theta$ generally.
2. The more concentrated the load, the higher the maximum stress couple.

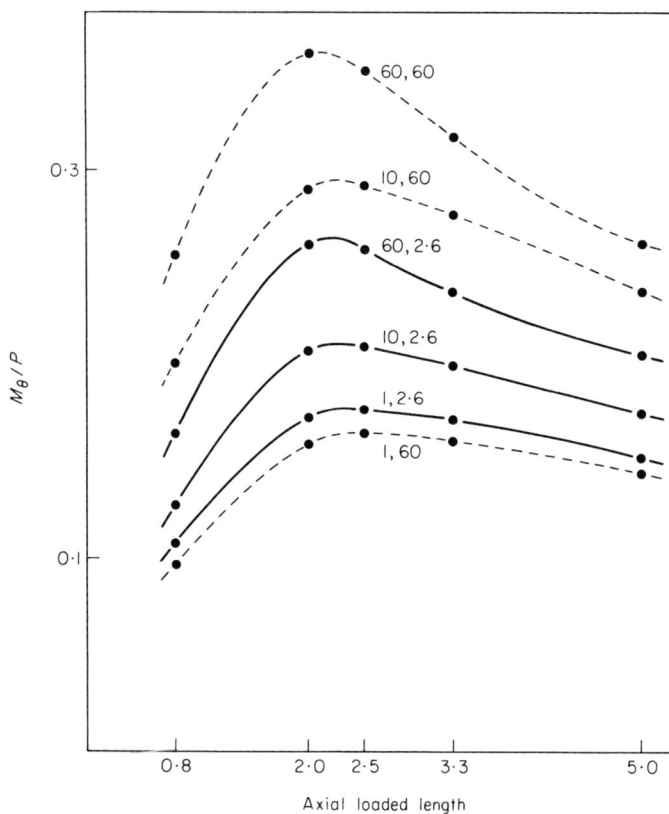

Fig. 7.8. Dimensionless circumferential stress couple parameter as a function of axial loaded length. Loaded area $= 4$ *in.*^{2} $h/R = 1/15$. *For* $E_i > E_j$ $(i, j = x, \theta)$.
————, $E_i/G_{iz} = 2 \cdot 6$; ------, $E_i/G_{iz} = 60$.

3. M_x increases as h/R increases.
4. As ϕ_1 increases N_θ increases.
5. When $(l_1 - l_2) > 2\phi_1$, N_x is insensitive to load geometry.
6. Maximum values of M_x and M_θ occur for loadings nearly square.
7. As E_i/E_j increases, N_i and M_i increase, while N_j and M_j decrease $(i, j = x, \theta)$.
8. Transverse shear deformation effects are significant and cannot be ignored in analysing shells of composite materials subjected to localised loads.

7.5 LAMINATED CYLINDRICAL SHELLS OF PYROLYTIC GRAPHITE TYPE MATERIALS

The solutions of this Section were obtained by Zukas[45,62] and are included herein because they represent the simplest example that can be included that incorporates thermoelastic effects including 'thermal thickening', and a laminated construction, in order to determine rationally the joint stresses.

Transversely isotropic materials are discussed in Chapter 5 and the elasticity matrix is given by eqn. (5.7). The stress–strain relations for an orthotropic material are given by (5.9). These equations will now be modified to conform to the problem of this Section. Consider a cylindrical shell utilising a material system that has identical properties in the x and θ direction, but different properties in the z, or thickness direction. Under those conditions $S_{11} = S_{22}$, $S_{13} = S_{23}$ and $S_{44} = S_{55}$. Inserting specific physical properties, as discussed in Section 5.2, (5.9) is written for the case of the circular cylindrical shell under axially symmetric loads as

$$\epsilon_x = \frac{1}{E}\left[\sigma_x - \nu\sigma_\theta - \nu_c\sigma_z\right]$$

$$\epsilon_\theta = \frac{1}{E}\left[\sigma_\theta - \nu\sigma_x - \nu_c\sigma_z\right]$$

$$\epsilon_z = \frac{\sigma_z}{E_c} - \frac{\nu_c}{E}\left[\sigma_x + \sigma_\theta\right]$$

$$\epsilon_{xz} = \sigma_{xz}/2G_c$$

In the above

$$E_x = E_\theta = E, \qquad G_{x\theta} = E/2(1+\nu), \qquad \nu_{x\theta} = \nu_{\theta x} \equiv \nu,$$

$$\nu_c \equiv \nu_{xz} = \nu_{\theta z}, \qquad G_c \equiv G_{xz} = G_{\theta z}, \qquad E_c \equiv E_z$$

Hence, there are five material constants involving E, E_c, G_c, ν and ν_c, and the following relationship still holds: (see eqn. (5.21)).

$$E/\nu_c = E_c/\nu_{zx}$$

If the cylindrical shell is subjected to changes in temperature, the above stress–strain relations must be modified to include 'thermal strains' as well as the isothermal strains shown above. These thermal strains at any material point are the product of the coefficient of thermal expansion of the material and the *difference* between the temperature of the material point at that time and the temperature at which the elastic body is considered to be stress-free when not loaded. This is discussed in great detail in Chapter 3 of Reference 63 as well as many other texts. For the transversely isotropic material considered here there is one coefficient of thermal expansion in the x-θ plane, labelled α, and another in the thickness direction, called α_c. (In an orthotropic material, there can be three differing coefficients of thermal expansion.) Thus, the thermoelastic stress–strain relations are as follows when T is the temperature difference between the temperature of the material point (x, θ, z) and the stress-free temperature

$$\epsilon_x = \frac{1}{E}[\sigma_x - \nu\sigma_\theta - \nu_c\sigma_z] + \alpha T \tag{7.59}$$

$$\epsilon_\theta = \frac{1}{E}[\sigma_\theta - \nu\sigma_x - \nu_c\sigma_z] + \alpha T \tag{7.60}$$

$$\epsilon_z = \frac{\sigma_z}{E_c} - \frac{\nu_c}{E}(\sigma_x + \sigma_\theta) + \alpha_c T \tag{7.61}$$

$$\epsilon_{xz} = \sigma_{xz}/2G_c \tag{7.62}$$

Since thermal strains are only dilitational in nature, the stress–strain shear relations (7.62) are not affected.

For the problem considered herein, consider the cylindrical shell to be sufficiently thin that Love's First approximation can be employed, that is to say $h/R \ll 1$. The transverse normal stress σ_z is small compared to other stress components, and can be ignored in the stress–strain relations. A lineal element normal to the undeformed middle surface undergoes at most a translation and a rotation, hence,

$$u(x, z) = u_0(x) + z\beta(x)$$

All of these assumptions are 'classical' as discussed in the derivations of the governing equations of Chapter 4, and earlier in this chapter.

However, it has been found that in pyrolytic graphite type materials[40] and also in some composite materials[2] the thermal expansion coefficient in the thickness direction can be an order of magnitude or more greater than that in the plane of the shell midsurface. Hence, a relatively new name appears in the analysis of plates and shells of pyrolytic graphite-type materials and composite materials: 'thermal thickening'. Mathematically, the lateral deformation must be of the form

$$w(x, z) = w(x) + \bar{w}(x, z) \tag{7.63}$$

where the first term on the right-hand side is the classical assumption, and

$$\bar{w}(x, z) = \int_0^z \alpha_c T \, dz \qquad (7.64)$$

With these assumptions, the stress–strain relations (7.59) to (7.62) become

$$\epsilon_x = \frac{1}{E} [\sigma_x - \nu\sigma_\theta] + \alpha T \qquad (7.65)$$

$$\epsilon_\theta = \frac{1}{E} [\sigma_\theta - \nu\sigma_x] + \alpha T \qquad (7.66)$$

$$\epsilon_z = \alpha_c T \qquad (7.67)$$

$$\epsilon_{xz} = \sigma_{xz}/2G_c \qquad (7.68)$$

The strain displacement relations for the circular cylindrical shell are analogous to those derived previously, but $\bar{w}$ of eqns. (7.63) and (7.64) must be included:

$$\epsilon_x = \frac{\partial u}{\partial x} = \frac{\partial u_0(x)}{\partial x} + z \frac{\partial \beta(x)}{\partial x} \qquad (7.69)$$

$$\epsilon_\theta = \frac{1}{R}(w + \bar{w}) \qquad (7.70)$$

$$\epsilon_z = \frac{\partial \bar{w}}{\partial z} \qquad (7.71)$$

$$\epsilon_{xz} = \frac{1}{2}\left(\beta + \frac{\partial w}{\partial x}\right) \qquad (7.72)$$

One now defines stress resultants N_i, stress couples, M_i, and shear resultants, Q_i, in the usual way for the ith lamina of a circular cylindrical shell, of thickness h_i

$$\begin{Bmatrix} N_{x_i} \\ N_{\theta_i} \\ Q_i \end{Bmatrix} = \int_{-h_i/2}^{+h_i/2} \begin{Bmatrix} \sigma_{x_i} \\ \sigma_{\theta_i} \\ \sigma_{xz_i} \end{Bmatrix} dz_i \qquad (7.73)$$

$$\begin{Bmatrix} M_{x_i} \\ M_{\theta_i} \end{Bmatrix} = \int_{-h_i/2}^{+h_i/2} \begin{Bmatrix} \sigma_{x_i} \\ \sigma_{\theta_i} \end{Bmatrix} z_i \, dz_i \qquad (7.74)$$

We also define thermal stress resultants N_i^* and thermal stress couples M_i^* as

$$\begin{Bmatrix} N_{x_i}^* \\ N_{\theta_i}^* \end{Bmatrix} = \int_{-h_i/2}^{h_i/2} E_i \alpha_i T \, dz_i \qquad (7.75)$$

$$\begin{Bmatrix} M_{x_i}^* \\ M_{\theta_i}^* \end{Bmatrix} = \int_{-h_i/2}^{h_i/2} E_i \alpha_i T z_i \, dz_i \qquad (7.76)$$

Retaining the surface shear and surface normal stress terms, as in eqns. (4.25) to (4.29), the integrated shell equilibrium equations for the shell, (4.125) through (4.127), become for the ith lamina

$$\frac{dN_{x_i}}{dx} + \tau_{1i} - \tau_{2i} = 0 \qquad (7.77)$$

$$\frac{dM_{x_i}}{dx} - Q_i + \frac{h_i}{2}(\tau_{1i} + \tau_{2i}) = 0 \qquad (7.78)$$

$$\frac{dQ_i}{dx} - \frac{N_{\theta_i}}{R} + p_{1i} - p_{2i} = 0 \qquad (7.79)$$

In the above $\tau_{1i} = \sigma_{xz}(+h_i/2)$, $\tau_{2i} = \sigma_{xz}(-h_i/2)$, $p_i = \sigma_z(+h_i/2)$ and $p_2 = \sigma_z(-h_i/2)$.

Substituting (7.69) and (7.70) into (7.65) and (7.66), multiplying each by dz_i and integrating the equations across the lamina thickness results in eqns. (7.80) and (7.81). Then multiplying them by $z_i \, dz_i$ and integrating across the lamina thickness provides (7.82) and (7.83). In both cases the definitions (7.73) to (7.76) are made use of.

$$N_{x_i} = \frac{E_i h_i}{(1 - \nu_i^2)}\left(\frac{\partial u_{0_i}}{\partial x} + \frac{\nu_i w_i}{R_i}\right) - \frac{N_{x_i}^*}{(1 - \nu_i)} + \frac{E_i \nu_i}{R_i(1 - \nu_i^2)}\int_{-h_i/2}^{h_i/2} \bar{w}_i \, dz_i \qquad (7.80)$$

$$N_{\theta_i} = \frac{E_i h_i}{(1 - \nu_i^2)}\left(\nu_i\frac{\partial u_{0_i}}{\partial x} + \frac{w_i}{R}\right) - \frac{N_{\theta_i}^*}{(1 - \nu_i)} + \frac{E_i}{R_i(1 - \nu_i^2)}\int_{-h_i/2}^{h_i/2} \bar{w}_i \, dz_i \qquad (7.81)$$

$$M_{x_i} = \frac{E_i h_i^3}{12(1 - \nu_i^2)}\frac{\partial \beta_i}{\partial x} - \frac{M_{x_i}^*}{(1 - \nu_i)} + \frac{E_i \nu_i}{R_i(1 - \nu_i^2)}\int_{-h_i/2}^{h_i/2} z\bar{w}_i \, dz_i \qquad (7.82)$$

$$M_{\theta_i} = \frac{E_i h_i^3}{12(1 - \nu_i^2)}\frac{\partial \beta_x}{\partial x} - \frac{M_{\theta_i}^*}{(1 - \nu_i)} + \frac{E_i}{R_i(1 - \nu_i^2)}\int_{-h_i/2}^{h_i/2} z\bar{w}_i \, dz_i \qquad (7.83)$$

Following the procedure of McDonough[40] a transverse shear resultant expression is derived using a weighting procedure. The result is:

$$Q_i = \frac{m_i}{6} + \frac{5}{6} h_i G_{c_i}\left(\beta_i + \frac{\partial w_i}{\partial x}\right) + \frac{5}{4}\int_{-h_i/2}^{h_i/2}\left[1 - \left(\frac{z_i}{h_i/2}\right)^2\right] dz_i \qquad (7.84)$$

wherein

$$m_i = \frac{h_i}{2}(\tau_{1i} + \tau_{2i})$$

Considering now a two-layered shell wherein the outer lamina is labelled a and the inner lamina labelled b, the unknown joint shear stress and unknown joint normal stress are defined as

$$\tau_j \equiv \sigma_{zx}(-h_a/2) = \sigma_{zx}(+h_b/2) \qquad (7.85)$$

$$p_j \equiv \sigma_z(-h_a/2) = \sigma_z(+h_b/2) \qquad (7.86)$$

Before proceeding to obtain the governing equations, the joint boundary conditions are specified, namely that no slippage and no separation are allowed. The first condition can be written as

$$u_a(-h_a/2) = u_b(+h_b/2)$$

while the latter is

$$w_a(-h_a/2) = w_b(+h_b/2)$$

Substituting the expressions for the displacements into the above, one obtains

$$w_a = w_b + \bar{w}_b(x, h_b/2) - \bar{w}_a(x, -h_a/2) \qquad (7.87)$$

$$u_{0_a} = u_{0_b} + \left\{ \frac{h_a}{2}\beta_a + \frac{h_b}{2}\beta_b \right\} \qquad (7.88)$$

Other useful expressions are the extensional stiffness K_i and the flexural stiffness D_i of the lamina about its own midsurface

$$K_i = \frac{E_i h_i}{(1 - \nu_i^2)}, \qquad D_i = \frac{E_i h_i^3}{12(1 - \nu_i^2)} \qquad (i = a, b)$$

Solving for Q_i from (7.78) and substituting its first derivative into (7.79) and making use of (7.81) and (7.82) results in the following two equations, in which $D \equiv \partial(\;)/\partial x$, and use of (7.87) and (7.88) is made

$$D_a D^3 \beta_a - \frac{K_a \nu_a h_a}{2R} D\beta_a - \frac{K_a \nu_a h_b}{2R} D\beta_b + \frac{h_a}{2} D\tau_j -$$

$$- p_j - \frac{K_a \nu_a}{R} Du_{0_b} - \frac{K_a}{R^2} w_b = \alpha_{17} + \pi_1 \qquad (7.89)$$

$$D_b D^3 \beta_b + \frac{h_b}{2} D\tau_j + p_j - \frac{K_b \nu_b}{R} Du_{0_b} - \frac{K_b w_b}{R^2} = \alpha_{27} + \pi_2 \qquad (7.90)$$

where

$$\alpha_{17} = \frac{N_{\theta_a}^*}{R(1 - \nu_a)} - \frac{K_a \hat{E}}{R^2} - \frac{E_a}{R^2(1 - \nu_a^2)} \int_{-h_a/2}^{h_a/2} \bar{w}_a \, dz_a +$$

$$+ D^2 \left\{ \frac{M_{x_a}^*}{(1 - \nu_a)} - \frac{E_a \nu_a}{R(1 - \nu_a^2)} \int_{-h_a/2}^{h_a/2} z_a \bar{w}_a \, dz_a \right.$$

$$\alpha_{27} = \frac{N_{\theta_b}^*}{R(1 - \nu_b)} - \frac{E_b}{R^2(1 - \nu_b^2)} \int_{-h_b/2}^{+h_b/2} \bar{w}_b \, dz_b +$$

$$+ D^2 \left\{ \frac{M_{x_b}}{(1 - \nu_b)} - \frac{E_b \nu_b}{R(1 - \nu_b^2)} \int_{-h_b/2}^{h_b/2} z_b \bar{w}_b \, dz_b \right.$$

$$\pi_1 = -p_{ia} - \frac{h_a}{2} D\tau_{1a}$$

$$\pi_2 = p_{2b} - \frac{h_b}{2} D\tau_{2b}$$

$$\hat{E} = \bar{w}_b(x, h_b/2) - \bar{w}_a(x, -h_a/2)$$

Substituting eqns. (7.80) to (7.88) into (7.77), the first equilibrium equation, results in two more useful equations

$$\frac{K_a h_a}{2} D^2 \beta_a + \frac{K_a h_b}{2} D^2 \beta_b - \tau_j + K_a D^2 u_{0_b} + \frac{K_a \nu_a}{R} Dw_b = \alpha_{37} - \tau_{1a} \quad (7.91)$$

$$\tau_j + K_b D^2 u_{0_b} + \frac{K_b \nu_b}{R} Dw_b = \alpha_{47} + \tau_{2b} \quad (7.92)$$

where

$$\alpha_{37} = D \left\{ \frac{N_{x_a}^*}{(1 - \nu_a)} - \frac{K_a \nu_a \hat{E}}{R} - \frac{E_a \nu_a}{R(1 - \nu_a^2)} \int_{-h_a/2}^{h_a/2} \bar{w}_b \, dz_a \right\}$$

and

$$\alpha_{47} = D \left\{ \frac{N_{x_b}^*}{(1 - \nu_b)} - \frac{E_b \nu_b}{R(1 - \nu_b^2)} \int_{-h_b/2}^{h_b/2} \bar{w}_a \, dz_b \right\}$$

Finally, manipulating (7.84), (7.78), (7.87) and (7.88) two final expressions are obtained

$$D_a D^2 \beta_a - \tfrac{5}{6} G_{c_a} h_a \beta_a + \tfrac{5}{12} h_a \tau_j - \tfrac{5}{6} G_{c_a} h_a Dw_b = \alpha_{57} + \pi_5 \quad (7.93)$$

$$D_b D^2 \beta_b - \tfrac{5}{6} G_{c_b} h_b \beta_b + \tfrac{5}{12} h_b \tau_j - \tfrac{5}{6} G_{c_b} h_b Dw_b = \alpha_{67} + \pi_6 \quad (7.94)$$

where

$$\alpha_{57} = \Delta_a + D \left\{ \hat{E} + \frac{M_{x_a}^*}{(1 - \nu_a)} - \frac{E_a \nu_a}{R(1 - \nu_a^2)} \int_{-h_a/2}^{h_a/2} z_a \bar{w}_a \, dz_a \right\}$$

$$\alpha_{67} = \Delta_b + D \left\{ \frac{M_{x_b}}{(1 - \nu_b)} - \frac{E_b \nu_b}{R(1 - \nu_b^2)} \int_{-h_b/2}^{+h_b/2} z_b \bar{w}_b \, dz_b \right\}$$

$$\pi_5 = -\tfrac{5}{12} h_a \tau_{1a} \qquad \pi_6 = -\tfrac{5}{12} h_b \tau_{2b}$$

and

$$\Delta_i = \frac{5}{4} \int_{-h_i/2}^{+h_i/2} \left[1 - \left(\frac{z_i}{h_i/2}\right)^2 \right] G_{c_i} \frac{d\bar{w}_i}{dx} \, dz_i \qquad (i = a, b)$$

These six governing equations describe the problem to be solved. To continue the solution in explicit detail requires far too much space. What is important to describe here is the kind of governing equations that result, the techniques of solution used, and the form of the resulting solution. Anyone interested in more detail can consult either Reference 45 or 62. Using Zukas' terminology, but not defining the lengthy constants, the governing equations (7.89) to (7.94) can be written as

$$(g_1D^7 + g_2D^5 + g_3D^3 + g_7D)w_b = L_1(x) \qquad (7.95)$$

$$(b_{11}D^4 + b_{12}D^2 + b_{13})\beta_b = L_{II}(x) - (b_{14}D^3 + b_{15}D)w_b \qquad (7.96)$$

$$a_{23}D^2\beta_a = L_{III}(x) - (a_{21}D^2 + a_{22})\beta_b - a_{24}Dw_b \qquad (7.97)$$

$$D^2u_{0_b} = L_{IV}(x) - k_1D^2\beta_a - k_2D^2\beta_b - k_3Dw_b \qquad (7.98)$$

$$\tau_j = L_V(x) - k_4D^2u_{0_b} - k_5Dw_b \qquad (7.99)$$

$$p_j = L_{VI}(x) - k_6D^3\beta_b - k_7D\tau_j + k_8Du_{0_b} + k_9w_b \qquad (7.100)$$

In the above $D \equiv \partial(\ )/\partial x$, $L_i(x)$ are known functions, the g_i, k_i, b_{ij}, and a_{ij} are all known constants.

Therefore (7.95) can be solved independent of the other equations to obtain w_b. Then (7.96) can be solved to obtain β_b; and (7.97) can be solved to obtain β_a; then (7.98) provides u_{0_b}. Knowing all the above, τ_j and p_j, the joint shear and normal stresses are provided explicitly and exactly by (7.99) and (7.100).

Looking at (7.95), the homogeneous solution is found by assuming a solution of the form $w_b = e^{sx}$, and defining $y = s^2$, it can be written as

$$y^3 + \frac{g_2}{g_1}y^2 + \frac{g_3}{g_1}y + \frac{g_4}{g_1} = 0 \qquad (7.101)$$

where the g_i are constants.

There are three forms of solution for (7.101) depending on the values of the constants g_i:

I. two conjugate imaginary roots and one real root,
II. three real and unequal roots,
III. three real roots of which at least two are equal.

Case I type solution can be written as

$$w_{b_H} = A_1e^{s_1x} + A_2e^{-s_1x} + e^{s_2x}(A_3\cos s_3x + A_4\sin s_3x) +$$
$$+ e^{-s_2x}(A_5\cos s_3x + A_6\sin s_3x) \qquad (7.102)$$

Here A_i are the six boundary value constants and s_i are the three roots of (7.101).

The Case II solution can be of several forms depending on whether the roots are positive or negative. The forms below are for the case of one, two, and three positive real roots respectively

$$w_{b_H} = A_1 e^{s_4 x} + A_2 e^{-s_4 x} + A_3 \cos s_5 x + A_4 \sin s_5 x +$$
$$+ A_5 \cos s_6 x + A_6 \sin s_6 x \qquad (7.102a)$$
$$w_{b_H} = A_1 e^{s_4 x} + A_2 e^{-s_4 x} + A_3 e^{s_5 x} + A_4 e^{-s_5 x} +$$
$$+ A_5 \cos s_6 x + A_6 \sin s_6 x \qquad (7.102b)$$
$$w_{b_H} = A_1 e^{s_4 x} + A_2 e^{-s_4 x} + A_3 e^{s_5 x} + A_4 e^{-s_5 x} +$$
$$+ A_5 e^{s_6 x} + A_6 e^{-s_6 x} \qquad (7.102c)$$

Case III represents the degenerate forms of (7.102) wherein two of the roots are equal.

$$w_{b_H} = A_1 e^{s_4 x} + A_2 e^{-s_4 x} + (A_3 + A_5 x) \cos s_5 x +$$
$$+ (A_4 + A_6 x) \sin s_5 x \qquad (7.103a)$$
$$w_{b_H} = (A_1 + A_3 x) e^{s_4 x} + (A_2 + A_4 x) e^{-s_4 x} + A_5 \cos s_6 x +$$
$$+ A_6 \sin s_6 x \qquad (7.103b)$$
$$w_{b_H} = (A_1 + A_5 x) e^{s_4 x} - (A_2 + A_4 x) e^{-s_4 x} + A_5 e^{s_6 x} + A_6 e^{-s_6 x} \qquad (7.103c)$$

In solutions for single-layered shells the s_i roots can be written explicitly in terms of material properties and geometry. In the case of the two-layer, transversely isotropic laminae problem the expressions for the g_i are very long, and obtaining explicit expressions for the roots is very involved. It should be noted that only in Case I do terms involving products of exponential functions and trigonometric functions appear, which of course are the types of terms in the solution of the single-layer isotropic shell (see eqn. (4.149)). In that case both exponential and trigonometric functions involve a single root ϵ, while herein the roots differ, that is to say $s_2 \neq s_3$.

To continue, eqn. (7.98) can be integrated to determine u_{0_b}. In doing this one sees that two additional constants of integration, say A_7 and A_8, are introduced. Thus we have eight boundary value constants, four at each edge, which are sufficient. Hence, solutions for β_a and β_b from equations (7.96) and (7.97) need only be particular solutions, since any constants derived from homogeneous solutions would subsequently be set equal to zero in solving any problem.

Analogous to eqns. (4.119) and (4.120) the natural boundary conditions at the ends of a single-layer circular cylindrical shell under axially symmetric loading conditions, and including transverse shear deformation are

at $\qquad x = 0 \quad \text{and} \quad x = L$

either

$$N_x = K\left[\frac{\mathrm{d}u_0}{\mathrm{d}x} + \frac{vw}{R}\right] = 0 \quad \text{or } u_0 \text{ is prescribed,}$$

and either

$$Q_{x^2} = D\frac{\mathrm{d}^3 w}{\mathrm{d}x^3} = 0 \quad \text{or } w \text{ is prescribed,}$$

and either

$$M_x = -D\frac{\mathrm{d}^3 w}{\mathrm{d}x^3} = 0$$

or β is prescribed.

However, for the laminated shell the stress quantities above need to be modified as follows

$$N_{\mathrm{TOT}} = N_a + N_b$$

$$Q_{\mathrm{TOT}} = Q_a + Q_b$$

$$M_{\mathrm{TOT}} = M_a + M_b + \frac{h_a + h_b}{2}N_a$$

Zukas solved the problem of a two-layer cylindrical shell, free at each end, subjected to a temperature change of $T = -1000\,°\mathrm{F}$, that is to say cooling from $3000\,°\mathrm{F}$ to $2000\,°\mathrm{F}$. The outer layer is pyrolytic graphite and the inner layer is ATJ graphite. The problem is important because the customary procedure in manufacturing pyrolytic graphite is to deposit it on an ATJ graphite mandrel. The properties of these two materials is given in Table 7.1.

Sample results are shown in Figs. 7.9 and 7.10, for the joint shear stress and joint normal stress as one varies with thickness of the ATJ graphite mandrel. First one sees that both stresses increase rapidly at the ends of the shell, and are virtually zero away from the edges (beyond the bending boundary layer). In these Figures $h_a = 0\cdot50$ in. $(1\cdot27\,\mathrm{cm})$, $E/G_c = 20$ and $5 \geqslant h_a/h_b \geqslant 1$. The curves indicate that a high h_a/h_b ratio is desirable, that is to say a thin mandrel, in order to keep joint shear and normal stresses at a minimum.

In order to determine some characteristics of the occurrence of the

TABLE 7.1

AVERAGE MATERIAL PROPERTIES FOR PG AND ATJ GRAPHITE
BETWEEN 3000°F AND 2000°F

	PG	ATJ graphite
E	$3 \cdot 1 \times 10^6$ psi $= 2 \cdot 18 \times 10^5$ kg/cm^2	$2 \cdot 26 \times 10^6$ psi $= 1 \cdot 59 \times 10^5$ kg/cm^2
ν	$-0 \cdot 21$	$+0 \cdot 30$
ν_c	$+0 \cdot 90$	$+0 \cdot 25$
R	varies with case	30 in. $= 76 \cdot 2$ cm
L	40 in. $= 101 \cdot 6$ cm	40 in. $= 101 \cdot 6$ cm

various forms of the solutions (7.102) to (7.103), Zukas investigated solutions in the following range of variables, for the PG–ATJ graphite cylinder

$$1000 \geqslant h_a/h_b \geqslant 0 \cdot 001$$

$$400 \geqslant L/h \geqslant 26$$

$$0 \cdot 05 \geqslant h/R \geqslant 0 \cdot 0033$$

$$E/G_c \text{ (pyrolytic graphite)} = 50, 20 \text{ and } 2 \cdot 6$$

The results are that the Case I solution (7.102), occurs whenever $h_a/h_b \geqslant 1$ regardless of the E/G_c ratio. The roots s_1, s_2 and s_3 are all affected by the E/G_c ratio, and s_1 is far more sensitive to changes in

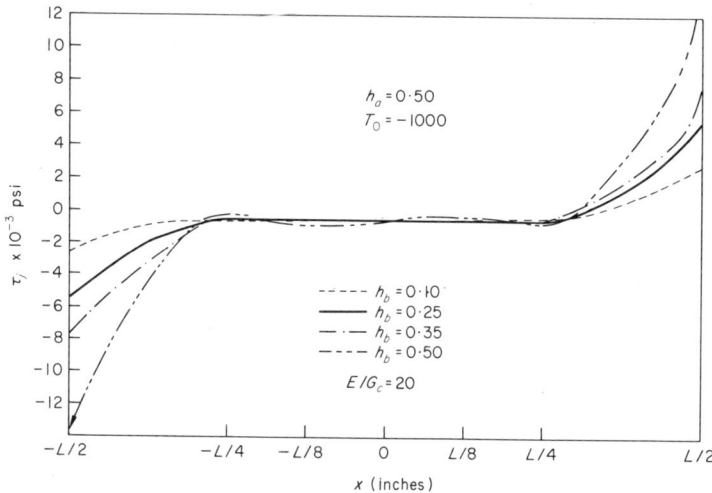

Fig. 7.9. *Behaviour of joint shear stress with varying mandrel thickness.*

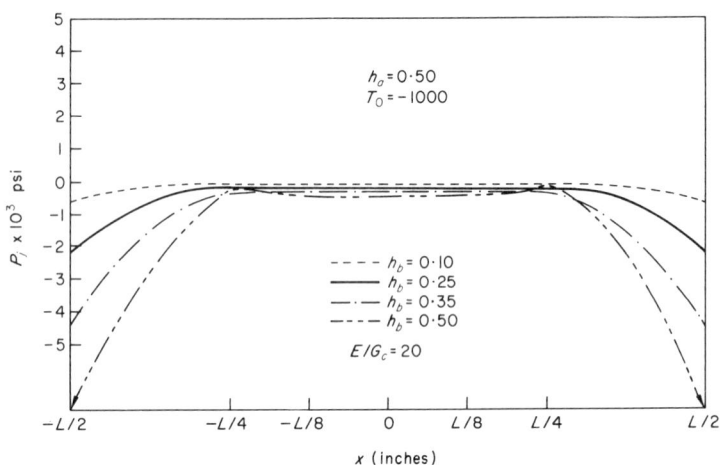

Fig. 7.10.   Behaviour of joint normal stresses with varying mandrel thickness.

this ratio than are s_2 or s_3. Furthermore, the first two terms of eqn. (7.102) are of primary importance at the edges only. It was observed that s_1 becomes so large that A_1 and A_2 become very small in satisfying the boundary conditions, and as a first approximation these first two terms can be omitted in the solution, leaving the remaining terms which have the same form as the classical shell solutions if $s_2 = s_3$.

This Section has demonstrated a technique of solution and a formulation of a problem wherein joint stresses are retained as unknown dependent variables that should be and could be carried out for various layered shells of composite materials. In 1974, Waltz[66] obtained solutions for interlaminar shear and normal stress, in laminated cylindrical shells of composite materials of arbitrary orientation, using this same technique.

More recently Zukas[64] has developed a theory for the layer deflection, non-linear response of laminated orthotropic shells of revolution. The equations of motion were derived using Hamilton's Principle (see Section 6.3.2) and include effects of transverse shear deformation, transverse normal stress (both isothermal and thermoelastic), rotatory inertia and finite deflections. Again he employed the theory to study thermal stress effects in cylindrical shells of pyrolytic graphite. Through extensive parametric studies, Zukas found that transverse shear effects and thermal expansion through the thickness significantly affect stresses even for shells considered geometrically thin in the sense of Love's First

Approximation. However, including isothermal normal stresses and non-linear effects only marginally affect the stresses calculated, but complicate the computation process considerably.

7.6 PROBLEMS

7.1. Write (7.23) explicitly for a three-layer, cross-ply shell $(0°, 90°, 0°)$ where each layer has the same thickness.

7.2. Write eqn. (7.23) explicitly for a two-layer, angle-ply construction $(+45°, -45°)$ where each layer has the same thickness.

7.3. For a cylindrical shell of $h/R = 1/15$, $E_i/G_{iz} = 60$, what is the maximum axial and circumferential stress, σ_x and σ_θ when the shell is subjected to a unit load of 1 lb. $(P = 1)$ over a 4 in.^{2} loaded area (knowing the stresses for a 1 lb. load provides an easy set of numbers with which to determine the maximum load that can be handled for a given allowable stress in the material).

7.4. Looking at eqn. (7.49), for a specified constant pressure p_0, derive the Euler coefficients in the Fourier series representation of the load given by (7.55).

7.5. Derive the right hand side of eqn. (7.23) for the shell discussed if it is subjected to static loads, both normal and shear stresses, on its outer and inner surfaces, i.e. $\partial^2(\ )/\partial t^2 = 0$, but p, q_x, q_θ, m_x and m_θ are not equal to zero.

REFERENCES

1. Wu, Cheng, Ih, 'On Vibrations of Laminated Anisotropic Plates and Shells', Ph.D. Dissertation, University of Delaware, June, 1971.
2. Daugherty, R. L., 'Stresses and Displacements in Shells of Revolution of Composite Materials', Ph.D. Dissertation, University of Delaware, 1971.
3. Shtaerman, I., 'On the Theory of Symmetrical Deformation of Anisotropic Elastic Shells', *Izv. Kievsk. Polit. I. Selkhoz. Institute* (1924).
4. Chakravorty, J., 'Vibrations of Spherically Aeolotropic Shell', *Bulletin of the Calcutta Mathematical Society*, **47**, No. 4 (1955), p. 235.
5. Bert, C. and Egle, D., 'Dynamics of Composite, Sandwich, and Stiffened Shell Type Structures', *J. Spacecraft and Rockets*, **6**, No. 12 (1969), p. 1345.
6. Ambartsumyan, S., 'Contributions to the Theory of Anisotropic Layered Shells', *Applied Mechanics Reviews*, **15**, No. 4 (1962), p. 245, also a revised version in *Applied Mechanics Surveys*, Spartan Books, Washington, D.C. (1966).
7. Ambartsumyan, S., 'The Calculation of Laminated Anisotropic Shells', (in Russian), *Izv. Akad. Nau. Arm. SSR, Seriga Fiziko—Mathematischeskikh I. Nauk*, **6**, No. 3 (1953), p. 15.
8. Ambartsumyan, S., 'Theory of Anisotropic Shells', NASA TTF–118, May 1964.
9. Dong, S., Pister, K., and Taylor, R., 'On the Theory of Laminated Anisotropic Shells and Plates', *J. Aerospace Sciences*, **29**, 969 (1962).

10. Becker, H. and Gerard, G., 'Elastic Stability of Orthotropic Shells', *J. Inst. Aero. Sci.*, **29**, 505 (1962).

11. Cheng, S. and Kuenzi, E., 'Buckling of Orthotropic or Plywood Cylindrical Shells Under External Pressure', *Proc. 5th Int. Symp. of Space Technology and Science*, Tokyo (1963).

12. Hess, T., 'Stability of Orthotropic Cylindrical Shells Under Combined Loading', *ARS. J.*, **31**, 237 (1961).

13. Hedgepeth, J. and Hall, D., 'Stability of Stiffened Cylinders', *AIAA J.*, **3**, 2275 (1965).

14. Thielemann, W., Schnell, W., and Fischer, G., 'Beul und Nachbeulverhalten Orthotroper Kreiszylinderschalen Unter Axial Und Innendruck', *Zeitschrift Für Flugwissenschaften*, **8**, 284 (1960).

15. Tasi, J., Feldman, A., and Strang, D., 'The Buckling Strength of Filament-Wound Cylinders Under Axial Compression', NASA CR–266, July 1965.

16. Tasi, J., 'Effect of Heterogeneity on the Stability of Composite Shells Under Axial Compression', *AIAA J.*, **4**, 1058 (1966).

17. Cheng, S. and Ho, B., 'Stability of Heterogeneous Aeolotropic Cylindrical Shells Under Combined Loading', *AIAA J.*, **1**, 1 (1963).

18. Ho, B. and Cheng, S., 'Some Problems in Stability of Heterogeneous Aeolotropic Cylindrical Shells Under Combined Loading', *AIAA J.*, **1**, 1603 (1963).

19. Dong, R. and Dong, S., 'Analysis of Slightly Anisotropic Shells', *AIAA J.*, **1**, 2565 (1963).

20. Vinson, J. and Brull, M., 'New Techniques of Solution for Problems in the Theory of Orthotropic Plates', *Proceedings 4th U.S. National Congress of Applied Mechanics* (1962).

21. Kingsbury, H., 'Stresses and Deformations in Anisotropic Shells', Ph.D. Dissertation, University of Pennsylvania, 1965.

22. Bert, C., 'Structural Theory for Laminated Anisotropic Elastic Shells', *J. Comp. Mater.*, **1**, No. 4 (1967), p. 414.

23. Kunukkasseril, V., 'Vibration of Multi-layered Anisotropic Cylindrical Shells', Report WVT-6717, Feb. 1967 (AD-649662), Watervliet Arsenal, Watervliet, N.Y.

24. Franklin, H. G., 'Membrane Solution of Fiber-Reinforced Corrugated Shells of Revolution', *J. Comp. Mater.*', **1**, 382–8 (1967).

25. Bessarabov, Y. D. and Rudis, M. A., 'On the Symmetrical Deformation of an Orthotropic Toroidal Shell', *Proceedings of the 4th All-Union Conference on Shells and Plates*, (1962), pp. 207–15.

26. Kraus, H., *Thin Elastic Shells*, John Wiley and Sons, Inc. (1967).

27. Bert, C., Baker, V., and Egle, D., 'Free Vibrations of Multilayer Anisotropic Cylindrical Shells', *J. Comp. Mater.*, **3**, No. 3 (1969), p. 480.

28. Dong, S., 'Free Vibrations of Laminated Orthotropic Cylindrical Shells', *J. Acoustical Soc. Amer.*, **44**, No. 6 (1968), p. 1628.

29. Reuter, R., Jr, 'Shear-Coupled Waves in Thin Helically Wrapped Cylindrical Shells', *J. Comp. Mater.*, **3**, No. 4 (1969), p. 676.

30. Matin, R., 'Free Vibrations of Anisotropic Conical Shells', *AIAA J.*, **7**, No. 5 (1969), p. 960.

31. Calcote, L., *The Analysis of Laminated Composite Structures*, Van Nostrand–Reinhold Company, New York (1969).

32. Librescu, L., *The Elasto-Statics and Kinetics of Anisotropic and Heterogeneous Shell Type Structures*, Publishing House of the Academy of the Socialist Republic of Romania (1969).
33. Reuter, R., Jr, 'Membrane Motion of a Thin Single Layer Generally Orthotropic, Cylindrical Shell', *J. Comp. Mater.*, **4**, 254 (1970).
34. Hildebrand, F., Reissner, E., and Thomas, G., 'Note on the Foundations of the Theory of Small Displacements of Orthotropic Shells', NACA TN-1833, 1949.
35. Kalnins, A., 'Analysis of Shells of Revolution Subjected to Symmetrical and Nonsymmetrical Loads', *J. Appl. Mech.*, **31**, 467 (1964).
36. Mirsky, I., 'Vibrations of Orthotropic, Thick, Cylindrical Shells', *J. Acoustical Soc. Amer.*, **36**, No. 1 (1964), p. 41.
37. Gulati, S. and Essenburg, F., 'Effects of Anisotropy in Axisymmetric Cylindrical Shells', ASME Paper No. 67-APM-28 (1967).
38. Ahmed, N., 'On the Axisymmetric Vibrations of Orthotropic Cylindrical and Conical Shells', Ph.D. Dissertation, Cornell University, 1967.
39. Garber, A. M., 'Pyrolytic Materials for Thermal Protection Systems', *Aerospace Engineering*, **22**, 126–37 (1963).
40. McDonough, T. B., 'Thermal Stresses in Transversely Isotropic Shells of Revolution', Ph.D. Thesis, University of Pennsylvania, 1965.
41. Kliger, H. and Vinson, J., 'Truncated Conical Shells of Pyrolytic Graphite Material', AIAA Paper 68-295.
42. Raju, P., 'Shallow Shells of Pyrolytic Graphite Type Materials', Ph.D. Dissertation, University of Delaware, 1968.
43. Daugherty, R. and Vinson, J., 'Asymptotic Solution for Noncircular Cylinders of Pyrolytic Graphite', *Developments in Mechanics*, **5**, 563–76 (1969).
44. Daugherty, R., Kliger, H., and Vinson, J., 'Governing Equations for Shells of Pyrolytic Graphite Type Materials', *AIAA J.*, **9**, 508–10 (1971).
45. Zukas, J. and Vinson, J., 'Laminated Transversely Isotropic Cylindrical Shells', *J. Appl. Mech.*, **38**, 400–407 (1971).
46. Pagano, N., 'Analysis of the Flexure Test of Bidirectional Composites', *J. Comp. Mater.*, **1**, No. 4 (1967), p. 336.
47. Schipper, J., 'An Exact Formulation of the Linear Equation for Thick, Orthotropic Shells with Arbitrary Imposed Temperature and Force Fields and Temperature Dependent Parameters', *Developments in Theoretical and Applied Mechanics*, Vol. 3, Pergamon Press (1967), pp. 255–78.
48. Kalnins, A., 'On the Derivation of a General Theory of Elastic Shells', *Indian J. Mathematics*, **9**, 381–425 (1967).
49. Vasilev, V. V., 'Theory of Orthotropic Laminar Cylindrical Shells', *Mekhanika Polimeros*, **4**, 136–44 (1968).
50. Kliger, H. S., 'Stresses in Shallow Spherical Shells of Composite Materials Subjected to Localized Loads', Ph.D. Dissertation, University of Delaware, 1970.
51. Kliger, H. and Vinson, J., 'Response of Spherical Shells of Composite Materials to Localized Loads', to be published in the *ASME J. of Pressure Vessel Technology*, (1974).
52. Kliger, H. and Vinson, J., 'Stresses in Shallow Spherical Shells of Composite Materials Subjected to Localized Loads', AFOSR Technical Report 70-1046 TR (April 1970).

53. Kliger, H. and Vinson, J., 'Computer Program for Calculating the Stresses in Shallow Spherical Shells of Composite Materials Subjected to Localized Loads', AFOSR Technical Report 70-1047 TR (April 1970).

54. Kliger, H., Forristall, G., and Vinson, J., 'Stresses in Circular Cylindrical Shells of Composite Materials Subjected to Localized Loads', AFOSR Technical Report No. 73-0494 TR (January 1973).

55. Pagano, N. J., Halpin, J. C., and Whitney, J. M., 'Tension Buckling of Anisotropic Cylinders', *J. Comp. Mater.*, **2**, No. 2, (1968), pp. 154–67.

56. Pagano, N. J. and Whitney, J. M., 'Geometric Design of Composite Cylindrical Characterization Specimens', *J. Comp. Mater.*, **4**, No. 3 (1970), pp. 360–79.

57. Pagano, N. J., 'Stress Gradients in Laminated Composite Cylinders', *J. Comp. Mater.*, **5**, No. 2 (1971) pp. 260–65.

58. Whitney, J. M. and Halpin, J. C., 'Analysis of Laminated Anisotropic Tubes under Combined Loading', *J. Comp. Mater.*, **2**, No. 3 (1968), pp. 360–67.

59. Whitney, J. M., 'On the use of Shell Theory for Determining Stresses in Composite Cylinders', *J. Comp. Mater.* **5**, No. 3 (1971), pp. 340–53.

60. Whitney, J. M., 'Analytical and Experimental Methods in Composite Mechanics', *Journal of the Structural Division, ASCE*, (1973), pp. 113–29.

61. Bijlaard, P. P., 'Stresses from Radial Loads in Cylindrical Pressure Vessels', *The Welding Journal*, **33**, No. 12, (1954).

62. Zukas, J. A., 'Laminated Cylindrical Shells of Pyrolytic Graphite Type Materials', Master's Thesis, University of Delaware, 1969.

63. Vinson, J. R., *Structural Mechanics: The Behavior of Plates and Shells*, Wiley–Interscience, John Wiley and Sons, Inc. (1974).

64. Zukas, J. A., 'Nonlinear Response of Laminated Orthotropic Shells', Ph.D. Dissertation, University of Arizona, 1973.

65. Naghdi, P. M., 'The Effect of Transverse Shear Deformation on the Bending of Elastic Shells of Revolution', *Quarterly J. Applied Mathematics*, **15** (1958).

66. Waltz, T., 'Interlaminar Stresses in Laminated Cylindrical Shells of Composite Materials', MMAE thesis, University of Delaware, 1975.

CHAPTER 8

THE STRENGTH AND FRACTURE OF COMPOSITE MATERIALS

8.1 INTRODUCTION

The strength of composite materials is influenced by a number of factors. These include the anisotropic and non-homogeneous nature of the materials, the mechanical incompatibility of the constituent phases, the effect of interfacial bonding, the elastic and plastic behaviour of the matrices and the reinforcing materials, the volume fractions of the component materials, and the directions of applied load. In view of the complexity of the deformation process, it is not surprising that there is a lack of comprehensive knowledge of failure mechanisms in composite materials. However, due to the intense efforts made by researchers in the field of composites, it is now possible to describe some typical fracture modes in fibrous composites. Theories both qualitative and analytical in nature are introduced to estimate the composite strength and work of fracture. The problem concerning the ductile failure of fibrous composites could very well be one of the most challenging problems facing structural engineers involved in fracture-safe design. The solution of this problem inevitably calls for better and more fundamental understanding of the plastic behaviour of materials and the role of crystalline defects. The strength of particulate composites is discussed in Chapter 3. The discussion in this chapter is centred on fibrous composites.

8.2 THE STRENGTH OF FIBROUS COMPOSITE MATERIALS

In this section, attention is focused on unidirectionally reinforced fibre composite materials. The loading condition is restricted to simple tension at both room and elevated temperatures, while the effects of fatigue and creep are not included. The fibres in the composite materials are assumed to possess uniform strength. The distribution of load among the matrix and fibres, as well as the strength of a composite depend very much upon the length of the fibres. Therefore the following discussions are made according to the fibres being in their continuous or discontinuous form. Extensive reviews of the strength theories concerning fatigue and creep of fibrous composite materials can be found in the works of Kelly[1] and Hale and Kelly.[2]

8.2.1 Continuous Fibres

Figure 8.1 depicts a unidirectionally reinforced composite material. The fibres are aligned in the direction of tensile force. By assuming satisfactory bonding between the fibres and the matrix, it is reasonable

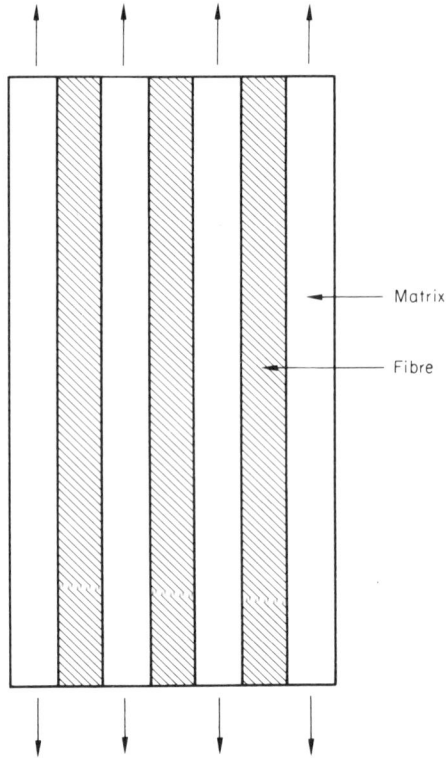

Fig. 8.1. *Fibre reinforced composite material.*

to assume that the strain of the composite, ϵ_c, is identical to that in the fibre, ϵ_f, and in the matrix, ϵ_m, namely

$$\epsilon_c = \epsilon_f = \epsilon_m \tag{8.1}$$

The engineering stresses are denoted by σ_c, σ_f and σ_m for composite, fibre and matrix, respectively. Since the applied load is distributed among the fibres and the matrix, it follows that

$$\sigma_c A = \sigma_f A_f + \sigma_m A_m$$

where A denotes the cross-sectional area of the composite and is composed of the fibre area A_f and the matrix area A_m. It is more convenient to express A_f and A_m in terms of the volumetric content of the fibre $V_f = A_f/A$, and the above equation can be written as

$$\sigma_c = \sigma_f V_f + \sigma_m(1 - V_f) \tag{8.2}$$

The response of composite materials to external load relies very much on the elastic and plastic behaviour of the component phases. Reinforcing materials such as glass fibres and tungsten wires at room temperature behave as linear elastic materials and fracture in a brittle manner. On the other hand, matrices made of plastics and metals can be deformed beyond the linear elastic range into a plastic or non-linear elastic range. A typical set of stress–strain curves for composite materials composed of brittle fibres and a ductile matrix is shown in Fig. 8.2. The ultimate tensile strength of the composite, following eqn. (8.2), can be expressed as[3–5]

$$\sigma_{cu} = \sigma_{fu} V_f + \sigma'_{mu}(1 - V_f) \tag{8.3}$$

where σ_{cu} and σ_{fu} denote the ultimate tensile strengths of the composite and the fibre, respectively. For a brittle fibre, σ_{fu} also stands for the fibre fracture strength. It should be pointed out that the ultimate tensile strength of the fibre in a composite may be affected by the fabrication process and is thus different from that of the virgin fibres. However, the effect is assumed to be small and the σ_{fu} used in Fig. 8.2 is the ultimate strength of the fibres in the composite. Another factor in eqn. (8.3) needed to be considered is the matrix strength σ'_{mu}. This is the stress in the matrix at the failure strain of the composite. Since the matrix is deformed in a ductile manner, the difference between σ'_{mu} and the ultimate tensile strength of the bulk matrix, σ_{mu}, denotes its capability of strain hardening.

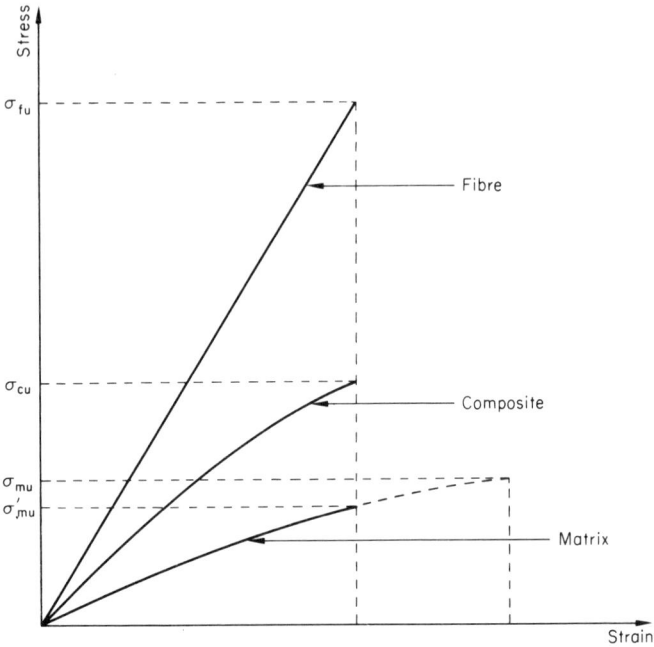

Fig. 8.2.   Stress–strain curves of fibre, matrix and composite.

The applicability of eqn. (8.3) has a lower limit with respect to fibre volume fraction. This occurs because at low V_f values, the fibres fail to significantly strengthen the ductile matrix material. The small amount of fibres dispersed in a ductile matrix cannot effectively constrain the deformation of the matrix and can be rapidly stretched to their fracture strain. Since the fibres do not contribute effectively to the load-carrying capacity of the composite, the composite strength is then given by

$$\sigma_{cu} = \sigma_{mu}(1 - V_f) \tag{8.4}$$

The minimum fibre content necessary for the strengthening effect to be realised is determined by equating eqns. (8.3) and (8.4)

$$V_{min} = \frac{\sigma_{mu} - \sigma'_{mu}}{\sigma_{fu} + \sigma_{mu} - \sigma'_{mu}} \tag{8.5}$$

It can be summarised that for a ductile matrix reinforced with brittle fibres, the ultimate tensile strength is given by eqn. (8.3) for $V_f \geqslant V_{min}$ and by eqn. (8.4) for $V_f < V_{min}$. These equations are indicated in Fig. 8.3

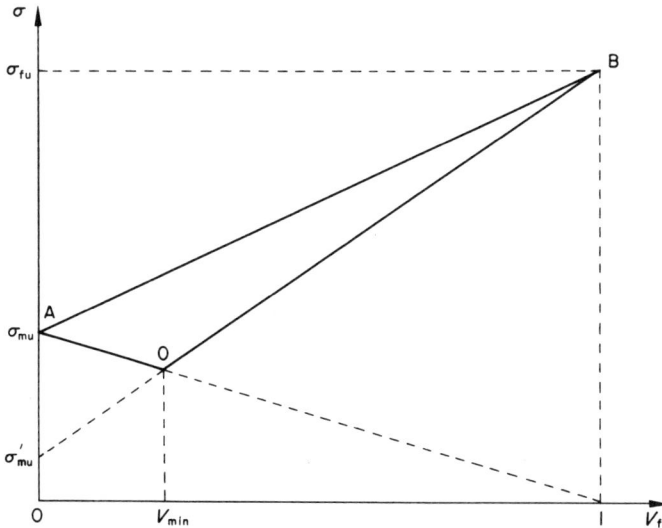

Fig. 8.3. Schematic diagram for strength–volume fraction dependence. (Reproduced from Reference 6 by permission of Chapman and Hall.)

by the line segments AO and OB. The failure mechanism proposed by Kelly *et al.* assumes that at $V_f > V_{min}$, the fibres and the matrix fail at the same time by a single fracture process. For $V_f < V_{min}$, multiple fracture of the fibres is accompanied by the plastic flow of the matrix. Failure of the composite then occurs due to the fracture of the matrix at its ultimate tensile strain. Experimental results of copper–tungsten systems[4] tested at room temperature (Fig. 2.10(b)) are excellent examples of the validity of this theory. The solid-line segments in Fig. 2.10(b) correspond to the prediction of eqns. (8.3) and (8.4). The minimum fibre volume fraction appears to be around 9 per cent.

Composite materials systems consisting of ductile fibres and a brittle matrix have less practical significance. However, their strengths can be predicted in the manner similar to the above case (Problem 8.1). Multiple fracture now occurs in the matrix rather than in the fibres.

When both the matrix and the reinforcing fibres have some ductility, the above strength theory has been modified by Mileiko.[6] The combination of the two components with different strengths and degrees of work-hardening is depicted by the stress–strain curves in Fig. 8.4. To examine the ultimate strength of the composite, it is necessary to first investigate the plastic instability of the component phases. The general

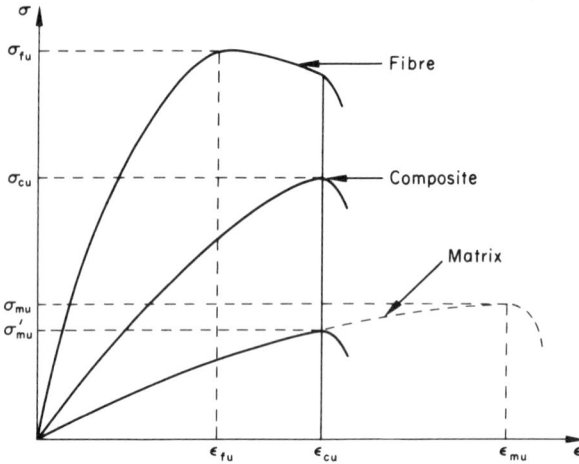

Fig. 8.4.   *Stress–strain curves for a composite with ductile fibres and ductile matrix. (Reproduced from Reference 6 by permission of Chapman and Hall.)*

relation between the true stress and true strain for ductile materials is given by the power function

$$\epsilon = \left(\frac{s}{s*}\right)^n \qquad (8.6)$$

The true strain $\epsilon = \ln(l/l_0)$ for a specimen is defined by the natural logarithm of the ratio of the instantaneous length l to the original length l_0. The true stress s is given by the ratio of applied load to the instantaneous load bearing area. $s*$ and n are constants to be determined from the material properties. When the engineering stress σ is used, the above equation can be expressed as

$$\sigma = s*\epsilon^{1/n}\exp(-\epsilon) \qquad (8.7)$$

The condition of incompressibility of the material is assumed in the derivation. At the moment of instability, necking of the specimen starts to develop and the load-carrying capacity of the specimen also starts to decrease. On the engineering stress–true strain curve, this is represented by the point of maximum load, or the condition of $d\sigma/d\epsilon = 0$. Using the condition for necking, the constants in eqn. (8.7) can be expressed in terms of material properties

$$n = 1/\epsilon_u$$

$$s* = \sigma_u\epsilon_u^{-\epsilon_u}\exp(\epsilon_u)$$

σ_u and ϵ_u are the ultimate tensile stress and strain, respectively. Equation (8.7) is now rewritten as

$$\sigma = \sigma_u(\epsilon/\epsilon_u)^{\epsilon_u} \exp(\epsilon_u - \epsilon) \tag{8.8}$$

The stress–strain relation of eqn. (8.8) is derived for ductile materials in general and is thus valid for the composite as a whole as well as its component phases. The fibre–matrix interface bonding is assumed to be ideal and is strong enough to prevent fibre necking without the necking of the composite. This assumption, physically reasonable, has greatly simplified the problem. The governing equations for the deformation process now consist of eqns. (8.1), (8.2) and (8.8). The stress–strain relation for a fibrous composite material composed of ductile fibres and matrix is given by

$$\sigma_c = V_f\sigma_{fu}(\epsilon_c/\epsilon_{fu})^{\epsilon_{fu}} \exp(\epsilon_{fu} - \epsilon_c) +$$
$$+ (1 - V_f)\sigma_{mu}(\epsilon_c/\epsilon_{mu})^{\epsilon_{mu}} \exp(\epsilon_{mu} - \epsilon_c) \tag{8.9}$$

Applying the criterion for necking, the relation between fibre content and composite ultimate tensile strain can be obtained

$$V_f\sigma_{fu}(\epsilon_{cu}/\epsilon_{fu})^{\epsilon_{fu}}(\epsilon_{fu}/\epsilon_{cu} - 1) \exp(\epsilon_{fu}) +$$
$$+ (1 - V_f)\sigma_{mu}(\epsilon_{cu}/\epsilon_{mu})^{\epsilon_{mu}}(\epsilon_{mu}/\epsilon_{cu} - 1) \exp(\epsilon_{mu}) = 0 \tag{8.10}$$

where ϵ_{cu} is the ultimate true strain of the composite. For most practical composite systems, $\epsilon_{fu} < \epsilon_{mu}$ and $0 < V_f < 1$. It can be shown from eqn. (8.10) that $\epsilon_{fu} < \epsilon_{cu} < \epsilon_{mu}$. Consequently, the instability of the fibres in the composite is delayed by the more ductile matrix and the composite fails at a strain intermediate between the ultimate tensile strains of the pure fibre and the pure matrix. This situation is illustrated in Fig. 8.4. The form of the fibre stress–strain curve implies that eqn. (8.8) is assumed to be valid at strains beyond the critical value ϵ_{fu}. Finally, the ultimate tensile strength of the composite material reinforced with ductile continuous filaments is given as

$$\sigma_{cu} = V_f\sigma_{fu}(\epsilon_{cu}/\epsilon_{fu})^{\epsilon_{fu}} \exp(\epsilon_{fu} - \epsilon_{cu}) +$$
$$+ (1 - V_f)\sigma_{mu}(\epsilon_{cu}/\epsilon_{mu})^{\epsilon_{mu}} \exp(\epsilon_{mu} - \epsilon_{cu}) \tag{8.11}$$

For the special case $\epsilon_{fu} = \epsilon_{mu}$, eqn. (8.11) is reduced to the simple form

$$\sigma_{cu} = V_f\sigma_{fu} + (1 - V_f)\sigma_{mu}$$

This relation is given by the line AB in Fig. 8.3. It has been suggested that for a composite with fibres and matrix that behave in a ductile manner, and for $\epsilon_{fu} < \epsilon_{mu}$, the tensile strength at various fibre volume

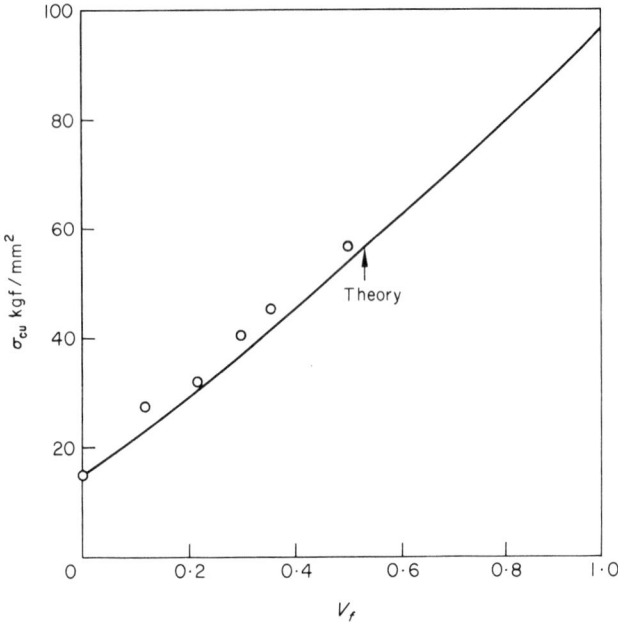

Fig. 8.5.   Strengths of nickel–tungsten composites at 400°C as functions of fibre volume fraction. (Reproduced from Reference 6 by permission of Chapman and Hall.)

fraction lies in the triangle region AOB (Reference 6). Some comparisons of this theory with experimental results are shown in Figs. 8.5 and 8.6. The agreement between theory and experiments is very satisfactory.

8.2.2   Discontinuous Fibres
It is common practice to strengthen a matrix material with discontinuous fibres. Chopped fibres and whiskers are typical examples, although continuous fibres may also be broken into segments during the manufacturing process. The effectiveness of strengthening by discontinuous fibres depends upon the length of the fibres. Figure 8.7 gives a typical example of this kind. It is seen that for a certain fibre content V_f and fibre diameter d, the strength of the composite increases as the fibres are made longer. This is due to the fact that the average tensile stress in a discontinuous fibre is always less than that in a continuous fibre. This average stress increases toward that of a continuous fibre

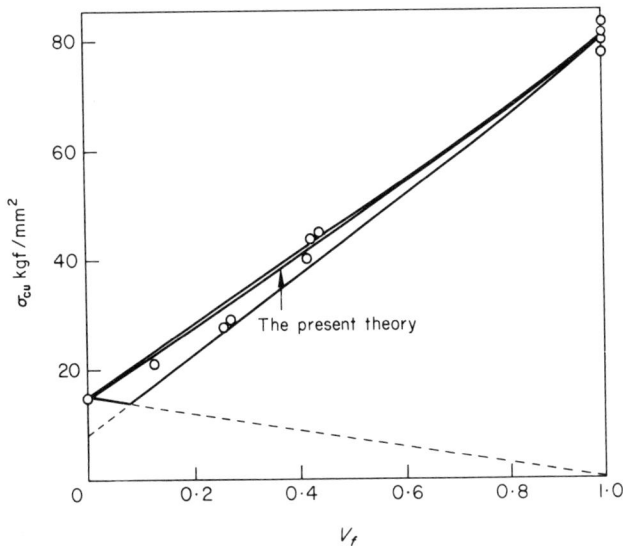

Fig. 8.6. *Strengths of silver–stainless steel composites at room temperature as functions of fibre volume fraction. (Reproduced from Reference 6 by permission of Chapman and Hall.)*

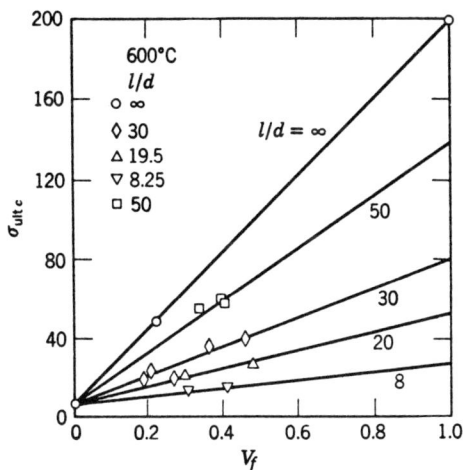

Fig. 8.7. *Tensile strength versus volume fraction of 0·2 mm diameter tungsten wires in copper tested at 600°C. Ratio of length-to-diameter of the wires is given by the value of l/d. (Reproduced from Reference 4 by permission of Wiley and Sons.)*

when the fibre length increases. To better understand the relationship between fibre average tensile strength and fibre length, it is necessary to examine the load-transference mechanism in the composite.

When unidirectionally aligned discontinuous fibres are dispersed in a ductile matrix, the isostrain condition of eqn. (8.1) is not maintained. As a result of the difference of the strain in the fibres and the strain in the matrix, shear stresses are induced around the fibres in the direction of the fibre axes. Plastic deformation may occur in the matrix or through the shearing of the fibre–matrix interface. These shear forces, acting near both ends of a fibre, stress the fibre in tension. It is through this transferring of stress that the applied load can be distributed among the fibres and the matrix. Figure 8.8 shows the variation of tensile stress in a fibre with its length. The maximum stress attainable in the fibre is the fibre tensile strength σ_{fu}. The minimum fibre length necessary for the tensile stress to build up to σ_{fu} is denoted as l_c, the *fibre critical length*. The length $\frac{1}{2}l_c$ is often referred to as the load transfer length. The building up of the tensile stress from the fibre ends is linear if the matrix material does not strain-harden and has a constant yield strength.

In order to find the average tensile stress in a fibre with length l, it is necessary to introduce the parameter β. β is defined as the ratio of the area under the stress distribution curve over the length $l_c/2$ in Fig. 8.8(c) to the area of $\sigma_{fu}l_c/2$. Then the average stress in the fibre $\bar{\sigma}_f$ can be expressed as[3–5]

$$\bar{\sigma}_f = \sigma_{fu}(1 - (1 - \beta)l_c/l) \tag{8.12}$$

For matrix materials with constant yield stress, $\beta = \frac{1}{2}$ and the above equation is simplified as

$$\bar{\sigma}_f = \sigma_{fu}(1 - l_c/2l) \tag{8.13}$$

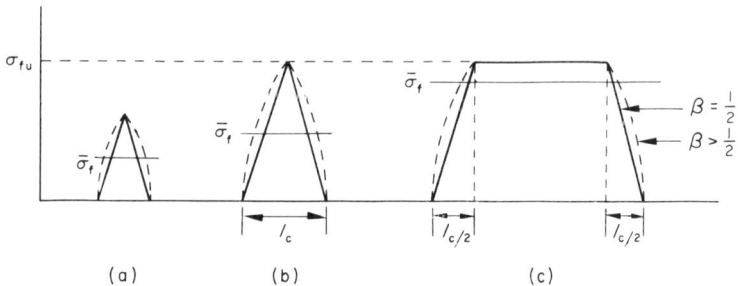

Fig. 8.8. *Variations of fibre tensile stress with fibre length.*

For matrix materials that show strain hardening behaviour, $\beta > \frac{1}{2}$, and the average stress is higher than that indicated in eqn. (8.13). It is also obvious from eqn. (8.12) that $\bar{\sigma}_f$ approaches σ_{fu} when $l \gg l_c$. The fibre critical length can be easily found by considering the balance of tensile and shear stresses as

$$\frac{l_c}{d} = \frac{\sigma_{fu}}{2\tau} \tag{8.14}$$

τ is the shear strength of either the matrix or the interface, whichever is least. The critical length of a fibre decreases if the matrix work-hardens. For discontinuous fibres embedded in a matrix it is now evident that fracture may occur if their lengths are greater than l_c. Measurements of tungsten wires extracted from a copper matrix[4] did indicate that the wires were broken into lengths of between l_c and $l_c/2$.

The strength of a composite material reinforced with discontinuous fibres is now examined. By assuming that the fibres are uniformly dispersed in the matrix and aligned in the direction of tensile force, the stress in a fibre at any cross-section can be approximated by the average stress $\bar{\sigma}_f$ in eqn. (8.12). Hence, the tensile strength of the composite for $\beta = \frac{1}{2}$, following eqn. (8.3), is

$$\sigma_{cu} = \sigma_{fu} V_f (1 - l_c/2l) + \sigma'_{mu} (1 - V_f) \tag{8.15}$$

for $l \geq l_c$. For $l < l_c$, the first term on the right-hand side of eqn. (8.15) should be written as $\sigma_{fu} V_f l/2l_c$. Also, if the fibre length is less than l_c, the matrix stress can be raised to higher than σ'_{mu}. Since the factor $1 - (1 - \beta)l_c/l$ in eqn. (8.12) is less than unity, it is obvious that for the same V_f value the strength of a composite reinforced with discontinuous fibres is always less than that of a composite with continuous fibres. However, the difference could be kept small. Consider, for instance, a composite with $\beta = 0.5$ and $l/l_c = 10$. It is seen that from eqn. (8.15), 95 per cent of the strength contribution from continuous fibres can be realised by using discontinuous fibres. Consequently, discontinuous fibres are feasible and effective in strengthening matrix materials. The experimental results in Fig. 8.7 on copper–tungsten systems tested at 600 °C demonstrated the linear relationship of σ_{cu} and V_f.

The ratio of l_c/d in eqn. (8.14), known as the *critical aspect ratio*, can be determined in the following manner. Differentiation of eqn. (8.15) with respect to V_f yields[4]

$$\frac{d\sigma_{cu}}{dV_f} = (\sigma_{fu} - \sigma'_{mu}) - \frac{1}{2} \sigma_{fu} \frac{l_c/d}{l/d} \tag{8.16}$$

The left-hand side of this equation can be measured from σ_{cu} versus V_f curves such as those in Fig. 8.7. A plot of $d\sigma_{cu}/dV_f$ versus $1/(l/d)$ at a constant temperature should give a linear relationship, and the critical aspect ratio can be obtained from the slope of the line. The critical aspect ratios for tungsten wires in a copper matrix at three temperatures are given in Table 8.1.[4] The values of the matrix shear strength τ, calculated from eqn. (8.14) are also listed. The τ values so obtained are nearly half of the ultimate tensile strength of bulk copper. For metal matrices and fibres that are well bonded, the high shear stress concentration near the fibre ends causes the matrix to deform plastically. Hence it is reasonable to assume the maximum shear stress in eqn. (8.14) to be one half the ultimate tensile strength of the matrix in estimating fibre critical length.

TABLE 8.1

FIBRE CRITICAL-ASPECT RATIO AND MATRIX SHEAR-YIELD STRENGTH OF THE COPPER–TUNGSTEN SYSTEM
(Reproduced from Reference 4 by permission of Wiley and Sons.)

Temperature °C	l_c/d	τ ksi (10^8 N/m²)
250	13·5	6·9 (0·476)
300	19·3	5·9 (0·407)
600	31·2	3·2 (0·221)

In resin matrix composites, shear fractures are often observed to occur along the interface. As a result, the strength of fibre–matrix interfaces should be examined. In Section 2.4 the determination of fibre critical length for a glass–epoxy resin composite[8] was discussed. The theoretical prediction of the composite strength was based upon eqn. (8.15). The experimental results matched with the theory for $l_c = 0.5$ in. (12·7 mm). Taking the measured values of σ_{fu}, d and l_c, eqn. (8.14) predicts that the interfacial shear strength τ_i should be 0·9 N/mm². The experimental measurement of interface bond[8] was carried out by first determining the average shear strength of the composite, τ_c. The shear failure of the composite was observed to occur through the simultaneous failure of the matrix and the interface.

Hence, τ_c can be expressed as

$$\tau_c = x\tau_i + (1-x)\tau_m \tag{8.17}$$

where τ_i and τ_m are the shear strengths of the interface and the matrix, respectively. x denotes the fraction of the fracture area occupied by the fibre interface and was determined by carefully following the actual fracture path in the specimen. Investigation of the fracture mechanism showed that the crack ran preferentially through the epoxy matrix. The actual value of the interfacial shear strength so determined is $9 \cdot 5 \, \text{N/mm}^2$. The discrepancy between the measured τ_i value and the calculated value is attributed to the fact that the fibres were not uniformly dispersed in the matrix. In fact the fibres were observed to behave as bunches with approximately fourteen fibres in a bunch. Consequently, the calculation based upon the fibre diameter of a single fibre was not realistic.

Finally, it is necessary to comment on the elastic modulus of the composite. Following McDanels et al.,[9] a typical stress–strain curve of a composite material consists of the following stages. During the initial loading of a composite, both the matrix and the reinforcement are in their elastic stage. The elastic modulus of the composite can be represented by a rule of mixtures type relation

$$E_c = E_f V_f + E_m (1 - V_f) \tag{8.18}$$

where E is the Young's modulus and the subscripts c, f and m denote the composite, fibre and matrix, respectively. The above relation is also given in Section 5.6. In the second stage, the ductile matrix material is undergoing plastic deformation. The elastic modulus of the composite is approximated by a different form

$$E_c = E_f V_f + \left(\frac{d\sigma_m}{d\epsilon_m}\right)(1 - V_f) \tag{8.19}$$

The term $(d\sigma_m/d\epsilon_m)$ measures the slope of the matrix stress–strain curve at a given strain. For matrices with a very low work-hardening rate, the contribution from the second term is negligible and the modulus of the composite is determined by the fibres. If the reinforcing fibres also have some ductility, the third stage of deformation is characterised by plastic deformations of the fibres and the matrix. The last stage of deformation is due to the fracture of fibres, which eventually leads to the failure of the composite.

The discussions of the strength of fibrous composite materials in this section is based upon the mechanical factors. These include the strength and elastic properties of the component materials, the fibre content, the fibre aspect ratio, and the orientation of the reinforcements. The fibres are assumed to be uniform in strength. A rigorous treatment of composite strength from the viewpoint of continuum mechanics can be found in the book by Spencer.[10] The dynamic aspect of composite strength can be found in Reference 11.

8.3 LINEAR ELASTIC FRACTURE MECHANICS

The elastic fracture mechanics deals with the prediction of fracture strength of relatively brittle materials. The linear elastic materials are assumed to be isotropic and to contain pre-existing flaws. The elastic field associated with a flaw or a crack depends upon the type of load applied. Three modes of deformation can be defined with respect to the crack geometry as shown in Fig. 8.9. The Mode I deformation is defined by the tensile stress applied normal to the plane of the crack. The Mode II and Mode III deformations are characterised, respec-

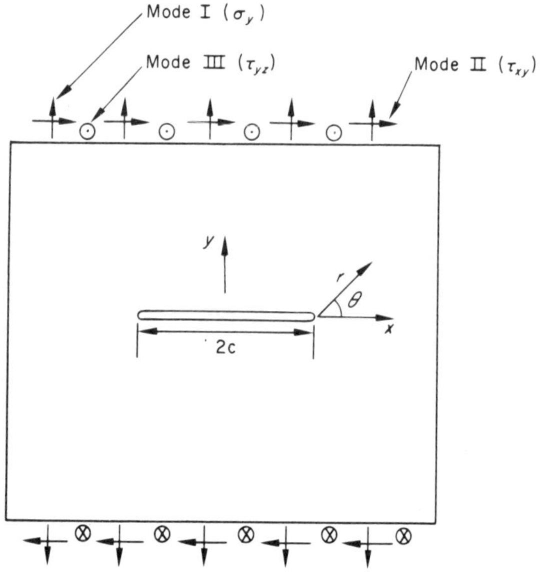

Fig. 8.9.   Three modes of deformation.

tively, by the shear stresses applied normal and parallel to the leading edges of the crack. Due to the nature of the deformation, Mode III is also known as anti-plane strain deformation.

8.3.1 Crack Tip Stress Concentration

The study of stress concentration at a crack tip dates back to 1913 when Inglis[12] obtained the stress solution in the vicinity of an elliptical hole in a flat plate. The major and minor axes of the hole are denoted by $2c$ and $2h$, respectively. Under a uniform applied stress σ normal to the hole major axis, a maximum stress concentration σ_{max} is induced at the apex of the major axis

$$\sigma_{max} = \sigma\left(1 + \frac{2c}{h}\right) \tag{8.20}$$

The elliptical hole can be degenerated into a crack by letting the radius of curvature, $\rho = h^2/c$, become very small. Then the crack-tip stress field found from eqn. (8.20) is

$$\begin{aligned}\sigma_{max} &= \sigma[1 + 2\sqrt{(c/\rho)}] \\ &\simeq 2\sigma\sqrt{(c/\rho)}\end{aligned} \tag{8.21}$$

Based upon the stress solution of Inglis for an elliptical hole, Griffith[13] in 1920 derived the now well-known criterion for the fracture of brittle solids. In his analysis, Griffith avoided making the assumption of the crack-tip radius, which is difficult to measure. Instead, the ingenious Griffith criterion is based upon an energy-rate consideration. A crack in a brittle solid will advance in an unstable manner if the rate of elastic energy released with respect to an infinitesimal extension of the crack is larger than the corresponding rate of increase in crack surface energy. The decrease in elastic energy due to the introduction of a crack of length $2c$ in a large plate under a tensile stress σ is

$$W_E = \frac{\pi c^2 \sigma^2}{E}$$

where E denotes the Young's modulus of the material. The sum of the surface energy γ_s of the crack surfaces is

$$W_s = 4c\gamma_s$$

Hence, the Griffith criterion requires

$$\frac{\partial}{\partial c}\left(\frac{\pi c^2 \sigma^2}{E}\right) \geq \frac{\partial}{\partial c}(4c\gamma_s)$$

The critical stress for fracture is

$$\sigma_F = \sqrt{(2E\gamma_s/\pi c)} \qquad (8.22)$$

The fracture stress is inversely proportional to $c^{1/2}$. Consequently, once the critical stress is reached a crack can propagate in an unstable manner.

The complete displacement and stress-field solutions for cracks in isotropic, anisotropic, homogeneous and non-homogeneous media are now widely available.[14,15] For the purpose of illustration, the crack-tip stress fields in an isotropic media associated with the three modes of deformation (Fig. 8.9) are given in the following

Mode I

$$\sigma_x = \frac{K_I}{(2\pi r)^{1/2}} \cos\frac{\theta}{2}\left(1 - \sin\frac{\theta}{2}\sin\frac{3\theta}{2}\right)$$

$$\sigma_y = \frac{K_I}{(2\pi r)^{1/2}} \cos\frac{\theta}{2}\left(1 + \sin\frac{\theta}{2}\sin\frac{3\theta}{2}\right)$$

$$\tau_{xy} = \frac{K_I}{(2\pi r)^{1/2}} \sin\frac{\theta}{2}\cos\frac{\theta}{2}\cos\frac{3\theta}{2} \qquad (8.23)$$

$$\sigma_z = \nu(\sigma_x + \sigma_y), \qquad \tau_{xz} = \tau_{yz} = 0$$

Mode II

$$\sigma_x = \frac{-K_{II}}{(2\pi r)^{1/2}} \sin\frac{\theta}{2}\left(2 + \cos\frac{\theta}{2}\cos\frac{3\theta}{2}\right)$$

$$\sigma_y = \frac{K_{II}}{(2\pi r)^{1/2}} \sin\frac{\theta}{2}\cos\frac{\theta}{2}\cos\frac{3\theta}{2}$$

$$\tau_{xy} = \frac{K_{II}}{(2\pi r)^{1/2}} \cos\frac{\theta}{2}\left(1 - \sin\frac{\theta}{2}\sin\frac{3\theta}{2}\right) \qquad (8.24)$$

$$\sigma_z = \nu(\sigma_x + \sigma_y), \qquad \tau_{xz} = \tau_{yz} = 0$$

Mode III

$$\tau_{xz} = -\frac{K_{III}}{(2\pi r)^{1/2}} \sin\frac{\theta}{2}$$

$$\tau_{yz} = \frac{K_{III}}{(2\pi r)^{1/2}} \cos\frac{\theta}{2} \qquad (8.25)$$

$$\sigma_x = \sigma_y = \sigma_z = \tau_{xy} = 0$$

In eqns. (8.23) and (8.24), the condition of plane strain is assumed and ν denotes the Poisson's ratio. These solutions can be easily rewritten for the plane-stress condition. The above expressions are obtained by neglecting the higher-order terms in r. Hence, they cannot be used to approximate the stress field at great distance from the crack tip, for instance, at $r \simeq c$.

The parameters K_I, K_{II}, and K_{III} are known as *stress-intensity factors* for the corresponding modes of deformation. The expression of the stress-intensity factor K_I for Mode I deformations for a through crack of length $2c$ in an infinite plate is

$$K_I = \sigma(\pi c)^{1/2} \tag{8.26}$$

where σ denotes the externally applied normal stress. Similar expressions of K_{II} and K_{III} are obtained by replacing the normal stress in eqn. (8.26) with the applied in-plane and anti-plane shear stresses, respectively.

8.3.2 Stress-Intensity Factors and Fracture Toughness

It is seen from eqns. (8.23–8.25) that the expressions for the crack tip stress fields consist of three distinct portions, namely, their dependence on θ, r and K. The stress components vary as $r^{-1/2}$ near the crack tips and have singularities at the crack tips. The stress-intensity factor signifies the redistribution of stress in an elastic body due to the presence of a crack. It also represents the deformation mode as well as the magnitude of stress near a crack tip. From eqn. (8.26) it is noted that the stress intensity factor depends only on the applied stress and the geometry of the crack. For instance, the stress intensity factor for a plate of width w containing a central crack of length $2c$, stretched in tension is approximately given by

$$K_I \simeq \sigma(\pi c)^{1/2} \left[\frac{w}{\pi c} \tan \frac{\pi c}{w} \right]^{1/2} \tag{8.27}$$

The elasticity solutions of stress intensity factors for various types of loadings and crack geometries can be found in References 14 and 15.

From the Griffith criterion of eqn. (8.22), it is noted that at the onset of unstable crack propagation the product of σ_F and $(\pi c)^{1/2}$ for a large plate is a constant and depends only on the material properties. The critical value of the stress-intensity factor at fracture is then defined as[16]

$$K_{Ic} = \sigma_F(\pi c)^{1/2} \tag{8.28}$$

K_{Ic} is known as the *fracture toughness*. The essence of the fracture toughness concept is that the stress intensity factor K for a certain material reaches a critical value K_c at the moment of fracture. The K_c value is independent of the crack geometry and can be considered as a material property. In the fracture-safe design of a structural part, the stress-intensity factor for a particular crack geometry can be determined by linear elastic stress analysis. The fracture toughness of the material used in the structure can be determined in laboratories with established procedures. Consequently, the fracture load for the particular structural part can be evaluated. The concept of fracture toughness has been successfully used in fracture-safe design for high-strength materials at all temperatures and in low and medium strength materials at low temperature. It needs to be borne in mind that the successful application of fracture mechanics to design relies on several factors.[17] These include the stress analysis for determining σ in the critical points of the structure, the knowledge of the stress intensity factor, the techniques for measuring the fracture toughness, and the estimation of crack length by non-destructive test means.

Finally, some comments need to be made on the limitation of the approach of fracture mechanics. It was pointed out previously that the stress expressions of eqns. (8.23–8.25) were valid at locations not far away from the crack tip. It was also noted that in these equations the stress components became unbounded near crack tips. Since no material can withstand a stress which is infinite in magnitude, the material in the immediate vicinity of a crack tip is inevitably deformed in a plastic manner. As a result, the stress expressions based upon elasticity theory break down in the plastic zones. The deformation process in the plastic zones as effected by the microstructures could be very complicated.[18] In spite of our ignorance of the exact nature of the plastic zones, the approach of fracture mechanics is valid for low nominal stress wherein the plastic zones are small relative to the crack size and specimen boundaries and are totally confined in the elastic regions. Consequently, the stress intensity factor provides a reasonably good approximation for the state of stress inside the fracture process zone near a crack tip and can be used as a criterion for unstable fracture. Information of fracture toughness values for structural metals is readily available.

The sizes of plastic zones R at crack tips have been determined for a Mode III crack[18]

$$R = c \left(\tau / \tau_Y \right)^2$$
$$= \left(K_{III} / \tau_Y \right)^2 \tag{8.29}$$

where τ and τ_Y denote the applied stress and the shear yield strength respectively. The plastic zone size also has been estimated for non-strain hardening material under Mode I deformations.[17] Results indicated that the size of the plastic zone is proportional to the square of the ratio of K_I to the tensile yield strength. Hence, knowing the fracture toughness and yield strength of a material, the extent of plastic deformation in front of a crack can readily be estimated. For example, for 7075-T6 aluminium alloy, which has a K_{Ic} value of 20 ksi $\sqrt{in}$. ($2\cdot2 \times 10^7$ N/m^2 $\sqrt{m}$) and a yield strength of 68 ksi ($4\cdot69 \times 10^8$ N/m^2), the plastic zone size under tensile stress is $0\cdot088$ in. ($0\cdot224$ cm).

Irwin[16,19,20] also examined the equilibrium and stability of cracks from an energy-rate analysis. The *strain-energy release rate* $\mathscr{G}$ of an elastic body under loading, and containing an extending crack, is defined as the energy released per unit of new crack area generated. The equivalence of the energy-rate and stress-intensity factor approaches has been verified.[14,21,22] For instance, the energy rates in isotropic media for the Mode I and II deformations are equal to the square of the stress intensity factors of the corresponding mode multiplied by the factor $(1 - \nu^2)/E$. The elastic constant factors are also available for anisotropic media. An example of the experimental measurement of energy release rate is given in Section 8.6.5.

8.3.3 Fracture-Toughness Testing

The measurement of fracture-toughness is one of the most important tests in fracture-safe design. The determination of the resistance of materials to unstable crack propagation provides designers with a material property in addition to the yield strength, ultimate tensile strength and elastic constants, commonly used in design. The testing of fracture-toughness for opening-mode cracks under plane-strain conditions has been discussed in a comprehensive manner by Brown and Srawley.[23] Some basic features of their work are reviewed here.

For an opening-mode crack, the stress-intensity factor for a through crack of length $2c$ in an infinite plate is, as given by eqn. (8.26)

$$K_I = \sigma(\pi c)^{1/2}$$

It is also understood that due to the stress concentration near the crack tips, regions in the immediate vicinity of crack tips are deformed plastically. The extension of the plastic zone is proportional to the parameter $(K_I/\sigma_Y)^2$. Hence, the effective crack dimension $2c$ in eqn. (8.26) can be considered as the summation of the actual crack length and the lengths of the plastic zones, which are usually very small compared to the actual crack size. The essential steps in fracture toughness testing consist of the measurement of crack extension and the load at the abrupt failure of the specimen. The extension of a crack is difficult to measure directly. Instead, the relative displacement of two points on the opposite sides of a crack plane is measured. This displacement can be calibrated and related to the actual extension of the crack front.

Typical plate specimens used in fracture-toughness testing contain either single and double edge cracks on the side surfaces or cracks at the plate centre. They can be loaded in tension or four-point bending. The relation of load and notch opening displacement depends upon the size of crack and the specimen thickness relative to the extent of plastic zones. When the actual crack length and specimen thickness are much larger than the quantity $(K_{Ic}/\sigma_Y)^2$, the load-displacement curve is of the type shown in Fig. 8.10(a). The load at brittle fracture is then clearly defined. When the specimen thickness is reduced to a certain range, a pop-in step occurs on the curve (Fig. 8.10(b)) and the specimen does not fracture completely at the load corresponding to K_{Ic}. The pop-in step indicates an increase in notch opening displacement without increasing the load. This phenomenon can be attributed to the fact that the crack front advances only at the centre of the plate where the material is constrained under a plane strain condition. However, near the free surface the plastic deformation is more extensive than in the centre and is close to a plane stress condition. Consequently, the plane strain fracture tunnels ahead in the central portion of the plate thickness and the portions of material near the specimen surfaces will eventually fail by shear rupture. When the specimen becomes even thinner, the plane stress condition prevails[17] and the load-displacement curve follows that of Fig. 8.10(c). In order for a valid plane strain fracture toughness measurement to be made, the constraint relieving influence of the free surface should be kept small and the crack tip should be far enough from the specimen boundary. This requirement enables the plastic deformation to be completely constrained by an elastic zone. Finally, the length of crack needs to be kept above a certain lower limit.

So far the requirements of specimen size and crack length have only been discussed in a qualitative manner. The lower limits on specimen width, thickness, and crack length all depend upon the extent of plastic deformation through the factor $(K_{Ic}/\sigma_Y)^2$. In view of our lack of

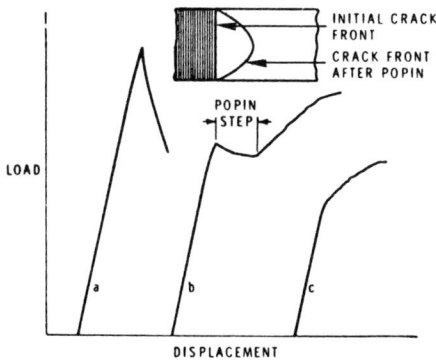

Fig. 8.10. Schematic load-displacement plots for tests of plate specimens. (Reproduced from Reference 23 by permission of the ASTM.)

knowledge of the exact size of the plastic zone for the opening-mode crack, it is not possible yet to determine the lower specimen size limits theoretically. These lower limits, for which K_{Ic} remains constant, have to be determined by a number of trial K_{Ic} tests. Specimens with sizes below these limits tend to overestimate the K_{Ic} value.

In carrying out the fracture-toughness test, the crack is preferably introduced by fatiguing from a starter notch in the specimen. The length of the fatigue crack should be long enough to avoid the interference of crack-tip stress field by the notch shape. Under applied load, the notch opening displacement can be measured between two points on the surfaces of the notch by various types of transducers. Figure 8.11 indicates the set-up for a double cantilever beam gauge used for displacement measurements. Electrical resistance measurements also have been used to detect the extension of cracks. Calibration curves are then used to convert the displacement and electrical resistance measurements to crack extension. The fracture load can be easily found from the load-displacement curve for a thick specimen (Fig. 8.10(a)). For thinner specimens, the idealised pop-in steps shown in Fig. 8.10(b) are seldom realised. The curves usually show a gradual deviation from linearity, and the pop-in step is extremely small (Fig. 8.12). Procedures used in analysing load-displacement records of this kind[24,25] can be explained using Fig. 8.12. The slope of the linear part of the load-displacement curve is denoted as OA. The line OP_5 is then drawn at a slope a certain percentage less than that of the line OA. The percentage offset depends upon the types of specimens and their sizes.

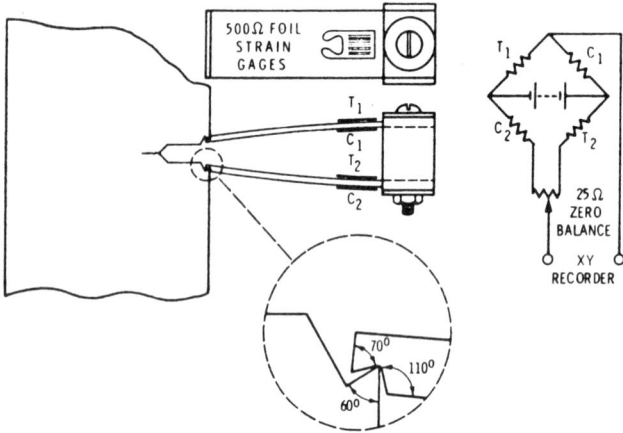

Fig. 8.11.   Double cantilever beam gauge and method of mounting on crack
notched specimen for displacement measurement (designed by J. E. Srawley).
(Reproduced from Reference 23 by permission of the ASTM.)

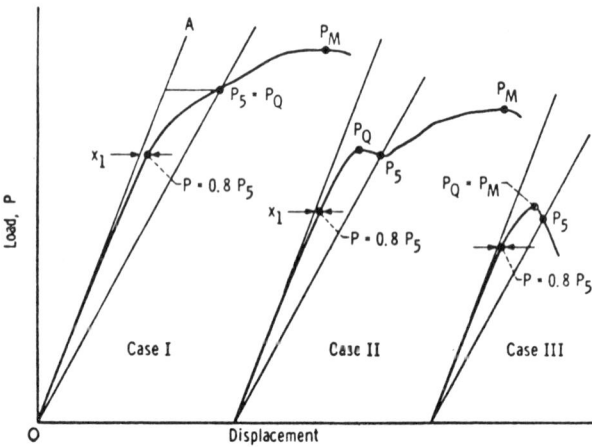

Fig. 8.12.   Types of load–displacement curves illustrating procedure for determination of K_{1c}. (Reproduced from Reference 24 by permission of the ASTM.)

A horizontal line is then drawn at 80 per cent of the load P_5. The distance x_1 between the tangent OA and the record can be measured. If this distance exceeds 25 per cent of the corresponding distance at P_5, the test is not considered to be valid. If the test has not been rejected, the load P_Q which is used to calculate a conditional value of K_{Ic}, called K_Q, is determined as follows. If the load at every point on the record which preceeds P_5 is lower than P_5, then P_Q is equal to P_5 (Fig. 8.12, Case I). If there is a maximum load preceding P_5 which exceeds it, then the maximum load is taken as P_Q (Fig. 8.12, Cases II and III). The conditional K_Q is calculated from P_Q according to the known equation for the particular specimen geometry used.[25] In order to ensure the validity of the test, it is necessary to check the crack length and specimen thickness with respect to the extension of plastic zone. If the calculated plastic zone length, which is proportional to the factor $(K_Q/\sigma_Y)^2$, is less than a certain fraction of the crack length and specimen thickness, K_Q is equal to the required K_{Ic}. Otherwise, the test needs to be repeated with a larger specimen. The basis for the procedures used in the analysis of load-displacement records can be found in the article by Brown and Srawley.[23] Some recent applications of the fracture toughness concept can be found in Reference 122.

8.4 MECHANISMS OF FRACTURE IN FIBROUS COMPOSITES

The dissipation of energy during the fracture of fibrous composites is controlled by the physical processes involved. These processes may include plastic deformation, crack branching, interface debonding, and fibre pull-out.[26,124] Unlike the fracture of reinforcements, heavy plastic deformation may occur in the matrix of a composite near a crack tip. The ductile matrix material may make an important contribution to the toughness of a composite. The crack branching in composites results in a larger fracture surface and more pronounced dissipation of energy. Debonding is due to the destruction of fibre-matrix bonds, and pull-out is the extracting of a fibre against the sliding friction of the matrix. Detailed discussions of the various failure modes in composites can be found in Reference 140.

The process of debonding and fibre pull-out are unique to the fracture of composite materials and deserve some detailed examinations. In the loading of fibres in a ductile matrix such as a metallic material, debonding of the fibre and the matrix is seldom observed. The shear stress resisting the fibre movement is equal to the shear strength τ of the matrix. Fibres with lengths below the critical length defined in eqn. (8.14) can be pulled out. The work done in extracting a fibre of length l and diameter d is

$$W_p = \int_0^l (\pi l d)\tau dl$$
$$= \pi d\tau l^2/2$$

(8.30)

A fibre of a length greater than its critical length will be fractured with an energy dissipation much smaller than that given in eqn. (8.30). When a fibre is embedded in a rather brittle matrix, debonding is first observed to occur along the fibre-matrix interface. The fibre then can be pulled out of the matrix when the debonding process is completed. The load on the fibre necessary for complete debonding in this case is higher than that for fibre pull-out. Some metal and glass-fibre reinforced epoxy systems have shown this kind of deformation process. The work done in debonding a fibre from the matrix may be considered to be stored in the fibres as elastic strain energy. For many brittle matrices the debonding energy is less than $100 \, \text{J/m}^2$. Kelly[26] has shown that the ratio of total work done in the pulling out of a fibre to that in debonding is $3(E/\sigma_f)$. E and σ_f are the modulus and strength of the fibre, respectively. Hence the energy dissipated in fibre pull-out is much higher than the work done in debonding.

As was pointed out previously, several mechanisms may contribute to the fracture of fibrous composite materials. These mechanisms seldom occur independently of one another. Instead, they are closely related and the fracture of a composite material may involve several processes occurring simultaneously or consecutively. In view of the wide variation of composite materials and the different fabrication methods employed in producing them, it is not yet possible to synthesise their behaviour and analyse them in a systematic fashion. In this section, only some typical failure phenomena of composites are reviewed. The role of plastic deformation, debonding and fibre pull-out in each example is examined. The strength and work of fracture of composite materials are then discussed according to the controlling failure mechanisms. Finally, the statistical nature of fibre reinforced composites is briefly discussed.

8.4.1 Notch Sensitivity of Composites Reinforced with Continuous Fibres

Cooper and Kelly[27] examined the notch sensitivity of metal matrix composites with continuous fibres. In this study, tungsten wires and silica fibres were used to reinforce the copper matrix. The composites were fabricated by casting and electrodeposition. When a notch is introduced transverse to the fibres, the stress concentration at the tip of the notch affects the mode of failure in the composite. The state of stress at the root of a notch has been examined by Cook and Gordon.[28] Although the discussion was for an isotropic medium, it does provide some qualitative indication as to the possible modes of composite deformation in front of a notch. For a crack lying in the xz plane, the shear stress σ_{xy} at the tip of a notch along the fibre direction tends to shear the matrix and the interface. This will result in blunting of the crack tip. It was also noticed that the maximum of the tensile stress σ_{xx} develops at a short distance ahead of the notch tip. In a composite material with weak interfacial bonding, this normal stress transverse to the fibre direction will cause splitting of the interface. The notch tip is again blunted when it runs into the splitted interface.

To identify the transition in failure modes, Cooper and Kelly carried out a series of tests on tungsten-wire reinforced copper. The specimens were all notched to remove 35 per cent of the original cross-section. The thickness of specimens varied from 120 μm to 250 μm. The transition of fracture mode occurs at the thickness of 180 μm. Below this thickness, delaminations were observed to occur first. Subsequent failure of the composite transverse to the fibres was due to the breaking of the remaining section as an unnotched specimen. The delamination phenomenon was attributed to the severe shear deformation in the matrix along the fibre direction and the very weak interface bonding. Above the transition thickness, the specimens failed by the propagation of a transverse crack. Some representative photomicrographs for these different modes of failure are shown in Plate XVIII. Because of the dissipation in energy during delamination, specimens below the transition thickness showed higher fracture load than those above the transition thickness.

When splitting does not occur, the crack tip stress concentration is not entirely removed by plastic deformation. This results in the propagation of the transverse crack. It was noticed that as the load was increased and the fibres started to fracture, plastic deformation of the

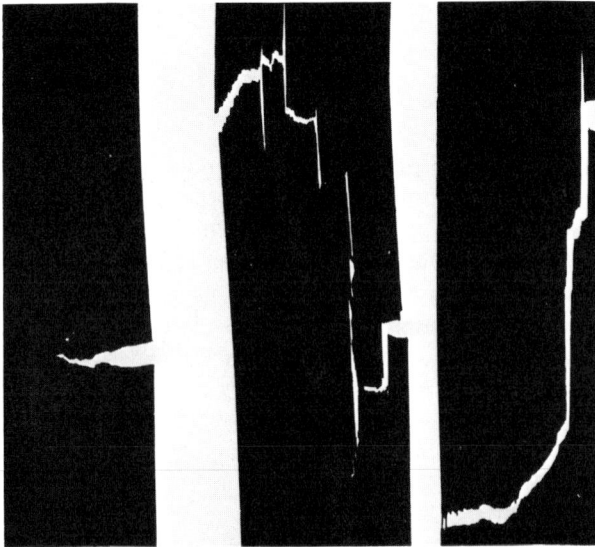

Plate XVIII. *Typical fracture appearance of specimens used in notch-sensitivity tests. Left to right, copper matrix (notch-sensitive); copper silica and copper tungsten (notch-insensitive). (Reprinted with permission of Microforms International Marketing Corp. exclusive copyright licensee of Pergamon Press journal backfiles.)*

matrix material developed near the crack tip. The unstable propagation of the crack took place when the combination of applied stress and crack length reached a critical value. Measurements of stress σ and crack length c at unstable crack growth fit very well into the relation $\sigma^2 c = $ constant. The $\mathcal{G}_c (= \pi \sigma_F^2 c / E)$ value was about $1.43 \times 10^8 \, \text{erg/cm}^2$.

At the microscopic level, the propagation of a transverse crack takes different forms, depending upon the strength of the fibre–matrix interface as well as the brittleness of the fibres. Plate XIX shows the fracture of a copper–tungsten composite made by electroforming. Because of the low interfacial bonding, delamination occurred by shearing the matrix and the interface. The delamination was not restricted to the region at the crack tip but extended to a distance several fibres ahead of the crack tip. The fibres were also observed to deform by necking. The stable crack growth was caused by the failure of the matrix and the subsequent breaking of the fibres ahead of the

Plate XIX. *The root of a crack in electroformed material; showing delamination at the fibre–matrix interfaces, and subsequent failure of the fibres some way away from the plane of the crack. (Reprinted with permission of Microforms International Marketing Corp. exclusive copyright licensee of Pergamon Press journal backfiles.)*

crack tip. The combination of tungsten wires and copper matrix by vacuum casting rendered an extreme combination of brittle and ductile components and strong interfacial bonding. Plate XX indicates the progressive growth of a crack from a surface notch. No delamination was observed. The fibres failed in a brittle manner not only in the immediate vicinity of the crack tip but also at some distance ahead of the crack. The matrix materials between the broken fibres were heavily deformed and subsequently failed by the advance of the notch. It is now obvious from the appearance of deformations associated with the notch (Plate XX) that the contribution to the work of fracture in the latter case is mainly due to the matrix. The work of fracture per unit area of the composite is determined by the extent of plastic zone, x, on either side of the crack plane and the matrix ultimate tensile strength.

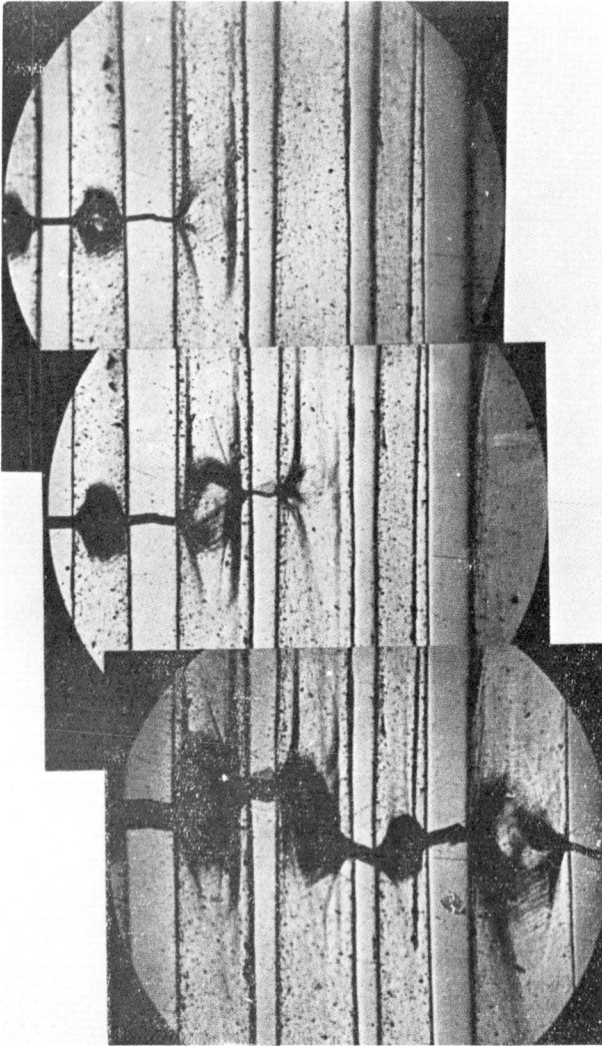

Plate XX. Vacuum-cast copper–tungsten; progressive slow advance of the crack. The three micrographs were taken in the order top, middle, bottom; showing growth of the crack. (Reprinted with permission of Microforms International Marketing Corp. exclusive copyright licensee of Pergamon Press journal backfiles.)

The contribution from fibre fracture is negligibly small. Cooper and Kelly estimated the width of the plastic zone as

$$x = \frac{(1 - V_f)}{V_f} \frac{\sigma_m d}{4\tau} \qquad (8.31)$$

where τ in a well-bonded composite may be assumed to be equal to the matrix shear strength, namely $2\tau = \sigma_m$. The basis of this relation is that the matrix material on both sides of the crack has to bear the load originally carried by the fibres.

Equation (8.31) also indicates a linear variation of x and, hence, of fracture work with fibre diameter. This has also been observed experimentally. The work of fracture for vacuum-cast copper–tungsten composites with 53 per cent of fibre is 10^7 erg/cm^2 at a fibre diameter of 20 μm and is 10^8 erg/cm^2 at 1000 μm. As to the fibre content, the work of fracture varies with V_f as the factor $(1 - V_f)^2/V_f$. An expression similar to eqn. (8.31) also has been used by Grenier and Cooper[29] for discussing an epoxy–tungsten system where the brittle matrix broke first and the load originally carried by the matrix across the crack plane was transferred to the ductile fibre over a distance x.

8.4.2 Fracture of Composites Reinforced with Continuous Fibres of Non-uniform Strength

Composite materials reinforced with continuous fibres can fail by fibre pull-out if the fibres contain flaws and are non-uniform in strength. To simplify the discussion, the matrix is assumed to have cracked completely. The fibres spanning the matrix are thus stressed more severely than those completely embedded in the matrix. Composites containing matrices with low failure strain can simulate such a situation. The fracture toughness of these composites is then affected by the mean length of fibres involved in the pull-out process. Cooper[30] examined the fracture toughness of composites reinforced with weakened fibres. In this study, weak points or flaws were introduced into the fibres in a controlled manner. The composite systems used were made of phosphor-bronze wires in plasticised epoxy resin. The reinforcing wires were very ductile. Plasticiser was used to increase the ductility of an otherwise very brittle matrix. The flaws were introduced by cutting the wires with a knife edge. The strength of a flaw was controlled by the depth of cut. The composite tensile specimens were designed in such a way as to promote the failure of the matrix along the mid-plane of the specimen. The fibre volume content

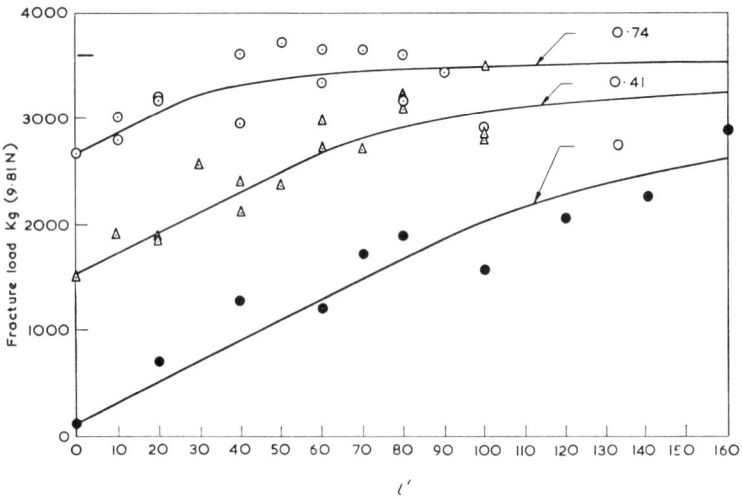

Fig. 8.13.   Variation of fracture load of 127-wire specimens with flaw spacing l′ (in mm) for values of σ*/σ$_{fu}$ of 0, 0·41, and 0·74. The solid lines are theoretical plots. (Reproduced from Reference 30 by permission of Chapman and Hall.)

was 30 per cent. When the strength of the flawed cross-section of the wire decreases, the fracture surface has more fibres pulled out of the broken specimen. The spacing between two neighbouring flaws on a fibre is denoted by l′. The variation of fracture load with l′ is shown in Fig. 8.13. σ* and σ$_{fu}$ in the figures are the strengths of the flawed region and the unnotched wire, respectively. The variation of fracture energy per wire due to fibre pull-out is shown in Fig. 8.14.

A theoretical treatment of the results also was developed by Cooper.[29,30] In this theory, it is assumed that the fibres in the composite have a uniform strength of σ$_{fu}$ except at the weakened points. A typical flaw may appear at a distance y from the plane of the matrix crack. The maximum stress in the fibre at the plane of matrix failure is

$$\sigma_{max} = \sigma^* + \frac{4\tau y}{d} \tag{8.32}$$

where τ is the shear strength of the interface and d is the fibre diameter. It is obvious that the contribution to fibre tensile stress from the second term is due to the shear stress distributed at the interface (see eqn. (8.14)). If σ_{max} reaches σ_{fu} failure of the fibre will occur at the

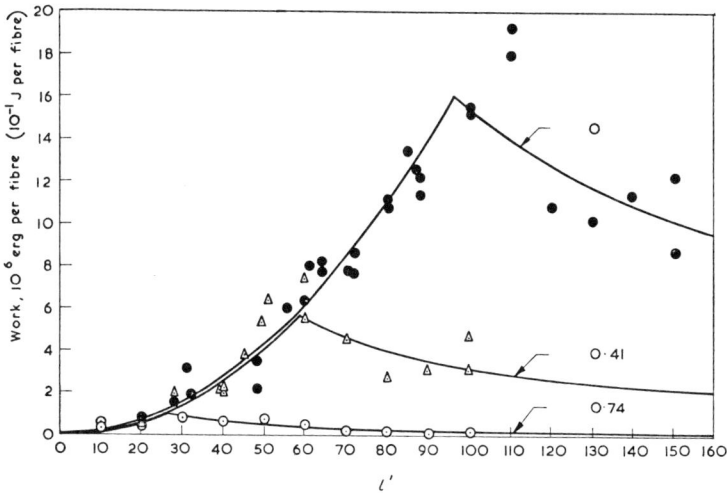

Fig. 8.14. *Variation of fracture work per wire for five-wire specimens with flaw spacing l' (in mm) for values of σ^*/σ_{fu} of 0, 0.41 and 0.78. The solid lines are theoretical plots. (Reproduced from Reference 30 by permission of Chapman and Hall.)*

plane of matrix crack. Failure at the flawed region occurs if the length y is insufficient to bring σ_{max} up to the level of σ_{fu}, namely,

$$y < \frac{(\sigma_{fu} - \sigma^*)d}{4\tau} = \frac{(\sigma_{fu} - \sigma^*)}{\sigma_{fu}}\frac{l_c}{2} \qquad (8.33)$$

where l_c is the fibre critical length. In this case, fibres break at the flawed regions and are pulled out of the specimen. The critical spacing of flaws, l'_c, that determines the occurrence of either one of these two events is given by

$$l'_c = \frac{(\sigma_{fu} - \sigma^*)}{\sigma_{fu}} l_c \qquad (8.34)$$

This result indicates that the critical length l'_c for a weakened fibre is less than the critical length of a discontinuous fibre, l_c, by the factor $(\sigma_{fu} - \sigma^*)/\sigma_{fu}$.

When fibre pull-out occurs, the stresses in the fibres at the plane of matrix crack range from σ^* to $[\sigma^* + (4\tau/d)(l'/2)]$. Hence the mean apparent breaking stress of these fibres, for $l' \leqslant l'_c$, is

$$\bar{\sigma} = \sigma^* + \frac{\tau l'}{d} = \sigma^* + \frac{l'\sigma_{fu}}{2l_c} \qquad (8.35)$$

Assuming that the shear friction stress at fibre pull-out is τ_s, then the work done in pulling out a single fibre of length l is $\pi d\tau_s l^2/2$. The mean value of the work for a distribution of fibre length is then $\pi d\tau_s l^2/6$. Since the lengths of fibres pulled out vary from 0 to $l'/2$ the mean work per fibre is

$$\bar{W}_p = \frac{\pi d\tau_s l'^2}{24} \tag{8.36}$$

When the fibre length is longer than l'_c, there is still a chance that fibres will be pulled out. This is because the length of the segments lying in between the crack plane and the flaws may be less than $l'_c/2$. Hence, a fraction l'_c/l' of the fibres can break at the weak points. These fibres have the mean apparent breaking stress of $(\sigma_{fu} + \sigma^*)/2$. The others are broken near the plane of the matrix crack at stress σ_{fu} and essentially make no contribution to the pull-out fracture toughness. As a result, the mean breaking stress of all the fibres is

$$\bar{\sigma} = \sigma_{fu} - \frac{l'_c}{2l'}(\sigma_{fu} - \sigma^*) \tag{8.37}$$

The work of fracture per fibre averaged over all the fibres in the composite is

$$\begin{aligned}\bar{W}_p &= \frac{l'_c}{l'}\frac{\pi d\tau_s}{24}l'^2_c \\ &= \left[\frac{\sigma_{fu} - \sigma^*}{\sigma_{fu}} \cdot l_c\right]^3 \frac{\pi d\tau_s}{24l'}\end{aligned} \tag{8.38}$$

The static shear strength of the interface, τ, was identified with the stress necessary to cause failure of the resin–fibre interface in pulling a single fibre from the matrix. This strength is then used to calculate the l_c value. The shear resistance to sliding, τ_s, can be determined from the same experiments during the subsequent pulling-out of the fibre. The values of l_c and τ_s are used for computations in eqns. (8.35) through (8.38). From eqns. (8.35) and (8.37), it is seen that the mean fibre breaking stress varies linearly with l' for $l' \leqslant l'_c$ and approaches σ_{fu} when l' becomes very large. This result is indicated by the solid lines in Fig. 8.13 for several σ^*/σ_{fu} values. The variation of work of fracture per fibre with l' is calculated from eqns. (8.36) and (8.38) and is shown in Fig. 8.14. The agreement of theoretical predictions with experimental results is very satisfactory.

8.4.3 Fracture of Composites Reinforced with Discontinuous Fibres

The strength and work of fracture for composites with discontinuous fibres can be easily deduced from the case of continuous fibres containing flaws. This is done by letting $\sigma^* = 0$ for equations discussed in Section 8.4.2. The expression of critical flaw spacing l'_c is then reduced to that of critical length l_c of discontinuous fibres. The fibre mean stresses of eqns. (8.35) and (8.37) then assume the following forms for discontinuous fibres

$$\bar{\sigma} = \frac{\tau l}{d} \qquad\qquad l \leq l_c$$
$$\bar{\sigma} = \sigma_{fu}\left(1 - \frac{l_c}{2l}\right) \qquad l > l_c \tag{8.39}$$

The critical length of the fibres is then given by the expression of eqn. (8.14). It needs to be emphasised again that the stress τ used to calculate l_c is usually assumed to be equal to the shear strength of the metal matrix material for fibres well-bonded to the matrix.[5] In resin matrices, τ is often determined by the strength of the fibre–matrix interface. The work done in pulling out a fibre can also be deduced from eqns. (8.36) and (8.38)

$$\bar{W}_p = \frac{\pi d\tau_s l^2}{24} \tag{8.40}$$

for $l \leq l_c$, and

$$\bar{W}_p = \frac{l_c}{l}\frac{\pi d\tau_s l_c^2}{24} \tag{8.41}$$

for $l > l_c$. The meaning of τ_s for metal matrix and resin matrix composites also has been discussed. The variations of fibre mean stress and work of fracture with fibre length for discontinuous fibres are also shown in Figs. 8.13 and 8.14 ($\sigma^*/\sigma_{fu} = 0$). The relative magnitude of fracture work of composites with weakened fibres and discontinuous fibres is discussed in Problem 8.3.

It is shown in Section 8.4.2 that the work of fracture of the composite with fibres well-bonded to a ductile matrix is proportional to the fibre diameter. The cause of energy dissipation is the plastic deformation of the matrix material. Furthermore, the work of fibre pull-out is also proportional to the fibre diameter as indicated by eqns. (8.40) and (8.41). The choice of the optimum size of fibres should also take several other factors into consideration.[26] These include the strength and stiffness of the fibres and the ease of handling and coating application during fabrication.

8.4.4   The Statistical Nature of Brittle Fibre-Reinforced Composites

High strength brittle fibres are non-uniform in their strengths. Glass and silica fibres as well as whiskers generally show considerable variation in strength. This is due to the high sensitivity of fibres to surface imperfections. For instance, under a stress of 100 ksi ($6 \cdot 89 \times 10^8$ N/m^2) glass breaks if there is a surface crack of only 1/10 000 in. ($2 \cdot 54$ μm) in depth. On the other hand, strong steels can tolerate notches of up to an inch ($2 \cdot 54$ cm) in depth at the same stress level.[31] The sensitivity of brittle fibres to surface imperfections is further reflected by their strength dependency on fibre length. This is merely because a longer fibre has a greater chance of being weakened by surface imperfections. The understanding of the mechanics of these very brittle fibres in composite materials is best approached by examining the statistical nature of their failures.

One of the earliest problems investigated was concerned with the strength distribution of unbonded fibre bundles. It was first pointed out by Peirce[32] that the strength of a simple parallel array of filaments is not equal to the average strength exhibited by the component filaments when tested separately. A rigorous statistical theory of strength distributions for fibre bundles was first developed by Daniels.[33] Coleman[34] further investigated the implication of the work of Peirce and Daniels regarding the relative magnitudes of the strength of a large bundle and the mean strength of the component filaments. The following assumptions were adopted: (a) the long, continuous fibres in a bundle are uniform in cross-sectional areas; (b) the fibre tensile strength is independent of the rate of loading; and (c) the fibres come from a common source and the statistical properties of the strength distribution are independent of distance along the yarn and thus free from long-range trends. Based upon these assumptions and, further, noticing that fibres break at their weakest cross-section, Coleman was able to show that the tensile strength distribution of separate long fibres obeys the Weibull distribution function.[35]

The Weibull statistical strength distribution function is given by:

$$G(\sigma) = 1 - \left[1 - \left(\frac{\sigma - \sigma_\mathrm{l}}{\sigma_\mathrm{u}} \right) m \right]^{\omega}$$

$G(\sigma)$ represents the probability of fracture of a fibre at a stress level equal to or less than σ. σ_u and σ_l denote, respectively, upper and lower strength limits of the fibres. σ_u is believed to be affected by fibre composition and thermal history. The effects of surface damage may be measured by the parameter m. The parameter ω reflects the size effect.[125] Coleman then considered a bundle composed of a very large number of filaments of equal length. The bundle strength was defined as the force at break per initial unit area. An expression was obtained for the ratio ϵ of the strength of the bundle to the mean strength of its component filaments as a function of the coefficient of variation in the strength of the component filaments, $\sigma_1/\bar{\sigma}$. σ_1 and $\bar{\sigma}$ are, respectively, the standard deviation in strength and the mean strength of the fibres. The result is presented in Fig. 8.15. It is seen that the ratio of strengths decreases monotonically with increasing dispersion in the strength of the constituent filaments. Unless there is no dispersion in fibre strength, the tensile strength of a large bundle is always less than the mean fibre strength. The decrease in bundle strength relative to the average strength of a group of filaments is attributed to the progressive failure of filaments in the bundle at surface flaws. It is then obvious that the experimental results of single filaments tend to overestimate the strength of a bundle made of the same filaments. Statistical analyses are then essential to correlate filament strength with strand strength.[125,126] Furthermore, the interaction of fibre and matrix needs to

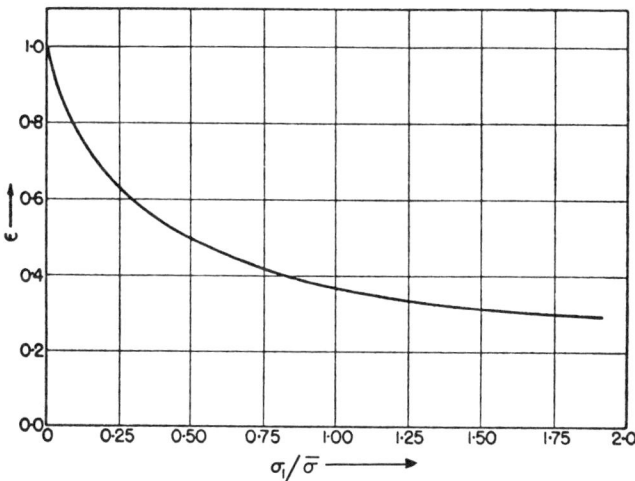

Fig. 8.15. *The strength efficiency, ϵ, for an infinite bundle versus the coefficient of variation in the strength of the component filaments, $\sigma_1/\bar{\sigma}$. (Reprinted with permission of Microforms International Marketing Corp. exclusive copyright licensee of Pergamon Press journal backfiles.)*

be taken into consideration when the strength of a composite is investigated.

The failure mode of composite materials as affected by the statistical nature of brittle fibre strength was first considered by Rosen.[36,37] As was stated previously, the strength of brittle fibres depends upon the degree of surface imperfection. The initial fracture of a composite under loading is most likely to occur in the brittle fibres in the regions of flaws. The manner of load redistribution near the fibre ends will affect the mode of fracture. Although it causes delamination and the propagation of transverse cracks, the stress concentration at the site of fibre breakage may not trigger fracture immediately. The increase in applied load may instead produce random fracture of the fibres at their weakened points. The final failure of the specimen will result from the accumulation of fibre breakage to such an extent that one weak cross-section is unable to carry the load transmitted to it. Consequently, the failure mode is one controlled by the statistical accumulation of fibre fractures. It should be noted that at the moment of incipient fracture, all the failure modes previously described may interact and result in the final failure.[37] In order to gain a clear understanding of this type of failure mode, it is helpful to refer to the experimental results of Rosen for direct observation of this type of fracture process. Plate XXI indicates the sequence of deformation of a single layer of E-glass–epoxy composite. The fibre volume fraction was about 50 per cent. The specimen was observed photoelastically. Under applied load, the fibres would brighten up while the broken region appeared black. It is seen that as the load increases, the fibres fracture in a random fashion instead of concentrated at the site of the initial crack. It was also concluded from the experiments that a strong and stiff matrix would enhance the possibility of the statistical failure mode.

A theoretical analysis of the 'cumulative weakening' failure mode was considered by Rosen (Fig. 8.16(a)). The strengths of composites are evaluated based upon the 'ineffective length' δ, namely, the length of a portion of a fibre over which the stress is perturbed by a fibre break. Rosen took δ to be that portion of the fibre in which the axial stress builds up from the fibre end to 90 per cent of the stress which would exist for infinite fibres. The ineffective length is determined from an analysis of the shear stress distribution along the fibre–matrix

Plate XXI.   *Failure sequence of the specimen at various percentages of the ultimate load. (Reproduced from Reference 37 by permission of the ASM.)*

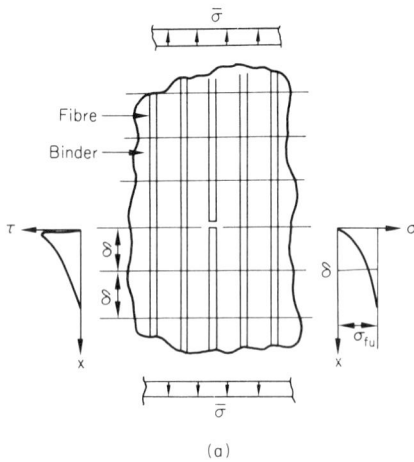

(a)

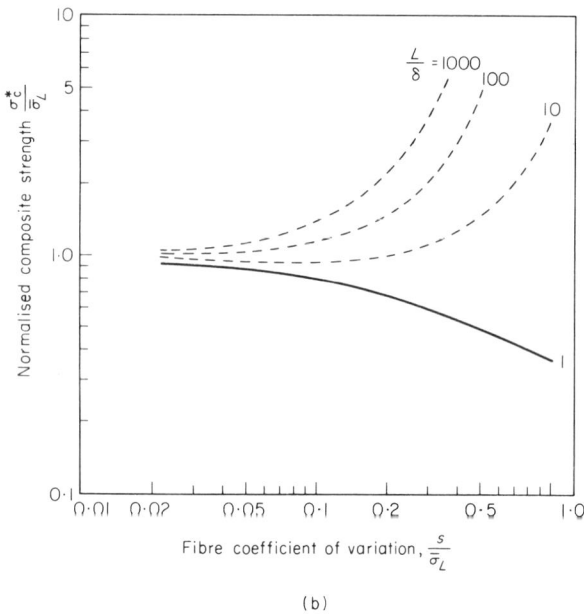

(b)

Fig. 8.16.   (a) *Tensile failure model*; (b) *effect of fibre coefficient of variation and reference fibre length upon the ratio of composite strength (the statistical mode) to mean fibre strength. (Reproduced from Reference 37 by permission of the ASM.)*

interface. The composite is then considered to be composed of layers of thickness equal to the ineffective length and the loading capacity of the matrix is neglected. The segment of a fibre in a certain layer may be considered as a link of a long chain. Each layer is then composed of a bundle of these links. When a segment in a layer breaks, it becomes ineffective in load carrying in that particular layer and the load in the layer is redistributed among the unbroken fibre segments. The strengths of the links in a fibre chain follow a statistical distribution. This distribution is determined from the experimental results of fibre strength–length relation. The distribution function for bundle strength can then be defined. Since the fibrous composite is composed of a series of bundles, the strength distribution of the composite is determined by applying the weakest-link statistical theorems.

The results of analyses for the 'cumulative-weakening' failure mode are shown in Fig. 8.16(b). The strength of the composite σ_c^* is normalised by dividing it by the mean strength $\bar{\sigma}_L$ of a set of individual fibres of length L. The fibre coefficient of variation is defined as the standard deviation S divided by the fibre mean strength at the same length. For composites with fibres of ineffective length δ, the strength is less than the mean fibre strength. When fibre lengths are greater than δ, the composite strength is less sensitive to the fibre coefficient of variation, and the composite strength is larger than the mean fibre strength. It is also seen that for a fibre coefficient of variation less than 15 per cent, the composite strength is close to the mean fibre strength. The experimental results on composites of glass-fibre reinforced epoxy resin follow the trend of the analytical result.

The effects on composite strength due to the statistical strength of fibres and the efficiency of the matrix in limiting the perturbation of local stresses at fibre ends have been demonstrated by the previous analysis. Further studies of similar nature have included the very important effects of stress concentrations in the vicinity of a fractured fibre. The consequence of the scatter in the monofilament strength and the progressive failure of the fibres is that the strength of some composites may deviate from the prediction of the rule of mixtures. Works concerning the stress concentration effect and other factors such as the size and distribution of internal cracks, compression failure and fatigue fracture in laminates can be found in the review articles of Rosen,[38] Hale and Kelly,[2] Rosen and Dow,[39] Argon,[40] and Rosen and Zweben.[128]

8.5 THE EFFECTS OF INTERFACES

It has been pointed out that the major function of the interfaces in composites is to transmit the load from the matrix to the reinforcing fibres. The behaviour of the interface determines not only the strength but also the mode of failure and the work of fracture of a composite. In order to optimise the performance of composite materials, it is necessary to gain some fundamental understanding of the nature of interfaces. The following discussion on interfaces in composites is made according to the metallurgical and mechanical considerations. The effects of crystalline defects such as interface dislocations are discussed in Chapter 3.

8.5.1 The Metallurgical Aspects of Interfaces

A desired interfacial region in a composite relies on several factors.[41] First, intimate contact between the fibre and matrix needs to be established through satisfactory wetting of reinforcing materials by the matrix. Secondly, good adhesion between the fibres and matrix is needed in order that adequate bonding can be developed. Thirdly, extensive diffusion between the component phases should be avoided so that the filaments will not be degraded by chemical reaction with the matrix phase. The choice of the fibre and matrix materials of a composite system often cannot satisfy these requirements at the same time as well as the requirements called for by the service conditions. One of the feasible ways of achieving a satisfactory interface while not having to sacrifice the high performance of the fibres is to apply a thin coating on the reinforcing materials. Besides achieving the purposes mentioned above, the coatings also prevent abrasion between filaments, and protect the filaments from corrosive environment in service. The method of using alloy additions also proved to be effective in reducing the diffusion of the component phases in a composite. In the following, the applications of coating and alloy addition in several composite systems are introduced.

Boron filaments, although excellent as reinforcing materials at low temperature, have shown poor oxidation properties and reactivity with metal matrices at elevated temperatures. Severe degradation of boron filaments in air was observed to occur at 500°C. The oxide layer formed on the surfaces of boron filaments tends to weaken the interface between the fibre and matrix materials. When boron fibres are used in a titanium matrix, the reaction product at the interface is a layer of titanium diboride. It was observed[42] that cracks that initiated in the brittle boride layer could extend into the boron filament. The tensile strength of the composites deterioriates as the thickness of the layer increases. In order to improve the high temperature performance of metallic materials reinforced with boron, the filaments are coated with a diffusion and oxidation inhibiting coating of β-silicon carbide.[43] Silicon carbide coated filaments are also known as Borsic filaments. Silicon carbide is chosen for its oxidation resistance, chemical inertness, and excellent strength retention at elevated temperatures. The coating is applied by thermally decomposing a halogenated silane compound, such as methyldichlorosilane, on the filament surfaces and is about 0·1 to 0·15 mil (2·54 to 3·81 μm) thick. The interfacial stability of the silicon-carbide coated boron in metal matrices is demonstrated by their room temperature ultimate tensile strength after prolonged heating at elevated temperatures.[44] The coated fibres were observed to retain all their room temperature strength of 500 ksi ($3·45 \times 10^9$ N/m^2) after 1000 hr of heating at 600 °C. On the other hand, the strength of uncoated boron filaments was reduced to nearly half of the room temperature value when heated for 1000 hr at 200 °C. The time required to cause degradation appears to be dependent on the time that boron takes to diffuse through the coated film.

When silicon carbide coated boron filaments were used to reinforce titanium, a significant improvement in tensile strength over the uncoated fibres was observed.[45] For instance, when the composite was subjected to heat treatment at 1600 °F (871 °C) for 10 hr the Borsic–titanium composite retained 95 per cent of its room temperature strength. However, under the same conditions the boron–titanium composite showed a strength retention of only 50 per cent. The transverse strength of the composite was relatively insensitive to the heat treatment time or the boride layer thickness.

The difficulty in wetting ceramic whiskers by molten metal matrices has posed a considerable problem in the fabrication process. Furthermore, some metal matrices may be corrosive to the filaments at the high infiltration temperature. The choice of the coating material should be such that it is chemically compatible with both the fibre and the matrix.[41] In some composite systems, the coating materials may be compatible with one component but is soluble in the other. In this situation a duplex coating consisting of two coatings needs to be applied to the fibres. The thin coatings of a wide range of materials can be satisfactorily applied by cathodic sputtering. In the sapphire whisker–silver system studied by Sutton and Chorńe,[46] a single coating of either platinum or nickel was applied to the whiskers to promote wetting by the matrix. When sapphire whiskers are incorporated into an aluminium matrix, either a single coating of nickel–chromium alloy or a titanium–nickel duplex coating can be used. The titanium–nickel duplex coating also has been used in one sapphire–silver system. It was found that the tensile strength of this system was nearly five times the strength of composites with uncoated whiskers.[41]

The addition of alloying elements in some matrices also proved to be an effective means of achieving compatibility between the filaments and matrix.[45] This has been demonstrated in boron composites by the alloying of titanium matrices with vanadium and a high vanadium content alloy (VCA). The two titanium alloys, Ti–13V–10Mo–5Zr–2·5Al and Ti–22V–3W–5Zr–2·5Al, were reported as having reaction rates with boron that were less than one-hundredth the rate of unalloyed titanium. Alloying of the matrix was also proved to be effective in eliminating reaction in the alumina–nickel system. A reaction layer of $NiAl_2O_4$ formed when the composite with a pure nickel matrix was heat-treated at elevated temperature. However, no visible reaction layer occurred when pure nickel was alloyed with 10 per cent of chromium. For a titanium matrix reinforced with alumina, a reaction layer also exists. Titanium alloys containing molybdenum and zirconium are believed to be effective in lowering the reaction rate in this case.

It may be concluded that satisfactory interfaces may be achieved in metal–matrix composites through the appropriate choice of matrix composition as well as the establishing of barriers to the reaction between the component phases. An equally important consideration is the appropriate choice of the fabrication techniques. Interfacial reaction and filament degradation can be significantly reduced if prolonged heating at elevated temperatures can be avoided.

The extensive uses of glass-fibre reinforced polymers have led to the study of the behaviour of interfaces and their effects on the mechanical properties. Considerable work has been done in the development of surface-treating compounds for glass fibres. The major functions of these compounds are to enhance the wetting of glass fibres by polymeric matrices and to develop high bond strength on exposure to water. The chemistry of the surface-treating compounds or coupling agents can be found in several review articles.[47–50] Two types of coupling agents are commonly used for glass. The first type consists of organometallic compounds such as organo complexes of cobalt, nickel and chromium. The second group consists of organosilane compounds. These coupling agents are believed to be able to develop bondings with both the glass substrate and the resin matrix. It also has been suggested that the coupling agents may form a barrier to water molecules and may also protect glass fibres against abrasion.[51] The thermal stresses resulting from the difference in thermal expansion coefficients of the fibres and the matrix and the shrinkage stresses due to resin polymerisation may also be relieved by the layers of coupling agents.

The failure of glass-fibre-reinforced plastics due to water damage has been studied by Ashbee, et al.,[52,53] among others. They first examined the dimensional change of the resin matrix due to water immersion at temperatures of 20, 60 and 100 °C. At each temperature, the first response to diffused water is resin swelling. In hot water, the swelling was then followed by shrinkage. The magnitude of the shrinkage increased as the duration of immersion was prolonged. It is believed that the shrinkage is due to the leaching of low molecular weight materials from the resin. Thin films of polyester resin containing glass fibres were then exposed to water at these same temperatures. The fibre debonding arising from resin shrinkage was studied from the observed optical anisotropy. When the composite specimens were immersed in water, the interfacial bond was rapidly destroyed by diffused water. However, if a coupling agent was coated on to the fibres, debonding occurred only in hot water after a long time of immersion. The mechanisms for debonding are believed to depend upon the compositions of the glass fibres. For E and C glass fibres, bond fracture was attributed to the osmotic pressure generated at the interface when water soluble constituents were leached from the fibre. For resins reinforced with fused silica fibres, debonding occurred at the fibre ends after much longer immersion time. This happened because the impurity content was small and debonding was mainly induced by the high interfacial shear stress.

Another consequence of interfacial debonding was the formation of indentation cracks in the resin. The fibre and resin could slide relative to each other after interface debonding. The glass fibres embedded in a shrinking matrix then acted as rigid indenters. Consequently, conical-shaped cracks were observed at fibre ends. A series of pictures showing the progressive penetration of a fibre into the matrix is given in Plate XXII. It is seen that as the first crack initiates and the resin continues to shrink, the fibre penetrates into the nearest crack surface

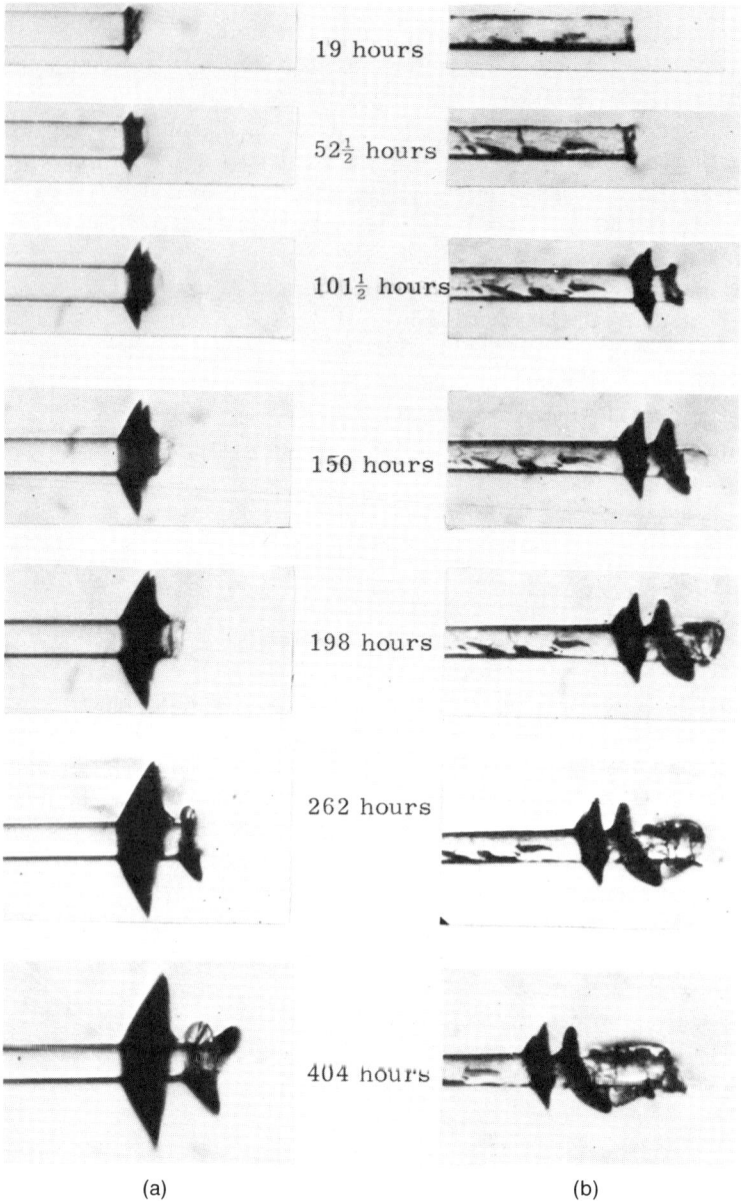

(a)   (b)

Plate XXII. *Growth of indentation cracks following debonding. E-glass fibres; (a) untreated and of length* 1·75 mm *and (b) treated with a coupling agent and of length* 2·25 mm. (*Reproduced from Reference* 53 *by permission of the Royal Society.*)

but also pushes against the outer one. Eventually, the outer crack surface is pierced by the fibre, and the crack moves toward the centre of the fibre. A second crack is then nucleated at the same fibre end. This process can repeat itself and results in multiple cracks at a fibre end and severe deformation in the matrix. The extent of indentation cracking diminishes when the water temperature and hence, the extent of interfacial debonding are decreased.

8.5.2 The Mechanical Aspects of Interfaces

The load transfer across an interface in fibrous composite materials has been examined by Cox[54] and Dow.[55] The composite model of Dow was made of two concentric circular components, simulating a fibre surrounded by the matrix material. Analytical solutions were obtained for the axial load transferred across the interface when either the fibre or the matrix was loaded in uniaxial tension. The interfacial load transferring phenomenon also has been studied by photoelastic experiments. For instance, Tyson and Davies,[56] and Schuster and Scala[57] examined the stress patterns around the reinforcing fibres. Schuster and Scala[58] also studied the interaction of stress patterns at whisker ends. Mullin *et al.*[59] were able to establish experimentally the critical fracture modes in boron fibre reinforced epoxy resin. It was further confirmed that the behaviour of interfaces played a significant role in affecting the strength of composites. In this section, the significance of mechanical compatibility of the component phases and its implication to the strength of composites is discussed in terms of some simple models.

The first model to be introduced is a bimaterial couple under uniformly applied stresses. Although the planar interface deviates from the situation in fibrous composites, the solution of the interfacial stress should provide us with insights of interface deformations in general. Figure 8.17(a) indicates a bimaterial couple. Some possible deformation modes of this couple was discussed by Hirth.[60] Under a uniform tensile stress parallel to the interface, several incompatibility effects are possible. One possibility is that the normal stress induces shear strains in the anisotropic component materials (Fig. 8.17(b)). If the component materials are free to rotate, such strains will induce rotation instead of stresses at the interface (Fig. 8.17(c)). If they are not free to rotate, as in compression tests, shear strains and shear stresses are introduced (Fig. 8.17(d)). Another possibility is that if the materials have different compliance constants along the tensile axis and they are unwelded, the response to tensile load would differ (Fig. 8.17(e)).

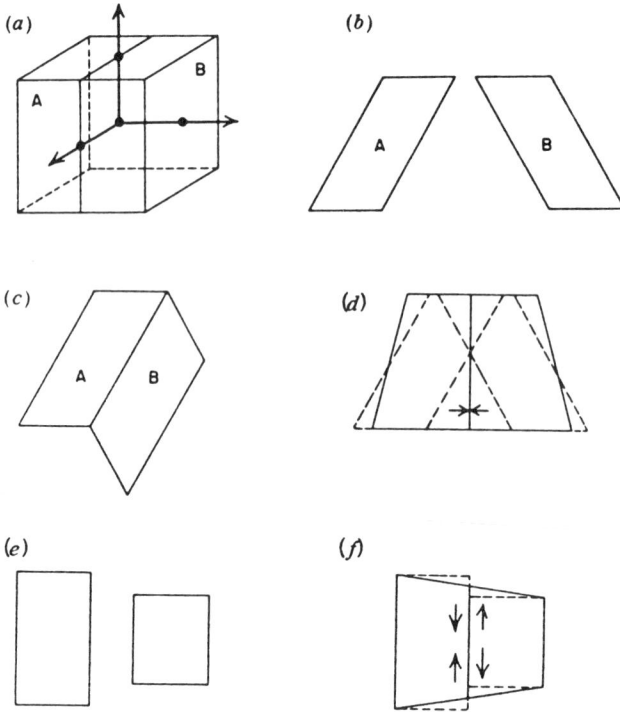

Fig. 8.17.   (a) Bicrystal; response to axial loading (b) unwelded, (c) with no end constraints, and (d) constrained. Cases (e) and (f), alternate response to axial loading. (Reproduced from Reference 60 by permission of the AIME.)

The compatibility requirement in this case when the materials are welded together, again leads to additional stresses at the interface.

To examine the stresses induced at the interfaces due to the compatibility requirement, Chou and Hirth,[61] and Chou and Olson[62] considered a bimaterial plate of thickness $2h$, composed of two elastic media welded together at $x = 0$. The origin of co-ordinate axes is at the centre of the interface and the media are infinite in extent in the x and z directions. The elastic constants are denoted by c'_{ij} and c''_{ij} for $x > 0$ and $x < 0$, respectively. Assuming that each phase possesses two orthogonal symmetry planes at $x = 0$ and $z = 0$, the elastic constants of the two adjacent phases assume the form of eqn. (5.6). First consider a uniformly applied antiplane shear stress $\sigma_{yz} = \sigma_a$ on the plate surfaces. The shear component σ_{xz} is induced throughout the bimaterial couple due to the compatibility requirement of the welded interface. At the interface, in particular, this stress takes the form

$$\frac{\sigma_{xz}}{\sigma_a} = \frac{2\bar{K}}{\pi\sqrt{c_1}} \ln \left[\sec \left(\frac{\pi y}{2h}\right) + \tan \left(\frac{\pi y}{2h}\right) \right] \tag{8.42}$$

where

$$\bar{K} = (c''_{44} - c'_{44})/\{c''_{44} + c'_{44}\sqrt{[(c'_{55}c''_{44})/(c''_{55}c'_{44})]}\}$$

and

$$c_1 = c'_{44}/c'_{55}$$

Equation (8.42) indicates that a high stress concentration occurs at both ends of the interface. Plastic relaxation or cracking will occur in regions near these singular points. Hence, the highly concentrated σ_{xz} near the surfaces of the plate can be the primary cause of plastic yielding and/or splitting of the interface under the specified load.

Both stress components σ_{yz} and σ_{xz} decrease exponentially with distance from the interface under the shear loading. On the middle plane of the plate, σ_{yz} can be expressed in closed form as

$$\frac{\sigma_{yz}}{\sigma_a} = 1 - \bar{K}\frac{4}{\pi}\tan^{-1}\left\{\exp\left[-\frac{\pi\sqrt{(c_1)}x}{2h}\right]\right\} \qquad y = 0, x > 0$$

$$\frac{\sigma_{yz}}{\sigma_a} = 1 - \tilde{K}\frac{4}{\pi}\tan^{-1}\left\{\exp\left[\frac{\pi\sqrt{(c_2)}x}{2h}\right]\right\} \qquad y = 0, x < 0, \tag{8.43}$$

where $c_2 = c''_{44}/c''_{55}$ and $\tilde{K}$ is obtained from $\bar{K}$ by interchanging c'_{ij} with c''_{ij}. It is seen that at locations far away from the interface the stress component σ_{yz} reduces to that of applied stress. When both components are isotropic, eqns. (8.42) and (8.43) can be simplified by noting that $c_1 = c_2 = 1$ and $K = -\tilde{K} = (G_2 - G_1)/(G_2 + G_1)$. G_2 and G_1 denote the shear moduli of the materials. A further implication of the stress distributions in the bimaterial system has to do with the movement of dislocations. By summing up the forces acting on a dislocation due to the stress components σ_{xz} and σ_{yz}, it can be shown that dislocations tend to form pile-ups against the interface in the component with the higher shear modulus. These dislocation pile-ups will certainly enhance the chance of fracture and plastic deformation in the bimaterial systems.

When the isotropic bimaterial couple is subjected to a uniform compression σ_{yy} on the surfaces, the compatibility requirement at the interface also causes additional stresses.[61] Among them, a normal component σ_{xx} is induced. At the interface, the concentration of σ_{xx}

occurs near the plate surface and is tensile in nature. With regard to mechanical properties of bimaterial systems, or by extrapolation to composite materials, the stress σ_{xx} would tend to cause debonding at interfaces. The other stress component also existing at the interface is the shear component σ_{xy}. In composite materials with weakly bonded interfaces, this shear component can be the primary cause of splitting of phase boundaries. Consequently, this could lead to crack blunting at the interface in the manner described by Mullin, et al.[59] and by Tyson and Davies.[56] The magnitudes of both the normal and the shear stresses decrease exponentially with distance from the interface. The effect of interfacial stresses becomes more pronounced as the difference between the elastic constants of the two constituent phases increases.

The above discussions on the effects of interfaces are centred on the phase boundary between the component materials. Another type of interface in fibre-reinforced composites is found at the interface between laminates. Since delamination in composites is an important failure mode, it is necessary to look into the interlaminar stress distribution and its implication on fracture. Figure 8.18(a) depicts a composite specimen consisting of two laminae with fibres oriented at $+\theta$ and $-\theta$. In order to understand the response of the laminate to tensile loading, it is most convenient to examine the behaviour of the individual lamina. Under external force, the two laminae tend to deform in the manner shown in Fig. 8.18(b), if each one is stressed independently of the other. The shape of the deformed lamina results from the fact that the reinforcing fibres tend to rotate to the orientation aligned with the tensile axis. In an angle-ply composite, an individual lamina is not free to deform due to the constraint from the neighbouring one. As a result, shear stresses are induced at the interfaces. The interfacial shear stress on lamina II due to lamina I is indicated in Fig. 8.18(b) by the arrows. The direction of the interfacial shear stress on lamina I is opposite to that on lamina II. The magnitude of the interfacial shear stress is affected by the strength of the individual lamina and the orientation of the fibres. These shear stresses acting at the interface are responsible for the delamination of the composite if they exceed the strength of the matrix or the interface.

Another consequence of the interfacial shear stress has to do with the coupling between stretching and twisting of the angle-ply laminate discussed in Section 5.4. The interfacial shear stresses on lamina II in Fig. 8.18(b) can be resolved into that shown in Fig. 8.18(c). A free body

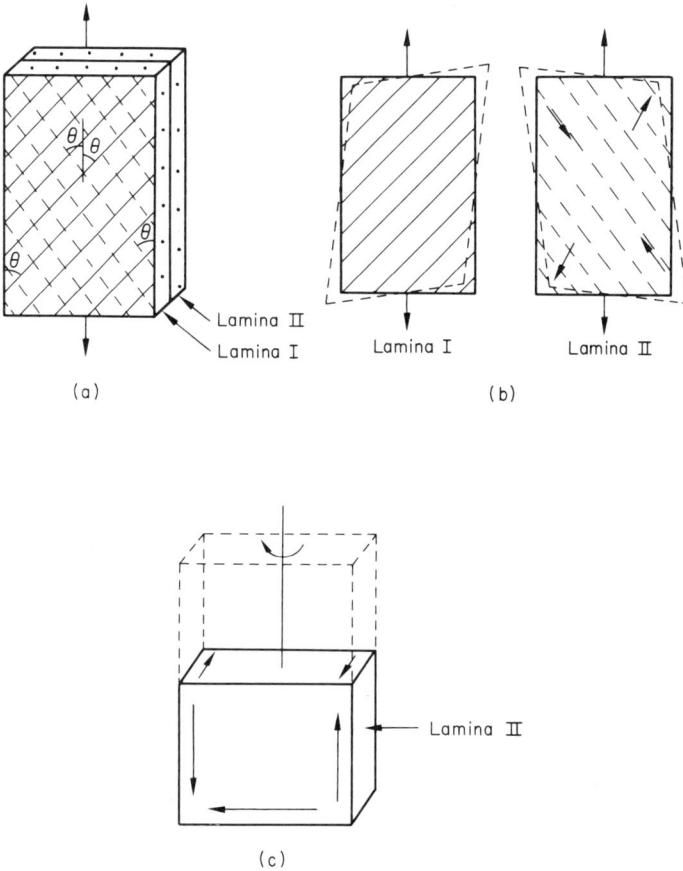

Fig. 8.18.   Deformation of a cross-ply laminate.

of this lamina is then taken to show the horizontal section. The directions of the shear stresses acting on this section are also indicated. It is then obvious that these shear forces tend to rotate the lower half of the lamina II in the direction indicated. A similar effect also occurs on lamina I. Consequently the cross-ply laminates are twisted by the axial tension. A general discussion of the influence of interlaminar stresses on the failure characteristics of composite laminates can be found in Pagano and Pipes.[117]

8.6   FRACTURE ANALYSIS OF FIBROUS COMPOSITES

8.6.1   Theory of Strain–Energy Density

The stress-intensity factors introduced in Section 8.3 for the various modes of deformation were derived from the analysis of the stress field near a crack tip. Unstable crack extension under tensile loading occurs when the stress intensity factor K_I reaches a critical value K_{Ic}, also known as the plane-strain fracture toughness. The concept of fracture toughness has become one of the most important tools in the fracture-safe design of high strength–low toughness materials. For the case when both the Mode I and Mode II deformations are existing, it has been suggested[63,64,69,136] that the failure criterion should be that the summation of energy-rates $G_1(= (1 - \nu^2)K_I/E)$ and $G_2(= (1 - \nu^2)K_{II}/E)$ reaches a critical value G_c at the onset of unstable crack propagation. However, the obvious deficiency of this criterion is that the crack is assumed to propagate in the direction collinear with the original crack plane. Experimental results indicate that cracks inclined to the tensile axis do not advance along the original crack plane.[92] The inadequacy of the classic concept of fracture mechanics has compounded the difficulty in the fracture analysis of fibrous composites. This is because the fibres in a laminated structure are oriented in directions not necessarily coinciding with the principal stress directions. As a result, cracks propagating in the direction parallel to the fibre orientation present a problem that cannot be solved by the classic theory.

The prediction of the direction of crack propagation and the criterion for incipient fracture have always been challenging problems to

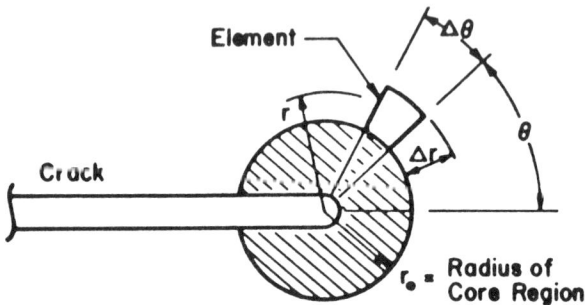

Fig. 8.19.   *Element outside core region around the crack tip. (Reproduced from Reference 68 by permission of the Technomic Publishing Co.)*

structural engineers. The development of the strain–energy density theory, recently made by Sih,[65-8,137,138] has not only greatly broadened the capability of linear fracture mechanics but has also made possible some fundamental understandings of the fracture processes in composite materials. The basic concept of the strain-energy density theory is introduced in the following. Figure 8.19 indicates a crack and a region associated with the crack tip. This region is called the core region, which has a radius of r_0. The size of the core region is of the same order of magnitude as the crack tip radius and this region is assumed to be free of mechanical defects of size comparable to r_0. It is also assumed that the solution of continuum mechanics is valid outside of this core region. The amount of energy stored in an incremental area $\Delta A = r\Delta\theta\Delta r$ outside of the core region, under plane strain, is (see Problem 8.4)

$$\frac{dW}{dA} = \frac{1+\nu}{2E}(\sigma_x^2 + \sigma_y^2 - \nu(\sigma_x + \sigma_y)^2 + 2\tau_{xy}^2) \tag{8.44}$$

dW/dA is called the local strain-energy density function. E and ν denote the Young's modulus and Poisson's ratio, respectively. The stress expressions in the above equation are readily obtained from eqns. (8.23) to (8.25). When the crack is deformed under more than one fracture mode, the summation of the same stress component from each mode is used in eqn. (8.44). The resulting expression of the strain-energy function is

$$\frac{dW}{dA} = \frac{1}{r}(a_{11}K_1^2 + 2a_{12}K_1K_2 + a_{22}K_2^2 + a_{33}K_3^2) \tag{8.45}$$

K_1, K_2 and K_3 are the stress-intensity factors for Mode I, II and III deformations, respectively. Their expressions are obtained by dividing the expression for K in eqn. (8.26) by the factor $\pi^{1/2}$. The coefficients a_{11}, a_{12}, a_{22} and a_{33} are given by

$$a_{11} = \frac{1}{16G}[(3 - 4\nu - \cos\theta)(1 + \cos\theta)]$$

$$a_{12} = \frac{1}{16G}(2\sin\theta)[\cos\theta - (1 - 2\nu)]$$

$$a_{22} = \frac{1}{16G}[4(1 - \nu)(1 - \cos\theta) + (1 + \cos\theta)(3\cos\theta - 1)] \tag{8.46}$$

$$a_{33} = \frac{1}{4G}$$

where G is the shear modulus. The strain–energy density varies inversely with r and is a maximum at the boundary of the core region. The intensity of this strain energy which varies along the periphery of the region $r = r_0$ is denoted by

$$S = a_{11}K_1^2 + 2a_{12}K_1K_2 + a_{22}K_2^2 + a_{33}K_3^2 \qquad (8.47)$$

S is called the *strain–energy density factor*. The significance of the strain–energy density factor is that it represents the variation of local strain–energy density around the region of fracture initiation.

The criterion for the direction of crack extension is determined by the stationary value of the strain–energy density factor, namely

$$\frac{\partial S}{\partial \theta} = 0 \quad \text{at} \quad \theta = \theta_0 \qquad (8.48)$$

The extension of the crack occurs when the strain–energy density factor reaches a critical value S_c, namely

$$S = S_c \quad \text{for} \quad \theta = \theta_0 \qquad (8.49)$$

From eqn. (8.47) it is seen that the critical strain–energy density factor depends only on the stress-intensity factors. Hence, the knowledge gained in the classical theory of linear fracture mechanics can still be utilised. On the other hand, this new concept of strain–energy density factor has the added advantage of determining the condition for crack extension and the direction of crack initiation simultaneously. Since S_c is again a material property, it can be determined from our knowledge of fracture toughness tests.

8.6.2   Fracture Mechanics of Unidirectional Composites

The strain–energy density theory has been employed by Sih et al.[67,68] to analyse the fracture of unidirectionally reinforced fibrous composites. The composite material they considered is the Scotchply 1002, a glass-fibre reinforced epoxy tape. The mechanical properties of this tape are given in Table 2.6. Under tensile loading, cracks in the tape were observed to start from defects such as voids and air bubbles in the matrix. These cracks tend to propagate in the matrix along the direction of the reinforcing fibres. The geometry of the composite specimen along with the crack is shown in Fig. 8.20.

Since the applied tensile stress can be resolved into stresses both normal and parallel to the crack plane, the strain–energy density factor of eqn. (8.47) is then reduced to

$$S = a_{11}K_1^2 + 2a_{12}K_1K_2 + a_{22}K_2^2 \qquad (8.50)$$

with the coefficients given by eqn. (8.46). The Scotchply 1002 tapes have the following orthotropic elastic properties: $E_1 = 5 \times 10^6$ psi $(3 \cdot 45 \times 10^{10} \text{ N/m}^2)$, $E_2 = 1 \cdot 67 \times 10^6$ psi $(1 \cdot 15 \times 10^{10} \text{ N/m}^2)$, $G_{12} = 7 \cdot 04 \times 10^5$ psi $(4 \cdot 85 \times 10^9 \text{ N/m}^2)$ and $\nu_{12} = 0 \cdot 05$,[69] where the 1 and 2 directions are parallel and transverse to the fibres, respectively. The matrix material is isotropic with $E = 4 \cdot 5 \times 10^5$ psi $(3 \cdot 10 \times 10^9 \text{ N/m}^2)$ and $\nu = 0 \cdot 35$. The stress-intensity factors in eqn. (8.50) have been obtained by simplifying the crack geometry in Fig. 8.20. It is assumed that the crack situated in the isotropic matrix material is sandwiched by two orthotropic media with the same elastic property as the composite materials. For the case that the matrix layer thickness is much smaller

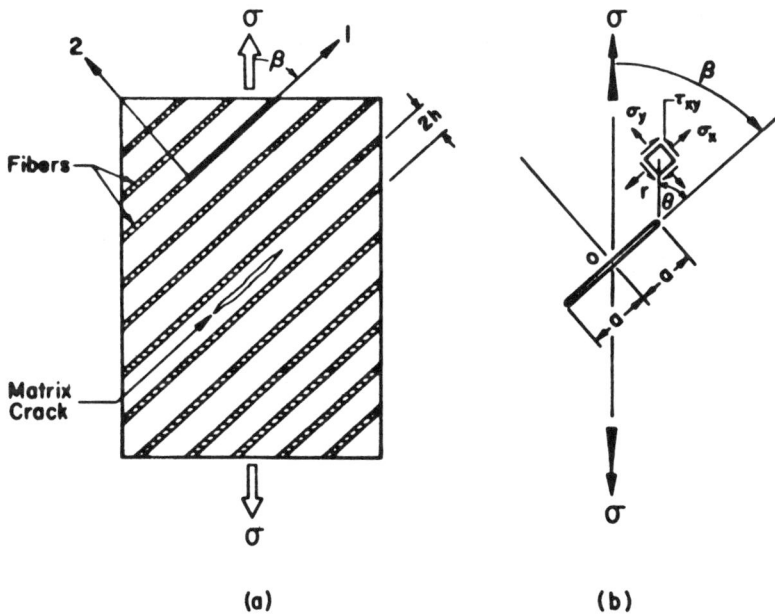

(a)   **(b)**

Fig. 8.20.   *Fibre composite in tension: (a) crack parallel to fibre, (b) idealised line crack. (Reproduced from Reference 68 by permission of the Technomic Publishing Co.)*

than the crack length, the stress-intensity factors are,[70]

$$K_1 = 0 \cdot 29 \sigma a^{1/2} \sin^2 \beta$$
$$K_2 = 0 \cdot 17 \sigma a^{1/2} \sin \beta \cos \beta \qquad (8.51)$$

Substituting eqns. (8.46) and (8.51) into eqn. (8.50), the general expression of the strain–energy density factor is

$$S = \sigma^2 a (0 \cdot 084 a_{11} \sin^4 \beta + 0 \cdot 098 a_{12} \sin^3 \beta \cos \beta + 0 \cdot 029 a_{22} \cos^4 \beta) \qquad (8.52)$$

For each fibre orientation, β, the direction of crack extension can be found by taking the derivative of S with respect to the angle θ. The corresponding value of S_c is then obtained from eqn. (8.52).

Comparisons of the analytical results with experimental data are shown in Fig. 8.21. The solid curves were obtained for various half crack lengths, a, by holding S_c constant. The experimental results[69] on cracked specimens with $a = 0 \cdot 51$ in. $(1 \cdot 30 \text{ cm})$ agree well with the theory especially for β in the range of $30°$ to $90°$. The theoretical curve for $a = 0 \cdot 02$ in. $(0 \cdot 05 \text{ cm})$ is close to the experimental result on uncracked specimens (dotted line). It is supposed that the uncracked specimens might actually contain cavities and air bubbles[69] of size comparable to the assumed flaw size. The outer broken curve was obtained by taking $(\sigma_c)_{\beta = \pi/2} = 6 \cdot 8$ ksi $(4 \cdot 69 \times 10^7 \text{ N/m}^2)$ and then transforming this stress to other values of β using a Mohr's circle analysis. The variations of composite critical stress with crack length at constant S_c are given in Fig. 8.22. The agreement between the theory and experiment is good for larger β values.

The discrepancy in the theoretical and experimental results is due to the increase in fibre fracture at a smaller misalignment between the fibre and tensile axis. The fracture of the matrix is no longer the only controlling factor in energy dissipation. It is evident from the discussion of Section 8.3 that the modes of fracture in fibrous composites are extremely complicated. It is not yet possible to predict the deformation of composites involving several failure modes, even on a qualitative basis. Nevertheless, the advancement made in the strain–energy density theory is significant toward the understanding of fracture processes in laminated fibre composites.

8.6.3   Failure Analysis by Acoustic Emission

The cracking of fibres which eventually leads to the failure of a composite is a process that cannot be readily detected. The observa-

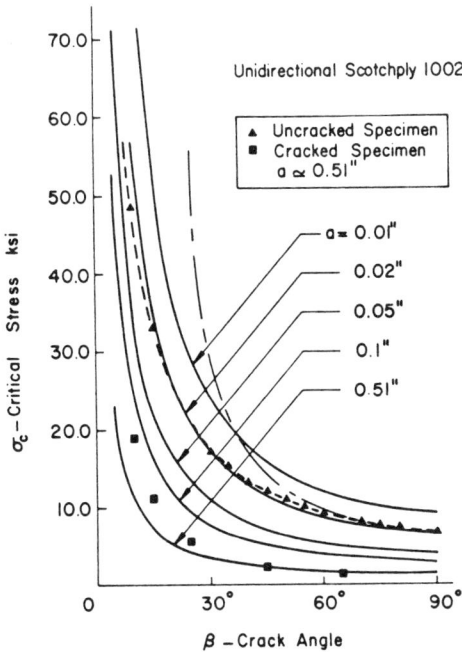

Fig. 8.21. *Critical stress versus crack angle. (Reproduced from Reference* 68 *by permission of the Technomic Publishing Co.)*

tion of fibre cracking may be made on the surface of a specimen using metallographic techniques or through photoelastic analyses.[71,118] In this section, the application of acoustic emission measurements to composite materials is introduced. Acoustic emission is the low-level sound emitted by a body under deformation.[72] The source of acoustic emission is the release of stored strain energy of a material. The sounds or stress waves emitted by materials provide information on the characteristics of the deformation and can be used to detect the possibility of impending failure of structures. The monitoring of stress waves of the earth by seismometers used for detecting and predicting earthquakes is a familiar example. Acoustic emission techniques have been applied for the purposes of materials research and structural integrity studies. One of the important characteristics of acoustic emission is its irreversible nature. A deformed material will not generate acoustic emission during reloading until the stress level

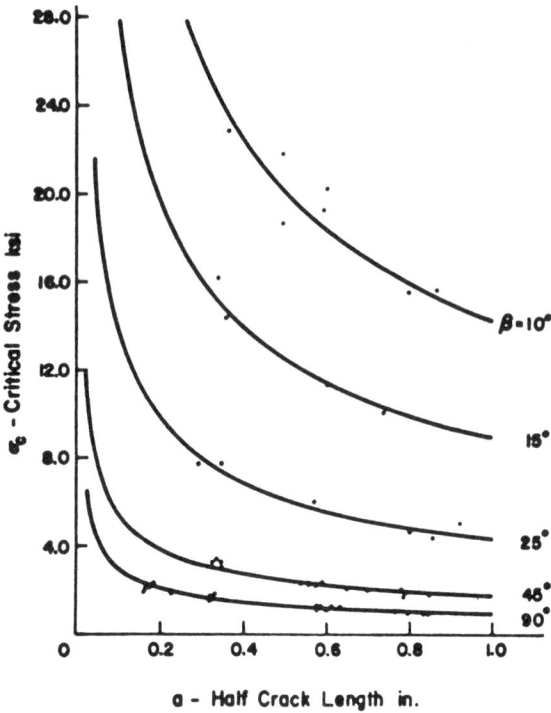

Fig. 8.22.   Variations of critical stress with half crack length. (Reproduced from Reference 68 by permission of the Technomic Publishing Co.)

exceeds its previous high. Acoustic emissions in the form of discrete pulses can be detected by a transducer when the emission is above its threshold level. Both piezoelectric and magnetostrictive type transducers can be used. Applications of this technique to the study of plastic deformations, phase transformation and fracture of solids can be found in various review articles.[72–4,119]

In composites, Harris, *et al.*[75] examined fibre cracking in Al_3Ni whisker reinforced aluminium by use of acoustic emission. They related the acoustic emission to the micromechanics of the deformation processes in a quantitative manner and thus demonstrated the feasibility of non-destructive evaluation of the composite integrity. The composite specimens were prepared by a unidirectional solidification of the Al–Al_3Ni eutectic alloy. Rod-like Al_3Ni whiskers were produced. The regions in the specimens near the loading pins were preloaded to values greater than the fracture load of the specimen. Thus, noise from these highly stressed regions can be eliminated. The cracking of fibres was first recorded by direct observation. The specimens loaded to various strain levels were unloaded and the number of cracked fibres was determined by using an optical microscope. The measured percentage of broken fibres, $\omega(\epsilon)$, as a function of strain, ϵ, was curve-fitted into the following equation

$$\omega(\epsilon) = a[1 - \exp(-b\epsilon^c)] \tag{8.53}$$

where $a = 0.427$, $b = 2.590$ and $c = 0.5385$. This is shown in Fig. 8.23.

The acoustic emission detected from the specimen was amplified, filtered and fed into a counter. The voltage of the amplified sensor signal was assumed to be a damped sinusoid of the form

$$V = V_0 e^{-\beta t} \sin \lambda t \tag{8.54}$$

In this equation, V_0 is the initial voltage. λ and β are constants and t denotes the time. Only the sensor signals that exceeded the trigger voltage V_t of the counter were counted. Consequently, from eqn. (8.54), it is obvious that a single event in a specimen will result in several counts before the voltage rings down to values below V_t. A

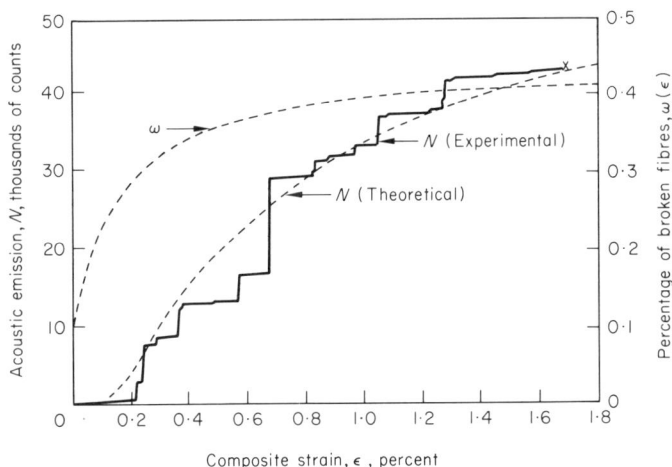

Fig. 8.23. *Summation acoustic emission and percentage of broken fibres as functions of composite strain. (Reproduced from Reference 75 by permission of ASTM.)*

larger event thus naturally gives more counts than a smaller one. The measured summation of acoustic emission counts as a function of composite strain is also shown in Fig. 8.23. From eqn. (8.54), the number of counts n from a single event can be expressed as

$$n = \frac{\lambda}{2\pi\beta} \ln \frac{V_0}{V_t} \qquad (8.55)$$

In view of the low strength and high purity of the aluminium matrix, it was assumed that all the emissions originated from fibre cracking. It was further assumed that the energy released during a fibre cracking is proportional to the square of the fibre fracture stress and also to the square of the initial voltage V_0. Based upon these assumptions, eqn. (8.55) can be rewritten as

$$n = \frac{\lambda}{2\pi\beta} \ln \frac{\epsilon}{\epsilon_0} \qquad (8.56)$$

where ϵ is the strain at fibre breaking and ϵ_0 is the strain corresponding to the threshold condition. Finally, the total number of counts N as a function of strain is obtained by integrating the above expression

$$N(\epsilon) = B \int_{\epsilon_0}^{\epsilon} \frac{d\omega}{d\epsilon} \ln \frac{\epsilon}{\epsilon_0} \, d\epsilon \qquad (8.57)$$

where B is a constant. By using the expression for ω in eqn. (8.53) and the appropriate choice of the values of constants B and ϵ_0, a relation between N and composite strain ϵ was obtained. This is given in Fig. 8.23.

The implications of this investigation are as follows. It demonstrates that the acoustic emission of a composite specimen due to fibre breaking can be predicted if the fibre cracking characteristics of the material are known. It is often easier to detect fibre cracking from acoustic emission measurements than by direct observation. Hence, a desirable application of this model will be to predict $\omega(\epsilon)$ by using the data of N from acoustic emission tests.

Acoustic emission studies were also made by Fitz-Randolph et al.[76] on epoxy reinforced with continuous boron filaments. The number of broken fibres in a specimen at each stage of fracture was determined by both electrical resistance measurements and a compliance calibration technique. Their results suggested that in this boron–epoxy system each fibre failure is accompanied by the same amount of emission and that acoustic emission monitoring can be used to determine the number of broken fibres in this composite. Studies of acoustic emission produced during burst tests of filament-wound bottles can be found in Reference 120.

The acoustic emissions, in the studies introduced above, were all attributed to the fracture of reinforcing fibres. However, efforts also have been made to discriminate between the energy released from specific events in the failure processes such as fibre fracture, matrix cracking and interfacial debonding.[77] The transmission of sound from a specific local event through a composite is a complicated process. Thus, a sound pulse monitored by the transducer may not be exactly the same as that which occurred some distance away. However, since it is possible to control the local failure mechanisms in composites, the specific acoustic signature associated with each event can also be identified. Consequently, if each event gives the same response whenever it occurs, it is possible to analyse the acoustic emission from a composite and to reconstruct the failure process based upon the characteristics of each individual failure event.

8.6.4 Cracks in Non-homogeneous Media

The solution of the elastic fields associated with cracks in non-homogeneous media is important to the understanding of fracture processes in composite materials. The presence of the second-phase material considerably complicates the elasticity problem. Earlier solutions for problems in non-homogeneous media mostly considered the bimaterial system made of two isotropic elastic half planes. The cracks were assumed to be either normal or parallel to the phase boundary. More recently, the solutions have been expanded to take into consideration the layered structures. Solutions of this kind are important to the understanding of fracture in laminated materials and in adhesive joints. These solutions may also provide some insight into the crack problems of unidirectionally reinforced composite tapes. Both isotropic and anisotropic media have been assumed in the layered structures. The problem of a unidirectional composite with a broken fibre also has been attempted. In view of the complexity of fibre interaction, exact solutions have only been obtained for very idealised cases. A review of the elasticity solutions concerning cracks in non-homogeneous material systems are given in the following.

The simplest geometry of a non-homogeneous system may consist of two elastic half-planes. The bonding is usually considered to be perfect along the interface. Williams[78] first examined the stresses around a crack lying in the interface. It was found that under Mode I and II deformations, the singularity of the stress field is of the type inversely proportional to the square root of the distance r from the crack tip, just as in a homogeneous medium. It was also realised that the stresses possess a sharp oscillatory character of the type $r^{-1/2} \sin(b \log r)$ or $r^{-1/2} \cos(b \log r)$ where b is a function of material constants. However, this behaviour is believed to be confined quite close to the crack tip.

Zak and Williams[79] later examined the crack point stress singularity where the plane of the semi-infinite crack is normal to the bimaterial interface. Under Mode I deformation, the crack tip stresses at a distance r are of the type $r^{\lambda-1}$. It is noted that as the cracked phase becomes harder than the uncracked phase, the order of singularity, $1 - \lambda$, increases. By assuming that the Poisson's ratios are 0·3 for both phases, the value of λ changes from 0·5 for two identical phases to 0·25 when the ratio of shear moduli of the cracked phase to the uncracked phase reaches 10. In the limiting case where the cracked component behaves as a rigid phase, λ approaches zero. The behaviour of the crack tip stress field in this case is characteristically different from the case where the crack lies along the interface.

Bogy[139] examined a more general case where a crack loaded on the surfaces terminates at a material interface. He was interested in finding out at what angle between the bimaterial interface and the crack plane the stress singularity at the crack tip is most severe. It has been found that when the crack is in the softer constituent the stress singularity is most severe if the crack lies in the interface; whereas, when the crack is in the stronger constituent a particular angle between the crack plane and the interface is associated with the most severe stress singularity for each composite.

The solution of crack problems in non-homogeneous materials also can be achieved if the elastic fields of dislocations are known. The essence of this method is that the Burgers vector of a dislocation represents a discontinuity in displacement of the elastic medium. Consequently, the crack opening displacements associated with the three modes of cracks can be simulated by appropriate arrays of dislocations. Each dislocation in an array acts as a Green's function of the elastic field. Chou[80] employed this method to obtain the stress singularity at the tip of a crack normal to a bimaterial interface under Mode III deformation. The stress singularity is of the type r^{-a} where $a = (2/\pi) \sin^{-1} [(1 - k)/2]^{1/2}$ and $k = (G_2 - G_1)/(G_2 + G_1)$. G_1 and G_2 are the shear moduli of the cracked and uncracked phases. The value of a, and hence, the order of singularity increases from zero at $k = 1$ to $0\cdot5$ at $k = 0$. The value of a approaches unity when k approaches -1, namely, the case that the cracked phase is nearly rigid compared to the uncracked phase.

The problem of two semi-infinite elastic planes with different elastic properties bonded to each other along a finite number of straight-line segments was examined by Erdogan,[81] and Rice and Sih.[82] As a result of this type of bonding, cracks finite or semi-infinite in length appear at the interface. Under Mode I and II deformations, the crack-tip stress singularities are again of the type of $r^{-1/2}$ and possess a pronounced oscillatory character. However, this oscillatory behaviour of the stresses is confined to a small region around the crack tip in which the material is beyond the elastic limit and where the linear theory of elasticity would not apply. Hence, for all practical purposes, this oscillatory character of the local stress can be ignored. The concept of stress intensity factor has also been extended to cracks in non-homogeneous systems. The expression for the stress-intensity factors for some crack geometries have been derived. It is noted that the stress intensity factors generally depend upon the elastic properties of the constituent phases. Rice and Sih found that for a finite crack at an interface, both the symmetric and skew-symmetric loadings were intermixed in the expressions for K_I and K_{II}. As a result, these stress intensity factors do not have the simple physical interpretation as in the homogeneous case. Furthermore, for the two half-planes with different elastic constants, the application of either a symmetric tensile loading or a skew-symmetric shear loading results in stress intensity factors of both K_I and K_{II}. Consequently, in the application of the Griffith–Irwin theory of fracture, it is necessary to assume that a function of K_I and K_{II} reaches some critical value when the crack grows in an unstable manner. The problem of bending of plates of dissimilar materials with interface cracks also was examined by Sih and Rice.[83]

Erdogan[84] furthered the study of stress distributions in bonded dissimilar materials with cracks by taking into consideration several other forms of external loads. These loads include the tractions on the crack surfaces, in-plane moments, residual stresses due to temperature changes, a concentrated load and couple acting at an arbitrary location in the plane, and a concentrated load on one side of the crack. The importance of the solution involving a concentrated load is that it may be used as a Green's function to generate solutions of problems with arbitrary loadings or crystalline defects. The crack problems introduced so far are two-dimensional. One of the three-dimensional problems considered involves a series of concentric ring-shaped cracks at a bimaterial interface.[85] The stress distributions around cracks have been obtained for some special cases. These include a penny-shaped crack, and an external circumferential crack on the interface. The problem of an interfacial crack also has been examined by Willis,[86] Lowengrub and Sneddon,[87,132] and Lowengrub.[88] In References 87 and 88, one of the elastic media was considered to be rigid. England[89] considered the problem of a single Griffith crack opened up by normal pressures between two bonded dissimilar half planes. He showed that the solution is physically inadmissable since it predicts that the upper and lower surfaces of the crack should wrinkle and overlap near its ends.

As a generalisation of the situation where two semi-infinite planes of dissimilar materials bonded along a straight line, England,[129] and Perlman and Sih[90] considered curvilinear cracks in bimaterial systems. The problem considered by Perlman and Sih dealt with an infinite plane containing an opening, the boundary of which is joined along circular-arc segments to an insert with different elastic properties. Exact solutions to several cases of practical importance were obtained. These included the cases of uniaxial tension applied at infinity and concentrated forces located in the insert or in the surrounding material. It was found that the interfacial stresses near the tips of a curved crack possess the same character of singularity as the one obtained for a straight crack between dissimilar media. For a crack with sufficiently small arc length, the crack-like imperfection has little influence on the stresses along the bond. An examination of the particle-matrix interfacial stresses indicates that both the normal and shear components are greater when the inclusion is made more rigid than the outer phase as compared to the case where the inclusion is less rigid than the surrounding material. The stress intensity factors for curvilinear cracks in non-homogeneous media were also derived by Perlman and Sih for the cases when the bimaterial system was under uniaxial tension and uniform tension in all directions. The results demonstrated the following facts. First, the stress intensity factors are functions of the elastic constants and the crack geometry. Secondly, a criterion of fracture for cracks in non-homogeneous media depends on a combination of K_I and K_{II} reaching some critical value regardless of the manner in which the external load is applied.

The treatment of cracks in non-homogeneous media also has been extended to anisotropic materials and to elastic-plastic materials. The problem of a crack between dissimilar anisotropic media has been treated by Gotoh[130] and Clements.[91] The problem of a crack in a layered structure treated by Hilton and Sih[70] also considered the effects of elastic anisotropy. The two elastic half planes were assumed to be orthotropic to simulate the anisotropy of fibre-reinforced laminates. The dislocation model of cracks has been extended to anisotropic media by Chou and Olson.[62] Erdogan[93] took into consideration the plastic deformation at the tips of interfacial cracks. The bimaterial system was subjected to Mode III antiplane shear deformation. Numerical examples were provided for the extension of plastic zones and crack opening displacements as functions of material properties and interface yield strength. The problem of plastic relaxation at the crack-tip in a ductile matrix material is of considerable practical importance. Chou and Tetelman[94] examined the extension of a crack normal to a phase boundary with plastic deformations in a comparatively softer material. The length of plastic zone, the crack opening displacement and the fracture strength in Mode III deformation were found to be sensitive to the yield strength, shear modulus and fracture toughness of the relatively softer material.

The stress analyses of multi-layered composites with flaws were initially performed by Erdogan and Gupta.[95] They considered the plane-strain and antiplane shear problems for a medium composed of homogeneous, isotropic layers with different elastic properties. The external loads were assumed to act on the crack surfaces. For the case that the loads are applied on the outer boundary, the stress solution can be obtained by a simple superposition of the solution from this problem and also the stress distribution in the flaw-free medium. One of the interesting results obtained is for the case of the plane-strain problem of an elastic layer bonded between two half planes with identical elastic properties. The crack of length $2c$ is located at the mid-section of the layer (Fig. 8.24). Uniform compressive stress $-\sigma_0$ and shear stress $-\tau_0$ are applied on the crack surfaces. The stress intensity factors are defined as

$$k_1 = \lim_{x \to c} (x^2 - c^2)^{1/2} \sigma_y(x, 0)$$

$$k_2 = \lim_{x \to c} (x^2 - c^2)^{1/2} \sigma_{xy}(x, 0)$$

The variation of stress intensity factors with layer thickness is shown in Fig. 8.24 for the case of a layer with stress-free boundaries and for

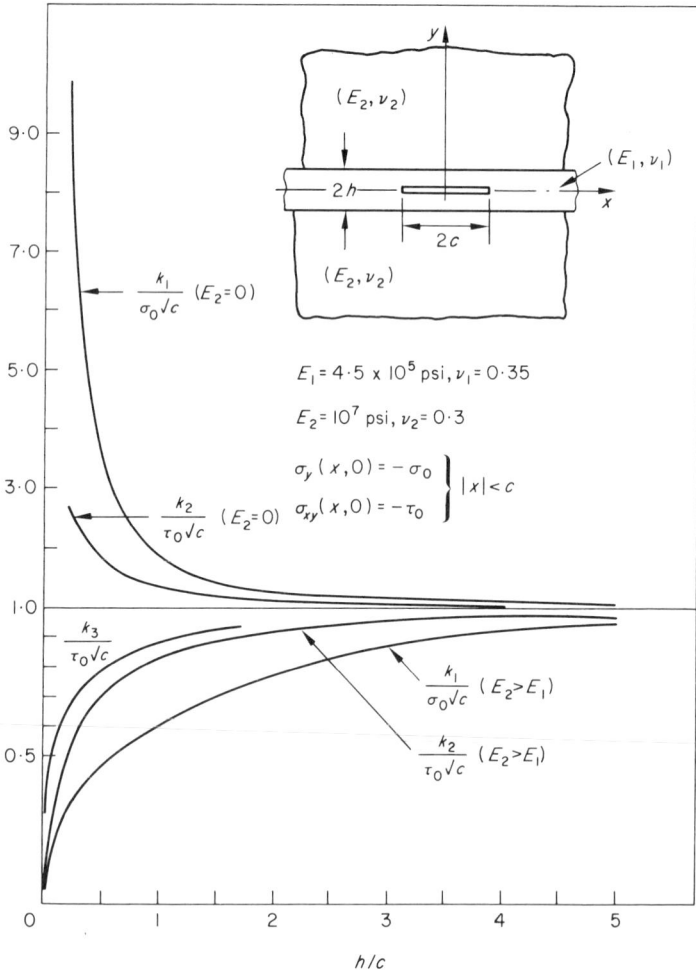

Fig. 8.24.   Stress intensity factors for the plane-strain case. (Reprinted with permission of Microforms International Marketing Corp. exclusive copyright licensee of Pergamon Press journal backfiles.)

$E_2 > E_1$. The values of the elastic constants are so chosen that they simulate a system of aluminium and epoxy. Several observations can be made from these results. First, the crack-tip stress intensities all approach that of a homogeneous medium when the layer thickness becomes very large compared to the crack length. Secondly, the crack-tip stress intensity in an elastic medium is reduced when second-phase materials with higher elastic moduli are introduced near to the crack tip. The effect is reversed if free surfaces are present near the crack tip. Thirdly, the fracture resistance of the layer increases as the h/c ratio decreases. The implication of this result will be discussed later. The solution for the same layered structure subjected to anti-plane strain was also discussed by Chen and Sih.[121]

Arin and Erdogan[96] later considered a penny-shaped crack in an elastic layer bonded to dissimilar half spaces. The crack was assumed to be parallel to the interfaces at an arbitrary location. For a crack situated on the middle section of the layer bonded by two identical half spaces the variation of the stress intensity factors with layer thickness and the relative magnitude of elastic constants follows the same trend as in the case of a through-crack just discussed. Hilton and Gupta[97] modified this result to consider a butt joint with a penny-shaped crack under a tensile stress σ. Let the radii of the circular cross-section of the joint and the crack be R and c, respectively. It is found that for $c/R = 0.6$ the ratio of $K_1/[(2/\pi)\sigma\sqrt{c}]$ is about 10 per cent higher than that for $c/R = 0$, namely, the case when the joint section is much larger than the crack area. Based upon the conclusions reached earlier,[95] it is easily seen that the stress intensity associated with a crack in a joint material is lower than that for the same material in bulk form. Naturally, it is assumed that the joint material is 'softer' than the materials it joints. Furthermore, the thinner the joint is, the higher the load-carrying capacity of the joint will be, provided the crack size and fracture toughness of the adhesive remain the same.

In layered composites as discussed above, the solutions of some special cases can be easily deduced. These cases include the degeneration of the systems to half spaces and strips. When the thickness of the layer containing the flaw vanishes, the problems reduce to those of interface flaws. However, since the nature of the stress singularities for a crack embedded in a homogeneous medium and for an interface crack is different, there is no smooth transition from one solution to the other as the distance of the crack from the interface approaches zero. A general discussion on layered composites composed of three different materials with interface flaws has been given by Erdogan and Gupta.[98] Numerical solutions are available for stress intensity factors as affected by layer thickness, crack length and elastic properties of the materials. The cases of interface cracks beneath an elastic layer and in sandwiched plates have been discussed by Erdogan and Arin.[99,100] The problem of coplanar cracks in layered materials was attempted by Dhaliwal.[101]

Another array of crack problems in layered composites is characterised by the crack planes normal to the interfaces. Hilton and Sih,[102] Gupta[134] and Bogy[135] considered the case of a single layer of material containing a crack which was sandwiched between two other layers of infinite height (Fig. 8.25(a)). To simulate the situation in layered composites, the elastic properties of the two half planes were assumed to possess the averaged properties of a large number of layers. The thickness of the layer and the crack length are $2h$ and $2c$, respectively. The origin of the co-ordinate axes is chosen at the centre of the crack with the x–y-plane as the crack plane. The stress intensity factors are defined as

$$k_1 = \lim_{x \to c^+} [2(x - c)]^{1/2} \sigma_y(x, 0)$$

$$k_2 = \lim_{x \to c^+} [2(x - c)]^{1/2} \sigma_{xy}(x, 0)$$

Under an internal pressure p, the stress intensity factor variation with layer geometry is given in Fig. 8.25(b). The subscript 1 denotes material properties of the layer. This result again indicates that the stress intensity is reduced when second-phase materials with higher elastic stiffness are present near the crack tip. Calculations were also carried out for approximating the stress intensity factor for a crack inclined at an arbitrary orientation in the sandwiched layer under biaxial loading. This was done by considering the extreme cases of cracks parallel[70] and normal to the interface. Upper and lower bounds to the stress intensity factors of cracks at arbitrary orientations were thus obtained. The effects of crystal anisotropy were also considered in Reference 70.

The treatment of cracks in layered structures by a dislocation model can be found in the work of Chou et al.[103]

Finally, the very important problem of cracks in fibrous composites is discussed. Muki and Sternberg[104–8] first examined the static transfer of load in a continuous, semi-infinite elastic filament that is embedded in an elastic matrix. Later, they coped with the more complicated problem of discontinuous elastic filaments in composites. They were concerned with the load absorption by a discontinuous filament that consisted of two disjoint but contiguous segments. These segments, circular in cross-section, were assumed to be well-bonded to an all-around infinite elastic matrix of distinct mechanical properties. The composite containing a single filament was subjected to uniform traction at infinity parallel to the filament. The variation of filament

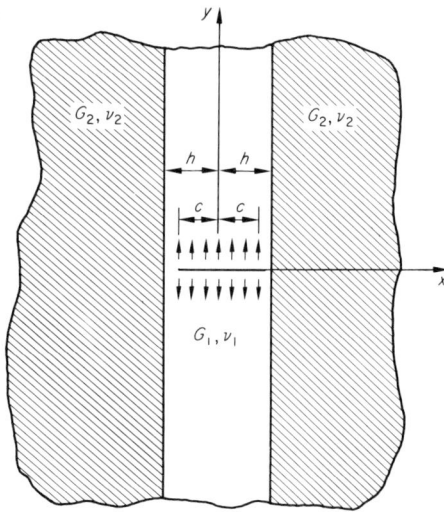

(a)

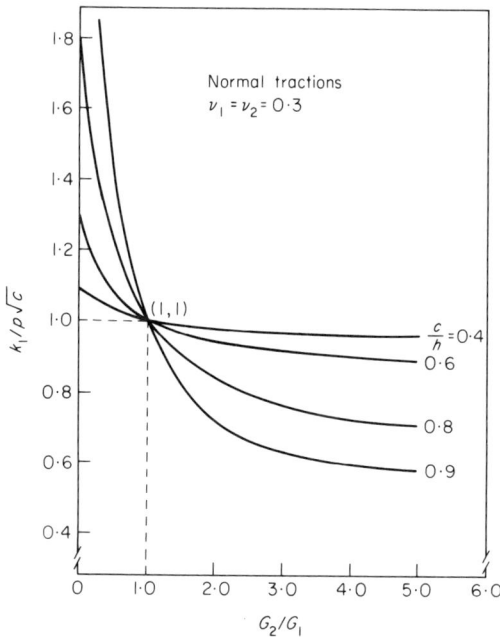

(b)

Fig. 8.25. (a) *A layer composite with a crack subjected to normal traction;* (b) *normalised stress-intensity factor* k_1 *versus* G_2/G_1 *for* $\nu_1 = \nu_2 = 0.3$. (Reprinted with permission of Microforms International Marketing Corp. exclusive copyright licensee of Pergamon Press journal backfiles.)

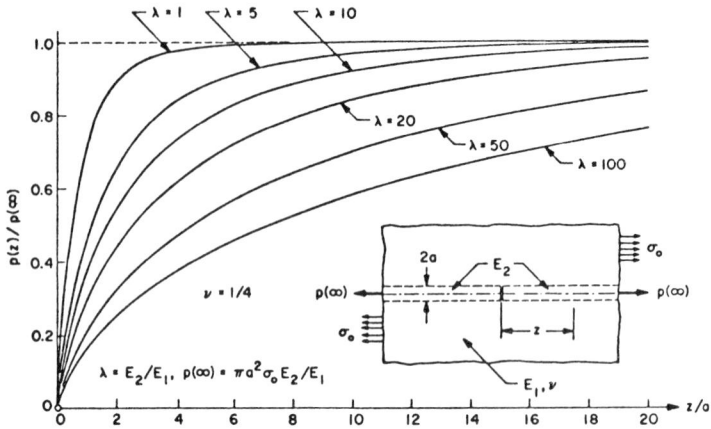

Fig. 8.26. Load-absorption by a discontinuous filament. Variation of filament-force. (Reproduced from Reference 108 by permission of Birkhaeuser Verlag.)

force $p(z)$ as a function of position along the axis of the discontinuous fibre is shown in Fig. 8.26 for various stiffness ratios. All these curves exhibit infinite initial slopes. The sudden build-up of stress in the fibre, and hence, along the interface near the fibre ends may cause interfacial debonding in these regions. Figure 8.26 also clearly indicates the trend of variation of the fibre axial stress with the stiffness ratio. It is noted that the axial stress builds up more slowly along the fibre length if the fibre material is much more 'rigid' than the matrix. This implies a lower efficiency of load absorption for a relatively stiff filament. It was also concluded that at the same stiffness ratio, the load absorption by the discontinuous fibres just discussed is more efficient than that of a semi-infinite continuous filament. The effect of the width of the gap between the fibre ends on stress concentration has been studied by Iremonger and Wood[131] using the finite element method and the results have been checked experimentally by means of a photoelastic model.

Concerning the behaviour of a single fibre embedded in an elastic matrix, Erdogan and Ozbek[109] considered another form of cracking. They examined the axially symmetric problem of interface cracks due to imperfect bonding between the fibre and the matrix. Numerical results were obtained for the stress intensity factors as well as the strain energy release rate for crack propagation along the interface. At high fibre volume contents, the fibres are close to one another and the interaction among them plays a rather significant role in affecting their behaviour. Exact elasticity solutions to take into consideration the fibre interaction would be extremely complicated, if not impossible. Hedgepeth and Van Dyke[110] examined the local stress concentration caused by broken fibres in filamentary composite materials. The model they used is common to shear-lag analyses. It is composed of tension-carrying elements connected to a matrix which carries only shear. Stress concentrations have been calculated in unbroken elements when various numbers of filaments are broken in the square and hexagonal filament arrays. The stress concentration at adjacent elements increases as the number of broken filaments becomes larger. For instance, in a square array, a stress concentration factor of 1·456 is reached at the fibre adjacent to nine broken fibres. This factor increases to 1·732 when a total of thirty-seven fibres are broken. Kulkarni et al.[133] also used a shear-lag analysis to examine the load concentration factors of circular holes in a unidirectional lamina subjected to uniform tension.

8.6.5 Fracture Mechanics of Anisotropic Elastic Media

At a macroscopic level, fibre-reinforced composite materials can be considered as homogeneous anisotropic media. When the dimension of a crack is much larger than the fibre spacings, one can neglect the microstructure of the composite and treat the problem by fracture mechanics. To achieve this purpose, it is necessary to employ the solution of cracks in anisotropic elastic media. Knowing the expressions of stress intensity factors from theoretical analyses and the fracture toughness data from experimental measurements, it is then a straightforward matter to determine the fracture load or the allowable crack length in service.

The general equations for crack-tip stress fields in homogeneous anisotropic media have been derived by Sih et al.,[111,127] using a complex variable approach. Just as in the case of an isotropic medium, the stress components at the tip of a crack are composed of three distinct portions. First, an elastic stress singularity of the order $r^{-1/2}$ is always present at the crack tip. The order of the singularity is the same as that in isotropic media. Secondly, the angular distribution of the stress components at a crack tip depends upon the material properties. This behaviour is not observed in isotropic media. Thirdly, the stress intensity factors, which represent the conditions for crack extension, can also be defined.

The stress intensity factors for a through crack in an infinite general anisotropic plate are given as follows[14,111]

$$K_I = \sigma(\pi c)^{1/2} \tag{8.58}$$

$$K_{II} = \tau(\pi c)^{1/2} \tag{8.59}$$

$$K_{III} = \tau(\pi c)^{1/2} \tag{8.60}$$

The three modes of deformation as defined earlier, are the inplane symmetric loading, inplane skew-symmetric loading and antiplane shear. σ and τ in the above equations denote respectively, the normal and shear loading on the specimen boundaries. The preceding equations show that the influence of the applied load and the size of the crack on the intensity of the local stresses is identical with the isotropic case. When the crack surface is subjected to unbalanced forces, the stress intensity factors are affected by the elastic properties just as in the case of an isotropic elastic medium.

The stress intensity factors in anisotropic materials can also be related to the energy release rates. When the crack is on one of the symmetry planes of an orthotropic material, the energy release rate for each basic mode of deformation depends only on the stress intensity factor of the same mode. They are given in the following

$$G_I = K_I^2 \left(\frac{b_{11} b_{22}}{2}\right)^{1/2} \left[\left(\frac{b_{22}}{b_{11}}\right)^{1/2} + \frac{2b_{12} + b_{66}}{2b_{11}}\right]^{1/2} \tag{8.61}$$

$$G_{II} = K_{II}^2 \frac{b_{11}}{2^{1/2}} \left[\left(\frac{b_{22}}{b_{11}}\right)^{1/2} + \frac{2b_{12} + b_{66}}{2b_{11}}\right]^{1/2} \tag{8.62}$$

$$G_{III} = K_{III}^2/2(c_{44} c_{55})^{1/2} \tag{8.63}$$

By assuming a plane-strain condition and that the x_2-x_3 plane is isotropic, such as the plane normal to the fibre direction in a composite material specimen, the elastic constant factors b_{ij} can be expressed in terms of the engineering constants as[67]

$$b_{11} = \frac{1}{E_1}\left[1 - \left(\frac{E_2}{E_1}\right)\nu_{12}'^2\right]$$

$$b_{22} = \frac{1}{E_2}(1 - \nu_{23}^2)$$

$$b_{12} = -\frac{\nu_{12}}{E_1}(1 + \nu_{23})$$

$$b_{66} = \frac{1}{G_{12}} \tag{8.64}$$

From the expression for the stress intensity factors above, it is seen that the stress fields near crack tips in anisotropic media are distributed in the same manner as in isotropic media, with probably the only difference being in their angular dependency. The intensity factors also depend upon the magnitude of the applied load and the crack geometry. Consequently, the concept of fracture mechanics widely used in isotropic medium can be applied equally well in anisotropic medium. The combination of stress-intensity factors will reach a critical value at the onset of unstable crack propagation.[111] The applications of this concept in fibrous composite materials are now demonstrated by a few examples.

The fracture toughness of two carbon–carbon composites and the ATJ-S graphite, a moulded polycrystalline anisotropic graphite, were determined by Guess and Hoover[112] from the load-deflection responses of single-edge, notched specimens under three-point bending. Two types of carbon composites were used. They included the CVD/FELT composite and the CVD-filament winding composite (CVD/FWW) where the specimens were taken along the axis of a filament wound cone. In the three-point bending tests, the ratios of crack length c to the beam depth W were kept in the range of $0\cdot2$ to $0\cdot5$. Under an applied load, P, the variation of load with the lateral deflection δ of the specimen at the mid-span was recorded. The maximum load determined from the P versus δ curve was used to calculate the K_{IC} value from the available formula for the particular beam geometry

$$K_{IC} = \frac{P_{max}S}{BW} [2\cdot9(c/W)^{1/2} - 4\cdot6(c/W)^{3/2} + 21\cdot8(c/W)^{5/2} -$$

$$- 37\cdot6(c/W)^{7/2} + 38\cdot7(c/W)^{9/2}] \quad (8.65)$$

where S and B are the span and width of the beam. The K_{IC} values so calculated were essentially independent of the crack and specimen geometry. The average values of K_{IC} are $1012\,psi\,\sqrt{in}$ ($1\cdot11 \times 10^6\,N/m^2\,\sqrt{m}$) for CVD/FELT, $530\,psi\,\sqrt{in}$ ($5\cdot82 \times 10^5\,N/m^2\,\sqrt{m}$) for CVD/FWW and $756\,psi\,\sqrt{in}$ ($8\cdot31 \times 10^5\,N/m^2\,\sqrt{m}$) for ATJ-S graphite.

In order to demonstrate the criterion of critical strain-energy release rate for unstable crack propagation mentioned in Section 8.3.2, the tests carried out by Guess and Hoover are introduced as an example. Under Mode I deformation, the rate of strain-energy release during crack extension, $\mathcal{G}_I$, can be shown as (see Problem 8.5 and References 14, 113 and 114)

$$\mathcal{G}_I = -\frac{\partial U}{\partial A} = \tfrac{1}{2}P^2 \frac{\partial(1/k)}{\partial A} \quad (8.66)$$

where U is the stored elastic strain-energy for the cracked elastic body and A represents the crack area. k is the spring constant of the elastic body and is measured from the initial slope of the P versus δ curve. The inverse of k is known as the compliance. The critical energy release rate, $\mathcal{G}_{IC}$, is determined experimentally by measuring the quantities in the above equation at the onset of unstable crack extension. The ways of measuring these quantities are now explained.

When the critical energy release rate is determined from the single-edge, notched bend specimens, it is desirable to write eqn. (8.66) in the following form

$$\mathcal{G}_{IC} = \frac{-1}{2BW} \left(\frac{P_{max}}{k} \right)^2 \frac{\partial k}{\partial (c/W)} \tag{8.67}$$

B and W again denote the width and depth of the beam specimen which has a pre-existing crack of length c. The curves of load versus beam deflection resemble the curve (a) of load versus notch opening displacement as shown in Fig. 8.10. The initial slope k of a P versus δ curve gives the value of the spring constant of the beam at a certain crack length. The P versus δ curves can be prepared for a series of crack lengths or c/W ratios and the initial slopes are measured. Figure 8.27 indicates the variation of k with c/W ratios. For a certain crack geometry or c/W ratio, the slope of the k versus c/W curve can be measured from the curves in Fig. 8.27, and the corresponding P_{max} value is also known from the P versus δ curve. Consequently the value of $\mathcal{G}_{IC}$ can be computed using eqn. (8.67). The average values of $\mathcal{G}_{IC}$ measured by the compliance technique are $1 \cdot 19$ in. lb/in.^{2} (212 joule/m^2) for CVD/FELT composite, $0 \cdot 189$ in. lb/in.^{2} ($33 \cdot 7$ joule/m^2) for CVD/FWW composite, and $0 \cdot 486$ in. lb/in.^{2} ($86 \cdot 6$ joule/m^2) for ATJ-S graphite.

Finally, it needs to be pointed out that, just as in the cases of isotropic media, either $\mathcal{G}_{IC}$ or K_{IC} can be calculated from eqn. (8.61) when the other is determined experimentally. The present results[112] indicate that the results of fracture toughness and strain-energy release rate measurements are in reasonably good agreement.

In glass-fibre reinforced composites, the fracture toughness and strain-energy release rate also have been measured. Sanford and Stonesifer[115] examined the fracture toughness of unidirectional glass-reinforced plastics. It was concluded that the toughness values were sensitive to the resin and hardener systems used. No significant improvement in fracture toughness was observed when E-glass was replaced by the higher strength S-glass. Composites with fine fibre

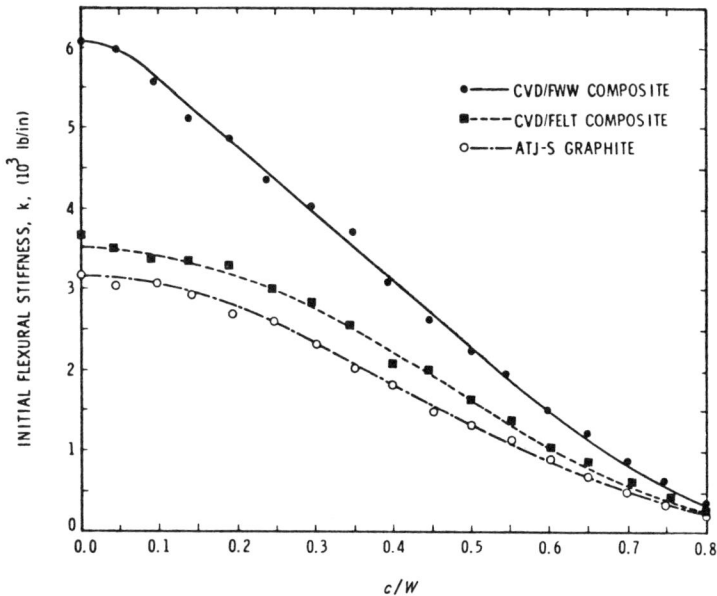

Fig. 8.27. *Initial flexural stiffness versus c/W ratio for the single-edge notched fracture toughness bend specimens of CVD/FWW, CVD/FELT and ATJ-S materials. (Reproduced from Reference 112 by permission of the Technomic Publishing Co.)*

seemed to have a higher fracture toughness than coarser-fibre composites. Furthermore, composites in their fully cured state showed the highest toughness values. Hancock and Swanson[123] measured the plane-strain fracture toughness for centre-notched specimens of 6061 aluminium containing 30 volume per cent of boron filaments. The fracture toughness value was found to be 30 ksi $\sqrt{\text{in}}$ ($3\cdot30 \times 10^7$ N/m^2 $\sqrt{\text{m}}$). It was noted that filament pullout contributed negligibly to the fracture toughness of this composite. So far, the fracture toughness measurements have been restricted to opening-mode deformation. Although in-plane shear is an important deformation mode in reinforced plastic, its measurement is rather difficult. The variation of $\mathcal{G}_{1C}$ with fibre content in E-glass reinforced composites has been examined by Sih.[67] The fracture toughness values were computed using eqn. (8.61). The application of fracture mechanics to laminated composites is discussed in Reference 116.

8.7   PROBLEMS

8.1.   Discuss the strength of composite materials composed of ductile fibres and brittle matrix. Derive an expression for the minimum fibre content at which the multiple fracture of the matrix occurs.

8.2.   The experimental results of Cooper in Figs. 8.13 and 8.14 indicate that the fracture load of a continuous fibre with flaw spacing l' is always larger than that of a discontinuous fibre of length l', provided the fibre diameters are identical. On the other hand, the work of fracture of continuous fibres is lower than that of discontinuous fibres. Furthermore, the work of fracture is a maximum at $l' = l_c'$. Discuss these results in a quantitative manner.

8.3.   Cottrell[141] has shown that the work of fracture per unit area of composite due to fibre pull out can be expressed as $W_p = \frac{1}{12} V_f \sigma_f l_c$. Reproduce this result and estimate the work of fracture for some composite systems discussed in Chapter 2. Compare your results for ductile and brittle matrices.

8.4.   Derive eqn. (8.44).

8.5.   Derive eqn. (8.66). To facilitate the derivation of this equation it is desirable to consider the plate with a pre-existing crack of length $2c$. Under an applied load P_1, the crack extends a length δc at each crack tip. As a result of the crack extension, the load on the plate drops to P_2. The ends of the cracked

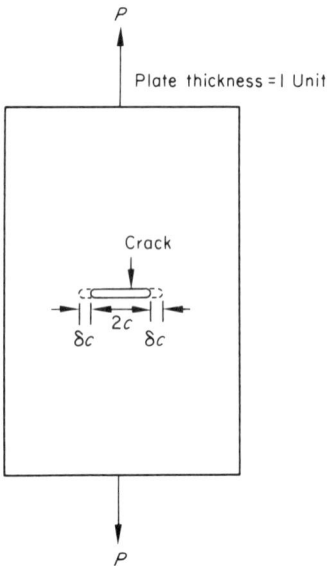

(a)

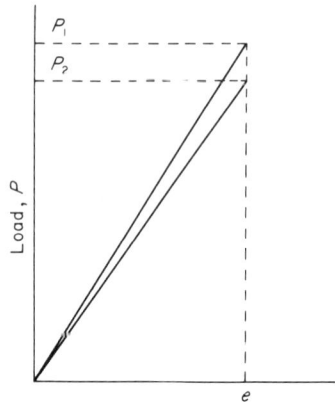

(b)

plate are assumed to be fixed during crack extension. The elastic energies stored in the plate are $U_1 = P_1 e/2$ and $U_2 = P_2 e/2$ before and after the crack extension, respectively. The rate of release of elastic energy with respect to crack area ($\delta A = 1 \times \delta c$) is defined as

$$\mathcal{G} = -\frac{\partial U}{\partial c}$$

The spring constant k in eqn. (8.66) is defined by the load–deformation relation $P = ke$.[19,113,114]

8.6. (a) A thin-walled cylindrical pressure vessel with radius R and wall thickness h is under internal pressure p. The cylinder axis is assumed to be the z-axis. The pressure vessel is made of 4340 steel with $K_{Ic} = 40$ ksi $\sqrt{\text{in}}$ (44×10^6 N/m^2 $\sqrt{\text{m}}$) and an ultimate tensile strength of 300 ksi (20.67×10^8 N/m^2). Show that the hoop stress and longitudinal stress are $\sigma_\theta = pR/h$ and $\sigma_z = pR/2h$, respectively.

(b) For a through crack of length $2c$ and oriented at an angle β with the θ axis, the stress-intensity factors are

$$K_1 = \frac{pR}{2h} \sqrt{c}\,(1 + \sin^2 \beta)$$

$$K_2 = \frac{pR}{2h} \sqrt{c}\, \sin \beta \cos \beta$$

Derive the expression of the strain-energy density factor.

(c) Find the S_c value of the material by assuming $\nu = 0.25$ and $E = 30 \times 10^6$ psi for steel.

(d) Find the values of the angle θ_0 of crack initiation for $\beta = 0°, 60°$ and $90°$.

(e) Find the critical internal pressure p_c in terms of the $h/2R$ ratio for $\beta = 0°$, $60°$ and $90°$, and a crack length of 0.2 in (see Reference 142).

The application of S_c-theory to pressure vessels made of unidirectional fibre composite is illustrated in Reference 143.

8.7. Based upon the experimental data of Fig. 8.22 find the S_c values of Scotchply 1002 for $\beta = 90°$ and $60°$. The crack lies in a layer of epoxy resin with $E = 4.5 \times 10^5$ psi (31.02×10^8 N/m^2) and $\nu = 0.35$.

REFERENCES

1. Kelly, A., 'Reinforcement of Structural Materials by Long Strong Fibres', *Met. Trans.*, **3**, 2313 (1972).

2. Hale, D. K. and Kelly, A., 'Strength of Fibrous Composite Materials', *Annual Review of Materials Science*, R. A. Huggins, ed., Annual Review, Inc., Palo Alto, Calif. (1972).

3. Kelly, A. and Davies, G. J., 'The Principles of the Fibre Reinforcement of Metals', *Met. Rev.*, **10**, 1 (1965).

4. Kelly, A. and Tyson, W. R., 'Fibre-Strengthened Materials', *High Strength Materials*, V. F. Zackay, ed., J. Wiley and Sons, Inc. (1965), p. 578.
5. Kelly, A. and Tyson, W. R., 'Tensile Properties of Fiber-Reinforced Metals: Copper/Tungsten and Copper/Molybdenum', *J. Mech. Phys. Solids*, **13**, 329 (1965).
6. Mileiko, S. T., 'The Tensile Strength and Ductility of Continuous Fibre Composites', *J. Mater. Sci.*, **4**, 974 (1969).
7. Piehler, H. R., 'Plastic Deformation and Failure of Silver–Steel Filamentary Composites', *Trans. Met. Soc. AIME*, **233**, 12 (1965).
8. Hancock, P. and Cuthbertson, R. C., 'The Effect of Fiber Length and Interfacial Bond in Glass Fiber–Epoxy Resin Composites', *J. Mater. Sci.*, **5**, 762 (1970).
9. McDanels, D. L., Jech, R. W., and Weeton, J. W., 'Stress Strain Behavior of Tungsten Fiber Reinforced Copper', NASA TN D-1881, National Aeronautics and Space Administration (October 1963).
10. Spencer, A. J. M., *Deformation of Fibre-reinforced Materials*, Clarendon Press, Oxford (1972).
11. Lee, E. H., ed., *Dynamics of Composite Materials*, The American Society of Mechanical Engineers, New York (1972).
12. Inglis, C. E., 'Stresses in a Plate Due to the Presence of Cracks and Sharp Corners', *Proceedings, Inst. Naval Architects*, **60** (1913).
13. Griffith, A. A., 'The Phenomena of Rupture and Flow in Solids', *Transactions, Roy. Soc. London*, **221** (1920).
14. Paris, P. C. and Sih, G. C., 'Stress Analysis of Cracks', *Fracture Toughness Testing and Its Applications*, ASTM STP 381, American Society for Testing and Materials (1965), p. 30.
15. Liebowitz, H., ed., *Fracture*, Academic Press, New York (1968–72).
16. Irwin, G. R., 'Analysis of Stresses and Strains Near the End of a Crack Traversing a Plate', *Trans. Am. Soc. Mechanical Engrs., J. Appl. Mech.* (1957).
17. Tetelman, A. S. and McEvily, A. J., Jr, *Fracture of Structural Materials*, John Wiley & Sons, Inc., New York (1967).
18. McClintock, F. A. and Argon, A. S., *Mechanical Behavior of Materials*, Addison-Wesley Publ. Co., Inc., Reading, Mass. (1966).
19. Irwin, G. R., 'Fracture', *Handbuch der Physik*, Vol. 6, Springer, Berlin (1958).
20. Irwin, G. R., 'A Critical Energy Rate Analysis of Fracture Strength', *Welding Journal* (Research Supplement) (1954).
21. Bueckner, H. F., 'The Propagation of Cracks and Energy of Elastic Deformation', *Trans. Am. Soc. Mechanical Engrs., J. Appl. Mech.* (1958).
22. Sanders, J. L., Jr, 'On the Griffith–Irwin Fracture Theory', *Trans. Am. Soc. Mechanical Engrs., J. Appl. Mech.* (1960).
23. Brown, W. F., Jr, and Srawley, J. E., 'Plane Strain Crack Toughness Testing of High Strength Metallic Materials', *Fracture Toughness Testing*, ASTM STP 410, American Society for Testing and Materials (1966), p. 1.
24. Srawley, J. E., Jones, M. H., and Brown, W. F., Jr, 'Determination of Plane Strain Fracture Toughness', *Mater. Res. Stand.*, **7**, 262 (1967).
25. ASTM Specifications E399-70T, American Society for Testing and Materials (1970).
26. Kelly, A., 'Interface Effects and the Work of Fracture of a Fibrous Composite', *Proc. Roy. Soc. Lond.*, **A319**, 95 (1970).
27. Cooper, G. A. and Kelly, A., 'Tensile Properties of Fiber-Reinforced Metal: Fracture Mechanics', *J. Mech. Phys. Solids*, **15**, 279 (1967).

28. Cook, J. and Gordon, J. E., 'A Mechanism for the Control of Crack Propagation in All-Brittle Systems', *Proc. Roy. Soc. London*, A282, 508 (1964).
29. Grenier, P. and Cooper, G. A., 'Some Observations on the Fracture of Composites with a Brittle Matrix', *Fibre Science and Technology*, 1, 219 (1969).
30. Cooper, G. A., 'The Fracture Toughness of Composites Reinforced with Weakened Fibres', *J. Mat. Sci.*, 5, 645 (1970).
31. Kelly, A., 'The Nature of Composite Materials', *Scientific American*, 217, 160 (1967).
32. Peirce, F. T., *J. Text. Inst.*, 17, T355 (1926).
33. Daniels, H. E., 'The Statistical Theory of the Strength of Bundles of Threads I', *Proc. Roy. Soc.*, A183, 405 (1945).
34. Coleman, B. D., 'On the Strength of Classical Fibres and Fibre Bundles', *J. Mech. Phys. Solids*, 7, 60 (1958).
35. Weibull, W., *Ing. Vetenskaps Akad. Handl.*, 151, 153 (1939).
36. Rosen, B. W., 'Tensile Failure of Fibrous Composites', *J. Am. Inst. Aero. Astron.*, 2, 1985 (1964).
37. Rosen, B. W., 'Mechanics of Composite Strengthening', *Fiber Composite Materials*, American Society for Metals, Metals Park, Ohio (1965), p. 37.
38. Rosen, B. W., 'Thermomechanical Properties of Fibrous Composites', *Proc. Roy. Soc. Lond.*, 319, 79 (1970).
39. Rosen, B. W. and Dow, N. F., 'Mechanics of Failure of Fibrous Composites', *Fracture*, H. Liebowitz, ed., Vol. 7, Academic Press, New York (1972).
40. Argon, A. S., 'Fracture of Composites', *Treatise on Materials Science and Technology*, H. Herman, ed., Vol. 1, Academic Press, New York (1972).
41. Noone, M. J., Feingold, E., and Sutton, W. H., 'The Importance of Coatings in the Preparation of Al_2O_3 Filament/Metal–Matrix Composites', *Interfaces in Composites*, ASTM STP 452, American Society for Testing and Materials (1969), p. 59.
42. Metcalfe, A. G., 'Interaction and Fracture of Titanium-Boron Composite', *J. Comp. Mater.*, 1, 356 (1967).
43. Basche, M., 'Interfacial Stability of Silicon Carbide-Coated Boron Filament Reinforced Metals', *Interfaces in Composites*, ASTM STP 452, American Society for Testing and Materials (1970), p. 130.
44. Basche, M., Fanti, R., and Galasso, F., 'Preparation and Properties of Silicon Carbide–Coated Boron Filaments', *Fibre Sci. Tech.*, 1, 19 (1968).
45. Lynch, C. T. and Kershaw, J. P., *Metal Matrix Composites*, Chemical Rubber Co. Press, Cleveland, Ohio (1972).
46. Sutton, W. H. and Chorńe, J., 'Potential of Oxide–Fiber Reinforced Metals', *Fiber Composite Materials*, S. H. Bush, ed., American Society for Metals, Metals Park, Ohio (1965), p. 173.
47. Johannson, O. K., Stark, F. O., Vogel, G. E., Lacefield, R. M., Baney, R. H., and Flaningam, O. L., 'Wetting, Adsorption, and Bonding at Glass Fiber-Coupling Agent–Resin Interfaces', *Interfaces in Composites*, ASTM STP 452, American Society for Testing and Materials (1969), p. 168.
48. Johannson, O. K., Stark, F. O., Vogel, G. E., Fleischmann, R. M., and Flaningam, O. L., 'The Physical Chemical Nature of the Matrix–Glass Fiber Interface', *Fundamental Aspects of Fiber Reinforced Plastic Composites*, Interscience Publishers, New York (1968), p. 199.
49. Wong, R., 'Mechanism of Coupling by Silanes of Epoxides to Glass Fibers',

Fundamental Aspects of Fiber Reinforced Plastic Composites, Interscience Publishers, New York (1968), p. 237.

50. Sterman, S. and Marsden, J. G., 'Bonding Organic Polymers to Glass by Silane Coupling Agents', *Fundamental Aspects of Fiber Reinforced Plastic Composites*, Interscience Publishers, New York (1968), p. 237.

51. Gutfreund, K., 'Interfacial Relationships in Fiber-Reinforced Plastic Composites', *Modern Composite Materials*, L. J. Broutman and R. H. Krock, eds., Addison-Wesley Publ. Co., Reading, Mass. (1967), p. 172.

52. Ashbee, K. H. G., Frank, F. C., and Wyatt, R. C., 'Water Damage in Polyester Resins', *Proc. Roy. Soc.*, A300, 415 (1967).

53. Ashbee, K. H. G. and Wyatt, R. C., 'Water Damage in Glass Fiber/Resin Composites', *Proc. Roy. Soc.*, A312, 553 (1969).

54. Cox, H. L., 'The Elasticity and Strength of Paper and Other Fibrous Materials', *British J. Appl. Phys.*, 3, 72 (1952).

55. Dow, N. F., 'Study of Stresses Near a Discontinuity in a Filament Reinforced Composite Metal', General Electrical Co. Report R63SD61 (1963).

56. Tyson, W. R. and Davies, G. I., 'A Photoelastic Study of the Shear Stresses Associated with the Transfer of Stress During Fiber Reinforcement', *British J. Appl. Phys.*, 16, 199 (1965).

57. Schuster, D. M. and Scala, E., 'The Mechanical Interaction of Sapphire Whiskers with a Birefringent Matrix', *Trans. Met. Soc. AIME*, 230, 1635 (1964).

58. Schuster, D. M. and Scala, E., 'Single and Multi-Fiber Interactions in Discontinuously Reinforced Composites', AIAA/ASME Eighth Structures and Materials Conference, Palm Springs, California, 1967.

59. Mullin, J., Berry, J. M., and Gatti, A., 'Some Fundamental Fracture Mechanisms Applicable to Advanced Filament Reinforced Composites', *J. Comp. Mater.*, 2, 82 (1968).

60. Hirth, J. P., 'The Influence of Grain Boundaries on Mechanical Properties', *Met. Trans.*, 3, 3047 (1972).

61. Chou, T. W. and Hirth, J. P., 'Stress Distribution in a Bimaterial Plate Under Uniform External Loadings', *J. Comp. Mater.*, 4, 102 (1970).

62. Chou, T. W. and Olson, D. F., 'Cracks and Screw Dislocation Arrays in Anisotropic Bimaterial Plates', *Met. Trans.*, 3, 2087 (1972).

63. Lauraitis, K., 'Tensile Strength of Off-Axis Unidirectional Composites', University of Illinois, TAM Report No. 344, 1971.

64. Lauraitis, K., 'Failure Modes and Strength of Angle-Ply Laminates', University of Illinois, TAM Report No. 345, 1971.

65. Sih, G. C., 'A Special Theory of Crack Propagation, Methods of Analysis and Solutions to Crack Problems', G. C. Sih, ed., Wolters–Noordhoff Publishing, Groningen (1972).

66. Sih, G. C. and Macdonald, B., 'What the Designer Must Know About Fracture Mechanics', IFSM-72-23, Institute of Fracture and Solid Mechanics, Lehigh University, 1972.

67. Sih, G. C., Chen, E. P., and Huang, S. L., 'Fracture Mechanics of Plastic-Fiber Composites', IFSM-72-23, Institute of Fracture and Solid Mechanics, Lehigh University, 1972.

68. Sih, G. C. and Chen, E. P., 'Fracture Analysis of Unidirectional Composites', *J. Comp. Mater.*, 7, 230 (1973).

69. Wu, E. M. and Reuter, R. C., 'Crack Extension in Fiber-glass Reinforced Plastics', University of Illinois, TAM Report No. 275, 1965.

70. Hilton, P. H. and Sih, G. C., 'A Sandwiched Layer of Dissimilar Materials Weakened by Crack-Like Imperfections', *Proceedings of the 5th Southeastern Conference on Theoretical and Applied Mechanics*, 5, 949 (1972).
71. Armenakas, A. E. and Sciammarella, C. A., 'Experimental Investigation of the Failure Mechanism of Fiber-reinforced Composites Subjected to Uniaxial Tension', *Experimental Mech.* (February 1973), p. 49.
72. Liptai, R. G. and Harris, D. O., 'Acoustic Emission—an Introductory Review', *Mater. Res. Stand.*, 11, No. 3, American Society for Testing and Materials (1971), p. 8.
73. Dunegan, H. L. and Harris, D. O., 'Acoustic Emission Techniques', *Experimental Techniques in Fracture Mechanics*, A. S. Kobayashi, ed., Society for Experimental Stress Analysis (1973), p. 38.
74. Palmer, I. G. and Heald, P. T., 'The Application of Acoustic Emission Measurements to Fracture Mechanics', *Mater. Sci. Engineering*, 11, 181 (1973).
75. Harris, D. O., Tetelman, A. S., and Darwish, F. A. I., 'Detection of Fiber Cracking by Acoustic Emission', *Acoustic Emission*, ASTM STP 505, American Society for Testing and Materials (1972), p. 238.
76. Fitz-Randolph, J., Phillips, D. C., Beaumont, P. W. R., and Tetelman, A. S., 'Acoustic Emission Studies of a Boron-Epoxy Composite', *J. Comp. Mater.*, 5, 542 (1971).
77. Mehan, R. L. and Mullin, J. V., 'Analysis of Composite Failure Mechanisms Using Acoustic Emissions', *J. Comp. Mater.* 5, 266 (1971).
78. Williams, M. L., 'The Stresses Around a Fault or Crack in Dissimilar Media', *Bull. Seismological Soc. of America*, 49, 199 (1959).
79. Zak, H. R. and Williams, M. L., 'Crack Point Stress Singularities at a Bi-Material Interface', *J. Appl. Mech.*, *Trans. ASME*, 30, 142 (1963).
80. Chou, T. W., 'Dislocation Pileups and Elastic Cracks at a Bimaterial Interface', *Met. Trans.*, 1, 1245 (1970).
81. Erdogan, F., 'Stress Distribution in Nonhomogeneous Elastic Plane With Cracks', *J. Appl. Mech.*, *Trans. ASME*, 30, 232 (1963).
82. Rice, J. R. and Sih, G. C., 'Plane Problems of Cracks in Dissimilar Media', *J. Appl. Mech.*, *Trans. ASME*, 32, 418 (1965).
83. Sih, G. C. and Rice, J. R., 'The Bending of Plates of Dissimilar Materials With Cracks', *J. Appl. Mech.*, *Trans. ASME*, 31, 477 (1964).
84. Erdogan, F., 'Stress Distribution in Bonded Dissimilar Materials With Cracks', *J. Appl. Mech.*, *Trans. ASME*, 32, 403 (1965).
85. Erdogan, F., 'Stress Distribution in Bonded Dissimilar Materials Containing Circular or Ring-Shaped Cavities', *J. Appl. Mech.*, *Trans. ASME*, 32, 829 (1965).
86. Willis, J. R., 'The Penny-Shaped Crack on An Interface', *The Quarterly Journal of Mechanics and Applied Mathematics*, 25, 367 (1972).
87. Lowengrub, M. and Sneddon, I. N., 'The Effect of Shear on a Penny-Shaped Crack at the Interface of an Elastic Half-Space and a Rigid Foundation', *Int. J. Engng Sci.*, 10, 899 (1972).
88. Lowengrub, M., 'Stress Distribution Due to a Griffith Crack at the Interface of an Elastic Half Plane and a Rigid Foundation', *Int. J. Engng Sci.*, 11, 377 (1973).
89. England, A. H., 'A Crack Between Dissimilar Media', *J. Appl. Mech.*, *Trans. ASME*, 32, 400 (1965).
90. Perlman, A. B. and Sih, G. C., 'Elastostatic Problems of Curvilinear Cracks in Bonded Dissimilar Materials', *Int. J. Engng Sci.*, 5, 845 (1967).

91. Clements, D. L., 'A Crack Between Dissimilar Anisotropic Media', *Int. J. Engng Sci.*, **9**, 257 (1971).

92. Erdogan, F. and Sih, G. C., 'On the Crack Extension in Plates Under Plane Loading and Transverse Shear', *J. Basic Engineering*, **85**, 519 (1963).

93. Erdogan, F., 'Elastic–Plastic Anti-Plane Problems for Bonded Dissimilar Media Containing Cracks and Cavities', *Int. J. Solids Structures*, **2**, 447 (1966).

94. Chou, T. W. and Tetelman, A. S., 'Dislocation Arrays and Elastic–Plastic Cracks in a Two-Phase System', *J. Comp. Mater.*, **4**, 162 (1970).

95. Erdogan, F. and Gupta, G., 'The Stress Analysis of Multi-Layered Composites with a Flaw', *Int. J. Solids Structures*, **7**, 39 (1971).

96. Arin, K. and Erdogan, F., 'Penny-Shaped Crack in an Elastic Layer Bonded to Dissimilar Half Spaces', *Int. J. Engng Sci.*, **9**, 213 (1971).

97. Hilton, P. D. and Gupta, G. D., 'Stress and Fracture Analysis of Adhesive Joints', *Application of Fracture Mechanics to Engineering Problems*, IFSM-72-23. Institute of Fracture and Solid Mechanics, Lehigh University (1972).

98. Erdogan, F. and Gupta, G. D., 'Layered Composites With an Interface Flaw', *Int. J. Solids Structures*, **7**, 1089 (1971).

99. Erdogan, F. and Arin, K., 'Penny-Shaped Interface Crack Between an Elastic Layer and a Half-Space', *Int. J. Engng Sci.*, **5**, 115 (1972).

100. Erdogan, F. and Arin, K., 'A Sandwich Plate With a Part-Through and a Debonding Crack', *Engng Fracture Mechanics*, **4**, 449 (1972).

101. Dhaliwal, R. S., 'Two Coplanar Cracks in an Infinitely Long Elastic Strip Bonded to Semi-Infinite Elastic Planes', *Int. J. Engng Sci.*, **11**, 489 (1973).

102. Hilton, P. D. and Sih, G. C., 'A Laminate Composite with a Crack Normal to the Interfaces', *Int. J. Solids Structures*, **7**, 913 (1971).

103. Chou, Y. T., Chou, T. W., and Li, J. C. M., 'Screw Dislocation Pileups and Shear Cracks in a Lamellar Composite', *J. Appl. Phys.*, **41**, 4448 (1970).

104. Sternberg, E. and Muki, R., 'Load-Absorption by a Filament in a Fiber-Reinforced Material', *Z. Angew. Math. Phys.*, **21**, 552 (1970).

105. Muki, R. and Sternberg, E., 'On the Diffusion of an Axial Load from an Infinite Cylindrical Bar Embedded in an Elastic Medium', *Int. J. Solids Struct.*, **5**, 587 (1969).

106. Muki, R. and Sternberg, E., 'Elastostatic Load-transfer to a Half-space from a Partially Embedded Axially Loaded Rod', *Int. J. Solids Struct.*, **6**, 69 (1970).

107. Sternberg, E., 'Load-transfer and Load-diffusion in Elastostatics', *Proc. Sixth U.S. Nat. Congress Appl. Mech.*, The American Society of Mechanical Engineers, New York (1970).

108. Muki, R. and Sternberg, E., 'Load-absorption by a Discontinuous Filament in a Fiber-reinforced Composite', *Z. Angew. Math. Phys.*, **22**, 809 (1971).

109. Erdogan, F and Ozbek, T., 'Stresses in Fiber-Reinforced Composites with Imperfect Bonding', *J. Appl. Mech., Trans. ASME*, **36**, 865 (1969)

110. Hedgepeth, J. M. and Van Dyke, P., 'Local Stress Concentrations in Imperfect Filamentary Composite Materials', *J. Comp. Mater.*, **1**, 294 (1967).

111. Sih, G. C., Paris, P. C., and Irwin, G. R., 'On Cracks in Rectilinearly Anisotropic Bodies', *Int. J. Fracture Mechanics*, **1**, 189 (1965).

112. Guess, T. R. and Hoover, W. R., 'Fracture Toughness of Carbon–Carbon Composites', *J. Comp. Mater.*, **7**, 2 (1973).

113. Wu, E. M., 'Fracture Mechanics of Anisotropic Plates', *Composite Materials*

Workshop, S. W. Tsai, J. C. Halpin and N. J. Pagano, eds., Technomic Pub. Co., Inc., Stamford, Conn. (1968), p. 20.

114. Corten, H. T., 'Influence of Fracture Toughness and Flaws on the Interlamina Shear Strength of Fibrous Composites', *Fundamental Aspects of Fiber Reinforced Plastic Composites*, R. T. Schwartz and H. S. Schwartz, eds., Interscience Publishers, New York (1968), p. 89.

115. Sanford, R. J. and Stonesifer, F. R., 'Fracture Toughness Measurements in Unidirectional Glass-Reinforced Plastics', *J. Comp. Mater.*, **5**, 241 (1971).

116. Waddoups, M. E., Eisenmann, J. R., and Kaminski, B. E., 'Macroscopic Fracture Mechanics of Advanced Composite Materials', *J. Comp. Mater.*, **5**, 446 (1971).

117. Pagano, N. J. and Pipes, R. B., 'Some Observations on the Interlaminar Strength of Composite Laminates', *Int. J. Mech. Sci.*, **15**, 679 (1973).

118. Sutliff, D. R. and Pili, H., 'Three-dimensional Scattered-light Stress Analysis of Discontinuous Fiber-reinforced Composites', *Expt. Mech.* (July 1973), p. 294.

119. Liptai, R. G., 'Acoustic Emission from Composite Materials', *Composite Materials: Testing and Design (Second Conference)*, ASTM STP 497, American Society for Testing and Materials (1972), p. 285.

120. Hamstad, M. A. and Chiao, T. T., 'Acoustic Emission Produced During Burst Tests of Filament-wound Bottles', *J. Comp. Mater.*, **7**, 320 (1973).

121. Chen, E. P. and Sih, G. C., 'Interfacial Delamination of a Layered Composite Under Antiplane Strain', *J. Comp. Mater.*, **5**, 12 (1972).

122. *Progress in Flaw Growth and Fracture Toughness Testing*, ASTM STP 536, American Society for Testing and Materials (1973).

123. Hancock, J. R. and Swanson, G. D., 'Toughness of Filamentary Boron/Aluminum Composites', *Composite Materials: Testing and Design (Second Conference)*, ASTM STP 497, American Society for Testing and Materials (1972), p. 299.

124. Tetelman, A. S., 'Fracture Processes in Fiber Composites Materials', *Composite Materials: Testing and Design*, ASTM STP 460, American Society for Testing and Materials (1969), p. 473.

125. Corten, H. T., 'Reinforced Plastics,' *Engineering Design for Plastics*, E. Baer, ed., Reinhold Publishing Corp. (1964), p. 869.

126. Tsai, S. W. and Schulman, S., 'A Statistical Analysis of Bundle Tests', Technical Report AFML-TR-67-351. April (1968).

127. Lekhnitskii, S. G., *Theory of Elasticity of An Anisotropic Elastic Body*, Chapter 3, Holden-Day, Inc., San Francisco (1963).

128. Rosen, B. W. and Zweben, C. H., 'Tensile Failure Criteria for Fiber Composite Materials', Materials Science Corp., Blue Bell, Pennsylvania, N72-30499 (1972).

129. England, A. H., 'An Arc Crack Around a Circular Elastic Inclusion', *J. Appl. Mech., Trans. ASME*, **33**, 637 (1966).

130. Gotoh, M., 'Some Problems of Bonded Anisotropic Plates With Cracks Along the Bond', *Int. J. Fracture Mechanics*, **3**, 253 (1967).

131. Iremonger, M. J. and Wood, W. G., 'Stresses in a Composite Material With a Single Broken Fibre', *J. Strain Analysis*, **2**, 239 (1967).

132. Lowengrub, M. and Sneddon, N., 'The Stress Field Near a Griffith Crack at the Interface of Two Bonded Dissimilar Elastic Half-Planes', *Int. J. Engng Science*, **11**, 1025 (1973).

133. Kulkarni, S. V., Rosen, B. W., and Zweben, C., 'Load Concentration Factors for Circular Holes in Composite Laminates', *J. Comp. Mater.*, **7**, 387 (1973).

134. Gupta, G. D., 'A Layered Composite With a Broken Laminate', *Int. J. Solids and Structures*, **9**, 387 (1973).
135. Bogy, D. B., 'The Plane Elastostatic Solution for a Symmetrically Loaded Crack in a Strip Composite', *Int. J. Engng Science*, **11**, 985 (1973).
136. Irwin, G. R., 'Fracture Mechanics', *Structural Mechanics*, Pergamon Press, New York (1960), p. 557.
137. Sih, G. C., 'Some Basic Problems in Fracture Mechanics and New Concepts', *Engineering Fracture Mechanics*, **5**, 365 (1973).
138. Sih, G. C., 'Application of Strain–Energy–Density Theory to Fundamental Fracture Problems', IFSM-73-49, Institute of Fracture and Solid Mechanics, Lehigh University (1973).
139. Bogy, D. B., 'On the Plane Elastostatic Problem of a Loaded Crack Terminating at a Material Interface', *J. Appl. Mech. Trans. ASME*, **40**, 911 (1971).
140. *Failure Modes in Composites*, ed., Metallurgical Society of AIME (1972).
141. Cottrell, A. H., Royal Society Discussion on New Materials, 1963, *Proc. Roy. Soc.*, A**282**, 2 (1964).
142. Sih, G. C. and Macdonald, B., 'Fracture Mechanics Applied to Engineering Problem—Strain Energy Density Fracture Criterion', *Engineering Fracture Mechanics*, to be published.
143. Sih, G. C., Chen, E. P. and Huang, S. L., 'Fracture Mechanics of Plastic–Fiber Composites', *Engineering Fracture Mechanics*, to be published.

CONVERSION FACTORS

Length
$1 \text{ ft} = 12 \text{ in} = 30 \cdot 48 \text{ cm}$
$1 \text{ in} = 2 \cdot 54 \text{ cm}$
$1 \text{ m} = 39 \cdot 37 \text{ in} = 3 \cdot 28 \text{ ft}$
$1 \mu\text{m} = 10^{-6} \text{ m}$
$1 \text{ Å} = 10^{-10} \text{ m}$

Mass
$1 \text{ kg} = 2 \cdot 2 \text{ lbm}$
$1 \text{ lbm} = 0 \cdot 4536 \text{ kg}$

Weight
$1 \text{ lb} = 16 \text{ oz} = 453 \cdot 59 \text{ g}$
$1 \text{ g} = 2 \cdot 2 \times 10^{-3} \text{ lb}$
$1 \text{ kg} = 2 \cdot 2 \text{ lb}$

Force
$1 \text{ N} = 10^5 \text{ dyn} = 0 \cdot 2248 \text{ lbf}$
$1 \text{ dyn} = 2 \cdot 2481 \times 10^{-6} \text{ lbf} = 1 \cdot 02 \times 10^{-3} \text{ g-wt}$
$1 \text{ lbf} = 4 \cdot 4482 \text{ N} = 453 \cdot 59 \text{ g-wt}$

Density
$1 \text{ lbm/ft}^3 = 5 \cdot 787 \times 10^{-4} \text{ lb/in}^3$
$= 16 \cdot 018 \text{ kg/m}^3$
$= 0 \cdot 016018 \text{ g/cm}^3$
$1 \text{ g/m}^3 = 0 \cdot 03613 \text{ lbm/in}^3$
$= 62 \cdot 43 \text{ lbm/ft}^3$

Stress
$1 \text{ psi} = 70 \cdot 307 \text{ g/cm}^2 = 7 \cdot 03 \times 10^{-4} \text{ kg/mm}^2$
$= 6 \cdot 8948 \times 10^3 \text{ N/m}^2$
$= 6 \cdot 8948 \times 10^4 \text{ dyn/cm}^2$
$1 \text{ N/m}^2 = 10 \text{ dyn/cm}^2$
$= 1 \cdot 01972 \times 10^{-2} \text{ g/cm}^2$
$= 1 \cdot 4504 \times 10^{-4} \text{ psi}$

Stress intensity
$1 \text{ psi} \sqrt{\text{in}}$
$= 1 \cdot 1 \times 10^5 \text{ (dyn/cm}^2) \sqrt{\text{cm}}$
$= 1 \cdot 1 \times 10^3 \text{ (N/m}^2) \sqrt{\text{m}}$
$= 3 \cdot 54 \times 10^{-3} \text{ (kg/mm}^2) \sqrt{\text{mm}}$

Energy

$\qquad$ 1 joule $= 10^7$ erg
$\qquad\qquad\quad = 1$ watt-sec
$\qquad\qquad\quad = 0.738$ ft-lb
$\qquad\qquad\quad = 2.389 \times 10^{-4}$ kcal
$\qquad\qquad\quad = 9.48 \times 10^{-4}$ BTU
$\qquad\qquad\quad = 3.725 \times 10^{-7}$ Hp-hr
$\qquad$ 1 BTU $= 0.252$ kcal
$\qquad\qquad\quad = 1055$ J
$\qquad\qquad\quad = 2.930 \times 10^{-4}$ kW-hr
$\qquad\qquad\quad = 3.929 \times 10^{-4}$ Hp-hr
$\qquad$ 1 cal $= 4.186$ J $= 0.003968$ BTU

Power

$\qquad$ 1 watt $= 1$ J/sec
$\qquad\qquad\quad = 0.0569$ BTU/min
$\qquad\qquad\quad = 0.01433$ kcal/min
$\qquad\qquad\quad = 1.341 \times 10^{-3}$ Hp
$\qquad\qquad\quad = 0.738$ ft-lb/sec
$\qquad$ 1 BTU/min $= 17.58$ W
$\qquad\qquad\qquad = 778$ ft-lb/min
$\qquad\qquad\qquad = 0.02357$ Hp
$\qquad$ 1 Hp $= 550$ ft-lb/sec
$\qquad\qquad\quad = 746$ W
$\qquad\qquad\quad = 0.707$ BTU/sec
$\qquad\qquad\quad = 10.69$ kcal/min

Temperature

$\qquad$ $\Delta 1° = \Delta 1°K$
$\qquad\quad = \Delta 1.8°F$
$\qquad$ $0°C = 273.15°K$
$\qquad\quad = 32°F$

INDEX